ב"ה יחי המלך המשיח!

To my friend Nathaniel :)

Shimon

AF779465

ספר שארית ישראל

ספר שארית ישראל

Scientific Thought in Messianic Times

ב"ה

יחי אדוננו מורנו ורבינו מלך המשיח לעולם ועד

ספר שארית ישראל

Scientific Thought in Messianic Times

Wisdom From Above...Wisdom From Below

SHIMON SILMAN, *Director*
Rabbi Yisroel Aryeh Leib Research Institute on Moshiach and the Sciences
Professor of Mathematics, Touro College

About the Cover

There are wonders of the Creator yet unseen—from the far reaches of space to deep inside the atom. This nebula (gas cloud) was only recently discovered by the Hubble Space Telescope. Yet this is only the "wisdom from below."

In the Era of Moshiach, "everything will be revealed—the true character and inner depth of the order of nature will be seen, together with the full revelation of the miraculous order. And beyond this—the revelation of that which is completely wondrous." —the Rebbe MHM

Rabbi Yisroel Aryeh Leib Research Institute on Moshiach and the Sciences,
Brooklyn, NY 11203 • 718.604.4701
 Published 2010
17 16 15 14 13 12 11 10 2 3 4 5 6 7 8 9

ISBN: 978-0-9827298-0-9

Library of Congress Cataloging-in-Publication Data

Silman, Shimon
Scientific Thought in Messianic Times:
Wisdom from Above...Wisdom from Below
Shimon Silman
p. cm.
Includes bibliographical references and index
LCCN 2010905211
ISBN-13: 978-0-9827298-0-9
ISBN-10: 0-9827298-0-4

1. Judaism and science. 2. Messianic era (Judaism)
I. Title.

BM538.S3S55 2010 296.3'75
QBI10-600221

Printed in the United States of America on acid-free paper
Book Design and outstanding general assistance: Asher Hecht
Cover Design: Asher Hecht, Yisroel Aryeh Silman and Chaya Mushka Silman
Transcribing: Tzivia Yaffee
Typesetting: Asher Hecht, Yisroel Aryeh Silman, Yitzchok Wagshul
Proofreading: Chana Shloush, Netel Silman and Yitzchok Wagshul
Editorial and publishing assistance provided by Final Word Consulting, Ltd.

This book is written in fulfillment of the instruction of the Rebbe MHM to me in Adar, 5753 (1993) to publish the proceedings of the first Moshiach and science conference on the theme *The Role of Science in the Era of Moshiach*, commemorating the 40th yartzeit of Rabbi Yisroel Aryeh Leib Schneerson, the youngest brother of the Rebbe MHM. It took place on 13-14 Iyar, 5752 (1992) in S. Paul, Minnesota.

On 11 Iyar, 5753 (1993), the Rebbe MHM gave his appproval and blessing to establish the *Rabbi Yisroel Aryeh Leib Research Institute on Moshiach and the Sciences*. Since then we have held annual conferences and seminars in United States, Canada and Israel. This book is based on the proceedings of these conferences.

It is my honor to dedicate this book to the memory of

Rabbi Yisroel Aryeh Leib Schneerson ז"ל

the chosid who inspired Moshiach & science research

In these days—the Era of Moshiach—which we are now in, all we need to do is open our eyes and then we see that the true and complete Geula is already here, literally.

—*the Rebbe MHM*

CONTENTS

Chapter IV
The Convergence of Science to Torah– *The Trend of Science*

Chapter V
The Quest for Unification– *Unifying Theories in Mathematics and Science*

Chapter VI
Special Lectures

Chapter VII
A Look to the Future

PREFACE

In late 1993 (a year and a half after the Rebbe MHM suffered a stroke) a Lubavitcher chosid wrote the following note to the Rebbe MHM with the above photo enclosed (emphasis in original):

"Will the Chassidim again see the Rebbe Shlita doing as in the enclosed picture, *with the same physical capabilities (standing, waving both arms joyously) and even more? We ask the Rebbe Shlita for his blessings that this occur soon.*"

The Rebbe MHM nodded in agreement.

INTRODUCTION

Scientific Thought ...

There are two searches—the search of the mind and the search of the soul.

The search of the mind is the search of the scientist who searches for wisdom from within the world in an attempt to describe the world he observes. He examines the physical world, experiments, and constructs mathematical models and theories to describe what he sees.

The search of the soul is a search for wisdom from above in an attempt to explain the world in which it finds itself. It looks to the Torah for truth and for instruction on how to live.

The search of the soul is a search for wisdom from above; the search of the mind is a search for wisdom from below. The ultimate goal is for the two searches to converge, since the same truth must be reached from whichever direction it is approached. But this goal will be attained only in the Era of Moshiach (the Messianic Era) when "the earth will be filled with the knowledge of G-d as water covers the sea,"[1] and all knowledge will converge to, and be covered by, the knowledge of G-d.

At first glance these two searches would seem to be in conflict with each other. After all, the conclusions of scientists have often been very different from the teachings of the Torah. In fact, historically, those who studied the sciences often did so with the intent of substituting human perceptions and reason for belief in G-d, and atheistic ideologies would draw upon science in their attempt to negate belief in G-d. Thus, the belief in the emergence of Moshiach, a personality who will bring the whole world to believe in G-d and to be concerned only with the knowledge of G-d, would seem to be diametrically opposed to everything that science represents.

Yet we find that some of the greatest scientists of the 20th century have said that their study of the universe brought them to an

1. Yeshaya 11:9.

awareness of a Supreme Being above and beyond the universe, or a Creator who created the universe with a definite purpose. Some have even described their research as an attempt to find out what G-d had in mind when He created the universe.

The greatest 20th century scientist, Albert Einstein, expressed this as follows: "I want to know how G-d created this world. I am not interested in this or that phenomenon, in the spectrum of this or that element. I want to know G-d's thoughts. The rest are details."[2] The famous theoretical physicist, Stephen Hawking, wrote, "If we do discover a [theory of everything] … it would be the ultimate triumph of human reason—for then we should know the mind of G-d."[3] In 1992, another famous theoretical physicist, Paul Davies, published a book discussing the relationship between science and religion called *The Mind of G-d*, which won him the Templeton prize of $1,000,000. The scientist who led the team that cracked the human genome, Francis Collins, director of the U.S. National Human Genome Research Institute, in his book, *The Language of G-d*, writes, "When you have for the first time in front of you this 3.1-billion-letter instruction book that conveys all kinds of information and all kinds of mystery about humankind … I can't help but look at those pages and have a vague sense that this is giving me a glimpse of G-d's mind."

Something must have happened to break the barrier between the search of the mind and the search of the soul and enable their convergence to begin. But *what* happened, and when did it happen?

The answer to this question is to be found in one of the *sichos* (talks)[4] of the Lubavitcher Rebbe MHM, Rabbi Menachem M. Schneerson, based on the *Zohar*, the classic text of Kabbalah, which was written nearly two thousand years ago by the great sage Rabbi Shimon Bar Yochai. The *Zohar* makes an amazing prediction based on a verse in the Torah about the flood in the time of Noach.

The Torah says, "In the 600th year of the life of Noach … all the fountains of the depths split open and the windows of heaven

2. Albert Einstein, *The Expanded Quotable Einstein*, ed. Alice Calaprice, 2nd ed. (Princeton, NJ: Princeton University Press, 2000), 202.

3. Stephen W. Hawking, *A Brief History of Time: From the Big Bang to Black Holes* (New York: Bantam, 1990), 193.

4. *Likutei Sichos*, 15:42–48.

opened up."[5] Referring to this, the Zohar makes the following statement: "In the six hundredth year of the sixth millennium [i.e., the year 5600 on the Jewish calendar, which corresponds to 1840 on the secular calendar], the gates of wisdom above will open and the fountains of wisdom below, and thus the world will be set to enter the seventh millennium"—i.e., the Messianic Era.

The "opening of the gates of wisdom above" refers to the tremendous revelation of the deeper levels of the Torah—the teachings of *Chassidus,* especially the publication of the works of the Alter Rebbe, Rabbi Schneur Zalman of Liadi (founder of Chabad *Chassidus*). These give us a deeper understanding and greater knowledge of G-d and prepare the world for the coming of Moshiach, when "the earth will be filled with the knowledge of G-d."[6]

The "fountains of wisdom below" refers to new discoveries in science and the development of new scientific theories. This requires some explanation: how could scientific developments actually *play a role* in preparing the world for Moshiach? Science has always been considered the enemy of religion. How could it come about that science not only becomes a friend of Judaism but actually helps prepare the world for the greatest Jewish religious event of all time—the coming of Moshiach?

In the *sicha,* known in our circle simply as "the *Sicha* of *Parshas Noach,*" the Rebbe MHM discusses three ways in which the sciences prepare the world for Moshiach. In describing the unique character of the knowledge of G-d in the Era of Moshiach, the prophet Yeshaya (Isaiah) says, "All flesh will *see* that the mouth of G-d has spoken,"[7] i.e., G-dliness itself will be *visible*. The first way in which modern science and technology prepare us for the Era of Moshiach is by enabling us to visualize certain concepts relating to G-d that could otherwise have been understood only abstractly. As an example, the Mishnah[8] tells us to be aware that there is "an eye that sees and an ear that hears." We believe that G-d is always watching everything we do. Now that we are familiar with communication technology, enabling people to be seen and heard instantly even on

5. Bereishis 7:11.
6. See sup., n. 1.
7. Yeshaya 40:5.
8. *Avos* 2:1.

the other side of the world, it is easy to comprehend—to actually visualize and not just believe in—the omnipresent watchfulness of G-d.

Second, by making use of the technology derived from the sciences. For example, communication technology can be used to spread knowledge of G-d—Torah—throughout the world. As we said, the Era of Moshiach is characterized as a time when "the earth will be filled with the knowledge of G-d." The electromagnetic waves which carry knowledge of G-d over radio or TV bring this information not only to listeners and viewers all over the world, but even to places where there are no listeners, since these waves spread throughout the whole world. Thus, such broadcasts literally *fill the world* with knowledge of G-d.

On the deepest level, the sciences themselves—the theories developed and the knowledge derived—prepare us for the Era of Moshiach in yet a third way: the ultimate knowledge of G-d that will be revealed in the Era of Moshiach is the absolute unity of G-d. The world as we perceive it appears to present an opposite picture of reality—***many*** different forces of nature and ***many*** different substances, rather than any kind of unity. In the last 170 years, however, scientific research has shown that the various forces of nature are just combinations of a few basic forces, and the various substances in the world are built from a few basic components—protons, neutrons, and electrons. This deeper understanding of the unity underlying the physical world, discovered through scientific research, helps prepare us for the revelation of the ultimate unity—the *Achdus Peshuta*,[9] or absolute unity of G-d, which is to be revealed in the time of Moshiach.

The unity that exists within nature is an expression of the unity of G-d, the Creator. Since He is one, His creation, with its multitude of materials and forces, has an intrinsic unity. Thus, our awareness of the unity within nature is essentially an awareness of the unity of G-d. The more that science reveals of the unity inherent within nature, the more the unity of G-d is revealed—from below. This sets the stage for the complete revelation of the unity of G-d— from above—in the Era of Moshiach.

9. *Achdus Peshuta* literally means "simple unity," which means a unity that does not consist of parts. There is only one G-d and He has no subdivisions.

Thus we may say that 20th century scientists, who came to an awareness of the Creator through their scientific research, were merely riding the crest of a new wave—a wave from the flood of a new type of knowledge flowing into the world. This was knowledge of a type that showed unity in the world and induced the scientist with an open mind and a sensitive heart to recognize the Supreme Power, the Master Planner Who designed and created the universe with a certain purpose in mind.

Theoretical physics, in particular, seems to have developed into a search for G-d. Earlier, we quoted Albert Einstein, Stephen Hawking, and Paul Davies, who described their quest to know the "mind of G-d." Davies once wrote that science offers "a surer path to G-d than religion."[10] Another prominent theoretical physicist, David Finkelstein, described his life's work as a quest to understand "how G-d might have thought the world into existence."[11]

An exception that proves the rule is the recent book, The *Trouble with Physics* by Lee Smolin, a prominent theoretical physicist. It is a survey of the history and current state of affairs of modern theoretical physics. It is not about science and religion. Yet the first sentence of the book reads, "There may or may not be a G-d." All of a sudden he's talking about G-d!

What happened here? Was he copying from the wrong manuscript? No. For Smolin, theoretical physics is nothing less than what we have described as the "search from below"; as he goes on to explain, "Yet there is something ennobling about our search for the divine … which is reflected in each of the paths people have discovered to take us to deeper levels of truth. Some seek transcendence in meditation or prayer.… Another way of engaging life's deepest questions is science.… If they are physicists, they want to know about space and time, and what brought the world into existence."

For Smolin, it seems that science itself is a search for G-d.

Similarly, it has been asserted that mathematicians whose field of interest is mathematical logic (like myself) are not like other mathematicians whose primary interest is the mathematics itself, or the

10. Quoted in *Albany Times Union*, March 9, 1995.

11. Quoted in Lee Smolin, *The Trouble with Physics* (Boston: Houghton Mifflin, 2006), 322.

applications of mathematics to other areas. Mathematical logicians, it is claimed, are engaged in a metaphysical quest—they are searching for G-d.[12]

An outstanding feature of Smolin's book is his emphasis on unification. As he traces the development of modern scientific theories, he explains how each new stage of development brought greater unification among the existing theories. (Perhaps Smolin's emphasis on unification is related to his feeling that science is a search for the divine.) This, of course, is of fundamental importance to us, since the unification of the forces and materials of nature is the aspect of modern science that prepares the world for the Era of Moshiach.

... In Messianic Times

The Zohar says that this new flood of knowledge is a prelude to the Era of Moshiach. But what exactly is the Era of Moshiach and who is Moshiach? The Era of Moshiach and Moshiach himself are discussed extensively in the Prophets, the Talmud, and Kabbala, and in many subsequent books throughout the generations. In chapter 1, sections 4–6, we present the basic information based on these sources. But for now, we want to make the point that we are currently in the Era of Moshiach. There are actually two points to address: one identifying the era, and one identifying Moshiach himself.

The emergence[13] of the great Jewish leader ultimately revealed as Moshiach is described by Rambam (Maimonides) in his work on Jewish Law, *Mishneh Torah.*[14] Rambam describes the sequence of events leading up to the revelation of Moshiach. He begins by stating that Melech HaMoshiach (the Messianic king, or King Moshiach), a descendant of King David, will arise and restore the dynasty of King David. He then describes the gradual emergence of this Jewish leader from the House of David, who will speak in Torah, deal in Mitzvos, return the Jews to the observance of Judaism, strengthen the observance of Judaism where it has been broken, and

12. David Leavitt (author of *The Man Who Knew Too Much: Alan Turing and the Invention of the Computer* (Norton, 2005)), radio interview with John Batchelor, *The John Batchelor Show*.

13. See the subsection, "Identity as Melech HaMoshiach" in Chap. I, sec. 5, for a further discussion of this.

14. Laws of Kings, chaps. 11–12.

fight the wars of G-d. This leader, states Rambam, is assumed to be Moshiach. Rambam's description has characterized the leadership of the Lubavitcher Rebbe since 1950, bringing many Jews to look upon the Rebbe as that leader who would eventually be revealed as Moshiach. While he was opposed to publicizing this until recently, over the years some people who addressed the Rebbe as Moshiach received a positive response, myself included.[15]

As the Persian Gulf War began to brew at the end of 1990, the world braced for an intense, high technology war expected to last for months, while the fear persisted that the international entanglements might naturally lead to a world war as had happened twice before. The Rebbe MHM, however, in his talks during those months, repeatedly quoted a Midrash from the *Yalkut Shimoni*[16] which foretold the war. The Midrash says that these events would take place in "the year that Melech HaMoshiach is revealed"; that the entire world, especially the Jewish people, would be worried; but that G-d would say (through a prophet) to the Jews, "My children, do not be afraid … the time of your redemption has arrived."

Thus, notwithstanding Saddam Hussein's stated intention to destroy Israel and his campaign of targeting that country with missiles, the Rebbe MHM said prophetically that Israel would be safe. He encouraged Jews to travel there, and indicated that the war would be over by the holiday of Purim. Everything happened exactly as he said. Even though 39 Scud missiles fell in Israel, no Jews were killed by them. And, indeed, the war ended on Purim. It was then clear that this was the revelation of Melech Moshiach. In his talks during 1991–92 the Rebbe MHM stated many times that this was in fact the case, and made it clear that he is, in fact,Melech HaMoshiach. On two occasions in May 1991, he publicly encouraged the singing of the declaration *"Yechi Adoneinu Moreinu V'Rabeinu Melech HaMoshiach L'Olam Voed!"* ("Long live our master, teacher, and Rebbe King Moshiach forever!") and subsequently encouraged many people who had been publicizing the fact that he is Moshiach to continue doing so.

---

15. In 1985, my family sent a gift to the Rebbe MHM for his birthday, 11 Nisan. It was a first edition of the *sefer* (book) *Beis Aharon V'Hosafos*, compiled by his maternal ancestor Rabbi Avraham Dovid Lavut. I wrote a cover letter in which I addressed the Rebbe as *Moshiach Tzidkeinu*, to which I received a positive response expressing thanks.

16. *Yalkut Shimoni*, Yeshaya, sec. 499.

It should be emphasized that this phrase—"The time of your redemption has arrived"—is the key phrase by which Melech HaMoshiach is identified, as we see from the continuation of the *Yalkut Shimoni*: "At the time that Melech HaMoshiach comes ... he announces to the Jews, 'O, humble ones, the time of your redemption has arrived!'" The Rebbe MHM made it clear that he was making the declaration that identified him as Moshiach.

In the summer of 1991, the Rebbe MHM said that with the miraculous victory in the Persian Gulf War, "the words of the *Yalkut Shimoni* ***have been fulfilled*** ... and since then, especially from the month of Nisan, the month of Geula (redemption), we are already 'at the time [*sha'ah*—literally "the hour"] that Melech HaMoshiach comes.'"[17] And again, "We have already actually seen wonders that confirm that this is ***'the year in which Melech HaMoshiach is revealed'*** and even ***'the hour when Melech HaMoshiach comes'*** and announces to the Jews, 'Humble ones, the time of your redemption has arrived' and even the announcement that 'Behold, this is Melech HaMoshiach who came'—***that he already came***."[18]

Later that summer he announced, "All aspects of the Geula have already begun and have already been brought into and accepted in this physical world.... So it will be no wonder when the Geula comes immediately."[19]

We cannot overemphasize the seriousness of a statement such as "All aspects of the Geula have already begun." It is, in fact, a matter of life and death. For many years the Rebbe MHM objected in the strongest terms to the idea expressed by some that the establishment of the State of Israel was the beginning of the Geula (*Aschalta D'Geula*) and in a letter written in 1968, after expressing his objection, he concludes by saying:

> I would not have written all the above had I not seen a danger in this declaration of "the beginning of the Geula"—for this is the only explanation that I have found for the "sacrifices" and deaths [from attacks] in the Holy Land. Among those who have fallen, there are even people who were saved

17. *Sefer HaSichos* 5751, 2:595.
18. Ibid., 690.
19. Ibid., 793.

> from the Holocaust in an openly miraculous manner. Nevertheless, they fell in the Holy Land—also in an unnatural manner, may G-d avenge their blood.
>
> The explanation is similar to what has been seen every time that a false Messiah arises and seduces people to follow him—it always ends with "sacrifices." The same thing applies to the idea that the Geula has already begun.[20]

Thus, when the Rebbe MHM said in 1991 that "all aspects of the Geula have already begun," it was a statement made with full certainty, and knowledge of the implications.

(In fact, the scholars who transcribed the *sichos* and prepared them for the Rebbe MHM to edit were so astonished by the strength of his statements about the Geula in this *sicha* that they weren't sure if they should write them up as statements of fact or as expressions of prayer. They decided to write the *sicha* both ways and to submit both versions. The Rebbe MHM rejected the "prayerful wish" version, and proceeded to edit and publish the "statement of fact" version.)

On 28 Nisan, 5751 (1991), in a very powerful speech addressed to the Jewish people,the Rebbe MHM began to encourage personal initiatives to bring Moshiach with the message, "Do everything in your power ... to bring about the actualization of Moshiach *Tzidkeinu* immediately."[21] Moshiach had already been revealed, and we were being instructed to bring this revelation to the world. Many projects to implement this were initiated with the Rebbe MHM's encouragement.

In the spring of 1992, the Rebbe MHM suffered a stroke, following which the activities to bring Moshiach intensified, with his continued encouragement. About two hundred rabbinical authorities worldwide signed a *Psak Din* (Halachic ruling) dated 11 Nisan, 5752 (1992) declaring that the Lubavitcher Rebbe has the status of "presumed Moshiach" (*B'chezkas Moshiach*) as defined by Rambam.

Throughout the year 5753 (1992–93), whenever the Rebbe MHM would enter "770" for *davening* (prayers)—sometimes three times a day—the service, in which thousands participated, would be fol-

20. *Igros Kodesh* 25:230–31.
21. *Sefer HaSichos 5751*, 2:474.

lowed by the singing of *"Yechi Adoneinu...."* The Rebbe MHM encouraged this enthusiastically and, on occasion, it was broadcast worldwide via satellite.

In the Talmudic writings, there is a description of a conversation that will take place between the Jews and Moshiach after he is revealed. The Jews approach him, addressing him by the title *Rabbeinu Melech HaMoshiach,* ("our Rebbe, King Moshiach") and ask him certain questions about the redemption.[22] This is the title by which countless Jews throughout the world now refer to the Lubavitcher Rebbe, and thus, in this book, we refer to him as "the Rebbe Melech HaMoshiach" or "the Rebbe MHM."

On 27 Tishrei 5754 (1993), I was part of a group that had the privilege to pass before the Rebbe MHM and receive from him a dollar (to be given to *tzedakah,* as customary).[23] As I passed before him, I greeted him saying, *"Shalom Aleichem, Melech HaMoshiach!"* and he nodded to me in acceptance.

From all the foregoing, we see that Melech HaMoshiach has emerged and has been clearly identified. It is the purpose of this book to examine various aspects of our current era, especially in the areas of science and technology, and to identify them as characteristics of the Era of Moshiach, following the statement of the Rebbe MHM himself on 19 Kislev, 5752 (1991) that we are now living in the Era of Moshiach: "In these days—the Days of Moshiach—which we are now in, all we need to do is open our eyes and then we see that the true and complete Geula is already here, literally."[24]

The idea that one "opens his eyes" to see something means that the thing is already there in front of him, but that he doesn't really *see* what he's looking at. This may mean that he must take a deeper, more penetrating look, possibly on a more abstract or spiritual level. But it could simply be that he is looking at something, or listening to something, and he doesn't grasp or notice certain aspects of it, even physical aspects. This may be because his senses are not attuned to it or because his mind is not focused on it or is distracted by some other aspect of it.

---

22. *Psikta Rabbasi,* Chap. 15, sec. 71.
23. This has been the only such event since the stroke.
24. *Sefer HaSichos* 5752, 1:173.

This happens frequently when one is so preoccupied with the details of what he is considering that he misses the overall picture. This is described by the famous metaphor of not being able to see the forest for the trees.

In recent years, a technique known as "magic eye" has been developed to generate certain computer images that appear to the viewer as a two-dimensional scene. For example, in the image reproduced below, it is a field of flowers. When the viewer deliberately "blanks out" his focus and "opens his eyes" he sees a three-dimensional image instead of a two-dimensional image, with new objects visible in the scene.[25] In this image, it is a deer standing in the middle of the field of flowers. The deer—a metaphor for Moshiach[26]—is hiding in plain view! The viewer need only open his eyes.

As another example, a few years ago I heard a report on NPR radio

A "magic eye" image: all the viewer need do is "open his eyes."

25. The magic eye technique, in addition to being fascinating and fun, it is also used in various vision therapies.

26. Midrash Shir HaShirim Rabbah 2:22.

by someone who was researching the chirping of birds. In one case he recorded the chirping of a certain bird that sounded to the human ear as a one-tone tweet. But when he played it at a much slower speed, one could hear that this single tweet was actually an entire song with different tones, high notes and low notes etc. All this was blurred at normal speed and thus inaudible to the human ear.

The story is the same for all our senses, as well as our intellectual perception. For that matter, the history of science is replete with the misperceptions of one generation being corrected by the revolutionary new theories of the next generation.

The Rebbe MHM is telling us that the same is true of the world as we are used to looking at it—a world of *Golus* (exile). But this is a misperception caused by our past experience and former ways of looking at things. He is now telling us to open our eyes and see Geula in the world. That is the purpose of this book.

Moshiach and Science

In the ensuing period (late 1991–92), the Rebbe MHM made stronger statements about our time being the Messianic Era, saying that all aspects of the Geula had already begun and discussing the fulfillment of some of the Messianic prophecies, such as "swords being beaten into plowshares," at great length. He instructed people to live their lives in accordance with this new era. He directed the Jewish people to strengthen their observance of Judaism and the study of the Torah—especially topics related to the Geula—and the non-Jews to increase in acts of goodness and kindness and the observance of the "Seven Noahide Laws."

One of his statements, from the beginning of 1992, made a strong impression on me personally as a mathematician: "We have to investigate the role that everything in the world will play in the Era of Moshiach."[27] This includes mathematics and science. My work in mathematics took on a new meaning, a new dimension. It was part of the wisdom from below that was preparing the world for Moshiach and was to play a role in the Messianic Era itself. This was very exciting.

The year 1992 also marked the 40th *yartzeit* (anniversary of the

27. *Sefer HaSichos* 5752, 1:250, n. 46.

passing) of the youngest brother of the Rebbe MHM, Rabbi Yisroel Aryeh Leib Schneerson, who was a great Torah scholar (especially in the area of Chassidic Philosophy) and a mathematician by profession. His life story had exerted a profound influence on my own development and thinking.[28] I was moved to commemorate the 40th *yartzeit* by organizing a conference on the theme, "The Role of Science in the Era of Moshiach." This conference took place in S. Paul, Minnesota, where I was then living. The keynote speaker was the world renowned Lubavitcher chosid and mathematician Professor Pesach (Paul) Rosenbloom, who had edited and published a well-known mathematical paper of Rabbi Yisroel Aryeh Leib.

In March of 1993, I received the approval and blessing of the Rebbe MHM to publish the proceedings of the Minnesota conference and to organize a similar conference in New York (my new place of residence), to mark the 41st *yartzeit*. During that conference, on May 2, 1993, I received the approval and blessing of the Rebbe MHM to establish the "Rabbi Yisroel Aryeh Leib Research Institute on Moshiach and the Sciences" to conduct research on the connection between developments in the sciences and the Messianic Era, and to organize annual conferences.

The RYAL Institute has sponsored annual conferences every year since then on the *yartzeit* of Rabbi Yisroel Aryeh Leib, as well as other conferences and seminars in major cities of the world including Tel Aviv, Beer Sheva, New York, Montreal, Toronto, and Chicago. The lecturers have included some of the most prominent religious Jewish scientists of our time, including Professor Alvin Radkowsky, designer of the United States nuclear navy (under Admiral Hyman G. Rickover) and the Radkowsky Thorium Nuclear Reactor; Dr. Naftali (Norman) Berg, Director Research of the Advanced Technology Office of the Army Research Laboratory of the Pentagon; and Professor Yirmiyahu (Herman) Branover, director of the Magnetohydrodynamics (MHD) Laboratories at the University of the Negev, Israel.

Structure of the Book

This book is based on those lectures, with much new material

28. I discuss this at length in chap. 2, sec. 3.

added to provide background or to complete topics so that the book will be self contained. Chapter I, "Wisdom From Above ... Wisdom From Below," provides historical background and introduces concepts—both in Chassidic philosophy and in modern science—necessary for a full appreciation of the lectures. Chapter II, "Rabbi Yisroel Aryeh Leib Schneerson," is an in-depth biography of the chosid who was the inspiration for the RYAL Institute, and Moshiach and Science research in general. (His famous mathematical paper, "The Location of Eigenvalues and Eigenvectors of Complex Matrices," and a commentary on it by a contemporary mathematician, are presented in the Appendix.) Chapter II also includes a report on the Minnesota conference. Chapter III, "Swords Into Plowshares," contains lectures on the transformation of military technology to peaceful uses. This worldwide phenomenon has been called "the trend of history" by some physicists. Chapter 4, "The Convergence of Science to Torah," contains lectures discussing ways in which the theories and attitudes of science are coming closer to the truth of the Torah. We call this "the trend of science." Chapter 5, "The Quest for Unification," contains lectures that describe scientific models for Torah concepts. Chapter 6, "Special Lectures," contains two self-contained lectures delivered by Professor Branover and myself, respectively. Chapter 7, "A Look to the Future," investigates the ultimate status of the sciences further into the Era of Moshiach. Finally, the Appendix contains the full story of the so-called Doctor's Plot (in connection with chapter1, section 5), Education Day USA (in connection with chapter 3, section 2) and the mathematical papers.

Notes

1) Many selections from *seforim*, especially *sichos*, *Igros Kodesh*, and *ma'amorim*, are translated here. The translations are my own "free translations," in which I try to convey the content and, at the same time, preserve the original language.

2) The vast majority of the sections are non-technical and accessible to the general reader. There are, however, a few sections for which technical background in mathematics or physics is necessary. In some cases, the reader can get the content of the technical section by reading a corresponding non-technical

section. For example, Dr. Berg's lectures on quantum mechanics and string theory are presented first in a non-technical manner (chapter 5, sec. 5a), then with full technical detail (sec. 5b). Similarly, the Radkowsky nuclear reactor is first discussed in a non-technical manner in the interview (chapter 3, sec. 7a), then described technically (sec. 7b).

3) References are given at the end of each chapter so that the reader can look further into the topics discussed. The book is thus suitable as a text for a university level course in science and religion.

4) Every section of the book was written by me unless otherwise indicated.

5) We quote Albert Einstein, a Jew from Germany, and Richard Feynman, a Jew from Brooklyn, as well as other scientists, frequently in this book for the scientific views they expressed-purely from the point of view of science, not Judaism. Their background in and commitment to Judaism was, in fact, very weak and had little or no influence on their philosophy of science. Their ignorance of Judaism becomes evident in their discussions on religion.

   We do not consider Einstein and Feynman examples of what a Jewish scientist should be like. Good examples of what a Jewish scientist should be like are our colleagues Professor Yirmiyahu Branover, the late Dr. Naftali Berg z.l., and Dr. Tsvi Saks z.l.

   (On a deeper level, however, the Jewish soul—the Jewish life—in these scientists certainly influenced their scientific thinking. At one of our Moshiach and Science conferences, Dr. Berg mentioned that when Einstein first published his theory of relativity, a certain Russian scientist who did not like the theory—and didn't like Jews either—commented, "It is the Jew in Einstein that makes him come up with such crazy ideas." As Dr. Berg pointed out, while this was an anti-Semitic comment,

29. Their background in and commitment to Judaism was, in fact, very weak and had little or no influence on their philosophy of science. Their ignorance of Judaism becomes evident in their discussions on religion.

30. Abbreviation for *zichrono livracha* (of blessed memory).

it is certainly true that it's a quality of the Jewish soul, the neshama, to sense deeper levels of reality.)

The sources of the concepts of the character of science presented in this book are the writings of the Rebbe MHM. These concepts may be new and may even appear radical to the reader. The average working scientist may not even be familiar with them. (He may have never even thought about the issues involved.) We therefore want to emphasize that these are also the views on science of the greatest masters of modern science such as Feynman and Einstein. Feynman's views on science in particular, are strikingly similar to those expressed by the Rebbe MHM.

It is for this reason that we quote Feynman, Einstein and others. Our point is that in these Messianic times, scientific thought is coming closer to the truth of the Torah.

In Appreciation

My deepest gratitude is to the Rebbe MHM himself for giving me his approval and blessing (*bracha v'haskama*) to establish the RYAL Institute and to publish the proceedings of the Moshiach and Science conferences.

I want to express my appreciation to the Rabbis, professors, and scientists who have spoken and lectured at our conferences since the Institute was founded. They include (in alphabetical order):

Rabbis: Shmuel Chefer, Moshe Feller, Levi Garelik, Yitzchok Ginsburgh, Zevulun Lieberman, Shloma Majesky, Nissan Mangel, Yitzchok Springer, and others.

Professors and scientists: Drs. Yitschok Block, Yirmiyahu (Herman) Branover, Aryeh (Arnie) Gotfried, Lawrence Lasher, Neil Miller, Levi Reiter, Dovid Wohl and others.

Professors and scientists, of blessed memory, who have since passed away: Drs. Naftali (Norman) Berg, Alvin Radkowsky, Pesach (Paul) Rosenbloom, and Tsvi (Victor) Saks.

Special thanks to Rabbi Yitzchok Springer for his consistent encouragement and advice throughout the project of writing this book; to HaRav Yakov Schwei for his instruction on Halachic issues related to the preparation of the book; to Rabbi Menachem Gerlitzky,

Director of Kollel Levi Yitzchok, Congregation Lubavitch "770" and Aron Mendelsohn for technical assistance; to Rabbi Zimroni Tzik for organizing our conference at Tel Aviv University and to Rabbi Yirmiya Kalifa and Rabbi Shlomo Dubnick for organizing our conference in Beer Sheva.

It is with deep respect that I express my honor to the memory of my father, HaRav Moshe DovBer Silman ז"ל, who set me on the path of Torah from the time I was born; and my honor to my mother, Rebbitzen Rochel Silman, may she have many long healthy years. Her encouragement and assistance has been strong and consistent throughout the years.

Finally, I want to express my deep appreciation to my family, especially to my *eshes chayil*[31] Netel, who stood with me and helped me throughout the years that the Moshiach and Science concepts were being developed and this book was being written.

My children grew to adulthood together with this book. Their father was always "working on the book." Each of them has helped me in his own way, each applying his own skills.

Finally, I would like to invite the reader to contact me with questions or comments on issues related to Moshiach and Science, or with his own views on scientific thought in Messianic times. I can be reached by email at: Moshiach.Science@gmail.com. Also, please visit our website: www.RYAL.org

May G-d grant that we merit to continue the work of the RYAL Institute until it brings "the revelation of the three levels of pure intellect … which is brought about only by the refinement and elevation of the sciences."[32]

Shimon Silman
Crown Heights
13 Iyar, 5770
58th *yartzeit* of Rabbi Yisroel Aryeh Leib

31. Lit., "woman of valor" (see Mishlei 31:10)
32. *Sefer HaSichos 5751*, 2:525, n. 73.

CHAPTER I

Wisdom From Above...Wisdom From Below

The innovation of Moshiach is that through him the two concepts of 'above' and 'below' will be united.

—the Rebbe MHM

CHAPTER 1

Wisdom From Above...Wisdom From Below

The innovation of Moshiach is that through him the two concepts of 'above' and 'below' will be united.

—the Rebbe MHM

In this chapter we investigate the duality inherent in the two searches for truth—of the mind and of the soul. We begin with the duality in the mind's search through intellect versus the soul's search through faith. Then we examine the wisdom attained by the search of the mind—modern science as it developed from the 1800s onward, and the wisdom revealed to the soul—the teachings of Chassidus since that time.

Finally we give a full exposition of the concept of Moshiach in three parts: general information about Moshiach from the prophets, Rambam, and the works of Kabbala and Chassidus; a description of the emergence of Melech HaMoshiach and his characteristics; an in-depth analysis of the concept of the everlasting life of Moshiach.

Faith and Intellect: From Conflict to Convergence

In the introduction, we described two searches for truth—one from above and one from below. The former is the search of the man of faith and the latter is the search of the man of intellect. By definition these two approaches are opposites and, historically, the tension between them was the source of great conflict. Yet, the ultimate goal—which is attained in the Era of Moshiach—is the convergence and unification of these two searches.

This issue is in fact one of the oldest in Jewish philosophy. It was addressed by Rav Saadia Gaon in his *Emunot V'Deot*, Rabbi Yehuda HaLevi in the *Kuzari*, and by the Rambam in *Moreh Nevuchim* (which was addressed to those who had been influenced by the philosophy of Aristotle). In recent times, the issue has been dealt with extensively in Chabad Chassidus. The Tzemach Tzedek especially, in his *Sefer HaChakira*, addresses these issues, unifying the discussions of the earlier Jewish philosophers with the concepts of Kabbala and Chassidus.

Before getting into our analysis of the relationship between faith and intellect, we will first make some general observations, beginning with a quote from the Rebbe MHM's first book, *HaYom Yom* (a compilation of Chassidic sayings of the Previous Lubavitcher Rebbe): "When Moshiach comes, we will see the advantage of faith and sincerity—that everyone believes in G-d, His Torah and mitzvos with pure faith. Regarding learning—understanding—the greatest understanding has a limit. But faith is a feeling with no limit. Melech HaMoshiach will give us the ability to understand the advantage of sincerity—heartfelt earnest service of G-d."[1]

We see, then, that the fundamental difference between faith and intellect is that intellect is finite, while faith is infinite.

The ability to have faith and the ability to understand are two separate faculties of the soul.[2] Faith is not just something held to be true despite lack of proof. It is the capacity of the mind to hold something to be true in a way that transcends any proof one may have for it. As the Previous Lubavitcher Rebbe wrote, "Faith is one of the faculties of the soul which are imbedded in the limbs of the body—in the inner depth of the brain and the heart."[3]

Something held to be true only because of one's understanding will be abandoned if the proof is found to be flawed or is contradicted. This is the normal progression of scientific thought, where new information or understanding overturns a previously held view of nature. Consider, for example, how the discovery of quantum mechanics overturned the previously held view that the laws of nature are deterministic, and replaced it with a view of indeterminism in natural law.[4] Understanding is finite.

Faith, on the other hand, transcends experience. Even if it is challenged by what one sees, one maintains one's faith. A young Israeli girl from a family expelled from Gush Katif was interviewed by a reporter. She said they had had faith that they would not be expelled.

---

1. *HaYom Yom,* 5 Teves.

2. See the section *Ha'amonas Elokus* ("Belief in G-d") in *Derech Mitzvosecha* of the Tzemach Tzedek.

3. *Sefer HaMa'amorim* 5702, 83. It is interesting to note that scientists have recently identified a region in the brain, separate from the parts that deal with intellectual activity, which handles faith. They maintain that one must have some kind of faith to be psychologically healthy.

"So your faith was wrong," the reporter said to her. "No," she replied immediately, "Our faith made us stronger!" Faith is infinite.

So if the question is raised regarding which path one ought to follow—the path of faith (wisdom revealed from above) or of intellect (wisdom developed from below), it appears that the issue should be decided in favor of faith.

Indeed, as we explain in the section "Wisdom from Below," later in this chapter, one is deceiving himself if he thinks that he can rely totally on intellect, because any system of logic or intellect is based on fundamental assumptions, or axioms, that cannot be proven. They must be accepted—on faith!

The dependence of intellect on faith actually goes much deeper. Logic, after all, is a creation and, like everything else, was created in accordance with the Torah. The *Zohar* says that G-d used the Torah as the blueprint for creation. Why is something logical in the first place? Why does it make sense to us? It is only because according to the Torah—the blueprint for our creation—it should make sense to us. For example, the commandment not to steal makes sense to us. However, it is not "logical" in any absolute sense. It is logical because G-d decided that this mitzva should make sense (it should be a *mishpat*) so he created man in a way that "Do not steal" would make sense to him. The commandment that we may eat only certain animals is not logical, precisely because we were created in such a way that this should not make sense to us; rather, it should remain a supra-rational mitzva (*chok*). But it could just as well have been the reverse. If G-d had created us differently—programmed our minds differently—eating Kosher would have been logical and not stealing would have been supra-rational. This profound concept was explained by the Rebbe MHM in a *sicha* on *Parshas Mishpatim*.[5]

At this point it seems quite clear that faith is the path to follow rather than intellect. But the issue is far from settled. Indeed, in the Torah itself we find imperatives for both faith and intellect. For example, when the Jewish people saw the splitting of the Red Sea and the destruction of the Egyptians, "they *believed* in G-d and in Moshe His servant."[6] The first of the Ten Commandments, "I am

5. *Sefer HaSichos* 5752, 2:366–67.

G-d, your L-rd, who took you out of the land of Egypt"[7] is a commandment to believe in G-d.[8] However, the Torah also instructs us to have *knowledge* of G-d: "You will *know* that I am your G-d"[9] and "You should *know* this day and take it to heart that G-d is the Power in the heavens above and on the earth below. There is none other."[10]

Chabad Chassidus, which certainly emphasizes and discusses faith, is at the same time focused on intellect. In fact, the very name *Chabad* is an acronym for *Chochma* (wisdom), *Bina* (understanding), and *Daas* (knowledge)—intellect! Faith alone does not produce a complete person. The Talmud gives an example of a thief who is breaking into a house to steal and who calls out to G-d to help him succeed in his theft. This example is analyzed in Chassidus, which points out a fundamental paradox in the thief's behavior. He certainly believes in G-d or else he wouldn't be calling out to Him. Yet G-d commanded us not to steal, so he's calling out to G-d to help him do something that G-d does not want him to do. What is he thinking?

The answer is that he's not thinking, he is just believing; his belief has not been internalized, so it does not affect his thought, speech, and action. This, then, is where intellect comes into the picture: it assimilates matters of faith into the person's mind and personality, and determines his behavior.

The Previous Lubavitcher Rebbe makes a further point on the importance of an intellectual understanding of matters of faith. He writes: "Dovid HaMelech says,[11] 'Know the G-d of your fathers,' meaning that you must involve your understanding in the faith in G-d that you received from your fathers. For in divinity there are things that can be understood. Understanding divinity illuminates one's faith and it becomes a pure faith. But if one only believes in something that can also be understood, in addition to the fact that he loses the mitzva of understanding divinity, his faith is not a pure faith because it is intermingled with matters that can be understood."[12]

---

6. Shemos 14:31.
7. Shemos 20:2.
8. Rambam, *Sefer HaMitzvos*, positive commandment 1; *Sefer HaChinuch*, sec. 25; Tzemach Tzedek, *Derech Mitzvosecha*, "Mitzvas Ha'amonas Elokus," chap. 1.
9. Shemos 6:7. The *Zohar* says that this is actually a mitzva to have knowledge of G-d.
10. Devorim 4:39.
11. In his instructions to his son Shlomo, Divrei HaYamim I, 69:9.
12. *Likutei Diburim*, *Sicha* of Shmini Atzeres 5697, sec. 24.

Now, since we have to do both, the question becomes where do we start, with faith or with intellect, and what should be the relationship between them? That will be the focus of the rest of this section.

Rambam and Aristotle

The answer to "where do we start" can be seen most clearly from an analogy drawn by Rabbi Shalom DovBer of Lubavitch, the fifth Rebbe of Chabad, contrasting Rambam with Aristotle: both Rambam and Aristotle were philosophers, but there was a fundamental difference between the way in which they developed their philosophical systems. Rambam first fixed a central point and then drew a circle around it. Aristotle first drew the circle and then groped to find the center. What does this mean? Rambam set the belief in G-d as the foundation—the central point—of his philosophical investigations. Thus he was able to successfully draw the circle of intellect around it. But Aristotle did the reverse. He started with the circle—human intellect. His subsequent attempt to identify the central fundamental point did not succeed.

We see from this that the search from below must be founded on faith. Otherwise, there is no telling in which direction it may go—it may go off on a tangent.

Corresponding to this fundamental difference between the *philosophy* of Rambam and Aristotle, we find that there is a fundamental difference between the *philosophers* themselves. The Tzemach Tzedek says that both Rambam and Aristotle worked on finding proofs of the existence of G-d. The difference between them was that whenever Aristotle would find a proof for the existence of G-d, he would become more arrogant because his new-found proof made him a greater philosopher. With Rambam it was the opposite. Every time he found a new proof for the existence of G-d he would become more humble. His new knowledge of G-d would inspire within him a deeper awe of G-d.[13]

There is a sequel to the Aristotle story. The *Seder HaDoros,* a book on the chronology of world history, relates that Aristotle himself came in contact with Shimon HaTzadik and was greatly influenced by him.

13. *Derech Mitzvosecha,* 16.

The influence of "the Shimoni" (as Aristotle called him) on Aristotle was so great that he said he was ready to revise his entire philosophical system to be based on what he learned from Shimon HaTzadik. However, Aristotle felt that he was too old to start over.

Levels of Interaction

Chassidus is the wisdom that brings unity to all knowledge and leads it to the Messianic knowledge of G-d. Thus, in Chassidus we find faith and intellect discussed, not as opposing forces, but rather as two complementary methods of attaining an awareness of G-d. In fact, the Rebbe MHM says that the barrier between faith and intellect was actually broken with the redemption of the Previous Rebbe on 12 Tamuz, 5687 (1927). The miracles of that redemption were of the type where the opposing forces of the supernatural and the natural worked together. So from then on, there was no real barrier between faith and intellect.[14]

Some of the Rebbe MHM's discussions on this topic focus on Rambam's statements concerning faith and knowledge.[15] Most prominently, Rambam begins his major work, *Mishneh Torah*, by saying, "The foundation of all foundations and the pillar of all wisdom is to *know* that there is a First Being" rather than by saying "to *believe* that there is a First Being."[16] Similarly, Rambam ends *Mishneh Torah* by discussing how the world will be filled with the *knowledge* of G-d in the Era of Moshiach,[17] without mentioning faith. To explain this, the Rebbe MHM identifies various levels in the relationship between faith and intellect.

First, there is the basic faith which is a prerequisite to the observance of the mitzvos. To do a mitzva—a commandment—one must first believe that there is a Commander who commanded the mitzva. (This is not just a matter of faith, however. The existence of G-d is a

---

14. *Sefer HaMa'amorim Melukat,* 5:325.

15. The ideas developed here are based primarily on the Rebbe MHM's *Hadranim* on the Rambam, 5746 and 5750. See *Toras Menachem, Hadranim al HaRambam,* 99–117, 187–200.

16. "Hilchos Yesodei HaTorah," chap. 1, sec. 1.

17. "Hilchos Melachim U'Melech HaMoshiach," chap. 12, sec. 5.

fact which is within the realm of knowledge.)[18]

This faith, which precedes knowledge, is analogous to the commitment to observe the mitzvos which must precede the actual observance of the mitzvos—the famous declaration that the Jews made before receiving the Torah, *Na'aseh V'Nishma* ("We commit ourselves to do it; then, we will [endeavor to] understand it").

Following this, there is actually a commandment to know G-d. It is one of the 613 mitzvos.[19] This is not just a general awareness (as in the first level), but involves detailed knowledge about G-d—all the information about G-d that the human intellect can comprehend. It is this knowledge that Rambam is describing at the beginning of *Mishneh Torah* when he says "The foundation of all foundations and the pillar of all wisdom is to know that there is a First Being and He brings all beings into existence. Everything that exists in the heavens and the earth and whatever is included in them, exists only from the truth of His existence." He continues to elaborate on this throughout the first four chapters of the book.

Attaining knowledge of G-d, however, is a continuous process. Each individual can understand only so much according to his capacity at any given point in time. Faith and knowledge now interact on a second level: what I have been able to learn and comprehend about G-d until now I *know*. That which is beyond my grasp now, I *believe* in. Later, as I learn more, things for which I had to rely on faith will be understood intellectually and will become part of my knowledge. This is an ongoing process—an upward spiral—with an elegant interplay between faith and knowledge. Today's faith becomes tomorrow's knowledge. A new horizon is then opened up and I begin to grasp higher things with my *faith*. Later, this too will become knowledge. Faith and intellect are both faculties of the soul, used in conjunction to attain truth.

This interaction between faith and intellect is described clearly in one of the letters of the Rebbe MHM written in response to the question of whether to follow faith or intellect:

---

18. *Derech Mitzvosecha,* 88–89.

19. "Hilchos Yesodei HaTorah," chap. 1, sec. 6.

> Regarding the questions in your letter...whether to believe with a simple faith or to research and investigate—they are both necessary. Not only is there no contradiction between them but, on the contrary, they complement each other. The beginning and the foundation of one's service of G-d is simple faith. Immediately following this, however, there applies the mitzva which is "the foundation of all foundations and the pillar of all wisdom—to know that there is a First Being" (as Rambam says). This is expressed in the *Zohar* as follows: "This mitzva is the first of all the mitzvos...to know G-d."

(Elsewhere the Rebbe MHM explains that this statement in the *Zohar*, a book of Kabbala, was the source of this Halacha in Rambam,[20] as Rambam himself was a Kabbalist. Rambam himself hid the fact that he had mastered the hidden wisdom of Kabbala, but this fact was later revealed by the Baal Shem Tov. Thus we find some laws in Rambam's *Mishneh Torah* whose only source is the *Zohar*. In fact, his philosophical work *Moreh Nevuchim* (Guide for the Perplexed) is actually a book of Kabbala written in philosophical language.[21])

The Rebbe MHM's letter continues: "One is obligated to exert his mind to know and understand as much as he is able to about G-d. Since one's ability is always limited, what is beyond his ability he will believe in with fundamental simple faith and this will give him life."

The mitzva of eating matza on Pesach has the effect of strengthening and increasing one's faith. In a letter[22] dated 13 Iyar, 5714 (Rabbi Yisroel Aryeh Leib's *yartzeit*), the Rebbe MHM expressed the prayer that everything that we now accept on faith—and for which we still need matza to strengthen that faith—should be comprehended and integrated into our intellect. Then we can apply our faculty of faith to matters of an even higher level. "For when one's intellectual capacity increases he comes to *understand* things that were previously only matters of faith for him. But there still remain matters that his intellect is not yet capable of understanding and he continues to *believe* them with simple faith." The Rebbe MHM compares this to a child who believes certain things that an adult actually *understands* very

20. See *Toras Menachem Tiferes Levi Yitzchok*, Shemos, 66.
21. See *Likutei Sichos*, 26:26–39.
22. *Igros Kodesh*, 9:35.

well. But an adult of even greater intellect will understand things that the first adult must still accept on faith.

The analogy between the faith-knowledge relationship and the commitment-observance relationship that we mentioned earlier can be extended to this second level: after our initial commitment to the mitzvos, we apply the full capacity of our intellect to understand the mitzvos as fully as possible. This is the in-depth study of the Torah, especially as it is developed in the Talmud and Chassidus.

There is a third, more intimate connection between faith and knowledge, in which faith applies to knowledge itself. Since human intellect is finite and G-d is beyond the finite (He is even beyond the infinite), when one uses human intellect to try to understand G-d, he is applying a finite tool to something that is infinite (and beyond). Only faith has the capacity to grasp the infinite. This contrast is expressed in the entry in *HaYom Yom* quoted earlier: "Regarding learning—understanding—the greatest understanding has a limit. But faith is a feeling with no limit."

Thus, when one attains an understanding of something about G-d, he must not deceive himself and think that he has seen the entire picture. This is expressed by a beautiful analogy in a *ma'amar* (discourse)[23] that the Rebbe MHM wrote for his father's birthday, the 18th of Nissan. In the *ma'amar,* he discusses the concept of a "yard" as it applies to the laws of Shabbos. There are two kinds of yard designed to contain animals. One is a yard that has a fence around it, but no roof; the other is a yard that has both a fence and a roof. The difference between them is a very practical one. The yard with the fence alone is good enough to keep an animal in it, say a cow or a bull, but it is not good enough to contain a bird; the bird will fly away. To contain a bird, the yard must have not only a fence but also a roof.

It may seem paradoxical that a bird, which is such a delicate animal, needs such a strong protection while the bull, such a powerful animal, needs only four walls without a roof. In the *ma'amar* the Rebbe MHM explains that the animal, the bird, and the roof have parallels in the spiritual realm—in the soul of a Jew. The animal corresponds to the "animal soul" which is interested only in physical

23. *Sefer HaMa'amorim Melukat,* 5:233–34.

pleasure, in what is down below. Therefore it needs to be watched and confined, but a regular confinement—a fence—is adequate. The bird, however, corresponds to the "intellectual soul." The intellectual soul is the intermediary between the "G-dly soul" and the animal soul. Its function is to communicate and explain to the animal soul the concepts of G-dliness known to the G-dly soul. The intellectual soul by its nature is not interested in lower, animalistic things, but rather in high-level intellect, in what is above. This is simply the nature of human intellect. It always wants to know something higher, and when it attains that, it wants to go higher again—like a bird which is always attempting to fly higher and higher.

Thus the Rambam, whose towering intellect was universally recognized as transcending that of the scholars of his generation as well as generations before him, was known as "the Great Eagle." It was said of him, "From Moshe (Rabbeinu) to Moshe (the Rambam) there arose none like Moshe (the Rambam)."[24]

Because of this tendency of the intellectual soul to go higher and higher, it needs to be watched very carefully, because when it starts to understand matters of G-dliness it will run into the error of thinking that what it understands is the whole story. It doesn't realize that it only understands a finite subset of the infinite knowledge of G-d. Therefore the intellectual soul needs a special confinement—a roof over it—to tell it, "You are only finite, and what you see is only a finite view of something that is infinite, so your knowledge is not a true and complete knowledge." (Recall the arrogance of Aristotle in his intellectual achievement as contrasted with the humility of Rambam that we mentioned above.)

In the words of the Previous Rebbe: "The G-dly intellect's manner of understanding is the opposite of that of human intellect. The human intellect is able to understand something only when that thing is delimited, so first it sets the boundaries of the thing and then it understands it. But the G-dly intellect, not only does it not set any prior boundaries to the thing that it is attempting to understand, rather it shows that even after the thing is understood it still remains unbounded and infinite."[25]

24. This saying is engraved on Rambam's tombstone.
25. *Likutei Diburim,* sicha of Simchas Torah 5690 (1929), sec. 26.

Thus, the Rebbe MHM explains in the *ma'amar* that even in those matters where we have a very clear *understanding*, it's not sufficient to rely on our intellect alone; we must have faith in those areas also. The point of this third level is that we need faith not only to pick up where intellect leaves off (since we do not yet have an understanding of the higher concepts, as at the second level), but even in those areas where we already have knowledge and understanding, we still need to have faith.

Here the parallel to the observance of the mitzvos is that even those mitzvos which we can understand intellectually—such as not to kill, not to steal etc.—must be observed primarily with an attitude of commitment to fulfill G-d's supra-rational will, rather than merely with the intention of doing what is rational. The mitzvos, which are the expressions of G-d's will, are infinite in scope. What we understand of them is only a finite view of something infinite. Even when the mitzva makes perfect sense to us, it's not the whole story.

Our Knowledge is Also Life

Having developed this multilevel relationship between faith and intellect, we are now in a position to answer our original question: Why does Rambam open his major work on Halacha with a discussion of the knowledge of G-d, with no mention of faith?

One may argue that it is reasonable to omit mention of the faith that is a prerequisite for the observance of the mitzvos (level 1) in a book devoted strictly to the laws and details of the observance of the mitzvos. But we have seen that faith goes hand in hand with knowledge at all levels. Shouldn't this faith be mentioned along with the knowledge of G-d?

The Rebbe MHM explains this. Faith transcends the definitions and limitations of the world. It is infinite, as we mentioned above. But this very advantage is also its disadvantage. The goal of the mitzvos is to affect the world as it exists in its natural state, and the Jew as he exists in this world. Intellect and knowledge are part of the structure of the finite natural world, so they are the *objects* of the mitzva of the awareness of G-d.

The correct path to follow lies in combining faith and intellect—but faith must be the foundation. The great Kabbalist Rabbi Levi

Yitzchok Schneerson, the father of the Rebbe MHM, discusses this issue in a letter[26] he wrote to the Rebbe MHM and Rebbitzin Chaya Mushka. He says that the *Etz HaChaim* (Tree of Life) and *Etz HaDa'as* (Tree of Knowledge) refer to faith and intellect, respectively. Faith is life, as the Prophet Habakuk says, "The tzadik lives through his faith." "If the faith precedes the knowledge," Rabbi Levi Yitzchok writes, "then the knowledge is also life....These two trees must be joined together....This is what it says [in the prophecy of Isaiah[27]] about Moshiach—that 'faith will be the belt of his body.' Only at the end of the prophecy does it mention knowledge—'The earth will be filled with the knowledge of G-d.'"

In the Era of Moshiach, faith and intellect become one. In general, the Era of Moshiach is characterized by the paradoxical unification of opposites, made possible by the revelation of *Atzmus,* the essence of G-d, which is above and beyond all definition and description. Relative to *Atzmus,* all opposites are equal. (Just as one would say that relative to a point in distant space, the top and the bottom of a tall building here on earth are equal.)[28] In particular, faith and intellect are unified. All those concepts which are higher than intellect and for which we now have to rely on faith, will be understood within the definitions and structures of human intellect.

The major issue in faith in our time is the faith in Moshiach—faith in the Rebbe Melech HaMoshiach himself and in his statements concerning the presence of the Geula. The correct path to follow is the same as in all matters of faith and intellect: We must start with the foundation of faith in the declarations and prophetic statements of the Rebbe MHM, in which he stated that we are in the Era of Moshiach and that the Geula is already here (level 1). Based on this foundation, we must develop a new intellectual outlook—a Geula mentality—to look at and analyze the world differently.

The Rebbe MHM has said, "All we have to do is open our eyes and then we see that the true and complete Geula is already here, literally."[29] Indeed, there are several major obvious Geula events that the

26. *Likutei Levi Yitzchok, Igros Kodesh,* 413–14.
27. Yeshaya, chap. 11.
28. This concept is discussed at greater length in the next section and in chap. 5, sec. 2b, "The Messianic World Order."
29. *Sicha* of 19 Kislev, 5752. Sefer HaSichos 5752, 1:173.

Rebbe MHM himself discusses in his *sichos,* such as the revelation of Melech HaMoshiach, the Jews leaving Russia peacefully to go to Israel ("Ingathering of the Exiles"), the "Swords Into Plowshares" transformation, etc.

For those aspects of the Geula that we are not yet able to identify—our eyes are not open to them—we have faith in the words of the Rebbe MHM that we will see them too, immediately (level 2). And even in those aspects that we do see, there is more to them than meets the eye (level 3).

Based on the faith that everything that is happening in the world and in our personal lives is part of the unfolding of the Geula, we come to understand the world and ourselves differently. That is the goal of this book.

In this way we truly live a Geula life and our "knowledge is also life."

The Wisdom From Above: Kabbala, Chassidus, and Moshiach

In the six hundredth year of the sixth millennium the gates of wisdom above will open up and the fountains of wisdom below, and the world will be set to enter the seventh.

—Zohar[1]

A Long Chain

In this prediction, the Zohar speaks of the flood of wisdom beginning in the year 5600 (1840) preparing the world for the Era of Moshiach. Wisdom descends from above and wisdom rises up from below. In the "Sicha of Noach" the Rebbe MHM explains that the wisdom from above is Chabad Chassidus which clearly prepares the world for the Era of Moshiach since it has flooded the world with the knowledge of G-d preparing it for the fulfillment of the Messianic prophecy that "the world will be filled with the knowledge of G-d as the water covers the ocean bed." The wisdom from below refers to the birth of modern science—the discoveries and unifying theories in the sciences that began at that time, which also play a role in preparing the world for Moshiach.

In this section we review some of the history of the revelation and development of the wisdom from above—Kabbala and Chassidus—

1. *Zohar* I:117a.

and the *tzadikim* who brought this about. In the next section we will discuss the wisdom from below.

We begin with a story.[2] Shortly after the Spanish Inquisition, a Spanish Jew named Joseph Jospa arrived in Krakow, Poland. He was a great scholar and a tzadik and greatly respected by the Jews of Krakow who called him "the Spanish Tzadik." Being fifty years of age and unmarried, he lived by himself.

Thirty years went by in this manner until a tragic event changed this. A young businessman from Krakow was killed during a business trip to Prague, leaving a widow with no children. Thus a *chalitza* ceremony was performed by the businessman's brother in the Beis Din of Krakow. It was the custom of Krakow in those days for the chalitza ceremony to be a community event after which the Rabbi of the Beis Din would arise and give the woman a blessing that she should soon marry and have children. Then an announcement would be made that if any man present wished to marry the woman he should present himself to the Beis Din.

No one responded on that particular occasion, but about five months later, Joseph Jospa, the Spanish Tzadik, suddenly came to the Beis Din and announced that he wished to marry the widow if she would agree. He explained that he had not intended to get married, but now, for certain reasons which he did not wish to reveal, he wished to marry in spite of his advanced years.

The Beis Din then sent for the widow. Immediately upon arriving in the Beis Din, even before she had a chance to ask why she was summoned, she burst into tears.

"Why are you crying?" they asked her.

"I have a terrible secret weighing me down, yet I can't make up my mind whether or not to tell you about it," she replied.

They continued to ask her what the matter was. All she would say is that it was connected with a dream in which her father, who had passed away many years before, appeared to her and asked her to do something, and she could not decide whether or not to listen to him. She said she was worried since she had had the dream several times and asked the Beis Din what to do.

---

2. *Lubavitcher Rebbe's Memoirs* 2:28–35.

After discussing the matter amongst themselves, the Rabbis of the Beis Din told her that it would be best for her if she would tell them what the dream was. She then said that in her first dream her father appeared to her dressed in his Shabbos clothes, put his hands over her head, blessed her and said, "And now I wish you *mazal tov* for it has been decreed that you marry the Spanish Tzadik, Joseph Jospa." She awoke from this dream trembling violently but then put it out of her mind. She had the dream again, but again did not take it seriously. Then her father appeared to her looking very serious and told her that there was no way out of it as it had been decided in the Heavenly Court. She must speak to someone to arrange the marriage. If she listened to him, he continued, she would be blessed with a son. But if she refused she would come to a bad end.

Three more times she had the dream when she finally decided to go to the Beis Din about it. She had just made the decision to go when the messenger arrived informing her that the Beis Din had sent for her.

When she finished her story the Rabbis of the Beis Din looked at each other in amazement and told the woman that Joseph Jospa himself had come to them and told them that he wanted to marry her. She now had no doubt that it was G-d's will that she marry the Spanish Tzadik and the marriage was arranged. The wedding was a great celebration for the whole community and everyone in Krakow felt that this was no ordinary wedding, but that there was an inner significance beyond their comprehension.

In the second year of their marriage they were blessed with a son whom Joseph Jospa named Eliyahu after *Eliyahu HaNavi* (Elijah the Prophet). From the age of two his father started teaching him Torah and he studied diligently. His father taught Eliyahu personally until he was of bar mitzva age.

About two weeks before his son's bar mitzva, Joseph Jospa told his wife that he felt that he was about to pass away. He told her that after their son's bar mitzva, Eliyahu would tell her that he wants to go out into the world and that she should not discourage him from doing this as he had been sent down to this world to fulfill a special mission. He told her that when her first husband had been killed, Joseph Jospa received a divine command to marry the widow, for a son of very high stature would be born to them with a special mission to fulfill for the Jewish people, to help them and uplift them. Eliyahu HaNavi himself had been studying with their son Eliyahu to prepare him for this mission. He was to be the first in a long chain of tzadikim leading up to the coming of Moshiach.

After concluding these instructions to his wife, Joseph Jospa passed away. A few weeks after Eliyahu's Bar mitzva he told his mother that he wanted to go out in to the world. Having been prepared for this, she did not object. She gave him her blessing and he left. Forty years later, in the year 5350 (1590), he appeared in the city of Wurms, Germany where he became known as a miracle worker and a healer. He also established a yeshiva there where he taught Kabbala—particularly the Zohar—in addition to the Talmud. He was the famous Rabbi Eliyahu Baal Shem.

Rabbi Eliyahu Baal Shem was indeed the first in the long chain leading up to the revelation of Melech HaMoshiach. Rabbi Eliyahu himself was the first of four "Baal Shem's." He was succeeded by his disciple Rabbi Yoel Baal Shem, then by Rabbi Adam Baal Shem who was succeeded by his disciple Rabbi Yisroel—the famous Baal Shem Tov.

With the Baal Shem Tov two major changes occurred that would affect the rest of the chain. Firstly, the earlier Baal Shem's and their followers were for the most part "hidden tzadikim" and they did their work secretly. The Baal Shem Tov, however, was instructed by his teacher the prophet Achiya HaShiloni[3] to reveal himself as a tzadik and Rebbe and to start doing his work openly. Secondly, as the Baal Shem Tov wrote in a famous letter,[4] one Rosh HaShana his soul ascended to the heavenly chamber of Moshiach and he asked Moshiach, "When is the master coming?" to which Moshiach replied, "When the wellsprings of your teachings spread to the outermost reaches." From that time on the spreading of the teachings of the Baal Shem Tov and his successors would become an integral

3. Achiya HaShiloni was a prophet who lived during the time of the first Beis HaMikdosh. He was a disciple of Dovid HaMelech and the master of Eliyahu HaNavi. Achiya HaShiloni appeared to the Baal Shem Tov many generations later to instruct him in the hidden dimension of the Torah. (Eliyahu HaNavi was famous for appearing to tzadikim throughout the generations to teach them.)

4. Printed at the beginning of *Kesser Shem Tov.*

component of the continuation of the chain, as it had been revealed that this would bring Moshiach.

The Baal Shem Tov was succeeded by his disciple Rabbi Dov Ber who was known as the "Maggid of Mezeritch." After Rabbi Dov Ber passed away the chain branched out in several directions as his disciples set up centers in different parts of Europe. The main link in the chain, however, was the Maggid's primary disciple, Rabbi Schneur Zalman of Liadi. A fellow disciple of the Maggid, Rabbi Levi Yitzchok of Berditchev, once said of Rabbi Schneur Zalman, "We all ate from the same bowl but he got all the cream."

Rabbi Schneur Zalman was a seventh generation direct descendant from the Maharal of Prague who was a direct descendant of King David.

Rabbi Schneur Zalman took a revolutionary approach to spreading the teachings of the Baal Shem Tov and the Maggid. He maintained that it was not sufficient to merely relate the teachings of Chassidus because the concepts could be fully understood only by scholars well versed in the Talmud and Kabbala. A system had to be devised whereby the average Jew could actually learn and understand the teachings of Chassidus. Thus he constructed the system of Chabad, a system of Torah study which was based primarily on Kabbala but which was actually a synthesis of Kabbala, Talmud and Halacha. It had its own set of definitions and concepts and could be studied by anyone in a progressive manner—starting from introductory and elementary concepts and working up to the most abstract and delicate ones. This made Chassidus accessible to everyone.

This new system encountered tremendous opposition not only from those who were opposed to the Baal Shem Tov and Chassidus but even from some of Rabbi Schneur Zalman's colleagues—fellow disciples of the Maggid who felt he was doing too much too fast. It should be noted that even the Gaon of Vilna, who was considered to be the head of the opposition to Chassidim in his time writes, "If one does not understand the secrets of the Torah (the *sode*), even the simple meaning (the *pshat*) is not clear to him. One who was able to comprehend the secrets of the Torah but didn't make the effort is judged severely."[5] Furthermore, he writes that the Geula is depend-

5. *Even Shleima*, chap. 8.

ent on this study of the deeper levels of the Torah.[6] (But he was against teaching it to the public.)

Some of Rabbi Schneur Zalman's opponents even tried to stop him by slandering him to the Russian government, saying that the charity that he was sending to *Eretz Yisroel* (the Land of Israel) was really being sent to the Turkish government (which controlled Eretz Yisroel at that time) to finance a war against the Czar. This resulted in Rabbi Schneur Zalman being arrested on charges of treason.

The great Kabbalist, Rabbi Levi Yitzchok Schneerson, the father of the Rebbe MHM, explains that whenever there is a new innovation in the world in the area of holiness there is opposition to it. The higher the level of the holiness being introduced into the world, the opposition itself will be that much greater and it will originate from a holier source i.e. the opposition comes from "closer to home." Thus, for example, when King David was first anointed king, thus establishing the dynasty that would lead up to Melech HaMoshiach, his main opposition was from a man who was a perfect tzadik—King Saul. Similarly, Rabbi Schneur Zalman encountered opposition from some of his fellow disciples of the Maggid.

While in prison Rabbi Schneur Zalman was visited by two illustrious visitors from Gan Eden, the Baal Shem Tov and the Maggid of Mezeritch. During their visit Rabbi Schneur Zalman asked them, "Why did this happen to me?" They explained to him that just as there was a controversy on earth regarding his innovation, there was a corresponding controversy in the heavenly court on the matter, resulting in the Alter Rebbe's imprisonment. He asked them if he should stop teaching Chassidus so extensively when he is released from prison. "On the contrary," they responded. "When you get out of here it will be an indication that the heavenly court has decided in your favor and you should teach Chassidus even more extensively." And so he did. After his miraculous release he developed Chabad Chassidus even more extensively and taught it more intensely.

Rabbi Schneur Zalman was succeeded by his son Rabbi DovBer who was succeeded by his son in law Rabbi Menachem Mendel Schneerson, known as the Tzemach Tzedek, who was also the grandson of Rabbi Schneur Zalman. The chain continued in a father to son

6. Quoted in *Igros Kodesh* 16:10.

succession for three generations after the Tzemach Tzedek: Rabbi Shmuel, Rabbi Shalom DovBer and Rabbi Yosef Yitzchok Schneerson. A major turning point came in 1941 when Rabbi Yosef Yitzchok made his famous announcement, "Immediate *Teshuva*—immediate *Geula*." The significance of this announcement was explained later by the Rebbe MHM. Until this announcement the immediate objective of Chabad was to develop and spread the teachings of Chassidus as a goal in itself, knowing that this would ultimately bring the Geula and Moshiach. But with this announcement the emphasis changed and the immediate objective of Chabad became bringing the Geula and Moshiach.[7]

The Final Link—Melech HaMoshiach

Upon the passing of Rabbi Yosef Yitzchok in 1950, the leadership of Chabad continued with his son in law Rabbi Menachem Mendel Schneerson, who is also a fifth generation descendant from the Tzemach Tzedek. In his first public Chassidic discourse[8]—on the first yartzeit of Rabbi Yosef Yitzchok, 10 Shevat, 5711 (1951)—he declared that it was the task of this generation to bring Moshiach. Those with "opened eyes" understood immediately that he was indicating to them that he himself is Moshiach and some of them actually addressed him as Moshiach. A letter sent to the Rebbe MHM by the chassidim of Kfar Chabad, Israel, on the occasion of his birthday that year was addressed him as Melech HaMoshiach and was well received. While this was accepted privately, it was not to be publicized. Rather, the chassidim were to stick to the work involved in bringing about the revelation of Moshiach by spreading Torah—especially the teachings of Chassidus—and the observance of mitzvos *throughout the world.*

This was the innovation of the Rebbe MHM which forged the final link in the chain—the emphasis that Torah and Chassidus must reach every *place* in the world, and every *person* in the world including non Jews who were to be taught the Seven Noahide Laws which G-d gave to all nations of the world through Moshe Rabbeinu, and that every *thing* in the world be involved in this. This included such

7. *Sefer HaSichos* 5750, 2:629.
8. See *Sefer HaMa'amorim Melukat* 1:3–10.

former enemies of religion as the sciences. Science and technology would now work hand in hand with Chassidus to prepare the world for Moshiach.

A major turning point came at the end of 1990 when the Rebbe MHM announced, "The time of your redemption has arrived," and this was to be publicized. It then became widely known that he is Moshiach. (We discuss this in detail later in this chapter, in the section "The Emergence of Melech HaMoshiach.")

Rabbi Joseph Jospa had told his wife that their son, Rabbi Eliyahu Baal Shem, would start a long chain of tzadikim leading up to Moshiach. Now the chain was complete.

The Concepts of Chassidus

"The main revelation that will take place in the future Geula is the revelation of the infinite Atzmus (essence of G-d) which is higher than comprehension."[9]

While the mysteries and secrets of the Torah—the subject matter of Kabbala and Chassidus—was always part of the Torah, it was the domain of the great scholars and was supposed to be kept hidden just as a king's greatest treasures are kept in his innermost chambers. On the day of the king's wedding however, he shows his full glory. Thus when G-d gave the Torah to the Jews at Mt. Sinai (the day of His "marriage" to the Jewish people) He revealed His glory and everyone saw His throne and all the mystical revelations that go along with it. But after that Kabbala was a private matter.

The Zohar, written by Rabbi Shimon Bar Yochai after the destruction of the Second Beis HaMikdosh, was the first book on Kabbala after the Torah was given. Rabbi Shimon was told by Eliyahu HaNavi that through the in depth study of his book the Jews would be able to leave their exile in a merciful manner.[10] Here too, as with Moshiach's statement to the Baal Shem Tov, we see how the study of the deeper dimension of the Torah—Kabbala and Chassidus—is connected with the coming of Moshiach. The "in depth study" is essential here since the intention is that the Jewish people live off of

9. Ibid. 3:109.
10. *Zohar* 3:124b.

Kabbala and Chassidus in the sense that they would understand it intellectually and this would affect their way of thinking and their emotions and ultimately their actions. This, again, was the motivation behind the Alter Rebbe's development of Chabad.

The Zohar, however, was a secret book and remained hidden for centuries. Similarly, many scholars who studied Kabbala kept it a secret and if they wrote about it they would clothe the exposition in other terminology. For example, Rambam who was famous as a great Talmudist and master of Jewish Law, was also a great Kabbalist and his Guide for the Perplexed was actually a book on Kabbala clothed in the terminology of philosophy.[11]

It was the great master of Kabbala, Rabbi Yitzchok Luria, the AriZal, who changed all this when he announced that, "In these later generations it is permissible and it is a mitzva to reveal this wisdom" of Kabbala. This was the first major step taken in the revelation of the hidden teachings of the Torah. The second major step was by the Baal Shem Tov and his successors who began to spread these teachings openly, as we mentioned above.

It is beyond the scope of this book to give a complete review of the concepts of Chassidus.[12] However, we will attempt to present some of the most fundamental concepts.

What does Chassidus teach about G-d and creation? There is, first of all, a fundamental question which must be answered: Before the creation of the world there was only G-d Himself, an absolute and infinite Existence. So where was there any "room" for a world to exist? Doesn't the absolute existence of G-d precludes the existence of anything else.

The AriZal explained this using the concept of *tzimtzum,* which means a contraction or a restriction. In the *Etz Chaim,* his fundamental work on Kabbala, he writes, "Before the creation, a simple exalted light filled all of existence and there was no free space—no empty space.... This is called the Infinite Light. When G-d wanted to create the worlds...He "restricted" Himself (*tzimtzum*).... After this tzimtzum...there was already space where there could be creations. He then drew, from the Infinite Light, one straight line."[13]

---

11. *Likutei Sichos* 26:27.
12. See the references at the end of this section for suggestions for further reading.
13. *Etz Chaim*, "Igulim V'Yosher," *anaf* (sec.) 2.

This is a very delicate concept and has been misunderstood even by those well versed in Kabbala. The Alter Rebbe explains firstly that this tzimtzum does not apply to G-d Himself—the *Atzmus*—Who never changes. It refers only to the infinite light emanated by G-d. Secondly, tzimtzum does not even imply a reduction in this light but only a hiding of it. He explains this with an example. Consider a person who has studied a certain topic in the Talmud thoroughly. As he thinks about it, the content and the words of the Talmud together with his own thoughts on the topic fill his mind. Then he turns to think about something else. At this point the words and concepts of the first topic are no longer out in the open in his mind. But they are certainly still in his mind in a hidden manner and he can go back to the first topic and bring all of it or part of it out in the open whenever he wants. This inward withdrawal of the topic into his mind is analogous to the tzimtzum. It is still there but it's just hidden.[14]

From this hidden light, the AriZal writes, G-d brought forth a short and thin line—a "thread"—from which the worlds were created. We say "worlds" instead of "world" because the physical world is actually the last stage in the process of creation, a process starting from a totally spiritual world, *Atzilus*, to our physical world, *Asiya*. Specifically, there are four worlds in the order of creation, distinguished from each other by the level of "reality" in each of them.[15]

The first world, *Atzilus*, is the world which is closest to G-d. The name comes from the Hebrew word a*etzel* which means near. At this level the real world does not yet have any reality. This world is "near" G-d and is characterized by the "absence of existence."

The second world is *B'riya* which means creation. At this level there is something but the something is as yet undefined. This world is characterized by the "possibility of existence" or potential existence.

The third world is *Yetzira* which means formation. At this level the created entity has a definite form and is said to exist. Thus this world is characterized by "existence."

14. *Likutei Torah*, VaYikra, 52c.

15. In this context it is interesting to note that one of the exotic, yet mainstream, interpretations of quantum mechanics, known as "many worlds theory," assumes the simultaneous existence of many parallel universes.

The fourth world is *Asiya* which means "made" or "done" i.e. the creation process is complete. The created entity attains its full existence with all its properties fully expressed. On a deeper level, it is in this world where not only the *act* of creation reaches its completion but also the original *intention* of creation reaches its completion—to make a physical world which will then be transformed into a holy place, a "residence" for G-d. This is attained fully in the Era of Moshiach.[16]

Thus the Era of Moshiach is the ultimate goal of creation as the Alter Rebbe says in Tanya, "The Era of Moshiach—in particular when the dead are revived—is the ultimate perfection of the creation of the world. It is for this that the world was created in the first place."[17]

These worlds and their names are alluded to in a verse from Isaiah about the creation, "I have *created* it, *formed* it even *made* it."[18] (The verse uses the Hebrew words *b'riya, yetzira, and asiya*. The world of *Atzilus* is indicated here indirectly.) The transition between each of these worlds and the next lowest world also requires a tzimtzum.

Actually, G-d the Omnipotent had the ability to create the world *directly* out of nothing, skipping the tzimtzum. But he did it through a process of tzimtzum, to provide structure to the process of creation, so that it would be more comprehensible to the human mind.

What is Chassidus?

One of the differences between Chassidus and Kabbala is that Chassidus not only teaches the knowledge of G-d, creation, the soul etc. found in the inner dimension of the Torah, but it also teaches—and is in fact focused on—the application of this knowledge in the service of G-d. Thus the Alter Rebbe, in his major work the *Tanya*, writes that there is a basic lesson to be learned from the fact that G-d created the world through this process of tzimtzum: "G-d put aside his great infinite light, so to speak, and hid it in three different *tzimtzumim* [between the worlds] because of His love for man—to elevate him to G-d.... Similarly—even more so—it is proper for man

16. This description of the four worlds is from *Sefer HaMa'amorim Melukat* 5:236–37.
17. *Tanya*, chap. 36.
18. Yeshaya 43:7.

to leave everything he has, from his soul to his flesh, and to give it all up in order to be attached to G-d with a strong willed desire and attachment. Nothing should stand in the way."[19]

So what exactly is Chassidus? This question was asked of Rabbi Shalom DovBer of Lubavitch, the 5th Rebbe of Chabad, in the summer of 1895 by a group of scholarly Jewish intellectuals who visited him and asked him various questions. Their question was expressed as follows:

> "What is Chassidus? On the one hand, it appears that Chassidus is a profound religious philosophy which deals with the deep issues of the existence of the Creator and the creation of the world, and finds all encompassing expla tions for them. On the other hand, Chabad Chassidus kindles a fiery flame in the heart of a person fulfilling a mitzva and brings him to the highest level of 'pouring out his soul' during prayer, and it motivates and inspires him to love his fellow Jew. Chabad Chassidus," they concluded, "appears to contain two opposite extremes—cold intellectual undstanding and fiery emotion."

Their question reminds one of the saying of the *Rishonim*: The question of a wise man is half the answer.

> The Rebbe responded as follows: "Indeed, that is Chassidus—fiery emotion within intellectual understanding, but with a certain criterion. When is the Chassidic understanding true? When one prays with the understanding that he has attained. A fiery prayer is the test of true understan ing. And when is the prayer truthful? When one acts with good character after the prayer. Proper conduct throughout the day is the test of the sincerity of the 'service of the heart' during one's prayer."[20]

The Role of the Mitzvos in the Order of Creation

The entire process of tzimtzum, as we said above, relates only to the infinite light emanated by G-d, but G-d *Himself*—the Atzmus—

19. *Tanya,* chap. 49.
20. *Likutei Diburim,* Sicha of 18 Elul, 5701.

is unaffected by it. The Atzmus, however, while omnipresent is not directly sensed in a coarse material world. Thus, there is a "gap" between the Creator *Himself* and the physical universe which He created. He is here but His presence cannot be felt. How is this gap to be bridged?

It is the role of the mitzvos—instructions from G-d that are to be performed with physical objects—to forge the connection between the physical world and G-d, and reveal the Atzmus in the physical world. The performance of the mitzvos refines the world from its coarse materialistic state to be more spiritual.

The famous question of the wise son of the Passover *Haggada*, "What are the *Edos, Chukim* and *Mishpatim* that G-d commanded?"[21] identifies the mitzvos by three different characteristics. *Edos*—testimony—refers to the fact that the mitzvos testify to something about G-d. *Chukim*—decrees—refers to the fact that the mitzvos have aspects that cannot be understood in human reason; they are just decreed by G-d. *Mishpatim*—rational laws—refers to the fact that the mitzvos have aspects that are understood in human reason. Every mitzva has these three characteristics and each of these characteristics has the effect of bringing G-dliness into the world in a different way.[22]

The rational aspect (*Mishpatim*) has the effect of revealing that aspect of G-dliness that can be understood rationally. The fact that G-d fills the world with life, making things move, grow and live can be understood rationally. One can become aware of this by his own observations just as he is aware of the soul within his own self that makes him live. The decreed aspect (*Chukim*) of the mitzvos reveals the aspect of G-dliness which is higher than the G-dly life in each individual creature. This is the G-dly life that encompasses the entire universe uniformly. We cannot understand the nature of such a life but we can come to the realization that there must be such a level of life, just as a complete intellectual system recognizes its own boundaries and the limitations of intellect itself, and brings one to the awareness that there is a level higher than intellect.

Testimony is something different. A known fact, or something that

21. Devorim 6:20.
22. See *Sefer HaMa'amorim Melukat* 4:216–18.

can be inferred from a known fact, does not require testimony. Testimony is really needed only when the fact in question is not known or not even knowable. The aspect of testimony (*Edos*) in the mitzvos reveals the unknowable Atzmus in the world.

More generally, all the 613 mitzvos can be divided into three categories: 1) The mitzvos that are primarily a testimony (such as Shabbos and the Jewish holidays which testify to the fact that G-d created the universe and performs miracles) are in the category of Edos; 2) the mitzvos that are primarily decrees (such as the mitzvos of Kosher food) are in the category of Chukim; 3) the mitzvos that are primarily rational (not to kill, not to steal etc.) are in the category of Mishpatim.

The mitzvos that are *Edos* have the special quality of revealing Atzmus in the world. In addition, mitzvos that deal directly with infinity have this quality. As an example, the mitzva of having children is an infinite process since one's children themselves have children etc. indefinitely, generation after generation. Furthermore, the conception and birth of a child is an act of creation, similar to the creation of the universe which can only be done by G-d Himself—the Atzmus.[23]

The revelation of the Atzmus in the world by fulfilling the mitzvos is an ongoing, cumulative process over the millennia whose climax is the total refinement of the world, making it a "dwelling place" for G-d, i.e. a place where the Atzmus is fully revealed. This is attained in the Era of Moshiach and is, in fact, fundamental to the Geula, as the Rebbe MHM writes, "The main revelation that will take place in the future Geula is the revelation of the infinite *Atzmus* which is higher than comprehension."

Another particularly powerful mitzva in this respect is the mitzva of Tefillin. Specifically, the Tefillin on the head "draws down" the Atzmus to the world of Atzilus, and the Tefillin on the hand brings the Atzmus down to the lower worlds. Many people who came to the Rebbe MHM for blessings for children, health or financial success were told to have their Tefillin checked to see if they were defective or, if they were not yet putting on Tefillin, to begin doing so. Indeed, having children is connected with the Atzmus, as we explained

23. Ibid. 2:81.

above, and the *neshama* (Jewish soul) which comes from G-d Himself must descend through all the higher worlds before it comes down to this physical world.

Prior to the Six Day War in Israel in 1967, the Rebbe MHM initiated his famous "Tefillin Campaign," promising that it would bring about a miraculous victory in the war. The miraculous victory is well known.

A professor of computer science, Dr. Abraham Polichenco, once visited the Rebbe MHM. They discussed various matters of Torah, science etc. and the Rebbe MHM told him that everything in the world has a source in Torah since G-d used the Torah as His "blueprint" to create the world. The professor asked, "What is the source of computers in the Torah?" and the Rebbe MHM responded, "Tefillin." He explained that the similarity between them is that computers connect to other computers and communication takes place between them even though the connection is not always visible. Similarly, Tefillin connects a Jew with G-d and with all other Jews even though one may not feel the connection taking place.

The fact that the Atzmus remains unchanged throughout the entire process of tzimtzum and creation is extremely important and has many implications. First and foremost it means that while there are vast differences between the higher more spiritual levels of creation and the lower material levels, all these levels are equal relative to the Atzmus. Indeed, the material and the spiritual are equal relative to the Atzmus which is neither material nor spiritual and is unaffected by their creation.How can two vastly different things be equal relative to a third thing? Consider, for example, two vastly different finite numbers such as 2 and 1,000,000,000. While one is much greater than the other, they are both equally insignificant relative to ∞ (infinity). In a mathematical calculation, any finite number relative to infinity is just as insignificant as any other.[24]

One important implication of this is that G-d maintains a personal relationship with the world. In the Kuzari, the famous classic book on Jewish philosophy written by Rabbi Yehudah HaLevi, the story is told of the king of an eastern nation called the Khazars who is searching for the true way to serve G-d. Repeatedly he has a dream in

24. *Tanya* chap. 48.

which he is told that his intentions are good but his pagan way of serving G-d is not correct. So he begins his search to find the truth. First he asks a philosopher what he believes about the Creator. The philosopher tells him, "The philosophers hold that He is above having any knowledge of the details of things because they change and the knowledge of the Creator does not change. He doesn't know you," the philosopher continues to say to the king of the Khazars, "and He certainly doesn't know your intentions or your actions and most certainly does not hear your prayers."[25]

The error of the philosopher was that he was missing just this point—that it is *because* G-d does not change throughout the act of tzimtzum and creation that He is omnipresent and omniscient just like before the tzimtzum. He is not "too high" to know what's going on in the world because the world is not "far away" from Him. In the process of creation the world never left Him and is therefore not removed from Him. The king of the Khazars and his entire nation ultimately converted to Judaism.

The Universe and the Jew

In the Talmud and Kabbala the Jew is called *"Olam Katan,"* a small universe, a microcosm. The universe was created for man and it is man's role to bring the universe to its perfection. Therefore man was created in such a way as to parallel the structure of the universe—the nature and structure of the universe are reflected in him. For example, a complete human act goes through three stages. First one thinks about an issue. Then he talks about it with other people to attain a better understanding of it. Finally, he acts on the matter. These three stages—thought, speech and action—are the three worlds of B'riya, Yetzira and Asiya on the human level.

This intimate connection between man and the universe has far reaching implications. It means that by making a change in oneself one can affect a change in the entire universe. Even the smallest act—good or bad—can affect the entire universe. Thus, the Zohar says, that when G-d created man the entire heavens and earth trembled. Rabbi Levi Yitzchok explained that this was because of the fear

25. *Kuzari,* part 1, beginning of sec. 1 (p. 20 in classic edition).
26. See *Toras Menachem–Tiferes Levi Yitzchok,* Bereishis, 28.

that man, the microcosm, might do something that would harm the entire universe.[26]

This is expressed in Rambam's famous statement that one should always look upon himself and the world as being perfectly balanced between good and evil so that the next act he does may tip the balance toward redemption or the opposite.[27] This again emphasizes the extreme importance of each individual observing the mitzvos, studying the Torah, doing acts of goodness and kindness etc.

The Rebbe MHM once said, while speaking about giving charity, that the next coin that a child somewhere in the world puts in a charity box may have the power to tip the balance of the world and bring the Geula to its completion.

27. Rambam, "Hilchos Teshuva," chap. 3, sec. 4.

Wisdom From Below— Elements of Modern Science

The fact that there are rules at all to be checked is a kind of miracle; that it is possible to find a rule, like the inverse square law of gravitation, is some sort of miracle. It is not understood at all.

—Richard Feynman[1]

Having discussed the wisdom from above in the previous section, we now turn to the wisdom from below. In the "Sicha of Noach," the Rebbe MHM explains that the scientific discoveries and the theories developed since 1840 show a high degree of unity within nature, and that this awareness of unity in the physical world prepares us for the revelation of the ultimate unity, the absolute unity of G-d, that will be revealed in the Era of Moshiach. Thus, our emphasis will be on unifying theories discovered and developed since the mid-1800s. We will begin our discussion with some general issues related to the methodology of science and the philosophy of science. Specifically, we will discuss the axiomatic method in mathematics and science (which is itself a unifying theory), and the logical structure and probabilistic nature of scientific theories. (In chapter IV, "The Convergence of Science to Torah," sections 3, 4 and 5, we develop these ideas further.)

We will then turn our attention to various unifying theories in physics developed at that time, first and foremost among them being

1. Richard Feynman, "The Uncertainty of Science," in *The Meaning of it All* (Cambridge, MA: Perseus Books, 1998), 23.

the theory of Electromagnetism. (In chapter V, "The Quest for Unification," sections 3, 4 and 5, we discuss quantum mechanics and string theory as well as unifying theories in general.)

The Logic of Science

Generally speaking, all scientific and mathematical thought uses two methods of reasoning, induction and deduction. In induction, first one makes observations, then draws a conclusion based on a generalization of those observations. In deduction, one makes assumptions and employs a system of logic to reason with those assumptions until he reaches a conclusion. Neither of these methods, however, gives us knowledge of the *truth*.

Induction does not arrive at the truth because we can only make a limited number of observations. No matter how many observations we make, we can never make all the possible observations so all we can ever say is that our conclusion is more or less probable. Induction is concerned with probability, not truth.

Deduction does not arrive at the truth because we cannot say that the assumptions which the conclusion is based on and the system of logic employed are true. All deductions are merely *"if / then"* statements which say that *if* you accept the assumptions and the system of logic, *then* the conclusion follows. If one does accept the assumptions and the system of logic, and the conclusion follows from the assumptions, then the conclusion is said to be *valid*. Deduction is concerned with validity, not truth.[2]

Do not be surprised that we cannot use science to reach the truth, because science was never designed as a tool for discovering truth. To try to use science to discover truth would be a misuse of science, perhaps even an abuse of science. Science is a type of wisdom, but with wisdom alone we cannot attain truth.

It would be a very closed mind indeed that thinks that there is no truth except what we learn from science. It reminds me of the famous world chess champion, Bobby Fisher, whose whole world

2. See Howard Eves, *Foundations and Fundamental Concepts of Mathematics*, 3rd ed. (New York: Dover Publications, 1997), 5–9, for a discussion of induction and deduction, including examples where valid reasoning may lead to conclusions that are certainly not true.

was chess. If anyone around him would talk about anything else, he would snap at him, "What does that have to do with chess!?"

The Role of Axioms

Modern science has come to the realization that no rational system of thought can be said to be absolute or "true." Every such system must be based on a set of axioms or assumptions which people are free to accept or reject. The most dramatic example of this concerned something as fundamental as the mathematical concepts of space—geometry. For many centuries everyone accepted the system of Euclid that was based on *his* axioms. But in the 19th century a few mathematicians proposed their own sets of axioms yielding different geometries and it was understood that these new non-Euclidean geometries were just as valid as Euclid's. One of these geometries was later used by Einstein in his development of general relativity. This will be discussed in more detail later.

This fundamental role of the axioms in any logical or scientific system, is expressed clearly and concisely by the great *Gaon* and Kabbalist, Rabbi Levi Yitzchok Schneerson in a letter written to his sons, the Rebbe MHM and Rabbi Yisroel Aryeh Leib:

"In mathematics, as well as in every intellectual matter, there must be two things: 1) elementary concepts and axioms, for example, that the whole is greater than any of its parts. 2) On this [foundation] all the mathematical ideas and theories are built. These two things are chochmah *and* binah. *Chochmah is the basic concepts and binah is the ideas that are built on the basic concepts."*[3]

Rabbi Levi Yitzchok continues the letter by explaining the Kabbalistic basis of this structure.

The point is that every logical system is of necessity based on fundamental assumptions—*axioms*—which themselves are not *logical* but are adopted by convention i.e. people who like these axioms agree to use them, but there is nothing necessarily compelling about them. One is free to take them or leave them and replace them with his own axioms which may lead to different conclusions. The *logical* part of intellect is the process of deriving new statements and ideas

3. *Likutei Levi Yitzchok,* "Igros Kodesh," 250.

from the axioms. In this process, however, we can only say that the conclusions are logically compelling *if one accepts the axioms*. The whole system has an if/then structure, as we explained above.

The Axioms of Science and the Axioms of Torah

A Jewish man I know was raised in a non-observant environment. His exposure to Judaism was minimal and the many questions that he had about Judaism went unanswered. Around the time of his Bar Mitzva, believing that there was nothing more to Judaism than what he had already seen, he made an important decision—after his Bar Mitzva he was going to leave Judaism and become an atheist. And so he did. But after studying the doctrine of the atheists for some time, a certain uneasiness began to overtake him. He realized that everything they said was based on certain assumptions which they could not prove. This made him feel very uneasy because the whole reason he left Judaism was because it had certain basic ideas and beliefs that he felt could not be proved. But now he was finding out that the doctrine of the atheists was not logically self contained either. They had their own set of assumptions upon which they based their philosophy and those assumptions could not be proved either. Rather it was a matter of faith. If you accepted the assumptions then the rest followed. But if you didn't then there was no system. "I am back where I started from!" he told himself. "Why are the non provable assumptions of the atheists any more valid than the fundamental beliefs of Judaism?" He then began to reexamine Judaism and after a period of time became a complete *Baal Teshuva*—and a Lubavitcher chosid.

The realization that any logical system is ultimately based on some assumptions or axioms that are not logical but must simply be assumed, takes the wind of absoluteness out of any system of thought—scientific, mathematical or philosophical.

This fundamental role of axioms has deep implications in understanding the difference between the sciences and Torah. The Rebbe MHM discusses this in various sichos and letters. For example, in one sicha he writes:

> It says,[4] "If one tells you that the nations of the world have

4. *Medrash Eicha Rabbah* 2:13.

wisdom, you should believe it; if one tells you that the nations of the world have Torah, you should not believe it." From this very statement of our Rabbis...we can understand the difference between wisdom and Torah.

In order to understand an intellectual concept one must have elementary concepts and rules of wisdom or, in the terminology of research, rules of logic. The rules of logic together with the elementary concepts necessarily imply the concept under consideration as a conclusion.

Thus one is forced to draw the conclusion because of the elementary concepts and rules of logic. But the elementary concepts and rules of logic themselves have no foundation. There are no prior concepts that necessarily imply these elementary concepts and rules of logic as a conclusion. It depends only on one's good will. If one wants to accept the elementary concepts then he accepts the conclusion. But if he doesn't want to accept the elementary concepts he is free to choose not to accept them and it follows that he will also not accept the resulting conclusion.

This is one of the ways in which intellect itself forces us to include in our system a level which is higher than intellect. Inside the intellectual process itself it is felt that the point of origin of intellect is not intellect itself because the starting point of the entire intellectual process—the elementary concepts—are not forced upon us by intellect. Rather we use them because we like them. This is a matter of faith,[5] an assumption.

This then is the difference between wisdom and Torah. Accepting a science (*chochmah*) depends on one's good will. If he likes the relevant rules of intellect, he draws a certain conclusion, but the science itself does not force him to. This is in contradistinction to Torah which means instruction.[6] Torah teaches a person what he must do, what he may do and what he may not do. These are the three categories of the pro-

5. Thus, ultimately, intellect itself is based on faith. See sec. 1 of this chapter, "Faith and Intellect."

6. *Zohar* 3:53b.

hibited, the permitted and *mitzva*. Furthermore, Torah requires that we transform the "permitted" into *mitzva*.

Torah teaches this in a manner independent of our willingness to accept it. Rather, Torah instructs and demands that one conduct himself in a certain manner, independent of one's own desire.

This is the meaning of the statement that "If one tells you that the nations of the world have wisdom, you should believe it; if one tells you that the nations of the world have Torah, you should not believe it." Wisdom, which is accepted because of one's will and desire to accept it, exists also by the nations of the world. If one accepts the axioms and rules of logic of the discipline, then he accepts the idea [which follows as a conclusion]. But Torah, which must be accepted regardless of one's agreement—such a thing does not exist among the nations."[7]

The Uncertainty of Science

Now, if you saw a notice for a lecture titled "The Uncertainty of Science" you might think that it would be a lecture attacking science by someone advocating mistrust of science. Actually, this was the title of a lecture delivered by a Nobel laureate in physics, one of the greatest physicists of the latter half of the 20th century, the man considered to have been the greatest theoretical physicist after Einstein—Richard Feynman.

Indeed, the title and the subject matter are surprising for a lecture of a scientist. But in fact, such powerful critical overviews of science, such as this lecture, are usually delivered by the greatest scientists, the ones on top.

While the statements made by Feynman which we quote below were made on a strictly scientific basis, we will see from the quotes from Feynman below, taken from "The Uncertainty of Science" and other lectures, some of the same ideas about science as those expressed by the Rebbe MHM in his sichos and letters.

---

7. *Sicha* of Shavuos, 5715, *Likutei Sichos* 2:561.

In quoting Einstein and Feynman, we will see the wisdom from above of the Rebbe MHM reflected in the wisdom from below of the greatest scientists of our times. We will see that these are actually the ideas of science itself, expressed by its greatest masters.

For example, in the sicha quoted above, the Rebbe MHM made the point that the conclusions of science are not absolute but are only conclusions based on axioms and the rules of inference that one is free to accept or reject.

Einstein was once asked how he came to develop the theory of relativity. He said, "By challenging an axiom."

And Feynman says it this way:

"If they say to you science has shown such and such, you might ask, 'How does science show it—how did the scientists find out—how, what, where?'...And you have as much right as anyone else, upon hearing about the experiments (but we must listen to all the evidence), to judge whether a reusable conclusion has been arrived at....The experts who are leading you may be wrong[8]....It is necessary and true that all of the things that we say in science, all of the conclusions, are uncertain, because they are only *conclusions.*"[9]

The main point in the Rebbe MHM's sicha was that general wisdom or science does not tell one what to do; only that certain conclusions follow from certain assumptions. Only Torah gives actual instruction to the person. This idea was also expressed by Feynman:

"Science as a method of finding things out...is based on the principle that observation is the judge of whether something is so or not.... The principle that observation is the judge imposes a severe limitation to the kind of questions that can be answered. They are limited to questions that can be put this way: "If I do this what will happen?" There are ways to try it and see. [But] questions like, 'Should I do this?' and 'What is the value of this?' are not of the same kind."[10] Elsewhere, he says:

> The common human problem, the big question, always is "Should I do this?" It is a question of action. "What should I

8. Richard Feynman, "What is Science?" in *The Pleasure of Finding Things Out* (Cambridge, MA: Perseus Books, 1999), 187. (See p. xxvii, note 5).
9. Richard Feynman, "Uncertainty of Science," 26.
10. Ibid., 15–16.

do? Should I do this?" And how can we answer such a question? We can divide it into two parts. We can say, "If I do this what will happen?" That doesn't tell me whether I should do this. We still have another part, which is, "Well, do I want this to happen?" In other words, the first question—"If I do this, what will happen?"—is at least susceptible to scientific investigation; in fact it is a typical scientific question....But the question "Do I want this to happen?"—in the ultimate moment—is not....At the end you must have some ultimate judgment.[11]

The Axiomatic Method

At this point, we should take a closer look at axiom systems in general, how they work and where the axiomatic method came from in the first place.

The reader is probably most familiar with the notion of axioms from high school geometry where certain axioms of geometry were stated, then mathematical statements were made and proved using the axioms and the methods of logical reasoning. Yes, this is a genuine example of the use of the axiomatic method. In fact this is one of the oldest examples—this is the geometry of Euclid who lived at the beginning of the Second Beis HaMikdosh. The reader who studied high school geometry should keep it in mind as an application of what we say about the axiomatic method. (Some of its terminology, however, is outdated. For example, Euclid divided his assumptions into two categories—"axioms" and "postulates"—and these terms are still used in high school geometry. Modern mathematics, however, does not make this distinction. Rather all the assumptions are called axioms.)

We mentioned Euclid's use of axioms in his development of geometry. This means that the axiomatic method goes back a long way. Aristotle, who preceded Euclid, also used this method. He wrote that "every demonstrative science must start from indemonstrable principles. Otherwise, the steps of demonstration would be endless."[12] But Aristotle did not invent the axiomatic method. In fact, the entire sys-

---

11. Feynman, "The Uncertainty of Values," in *Meaning*, 44–45.

tem of logic presented in Aristotle's famous "Organon," which was the authoritative text on logic in the secular world for the next two thousand years, did not originate with Aristotle. As one historian of logic, Prof. G.T. Kneebone, explained, "Concepts as general and as mature as those which Aristotle was able to use are only formed very slowly, and are normally the ultimate outcome of a prolonged process of inquiry."[13] Regarding the axiomatic method in particular, Professor Raymond Wilder, a major authority on the foundations of mathematics, wrote that "the origin of the axiomatic method is not known"[14]. Another authority on the foundations of mathematics, Professor Howard Eves, writes similarly, "We do not know with whom the axiomatic method originated."[15]

However, as we saw earlier from the quote from Rabbi Levi Yitzchok, axioms and the methods of reasoning applied to them are *chochma* and *bina* which have always been part of *Jewish* scholarship. For example, when Moshe Rabbeinu appointed judges and officers for the Jews, he looked for people who had both of these qualities.[16] Furthermore, he encourages the Jews to follow the Torah "for it is your *chochma* and *bina* in the eyes of the nations. They will hear all of these laws and they will say, 'This great nation is uniquely a nation of *chochma* and *bina*.' "[17] In the book of *Mishlei* (Proverbs), King Solomon refers to *chochma* and *bina* repeatedly.

Moreover, the Jewish tribe *(shevet)* of Yissachar wrote books on the mathematics of astronomy for use in determining the times of the new months and new years etc. in Jewish Law. The Greek mathematicians and astronomers first learned these disciplines from the Jewish scholars.[18]

Thus it is highly probable that the Greek mathematicians learned the axiomatic method from the Jews who were already using it. The *Seder HaDoros,* a book on the chronology of the history of the world,

12. Quoted in Raymond L. Wilder, *Introduction to the Foundations of Mathematics*, 2nd ed. (New York: Wiley, 1965), 3.
13. G.T. Kneebone, *Mathematical Logic and the Foundations of Mathematics* (New York: Dover Publications, 2001), 7.
14. Wilder, *Introduction,* 4.
15. Eves, *Foundations,* 15.
16. Devorim 1:13.
17. Ibid. 4:6.
18. *Likutei Sichos* 30:195, from *Kuzari* and *Seder HaDoros*.

relates that Aristotle himself came in contact with Shimon HaTzadik and was greatly influenced by him. The influence of "the Shimoni" (as Aristotle called him) on Aristotle was so great that he said he was ready to revise his entire philosophical system to be based on what he learned from Shimon HaTzadik. However, Aristotle felt that he was too old to start over.

What makes something an axiom? Can we just assume anything and call it an axiom? Ideally, an axiom should be a statement that, on the one hand, is intuitively clear such as the example in Rabbi Levi Yitzchok's letter "the whole is greater than any of its parts," but on the other hand, should be as basic and elementary as possible in the sense that it cannot be proved from any other statement. Consider, for example, Euclid's fifth postulate,[19] the so called "Parallel Postulate" or "Parallel Axiom" which can be expressed as follows: *"Through any point there passes only one line parallel to a given straight line."* This appears to be quite intuitively obvious. However, even the earliest commentators on Euclid held that this axiom was not sufficiently evident to be accepted without proof. "The question of whether Euclid's parallel postulate is an independent axiom or can be derived from other axioms had puzzled mathematicians for two thousand years,"[20] wrote one historian of mathematics. Many attempts were made to prove the Parallel Postulate from other axioms but they all failed. Some even tried to find a more elementary axiom from which the parallel postulate would follow but that failed too. It was finally realized that one had to just "take it or leave it."

This brings us to the year 5600 (1840) when the fountains of wisdom below broke open and the issue was settled as follows:

Some mathematicians "took it." They accepted the parallel postulate as an axiom. This was Euclidean geometry.

Some mathematicians such as Lobachevski and Riemann "left it" and replaced it with alternative axioms, giving rise to the non-Euclidean geometries. (In Lobachevski's geometry, *two* lines, parallel to a given straight line, pass through a point and in Riemann's

---

19. For a thorough, yet readable, discussion of Euclidean geometry, the issue of the fifth postulate, non-Euclidean geometry, and their relation to the axiomatic method, see Eves, *Foundations,* chaps. 2 and 3.

20. Dirk Struik, *A Concise History of Mathematics,* 4th ed. (New York: Dover Publications, 1987), 167.

geometry there are no parallel lines. This gave rise to three different geometries.)

Most importantly, all mathematicians realized that anyone was free to accept—or reject—any or all of these geometries as he pleased since none of them was absolute. The discovery of non-Euclidean geometry and the development of the modern axiomatic method "dealt a severe blow to the absolute truth viewpoint of mathematics."[21] It was all a matter of which axioms one wanted to accept, as the Rebbe MHM said in the sicha quoted above.

Aristotle had understood an axiom to be "a universal assumption that is so self-evident that no sane person would question it,"[22] (hence the use of the word "axiomatic" to mean "self-evident" or "obvious"). In the modern theory, however, this point of view was rejected. (So now, the correct meaning of "axiomatic" is merely—"assumed.") To summarize the above discussion in one phrase, we might say: *axioms are the rock bottom but they are not rock solid.*

We emphasize again that all this is due to the fountains of wisdom from below breaking open. Since the time of Euclid no progress at all had been made in this area "for the human mind had for two millennia been bound by the prejudice of tradition to the firm belief that Euclid's geometry was most certainly the only possible one."[23] Then, suddenly, there was an explosion of ideas on the foundations of geometry simultaneously by different mathematicians in different places. One well known mathematician and historian of mathematics, expressing his amazement at this, commented, "It is remarkable how the new ideas sprang up independently in Gottingen, Budapest and Kazan, and in the same period, after an incubation period of two thousand years."[24]

We mentioned above that an axiom should be intuitively clear. But, on the other hand, if it relies on intuition too much, we may be accepting as an obvious assumption something that really needs to be proved. This was the problem with the parallel postulate. It was found to be a problem with some more of Euclid's axioms as well. The mathematicians of the late 1800s, employing the modern

21. Eves, *Foundations,* 68.
22. Ibid., 31.
23. Ibid., 61.
24. Struik, *Concise History,* 170.

axiomatic method, were able to fix this and restructured classical Euclidean geometry in a more logical way. The most notable accomplishment in this area was David Hilbert's "Foundations of Geometry" published in 1899.

All this had far reaching implications. It resulted in an entirely new view of mathematics and science. No scientific theory, even a very intuitive theory such as geometry, could be considered absolute. It all depended on what axioms you wanted to set down. Furthermore, the axioms themselves did not necessarily have to be all that intuitive, as in the case of the new geometries. This became extremely important at the beginning of the 20th century when Albert Einstein developed his own (non-intuitive) theory of space, general relativity, for which he used Riemann's geometry.

The most non-intuitive theory of all time is, of course, quantum mechanics. Here too, after the experiments were done and the equations developed, an axiom system was presented. The axioms were statements that the scientists seemed to be compelled to accept as a result of their experiments. But again, every scientist was free to accept or reject these axioms and some scientists did reject them, most notably Albert Einstein who was opposed to the probabilistic aspects of quantum mechanics.

Professor Ernest Nagel, an authority on the philosophy of science, explained the role of axioms in scientific theory as follows: "Formal sciences are characterized by the fact...that their theorems are not 'about' some phase of the existing world but are 'about' whatever is *postulated* by thought."[25] In other words, scientific theories are not about the world itself but about what we assume about the world. Similarly, Neils Bohr once said, "It is wrong to think that the task of physics is to find out how nature is. Physics is concerned with *what we can say* about nature."

It's not about the world; it's about the axioms.

In the late 1800s the discovery of inconsistencies in mathematics and paradoxes in logic led to a crisis in the very foundations of mathematics. This crisis too was resolved by appeal to the axiomatic method, restructuring all of mathematics and logic to be based on the newly developed theories of mathematical logic and set theory.

25. Quoted in Wilder, *Introduction*, 6.

The Lesson from Geometry

The fact that all science and mathematics is based on axioms that cannot be proved is important for understanding the difference between science and Torah, as explained above in the letter of the Rebbe MHM. In another letter, written to a young Jewish engineer who had written to him about his life and background, the Rebbe MHM discusses this again while focusing specifically on geometry:

> There is a well-known saying of the Baal Shem Tov that we heard from [the Previous Rebbe]—that everything that a Jew sees or hears certainly has some instruction for him in his service of G-d.... I see from your letter that you are a *mehandes* [from the Hebrew word *handasa* which was used classically to refer to geometry. Thus a mehandes would be a geometer. In more modern usage, *handasa* refers to any kind of engineering so a *mehandes* would be an engineer.] But it is not exactly clear what professional work you do. Is it construction of buildings or various types of measurement—surveying etc.? In any case at the foundation of all of these is the discipline of *handasa*—geometry. What lesson can be learned from this discipline?
>
> Geometry has characteristics of an exact mathematical science and also of an applied science... On an infinitely higher level, the same thing applies to our Torah. On the one hand, it is the wisdom of G-d and thus the ultimate of truth and precision—"no man can fathom its worth and it is hidden from the eyes of the living." On the other hand, its ultimate purpose, as its name *Torah* indicates, is *instruction* in the daily life in this coarse material world. Therefore it is in [analyzing] the contrast between these two extremes that we can find the fundamental and infinite distinction between the Torah, which is called "our wisdom and understanding in the eyes of the nations" on the one hand, and the wisdom and understanding of the nations themselves or the intellect of the "animal soul" of the Jew, on the other.
>
> The distinction is as follows: Human intellect, even that of the sciences that are considered exact sciences, is based on foundations that science itself has nothing to say about. This is because science, especially exact science, accepts as con-

clusions only those things that have proof, and the *foundations* of all sciences and mathematics, including geometry, have no proof. Thus a person is free to accept them or to reject them. This is especially clear in the case of geometry in which, as is well known, there are three primary theories, each based on a number of assumptions—axioms, and the set of axioms of each theory contradict those of the others. In other words, science does not have the ability to make an absolute statement, only conditional statements—"If you accept these axioms as true and you also accept the methods of reasoning and proof, then you will get the following results...."

Two major points follow from this: 1) It is up to the person whether or not to accept the axioms. 2) Even if he does accept them, he cannot be compelled to do anything in accordance with the results that follow because the whole process merely says, "If you take the following course of action, the results will be as follows." If, for example, one does not care what harm may possibly come to him by following a certain course of action there is nothing preventing him from acting in that manner. In other words, science does not give instruction in life but rather "tells a story" or predicts the future and says, "Based on our experience until now and based on the assumptions which we now want to accept as true, the sequence of events will be as follows...."

In total contrast to this is our holy Torah. Being the wisdom of the True Existence—G-d—it is necessarily absolute. It is absolute truth—both in its basic statements and in the "principles of the Torah" that direct the manner in which the basic statements are discussed [the rules of reasoning]. Since this is the wisdom of the Creator of the entire universe, which includes man, it follows that all its conclusions compel[26] man to act in accordance with those conclusions and in no other manner at all.

This is one of the points that, as an engineer, should be embedded in your mind: that it is impossible to raise any

26. Elsewhere the Rebbe MHM explains that since the Torah is the law of the Creator,

> question whatever based on science against the Torah since the Torah is absolute truth and science, according to the way it defines itself, is not absolute but rather dependant on the assumptions that a person wants to make. Furthermore one has the freedom to establish contradictory theories which may all be maintained in accordance with the will of various people, such as the three theories of geometry—that of Euclid, that of Lobachevsky and that of Riemann.[27]

Here, again, we have a similar quote from Feynman: "Scientists take all those things that *can* be analyzed by observation...But there are some things left out, for which the method does not work. This does not mean that those things are unimportant. They are, in fact, in many ways the most important. In any decision for action, when you have to make up your mind what to do, there is always a "should" involved, and this cannot be worked out from 'if I do this, what will happen?' alone....That is the step the scientist cannot take."[28]

Finally, we note that the modern axiomatic method unified various mathematical structures that were thought to be diverse. For example, the *field* axioms showed that there is a unity among the many different mathematical entities, all of whom have the structure of a field. The axioms of measure theory unified analysis and probability etc.

To summarize the role that the axiomatic method plays in modern science:

1) It shows that no scientific conclusions can be considered absolute. They are only valid relative to the axioms on which they are based.
2) It shows that there is a limit to human knowledge. We can never claim to have penetrated the full depths of reality. All we can ever hope to do is to make observations and state our assumptions based on those observations. The scientific theories that we then develop are not about the world itself but rather about what we are assuming about the world.

---

the world would be forced to follow the laws of the Torah in the same way that it follows natural law—automatically—except for the fact that G-d wants there to be free will. But if one chooses to violate a law of the Torah, he is actually going against the natural order. See *Sefer HaMa'amorim Melukat* 4:275, n. 45.

27. *Igros Kodesh* 6:145–47.

28. Feynman, "Uncertainty of Science," 16–17.

3) It brings a high degree of unity to our picture of the physical world as there is now a uniform method for developing scientific theories.

Electricity and Magnetism

In the Sicha of Noach the Rebbe MHM explains that before the fountains of wisdom below broke open people would look at the various forces of nature as being diverse and unrelated but the new wisdom revealed the underlying unity among them.

One of the best examples of this is the unification of electricity and magnetism, two forces that were previously thought to be unrelated, under one unified concept: electromagnetism. We will discuss electricity and its unification with magnetism in some detail since the theory of electromagnetism was the first major unifying scientific theory after the fountains of wisdom below broke open.

Many phenomena due to electricity had been known for thousands of years but were not known to be manifestations of electricity. They were not even thought to have any connection with each other. For example, as far back as "the time of the Greeks" people knew of such phenomena as the "marvels concerning the attraction of amber" as Plato called it. When amber is rubbed with a cloth it would attract small lightweight objects such as straw. Another phenomenon observed was a certain pale glow sometimes seen on the tips of pointed objects such as ships' masts, especially during thunderstorms. The stunning effect of touching an electric eel was known and, of course, the most well known electric phenomenon of all—lightning. Since the Greek word for amber is *elektron,* these phenomena eventually became known as electricity. But in the time of the Greeks no connection was seen between them.

At this point we should comment on repeated reference to "the time of the Greeks." This is a common reference point in the history of science since their mathematical and scientific writings are available for reference. However, the Greeks were not necessarily the ones who discovered the phenomenon or originated the concept, as the historians of science themselves sometimes indicate. We have already seen that the Greeks learned mathematical astronomy from the Jews. The Rebbe MHM once explained that, generally speaking, all of the sciences can be traced back to the science of ancient Egypt (*Chochmas Mitzrayim*). Even those sciences that made their first appearances later in other countries can be traced back to ancient Egypt. Thus, when the prophet wishes to describe the great wisdom of Shlomo HaMelech, he says that it was even greater than the wisdom of Egypt.

For thousands of years not much more was discovered about electricity than what was known in the time of the Greeks. Then, in the late 1700s, investigations into the properties of electricity revealed some new information. But it was not until the mid 1800s, after the fountains of wisdom from below broke open, that the real nature of electricity was understood and its unification with magnetism took place.

It should be pointed out that "the 600th year of the 6th millennium"—1840—that the Zohar speaks of, is what we might call the breakthrough year. Prior to that year, both in the wisdom from above and in the wisdom from below, we find what we might call a buildup: Some of the major breakthrough developments in both of these areas actually began before 1840, as we now explain.

In the Sicha of Noach, the Rebbe MHM discusses the example of the publication of the ma'amorim (Chassidic discourses) of the Alter Rebbe. By the year 1837 the Tzemach Tzedek had prepared two volumes of the Alter Rebbe's ma'amorim for publication. They were to be published under the title Torah Or. At the time only the first volume was published; the second was published 10 years later under the title Likutei Torah. This was the breakthrough of wisdom from above. But these were Ma'amorim that the Alter Rebbe had said many years earlier, over the years 1796—1812.[29] The Tzemach Tzedek himself had spent 32,000 hours over a 30 year period (1804—1834) in intense study of the Alter Rebbe's Ma'amorim, beginning work on their publication in 1834. So there was a period of a buildup prior to the breakthrough year.

Thus, the Rebbe MHM identifies three stages in the revelation of the new wisdom:

The year 1740—the *beginning* of the 6th century of the 6th millen-

29. *HaYom Yom*, 3 Shevat.

nium (not the *600th year* of the 6th millennium). This is the year that the Or HaChaim (commentator on the Torah) calls "the dawn of the Geula." It corresponds to the time that the Baal Shem Tov was revealed and began teaching the wisdom of Chassidus. And around that time the mathematician Girolamo Saccheri *almost* discovered non-Euclidean geometry.[30]

The year 1800, following the release of Rabbi Schneur Zalman, the Alter Rebbe, from imprisonment. At this time he began spreading the wisdom of Chabad Chassidus far and wide. In the Chabad system, the Alter Rebbe developed Chassidus in a way that it could be understood by everyone and after 1800 it was made available to the public.

The year 1840—the breakthrough year when the new wisdom began to "flood the earth." At this time the Alter Rebbe's Chassidic discourses were published and the new unifying developments in the sciences proceeded rapidly.

(Indeed, it is interesting to note that in the case of the flood in the time of Noach, on which the statement in the Zohar is based, there was a preparatory period of 120 years before the flood itself—which was in "the 600th year of the life of Noach.")

In the wisdom from below, we find that the following important facts about electricity were revealed in the century prior to 1840:

1) It was found that while, on the one hand, charged pieces of amber *attracted* small lightweight objects, on the other hand, two such charged pieces of amber *repelled* each other. It was further found that there are two different types of electric charges which came to be called positive and negative. Each of these types attracted the opposite type and repelled its own type.

2) Devices for storing electric charge (what we would now call a battery) and for measuring electric charge were developed. Thus it became possible to talk about electric charge quantitatively.

3) It was determined that lightning was in fact an electric charge as was the charge of the eel.

30. Eves, *Foundations,* 58.

The Unification of Electricity and Magnetism

Magnetism was also a phenomenon which had been known for thousands of years. Around the time of the AriZal (late 1500s) it was discovered that the earth itself is a magnet which proved useful in the navigation of the oceans. One could find his direction using a magnetic compass. But while magnetic poles appeared to behave the same way as electric charges, no connection was known between them. From all observations, electric charges did not influence the magnets in any way nor did the magnets have any influence on electric charges.

Then, one day in the year 1820, Hans Oersted, a Danish physicist, was walking to his lecture at the University of Copenhagen. On the way, he had an idea. If static electricity did not affect magnets in any way, maybe things would be different if one tried electricity *moving* through a wire connecting the two poles of a battery. Arriving at the classroom filled with a crowd of young students, he placed a battery on the lecture table, connected its two opposite ends with a platinum wire and placed a compass needle close to it. The needle, which was supposed to orient itself in the north-south direction, turned around and came to rest in the direction perpendicular to the wire. The audience was not impressed—but Oersted was. The unification of electricity and magnetism had just taken place. The *electric* current flowing through the wire was generating a *magnetic* field around the wire and the magnet turned in the direction of that magnetic field. Electromagnetism, as Oersted called it, had just been discovered.

Oersted submitted his discovery to a prominent French journal of physics whose editors were rather skeptical of the results. They did publish it, but did so with a note of apology. They said that they don't usually publish such extraordinary findings for fear that they are fake. "But in regard to the paper by Mr. Oersted," they continued, "the results obtained by him, however singular they may appear, are accompanied by too many details to give place to any suspicion of error." This was a scenario which was to repeat itself several times over the coming century—extraordinary revelations of wisdom from below which were to be greeted with skepticism, and sometimes ridicule, by the scientific community. This may be attributed to the fact that these revelations were not simply the results of the natural progression of scientific research but rather the result of an infusion of wisdom—a flood of knowledge—that was being forced into the world. And it was coming faster than the world could handle it.

Oersted had shown that an electric current—moving electricity—could produce a magnetic field, demonstrating a unity between the two phenomena. About ten years later, after much experimentation, Michael Faraday showed that the reverse was also true—that a moving magnet could produce electricity. In the meantime Andre Ampere had postulated that magnetism itself was due to moving electric charge within the molecules of a magnetic material. Thus the unity of electricity and magnetism was almost a complete theory.

There was still one thing missing however, and that was a *mathematical* theory of electromagnetism i.e. a set of definitions of physical quantities and a set of mathematical formulas describing the relationship between those quantities. A deep mathematical theory of electromagnetism was developed by James Clerk Maxwell *after* the fountains of wisdom below broke open. We turn now to consider this, as well as the role of mathematical theories in general in the wisdom from below.

The Mystery of Mathematics

Einstein used to say that there was not much new in his special theory of relativity. He claimed that it was just an extension of the work of Maxwell, Lorentz and others. Maxwell's mathematical theory of electromagnetism was the greatest unifying theory in the era after the fountains of wisdom from below broke open.

In addition to developing the equations of electromagnetism, the equations themselves led him to theorize that light was also an electromagnetic wave. He thereby caused the whole science of optics to be included within the framework of electromagnetism, thus uniting two domains which had seemed entirely distinct. This new unifying theory was one of the most beautiful syntheses in the history of physics.

Maxwell's theory was further confirmed when Heinrich Hertz produced electromagnetic waves (radio waves) which were shown to have all the known properties of light such as reflection, refraction, interference and diffraction.

It was at this time that scientists began to realize the importance of mathematics in revealing the underlying unity in nature. As one of the great mathematicians of the era, Joseph Fourier, expressed it, "Mathematics compares the most diverse phenomena and discovers the secret analogies which unite them."[31]

Richard Feynman saw in the mathematical aspect of the laws of nature—and even in the very *existence* of laws of nature—something mystical:

"The fact that there are rules at all to be checked is a kind of miracle; that it is possible to find a rule, like the inverse square law of gravitation, is some sort of miracle. It is not understood at all...The rules that describe nature seem to be mathematical....It is not a characteristic necessity of science that it be mathematical. It just turns out that you can state mathematical laws, in physics at least, which work to make powerful predictions. Why nature is mathematical is, again, a mystery."[32]

And a contemporary theoretical physicist, Paul Davies, who has written extensively on science and religion, said, "What most impresses me is the existence of an underlying mathematical order, an order that led the astronomer Sir James Jeans to declare, 'G-d is a pure mathematician.'"[33]

From the time of Maxwell until now, mathematics and physics have gone hand in hand. New discoveries in physics gave rise to new discoveries in mathematics to describe them, and new discoveries in mathematics brought greater simplicity and unification to the theories of physics. Any improvement or unification in mathematics itself would have a further unifying effect in every area of science to which the mathematics was applied. If a certain scientific theory was stated in the terminology of a particular area of math, say calculus, and a fundamental improvement was made in calculus itself, the scientific theory then benefits from the same improvement or unification. This often leads to new predictions etc.

As an example, Maxwell's equations were first expressed as four partial differential equations (homogeneous equation), then later in a simpler form as two equations using vector calculus. Now the theory can be expressed simply as one equation using differential forms: $dF = 0$.

31. Quoted in James Stewart, *Precalculus* (Brooks/Cole, 2002), 335.
32. Feynman, "Uncertainty of Science," 23–24.
33. Reported in the *Albany Times Union*, March 9, 1995.

Mathematics, the great unifier, continued to develop and diversify, branching out in many directions. Eventually, mathematics itself was in need of unification. "By 1870 mathematics had grown into an enormous and unwieldy structure, divided into a large number of fields in which only specialists knew the way. Even the great mathematicians...at most could be proficient in only a few of these many fields."[34]

To achieve this unification, mathematicians tended toward more and more towards abstraction. As one mathematician wrote, "Generalization by abstraction permits us to see the essential sameness of what at first appear to be distinct mathematical developments. The unifying power of abstraction makes it possible to comprehend as one a number of theories which were initially presented at the level of concrete specialization...."[35] (Note the similarity to Fourier's statement above.)

Thus, new abstract theories, especially the theory of groups and other algebraic structures, the concepts of function and space, and especially, mathematical logic and set theory brought unification to the various fields of mathematics. The axiomatic method itself, which we discussed earlier, is, of course, the best example of this.

Now mathematics is very technical by nature so we will not describe the mathematical developments of this era in detail,[36] but (in addition to geometry which we discussed at length earlier) we will mention briefly three important areas of math where major developments were made, and the mathematicians involved:

1) Analysis. While differential and integral calculus had been developed by Newton and Leibniz in the late 1600s, its foundations were not understood, its definitions were not clear and its concepts of the infinitesimal and the infinite were often paradoxical. For over a century after Newton and Leibniz, the mathematicians who continued the development of calculus "intoxicated by the power and applicability of the new tool, failed to consider sufficiently the solidity of the base on which the subject was founded, so that instead of

34. Struik, *Concise History,* 177.

35. Angus Taylor, *Generalized Theory of Functions and Integration* (New York: Blaisdell, 1965), 90.

36. See chapter V, secs. 2 and 3, for extensive discussions of Mathematical Logic and Mathematical Infinity.

having demonstrations [i.e. proofs] justify results, results were used to justify demonstrations. With the passage of time, contradictions and paradoxes arose in increasing numbers, and a serious crisis in the foundations of mathematics became evident. It was realized more and more that the edifice of analysis [calculus] was being built on sand."[37]

This situation prevailed for over 100 years. A typical quote from this era, attributed to d'Alembert was, "Go forward and faith will come to you"[38] talking about math!

The foundations of calculus as we now know it—with precise definitions and rigorous proofs—began to be laid by Augustin Cauchy. "Cauchy offered at last a beginning of an answer to that series of problems and paradoxes which had haunted mathematicians since the days of Zeno [2,300 years earlier]...by creating a mathematical technique with which it was possible to do them justice." This was accomplished by replacing the murky method of infinitesimals with the precise method of limits.

Analysis was developed and further unified by Bernhard Riemann who brought considerations of topology into analysis, gave a comprehensive definition of the integral (the "Riemann integral") and developed a general theory of manifolds that unified all the various types of geometry—the Euclidean and non-Euclidean geometries and many new ones which became known as Riemannian geometries. The axioms of a topological space brought unity to these areas.

2) Algebra. A young and very brilliant mathematician named Evariste Galois, on the night before he was to participate in a duel in which he was killed at the age of 21, wrote a long letter to a friend in which he developed a unifying theory of algebra which came to be known as group theory. It is the key to modern algebra and geometry; groups are like "algebraic atoms" from which many algebraic systems can be constructed. Galois concluded his letter, obviously written under tremendous pressure, by saying, "There will be, I hope, some people who will find it to their advantage to decipher all this mess."

(I know a Jewish symphonic composer named Paul Schoenfield who had considered himself an agnostic. He would study modern

37. Eves, *Foundations*, 263.

38. Struik, *Concise History*, 151. (See pp. 151–60 as a reference for this discussion.)

abstract algebra as a hobby. When he came upon the area of abstract algebra known as Galois Theory, as he told me, he was very moved by its beauty and elegance and felt that if there was something so elegant in the world then there must be a G-d Who created it. Thus he began a search for Judaism, came to Chabad and eventually became a completely observant Jew, a *Baal Teshuva*. Another friend of mine became a Baal Teshuva from studying linear algebra. It was the concept of n-dimensional space that triggered it. The physical world as we see it has only 3 dimensions (or 4 if we include time). To my friend, the fact that the mathematical theory of linear algebra deals with any number of dimensions (n) just as easily, meant there are deeper levels of reality than we observe. In search of the deeper dimensions he began to study the Torah and eventually returned to Judaism.)

3) Mathematical logic and set theory. These were entirely new areas of mathematics that were developed by George Boole and Augustus DeMorgan (mathematical logic) and Georg Cantor (set theory). These two disciplines themselves were unified by the axioms of Boolean algebra. Three essential points must be made concerning mathematical logic and set theory as they relate to our present study:

They revolutionized the study of logic. After two thousand years of stagnation since Aristotle, suddenly, with the gushing forth of the fountains of wisdom from below, logic took on a whole new dimension compared to which Aristotelian logic is primitive. As Prof. Kneebone wrote, "Symbolic logic, as it exists today, has indeed supplanted and rendered wholly obsolete the traditional logic that persisted from the time of Aristotle." [39]

More than just surpassing Aristotelian logic, the modern study of logic showed that the Aristotelian scheme was not absolute but that many other logics were possible. As the great logician Alonzo Church wrote, "We do not attach any character of uniqueness or absolute truth to any particular system of logic. The entities of formal logic are abstractions, invented because of their use in describing and systematizing facts of experience or observation, and their properties...depend for their exact character on the arbitrary choice of the inventor."[40]

Just like the modern study of geometry produced non-Euclidean

39. Kneebone, *Mathematical Logic,* 133.

geometries based on their own axioms, with the new awareness that one was free to choose his own axioms and develop his own mathematical system, many alternative logics were developed. These became known as non-Aristotelian logics. Just like Einstein employed one of the non-Euclidean geometries in the development of general relativity theory, Hans Reichenbach suggested using one of these logics, the 3-valued logic, in the study of quantum mechanics.[41] (We present 3-valued logic in detail in chapter V, section 2c.)

c) In addition to being disciplines in their own right, these areas of math brought the most fundamental unity to modern mathematics as they included the study of the axiomatic method and the foundations of mathematics itself. They became the basic "mathematical language" in which modern mathematics was expressed.

Atomic Theory, Quantum Mechanics and Beyond

Atoms are to matter what axioms are to logic. Just as the axiomatic method brought a high degree of unity to scientific and mathematical theories, atomic theory brought a high degree of unity to our understanding of matter. Electromagnetic theory revealed the unity between two of the major forces of nature, electricity and magnetism; atomic theory revealed the unity among the various materials of the natural world.

Atomic theory appeared in two stages. Firstly, in the mid-1800s the chemists proposed the theory that all substances were in fact composed of atoms. They postulated that:

1) Elements are made of atoms

2) The atoms of an element are identical in their masses

3) Atoms of different elements have different masses

4) Atoms only combine in simple, whole number ratios.

Later, the internal structure of the atom was discovered. It was found that every atom consists of a nucleus with protons that had a positive electric charge and neutrons that had no electric charge. The nucleus is surrounded by electrons that have a negative electric

---

40. Quoted in Eves, *Foundations,* 258.

41. Hans Reichenbach, *Philosophic Foundations of Quantum Mechanics* (Berkeley and Los Angeles, 1944), esp. 142–48, 160–66.

charge. Thus all materials are basically the same thing—a bunch of protons, neutrons and electrons. The only difference between one material and another is the quantity and arrangement of these protons, neutrons and electrons. When one drinks a cup of coffee, the coffee in the cup is just a bunch of protons, neutrons and electrons—and so is the cup itself. These discoveries revealed a very high degree of unity in the physical world.

After the turn of the century, unifying theories appeared at an ever accelerating pace. Albert Einstein's theory of relativity showed the unity of space and time. Quantum mechanics (which Einstein was also one of the originators of) introduced many new concepts into physics. One of them, the wave—particle duality, demonstrated that every particle had a wave associated with it and that every wave (such as light) is also composed of particles (photons). Thus wave and particle phenomena were unified.

Einstein himself spent the last decades of his life trying to discover a unified field theory which would unify the entire view of nature under one theory. While he did not succeed, the quest for grand unifying theories (GUTs) became a major preoccupation of physicists.[42] This was only natural since, as we explained above, the whole reason this wisdom was introduced into the world in the first place was to reveal the underlying unity in the created universe as a preparation for the revelation of the absolute unity of G-d in the Era of Moshiach.

A major step towards a complete unifying theory was the development of quantum electrodynamics by Richard Feynman and others. This theory unified three of the four known forces of nature—electromagnetic, nuclear and the so-called "weak" forces. The only one left out was gravity. Currently, theoretical physicists are working on quantum gravity theories which attempt to unify gravity with quantum mechanics. This would unify all the force of nature into one theory.[43]

One of the most promising of these theories is string theory. We discuss quantum mechanics and string theory briefly in chapter V, "The Quest for Unification." For now we will just note, for comparative purposes, the difference between the ways that classical physics, quantum mechanics and string theory look at a particle moving in

---

42. See chap. 5, secs. 4 and 5.

space. In classical mechanics, a particle is always seen as having a very definite position and a very definite velocity. In quantum mechanics, because of the Uncertainty Principle, there is an inherent limitation in this knowledge. We cannot know both of these quantities precisely. The more accurately we know the position, the less accurately we know the velocity and vice versa.

In string theory, the basic entity is not even a particle but a string, which is a loop drawn in space. There is only one kind of string and the different kinds of particles are postulated to be different modes of vibration of these strings. All the different interactions between particles (forces) are understood to be splittings and joinings of these strings. String theory is the only known way of consistently unifying gravity with quantum theory and the other forces of nature. It unifies all particles and forces.

In the Sicha of Noach the Rebbe Melech HaMoshiach gave a characterization of the ultimate unification theory. He said that it would contain one concept of force and one concept of matter which carries the force. Perhaps string theory will be that theory.

The Character of Physical Law

We conclude this chapter with some important observations about the nature of science and scientific theories as they are now understood, especially since the development of quantum mechanics.

The issue here is the role that probability plays in *the character of* scientific theory. Probability and statistics have actually played a role in science in different ways for quite some time. As an example, when scientists were investigating the properties of gases in the 1800s they realized that one could not study a gas by applying the laws of mechanics to all the individual molecules of the gas in a container since there are many millions of molecules. Instead, they applied the laws of statistics to the gas, considering the average motion of its molecules rather than the motion of the individual molecules. This was very successful and gave rise to the science of statistical mechanics.

43. See Lee Smolin, *The Trouble with Physics* (Boston: Houghton Mifflin, 2006), part 1, for a review of current unifying theories.

Another area where statistics was applied was in measurement. In scientific research, one does an experiment and collects data, then tries to find a theory—an equation or a formula—to describe the observations. But no observation is perfect and slight errors might interfere with the development of an accurate theory. Often statistical methods can be used to effectively "smooth over" these errors.

Now, a third way in which probability applies to physical law gets closer to the issue at hand—*the character of* physical law. This is the observation that any scientific theory is based on a limited number of observations. If, for example, we want to find a formula for a bouncing ball, we will observe many bouncing balls, but we cannot possibly observe *every* bouncing ball. Thus, the best we could say is that our formula is probably true. This is the deficiency in the method of induction that we mentioned at the beginning of this section.

As Lee Smolin writes, "We are masters at drawing conclusions from incomplete information. We are constantly observing the world and then making predictions and drawing conclusions about it....We never have enough information to completely justify the conclusions we draw."[44]

It follows from this that a scientific theory can never be *proved* to be correct. But it can be proved to be false by finding examples where it fails. This is the famous criterion for scientific theories expressed by Karl Popper, the great 20th century philosopher of science: for a scientific theory to be considered valid it must be falsifiable i.e. there must be an experiment to test it to see if it can be proven false. "A theory is never proved right, but if it survives many attempts to prove it wrong, we can begin to have *faith* in it—at least until it is finally falsified."[45]

But the issue of probability brought to the fore by quantum mechanics goes much deeper than this. Shortly after the discovery of the Schroedinger Equation, the fundamental equation of quantum mechanics, in 1926, it was understood by most physicists that the physical meaning of this equation was that it gave the *probability* of the location of a particle at a given point in space at a given time. That was all that could be known. The uncertainty principle that we

---

44. Smolin, *Trouble with Physics,* 299.

mentioned above, which follows from the Schroedinger Equation, expresses the same idea.

It follows from this that no physical law can make a definite deterministic statement, only a probabilistic statement. This is in contrast to classical physics where it was assumed that one could get an exact description of a physical system because it was assumed that one could know all the relevant values exactly. But according to quantum theory this is not possible. The more precisely the variables related to the position of the system are known, the less precisely the variables related to dynamic change of the system can be known, and vice versa. So it is no longer possible to predict a future event precisely based on conditions in the present. Thus determinism—the concept that the future is determined by the present—is no longer valid.

As a result there has been a considerable change in the manner in which physicists understand the concept of prediction in physical theories. Having determined the values of the variables characterizing a system at a given time, they can only state what the probability is that these variables will have certain values at a later time.

This leads to a basic understanding of the position of science relative to Torah. Earlier, in the section on axioms, we quoted the Rebbe Melech HaMoshiach as saying that no challenge can be brought from science against Torah since any scientific theory is based on assumptions that one is free to accept or reject.

Here too, the role of probability in modern science shows that no challenge can be brought from science against Torah. As the Rebbe MHM writes in a letter to Chief Rabbi Herzog of Israel:[46]

"In general, regarding all the conclusions of science, there is the well known fundamental principle (named after Heisenberg) that all 'laws' in all branches of science are only estimates of what is most probable, the thing most likely to actually happen. This means that regarding the future it is impossible for a person to know anything with absolute certainty, only as an approximation."

The Rebbe MHM goes on to explain that according to scientific theory, the greater the number of events that are considered, the higher the percentage of them that will follow the law. But still, every

45. Ibid., 296. Italics are mine.
46. *Igros Kodesh* 13:143.

law is just a statement of the probability of what could happen, not a statement of what must happen. Thus any law of science may be violated.

"For example," he continues, "there would be no contradiction to science if a stone would be observed flying from below to above rather than the opposite [i.e. "falling up" instead of falling down]. It would only be contrary to what we are used to but not to the conclusions of science. This applies to all the laws of science without exception."

The fundamental uncertainty in science that we have discussed in this chapter—uncertainty due to the fact that scientific conclusions are merely conclusions drawn from assumptions, and uncertainty due to the issues of probability just discussed—was expressed frequently by Richard Feynman, as in the following quotes:

"To make progress in understanding, we must remain modest and allow that we do not know. Nothing is certain or proved beyond all doubt....And as you develop more information in the sciences, it is not that you are finding out the truth, but that you are finding out that this or that is more or less likely. That is, if we investigate further, we find that the statements of science are not of what is true and what is not true, but statements of what is known to different degrees of certainty."[47]

"Scientific knowledge is a body of statements of varying degrees of certainty—some most unsure; some nearly sure; none *absolutely* certain."[48]

"Scientists...are used to dealing with doubt and uncertainty. All scientific knowledge is uncertain. This experience with doubt and uncertainty is important. I believe that it is of very great value...You have to permit the possibility that you do not have it exactly right....This freedom to doubt is an important matter in the sciences...I feel a responsibility to proclaim the value of this freedom...I want to demand this freedom for future generations."[49]

Because of this fundamental uncertainty in physical law, no proof can be brought from science against the Torah.

---

47. Feynman, "The Relation of Science and Religion," in *Finding Things Out*, 248. (See p. xxvii, note 5).
48. Feynman, "The Value of Science," in ibid., 146.
49. Feynman, "Uncertainty of Science," 26–28.

A Day in the Life of Melech HaMoshiach

The world was created only for Moshiach.

—Talmud[1]

In this chapter we attempt to define who and what Moshiach is. It would certainly be impossible to describe the full greatness of Moshiach. To appreciate this, consider the statement of the Midrash[2] that Moshiach will be greater than Avraham and greater than Moshe. We would first have to have a full understanding of the greatness of Avraham and Moshe before we could start to understand the greatness of Moshiach!

As another example, consider the following prophecy of Zechariah about Moshiach: "On that day his feet will stand on the Mount of Olives."[3] In addition to its literal meaning, this is an allegory for the wisdom of Moshiach. Oil represents deep wisdom. Olives, the source of oil, represent an even greater wisdom. Olives on top of a mountain represent a higher wisdom yet. On this mountain of wisdom stand the *feet* of Moshiach, the wisdom of the *head* being so much greater.[4]

It is our intention then to present some of the basic fundamental characteristics of Moshiach based on which the reader will at least be able to formulate an image of Moshiach.

1. Sanhedrin, 98a.
2. *Midrash Tanchuma*, Toldos, sec. 14.
3. Zechariah 14:4.
4. *Likutei Sichos* 1:103.

The peace and perfection of the Era of Moshiach is the climax of creation. We find this goal stated at the outset, right at the beginning of the Torah. The second verse says, "The spirit of G-d hovered above the waters," on which the Midrash comments, "This is the spirit of Melech HaMoshiach."[5]

Throughout the Torah—and especially in the books of the prophets—there are countless prophecies about the Era of Moshiach, some explicit and some allegorical. Explanation of and elaboration on these prophecies as well as further information is found in the Talmud, Midrash, Zohar and books on Kabbala, Chassidus and Halacha throughout the generations.

Rambam, the great Jewish philosopher and codifier of Jewish law, concludes his major work on Jewish law, Mishneh Torah, with two chapters discussing the laws of Moshiach—the basic definitions of Moshiach himself and the Messianic Era. Elsewhere, in his famous list of the "13 Principles of Jewish Faith," the belief in the coming of Moshiach and the belief in the revival of the dead are the 12th and 13th on the list.

We begin with a prophecy of Isaiah which describes first Moshiach himself and then what the world will be like in the Era of Moshiach. This will be followed by selections from the last two chapters of Mishneh Torah which follow the same pattern—first describing Melech HaMoshiach (chap. 11), then the Era of Moshiach (chap. 12).[6] Finally, we will draw on these sources as well as selections from the Talmud, Midrash, Kabbala and Chassidus to sketch certain prominent characteristics of Moshiach.

Isaiah Chap. 11

> And a staff shall come forth from the stump of Yishai and a branch shall sprout from his roots.
>
> And a spirit of G-d will rest upon him [Moshiach], a spirit of wisdom and understanding, a spirit of counsel and

5. *Bereishis Rabbah* 2:4.

6. It is interesting to note the following anecdote in which the Rebbe MHM himself followed the same pattern: when he was presented with a painting of himself—Melech HaMoshiach—by the Chassidic artist Michel, he said, "And now, Michel, I would like you to make a painting of what the *world* will be like in the time of Moshiach."

power, a spirit of knowledge and fear of G-d.

And he will judge [with ease] as with his sense of smell because of his fear of G-d; neither with the sight of his eyes shall he judge, nor with the hearing of his ears shall he prove.

And he will judge the poor with justice and he will admonish with fairness the humble of the earth; and he will strike [the evil ones of] the earth with the rod of his mouth and with the speech of his lips he will put the wicked to death.

And righteousness shall be the belt of his waist and faith the belt of his loins.

And the wolf shall dwell with the lamb and the leopard shall lie with the young goat; and a calf and a lion and a fat ox shall lie down together and a small child shall lead them.

And the cow and even the bear shall graze, together their children will lie down; and the lion as the cattle will eat straw.

And the nursing child will amuse himself over the hole of a poisonous snake; and into the hole of a venomous snake shall a weaned child stretch his hand.

They shall do no evil nor shall they destroy throughout My holy mountain, for the land will be filled with the knowledge of G-d as the waters cover the ocean bed.

And the one who comes forth from the root of Yishai [Moshiach] who will stand on that day shall be for the nations as a banner carrier—the nations shall seek him; and his peace shall be an honor.

And it shall be on that day that G-d will continue to apply His hand a second time to acquire the rest of his people who will remain from Assyria and from Egypt and from Pasros and from Cush and from Elam and from Shinar and from Hamas and from the islands of the sea.

And He shall raise a banner to the nations to gather the exiles of Israel; and the dispersed of Judah He shall gather from the four corners of the earth.

And the envy of Ephraim will depart and the oppressors of

Judah will no longer be; Ephraim will not envy Judah and Judah will not oppress Ephraim.

And [Judah and Ephraim] shall swoop down upon the Philistines on the west side [of the land of Israel], together they shall plunder the people of the East; upon Edom and Moav they will stretch forth their hand and the children of Amon will obey them.

And G-d will cut off the gulf of the sea of Egypt and He shall wave His hand on the [Euphrates] river with the strength of His spirit; and He shall beat it into seven streams and cause [those returning from exile] to walk with their shoes [on the dry land].

And there shall be a road from Assyria for the rest of his people who will remain, just as there was for Israel on the day they went up from the land of Egypt.[7]

Laws of Kings and Melech HaMoshiach

Melech HaMoshiach will arise and restore the ancient dynasty of King David to its original sovereignty. He will build the Beis HaMikdosh and gather the dispersed of Israel [to the Land of Israel]. In his time all the laws of the Torah will be restored as they were in days of old: We will bring sacrifices and observe the Sabbatical and Jubilee years in accordance with all the Mitzvos stated in the Torah.

Anyone who doesn't believe in him or does not await his coming denies not only the prophets but also the Torah and Moshe Rabbeinu since the Torah itself gives testimony regarding Moshiach. It says, "G-d will return your exiles and will have mercy on you and gather you in.... even if your exiles are at the end of the heavens ... and G-d will bring you [to the Land of Israel]." These explicit words of the Torah contain within them all the things that all the prophets said on the matter.

---

7. Yeshaya chap. 11, translated by the author of this book based on *Metzudas Dovid* and *Metzudas Tzion,* with reference also to the translations of Rabbi Nisen Mangel (Kehot Publications) and the Judaica Press.

Also in *Parshas Bilam* there is a prophecy about the two Moshiachs—the first Moshiach which was King David who saved the Jews from their enemies, and the later Moshiach who arises from David's descendants who saves the Jews from the hands of the sons of Esau. In that prophecy it says, "I see him but not now"—this refers to David; "I behold him but it is not close"—this refers to Melech HaMoshiach. "A star will shoot forth from Yakov"—this is David; "And a staff will rise from Israel"—this is Melech HaMoshiach....

Also regarding the cities of refuge the Torah says, "If G-d will expand your borders you will add on three more cities." This has not yet happened. But G-d doesn't give an instruction for nothing. [Thus it will happen in the future—in the time of Moshiach.]

We do not, however, have to bring proofs from the prophets [about Moshiach] since all their books are filled with this matter.

Do not think that Melech HaMoshiach is required to perform signs or wonders, bring new things into the world or revive the dead or similar things....

If a king will arise from the House of David who teaches Torah and is occupied in the Mitzvos like David his ancestor in accordance with both the Written Torah and the Oral Torah, induces all the Jews to follow the Torah and to strengthen it where it has been broken and who fights the battles of G-d he has the assumed status of being Moshiach (*chezkas Moshiach*). If he succeeds further, builds the Beis HaMikdosh in its place and gathers the dispersed of Israel—then he is Moshiach with a certainty (*Moshiach vadai*) and he will direct the entire world to serve G-d together.

Do not think that in the time of Moshiach anything of the way of the world will be cancelled or that there will be innovations in the order of nature. But rather the world will follow its natural order.... Some *chachomim* (Rabbis of the Talmud) say that before Moshiach comes, Eliyahu HaNavi will come. Regarding all these matters and similar matters no one knows how they will happen until they happen.... [The

> main thing is] to await it and to believe in the matter in general as we explained above.
>
> The scholars and the prophets desired the Era of Moshiach ... only so that they would be free to pursue the study of the Torah and its wisdom without anyone oppressing or distracting them so that they will merit the "World to Come."... At that time there will be no famine or war, no jealousy or rivalry since there will be abundant good and all the delicacies will be as plentiful as the dust of the earth. The only occupation of the entire world will be to know G-d. Thus the Jewish people will be great scholars and will know the hidden secrets and will attain knowledge of the Creator to the full capacity of man as the prophet says, "For the earth will be filled with the knowledge of G-d as the waters cover the ocean bed."[8]

While Rambam writes that Moshiach is not required to perform wonders, this is only as a Halachic *requirement* i.e. the Halachic definition of Moshiach does not require him to perform wonders but, as we will see later from many sources, Moshiach will in fact have a miraculous nature. Rambam is only stating what the minimum Halachic requirements of Moshiach are. Similarly, his statement that the world will maintain its natural order applies only to the initial period of Moshiach's coming, the so called "First Era." Later on however, in the "Second Era," the order of nature will certainly change as the revival of the dead will occur—a very great change in the order of nature indeed.[9]

As mentioned before, the belief in the coming of Moshiach and the belief in the revival of the dead are the 12th and 13th principles of faith on Rambam's famous list of the Thirteen Principles of Faith. Similar to Rambam's statement here that one who does not believe in Moshiach denies the whole Torah, is his statement elsewhere that one who does not believe in the revival of the dead has no part in Judaism.[10]

---

8. *Hilchos Melachim U'Melech HaMoshiach*, chaps. 11–12, free translation by the author.
9. *Dvar Malchus* on *Hilchos Melachim*, *sicha* no. 3.
10. Rambam's commentary to the first *mishnah* in Sanhedrin, *Perek Chelek*.

"The innovation of Moshiach is that through him the two concepts of 'above' and 'below' will be united."[11]

The unification of opposites is a fundamental characteristic of Melech HaMoshiach which derives from his life source which is *Atzmus*, the essence of G-d. This is indicated by the verse in Isaiah, "And a spirit of G-d will rest upon him, a spirit of wisdom and understanding, a spirit of counsel and power, a spirit of knowledge and fear of G-d."[12] This verse is analyzed carefully in the writings of Chassidus where it is pointed out that even though wisdom and understanding are opposite intellectual powers (wisdom is the essential kernel of a concept, often symbolized by a fountainhead, and understanding is the broad expansion and development of a concept, symbolized by a wide river), the spirit of G-d the descends upon Moshiach will contain both of these opposites. This will enable him to unify opposing forces in the world.[13]

We will use this concept of the unification of opposites as the framework for our presentation of some of the primary characteristics of Moshiach stated in the works of the Talmud, Kabbala and Chassidus.[14]

Majesty and Modesty

Regarding Moshiach, the prophet Isaiah says,[15] "Behold, my servant shall have great wisdom. He will be elevated, exalted and of very high stature." In various sources this high exaltation is explained to mean that Moshiach will be greater than Adam, Avraham Avinu and even Moshe Rabbeinu.[16]

Melech HaMoshiach will be exalted, greater and more honored and more glorious than all kings who preceded him.[17] His reign will

---

11. *Sefer HaMa'amorim Melukat* 2:297.

12. Yeshaya 11:2.

13. *Sefer HaMa'amorim Melukat* 2:297–98. See also chapter V, sec. 2b, "The Messianic World Order," for a discussion of this in the wider context of Messianic phenomena.

14. It is significant to note that at a conference about the Rebbe MHM and his influence on the world held at New York University (NYU) in November 2005, one professor described him as "the embodiment of impossible contraries."

15. Yeshaya 52:13.

16. *Midrash Tanchuma*, Toldos, sec. 14.

17. *Likutei Sichos* 6:254.

extend over the entire world and will last forever.[18] His wisdom will surpass that of *Shlomo HaMelech* (King Solomon) and he will be a prophet of the caliber of Moshe.[19] He will reveal the deepest level of the Torah—both its Halachic content and its mystical secrets.[20] What we know of the wisdom of the Torah now is nothing compared to what he will reveal.[21]

At the same time he will be of the greatest humility concerned with the common man,[22] bringing this new wisdom and knowledge down to the level where anyone can understand it. He will show that the sincere faith of the simple Jew, who is totally dedicated to Judaism, is greater even than the deep understanding of the scholars.[23]

The Midrash tells a story about Moshe from the time that he was a shepherd. One of the sheep ran away from the flock and Moshe left the entire flock to care for this sheep and bring it back to the flock. This trait qualified him to be the leader of the Jewish people as he would be concerned about every individual Jew.[24] Similarly Moshiach will cherish every single Jew—even one who has run away from the flock.[25]

These two extremes of majesty and modesty are symbolized by two prophecies, one which says that Moshiach will appear on a cloud and one which says that he will appear as a poor man riding on a donkey.[26]

Wisdom and Prophecy

The *chacham* (man of wisdom) and the prophet are two distinct personalities with different characteristics and abilities. Each has an advantage over the other. Moshiach will attain the highest levels of perfection in both of these areas As Isaiah says in the prophecy quoted above, "And the spirit of G-d (prophecy) will rest upon him, a spirit of wisdom and understanding ... a spirit of knowledge...."

18. *Bamidbar Rabbah* 13:14.
19. See references in nn. 27 and 28.
20. *Sefer HaSichos* 5751, 566.
21. *Koheles Rabbah* 11:8.
22. *HaYom Yom*, 75.
23. *HaYom Yom*, 9, 56.
24. *Shemos Rabbah* 2:2.
25. "Sicha of Achron Shel Pesach 5743," in *Hisva'aduyos* 5743, 3:1314.
26. *Sefer HaSichos* 5749, 109.

Rambam[27] writes that Moshiach will be a chacham greater than King Solomon who was called "the wisest of all men," and will attain a level of prophecy close to that of Moshe Rabbeinu of whom the Torah says, "There arose none like Moshe who knew G-d face to face." The Midrash Tanchuma goes even further and indicates that Moshiach will be a greater prophet than Moshe.[28]

In fact, there is a close connection between these two characteristics of being greater than King Solomon and similar to Moshe, as we see from the continuation of Rambam's statement: "therefore he will teach Torah to all Jews." This refers to a deep level of the Torah which will be revealed by Moshiach. Now we know that the Torah has great depth. The greatest Rabbis and scholars—starting with Moshe Rabbeinu himself—have been studying it, analyzing it and delving into its depths for thousands of years. But the Torah being the wisdom of G-d has infinite depth so even the deepest analysis by the greatest *chachamim* for the longest time cannot reveal its full depth. Moshiach, however, will reveal in it a new dimension previously unattainable. So new will this revelation be that the Midrash (based on a verse in Isaiah) calls it *Torah Chadasha*—a new Torah. King Solomon refers to it at the beginning of his Song of Songs and symbolizes it by a kiss from G-d.[29]

To be able to accomplish this task Moshiach needs two qualities. Firstly, he needs to have the highest level of prophecy—similar to Moshe—to be able to receive this revelation from G-d, just as Moshe received the original revelation of the Torah from G-d at Mt. Sinai. But that is not yet enough. To be able to teach this new dimension of Torah to all Jews he must be able to bring it down to their level. To do this one would have to have the wisdom of King Solomon—literally. The greatness of the wisdom of Solomon consisted of his ability to take the greatest abstract concepts and, by giving one example after another, bring it down one level after another until even the common man could understand it. Moshiach, who will reveal a wisdom greater than Solomon's, must therefore have this ability to a greater degree than even Solomon had.[30]

27. Rambam, *Hilchos Teshuva,* chap. 9, sec. 2.
28. *Sefer HaSichos* 5751, 789.
29. Shir HaShirim 1:2 and commentary of Rashi.
30. See *Sefer HaSichos* 5751, 566–82.

There is another aspect of King Solomon's wisdom which is also essential to the wisdom of Moshiach. Regarding wisdom King Solomon says, "I saw that there is an advantage to wisdom over nonsense similar to the advantage of light over darkness." The Zohar asks a strong question on this: "Does it take King Solomon, the wisest of all men, to see that wisdom is better than nonsense and light is better than darkness? Doesn't everybody know this?" Chassidus explains that "wisdom" and "nonsense" here actually refer to two types of wisdom one which is termed "light" and the other which called "darkness." The light derives directly from the Torah—the search from above—while the darkness is the view of reality resulting from the groping search from below—the attempt to understand the world in its own terms, i.e. scientific research. The difference between these two types of wisdom is so great that the latter is called "nonsense" relative to the former. The point that Solomon was making was that when the sciences are used to explain a matter in Torah—for example, when astronomy and mathematics are used to explain the laws of the new moon—this scientific knowledge ceases to be secular. It becomes part of Torah. The darkness is thus transformed into a great light—greater than the light derived directly from the Torah. This is the "advantage" that he is talking about. The light resulting from transformed darkness has an advantage over the light derived directly from the Torah.[31]

The greatness of Solomon's wisdom was due in part to his having transformed the sciences.[32] This transformation of "nonsense" into wisdom is a major part of the refinement of the world that prepares it for and actually brings about the highest revelations in the Torah by Moshiach.[33] This process actually began with Joseph, who was the master over ancient Egypt and its wisdom, reached a very high level with King Solomon and will be completed by Moshiach himself.[34]

Thus Melech HaMoshiach will unify the concepts of 'above' and 'below'—including the wisdom from above and the wisdom from below.

Finally, the prevalence of peace in the time of Solomon (for which

31. *Sefer HaMa'amarim Melukat* 3:58–60.
32. "Sicha of Parshas Emor," sec. 6, in *Sefer HaSichos 5751*, vol. 2.
33. *Sefer HaSichos 5751,* 2:525, n. 73.
34. *Ma'amorei Admur HaEmtzai,* Devorim 1:9–10.

he was named, as the prophet said, "His name will be *Shlomo* and I will put *Shalom*—peace—and tranquility on Israel in his days"[35]) was a sample of the peace that will prevail in the time of Moshiach. The power of his monarchy was due to the power of his wisdom by which all the nations were awed. It was a great light which dispelled the darkness of war and brought about unity just as Isaiah says in the Messianic prophecy quoted above, "They will not cause destruction ... because the earth will be filled with the knowledge of G-d."[36]

Natural and Miraculous

Moshiach is a human being with human parents whose ancestry goes back to King David. Yet his soul is from the loftiest heights—from the essence of G-d Himself. "Moshiach, as a messenger of G-d, combines two extremes. On the one hand he is a being in and for himself, a soul in a physical body. But at the same time, being a messenger of *Adam HaElyon* (G-dliness), he is literally like—Him."[37] The essence of Moshiach is the same as the essence of G-d.

This essential fact about the life of Moshiach is connected with another essential fact: that he lives forever—an uninterrupted everlasting life. This fact is stated in the Talmud based on the verse in Psalms, "Life he [Moshiach] asked from You, and You granted it to him—length of days forever."[38] This is discussed extensively in the writings of Kabbala and Chassidus. The Mittler Rebbe, the second Rebbe of Chabad, in his explanation of this, contrasts Moshiach with other tzaddikim. He says that all other tzaddikim require some refinement of their bodies and therefore there has to be some sort of "passing away." This is true even of Moshe Rabbeinu of whom the Talmud and Zohar say, "Moshe did not die," and even regarding Dovid HaMelech (King David) of whom it is said, *"Dovid Melech Yisroel Chai V'kayam"*—David King of Israel is alive.

Moshiach, however, is on a much higher level. "No change or refinement takes place in his body at all. He will have a physical body similar to the bodies of all Jews of that time" which will eat, drink,

35. Divrei HaYamim I, 22:9.
36. See *Sefer HaSichos* 5751, 527–29.
37. *Sefer HaSichos* 5752, 1:107.
38. Psalms 21:5.

wear clothes and sit on a chair but "Moshiach will remain alive in his body together with the essence of his soul—down here just as it is above, literally. By Moshe Rabbeinu "there was no difference at all between his body and his soul ... and this will hold true of Moshiach in a much higher way. Therefore ... he will not need to ascend above."[39] (There is, however, a period of concealment between the time he attains the "assumed status of Moshiach" and when he attains the status of "Moshiach with a certainty."[40])

Regarding Dovid HaMelech the Mittler Rebbe writes, "His life came only from the ten *sefiros* ... of *Atzilus* (the highest spiritual world), but regarding Moshiach it says '*You* granted it to him' which means from *Atzmus*, the infinite."[41] When G-d is addressed in second person ("You") it refers to *Atzmus* i.e. G-d Himself, His essence. A life which comes from Atzmus is above all definitions and limitations of natural life and death as we know it. As the *Rebbe Rashab* says, "Relative to Atzmus, life and death are equal."[42] This will be discussed at greater length later in this chapter.[43]

In fact Atzmus transcends all limitations. The miraculous and the natural are equal relative to Atzmus and can coexist. It follows that the life of Moshiach is completely above the laws of nature, even as he lives in the physical world. As the Maharal of Prague writes "Moshiach will be ... outside of nature and the way of the world."[44] The Midrash says that G-d conceived of Moshiach even before He created the world because Moshiach is higher than the order of creation.[45] Yet at the same time, as we emphasized above, he lives within nature and the physical world. Because he contains these two opposites within himself, he will eventually bring the entire world to a new world order which will be characterized by the full revelation of G-d's infinite essence within the finite physical world,[46] a world order in which the natural and the miraculous combine and function simultaneously. This is expressed by the verse, "As in the days of your exodus from Egypt, I will show you wonders."[47]

39. *Ma'amorei Admur HaEmtzai*, "Drushei Chasunah," 131, 155.
40. *Bamidbar Rabbah,* chap. 11, sec. 2. *Likutei Sichos* 9:381.
41. *Ma'amorei Admur HaEmtzai*, Devorim 1:214–15.
42. *Sefer HaMa'amorim* 5670, 100.
43. See sec. 6, "The Everlasting Life of Melech HaMoshiach."
44. Quoted in *Sha'arei Geulah* 2:69, n. 29.
45. *Bereishis Rabbah* 1:4; *Or HaTorah,* Miketz, 2202.
46. *Sefer HaMa'amarim Melukat* 3:297–98.
47. Micha 7:15. See *Sefer HaMa'amarim Melukat* 4:225–34.

Soul Within a Soul

While the everlasting life of Moshiach transcends that of Moshe Rabbeinu and King David, there is, however, a deep connection between them. In fact, in the writings of Kabbala and Chassidus it says that the souls of both Moshe Rabbeinu and King David will "enter into" the soul of Moshiach.[48]

This concept of a soul entering into another soul or one soul containing another soul, is a deep topic in Kabbala. But one of the most important facts about Moshiach is that his soul is a *Neshama Klalis*—a general soul which contains the souls of all Jews. This means that the soul of every Jew is a part of Moshiach and thus there is a part of Moshiach in every Jew. This technical fact has a very important practical implication: since the soul of every Jew corresponds to a unique part of the soul of Moshiach, each Jew has a unique role to play in preparing the world for the coming of Moshiach.[49]

---

48. Likutei Sichos 35:206, n. 6.
49. *Ibid.* 20:522; *Sefer HaSichos* 5751, 590.

The Emergence of Melech HaMoshiach

Immediately we see that Moshiach is already present among us, and everyone points with his finger and says, "Behold, this is Melech HaMoshiach. He has already come."

—the Rebbe MHM[1]

The Birth of Moshiach

Rabbi Levi Yitzchok Schneerson, a fourth generation descendent of the Tzemach Tzedek, was known as a great Gaon and Kabbalist. On the 11th of Nissan, 5762 (1902), his first son was born in the city of Nikolayev in the Ukraine. At the bris, which took place on 18 Nissan—Rabbi Levi Yitzchok's birthday—the boy was named Menachem Mendel after the Tzemach Tzedek. This was the boy who would grow up to be Melech HaMoshiach.

The Talmud says that when Moshe Rabbeinu was born his home was filled with light. Already at the time of his birth there was an indication that an extraordinary person was born. So too it was with Melech HaMoshiach. On the day of his birth, the fifth Lubavitcher Rebbe, Rabbi Shalom DovBer sent six telegrams to Rabbi Levi Yitzchok and his wife, Rebbitzen Chana, regarding the birth of the baby which included special instructions on how he should be cared for.

It became clear very soon that the young boy had special qualities

1. *Sefer HaSichos 5751*, 2:548.

of intellect and strength of character. He was educated primarily by his father, studying intensely on his own also.

In 1905, after an attempted revolution against the Czar failed, a wave of anti-Semitic pogroms began sweeping through Russia. One such attack was directed against the Jews of Nikolayev. Groups of Jews hid in various hiding places during the attack. In one such shelter, several families with small children were gathered. The children were frightened and started crying, putting the entire group in danger of being discovered, which would mean certain death for all of them. The situation was becoming desperate when suddenly little Menachem Mendel, four years old at the time, got up and went over to each of the crying children, giving a candy to one, whispering something to another etc. calming down all the children—as well as their parents—and saving their lives.[2]

This young boy, who would grow up to be Moshiach, was already thinking about Moshiach and the Geula at this young age. He described these thoughts many years later in a letter to Yitzchok Ben-Tzvi, the president of Israel: "From the day that I started going to *cheder* (school), and even before then, the idea of the future Geula began to take shape in my mind—the redemption of the Jewish people from their final exile. It would be a redemption of such a nature that all the suffering and devastation of the exile would be understood. Part of that bright future, part of that Geula, would be the *Nasi*—the king ... who has no one above him except G-d."[3]

The Wisdom of Moshiach

In the previous section we discussed the characterization of Moshiach as combining two extremes. He will reach the loftiest heights of wisdom and prophecy, and at the same time have the greatest humility and care for the most humble people. We now look at some aspects of the Rebbe MHM's wisdom which began to blossom in his early youth.

While still a teenager, his father would consult with him on com-

2. For the stories mentioned here on the Rebbe MHM's youth, as well as many others, see Herman Branover and Avraham Naveh, *The Ultimate Jew* (Jerusalem: Shamir, 2003), chap. 6.

3. *Igros Kodesh* 12:414.

munity matters. There are also stories attesting to his genius during those years. Once, a dean of engineering at the university in Yekatrinislav was visiting the home of Rabbi Levi Yitzchok. He mentioned that they were working on a difficult mathematical problem. Rabbi Levi Yitzchok suggested that his sons, who knew mathematics, might want to try to solve the problem, which the dean then presented to them. A half hour later, Menachem Mendel came back and handed the dean a piece of paper. The dean, knowing that no one could solve the problem in half an hour, did not take it seriously, just stuck the paper in his pocket and forgot about it.

Sometime later, the dean told Rabbi Levi Yitzchok that he was astounded to find that Menachem Mendel had in fact given him the correct solution to the problem—which would have taken the best mathematicians days to solve.

Young Menachem Mendel was interested in astronomy as well as mathematics. At the age of sixteen he told one of his friends that there would be a solar eclipse on February 25 of that year. When that day came, his friend and some other young men went out to observe the eclipse but, to their surprise, they saw no eclipse. The sun shone in its full glory as usual. The friend came back to Menachem Mendel the next day and said that there was no eclipse. Menachem Mendel responded that there certainly was an eclipse. His friend explained that they all went out to look for the eclipse but saw no sign of it. "But I'm telling you," insisted Menachem Mendel, "that there was an eclipse. My calculations are very exact and they are correct."

A few days later an article appeared in a local periodical reporting on the eclipse that had in fact taken place and explained that it was visible in nearby areas but not in Yekatrinislav itself.

This was perhaps a foreshadowing of things to come—the Rebbe MHM knew things that no one else knew. When he would announce them, people who go only by what is visible to the eye at the moment, would be skeptical. But eventually, the truth of his words would be clearly visible. This is what happened in 1990-91 when he first announced that the time of the Geula had arrived. Since then we have seen many aspects of the Geula actualized, as we have been discussing in this book. As the Rebbe MHM himself said, "All we need to do is open our eyes and then we will see that the true and complete Geula is already here, literally."[4]

During the 1920s, the Rebbe MHM lived and learned in the household of the Previous Lubavitcher Rebbe in Petersburg (Leningrad). On 14 Kislev, 5689 (1928), he married the Previous Rebbe's daughter, Rebbitzen Chaya Mushka Schneerson.

Following their marriage, the Previous Rebbe sent them to Berlin where, at the instruction of the Previous Rebbe, the Rebbe MHM studied physics and other sciences at the University of Berlin (1928-1932). He attended the lectures of Professor Irwin Schroedinger who discovered the fundamental equation of Quantum Mechanics known as the Schroedinger Equation.[5] Some forty years later, the Rebbe MHM approved a suggestion by Professor Pesach (Paul) Rosenbloom that college students be given Schroedinger's philosophical work "What is Life" to read.

(It is interesting to note that Schroedinger was very sympathetic to the Jews. When the Nazis came to power and began firing Jewish academics from their positions, Schroedinger left Germany in protest. One may wonder to what extent Schroedinger's attitude may have been shaped by his relationship with the Rebbe MHM.)

Majesty and Modesty

Even while mastering the sciences at the University, the Rebbe MHM's primary occupation was learning Torah. Some of the gaonim of our generation, such as Rabbi Joseph B. Soloveitchik, Rabbi Yitzchok Hutner and Rabbi Chaim Heller, were also at the University of Berlin at that time and they attest to this fact.

In 1933, after the Nazis came to power, the Rebbe MHM and Rebbitzen Chaya Mushka moved to Paris where he continued studying mathematics, science and engineering at the Sorbonne University (1934–1938).

Rabbi Joseph B. Soloveitchik, later to become the leader of modern Orthodox Jewry and the *Rosh Yeshiva* of Yeshiva University, was a close friend of the Rebbe MHM in Berlin. It was a friendship that

4. *Sefer HaSichos* 5752, 1:173.

5. This was related to me by Prof. Rosenbloom, who heard it from the Rebbe MHM himself.

was to continue for decades to come. It was also a friendship that went back several generations. It went back at least as far as the friendship between Rabbi Shmuel, the fourth Rebbe of Lubavitch, and the elder Rabbi Joseph DovBer Soloveitchik (1820–1892), but was most notable in the next generation when Rabbi Shalom DovBer, the fifth Rebbe of Lubavitch, and Rabbi Chaim Soloveitchik were very close friends and worked together very closely in matters relating to the Jewish Community.

Much of what is known about the Rebbe MHM in the Berlin days comes from statements and stories by Rabbi Soloveitchik.[6] He once said that whenever he had a question on a matter in Torah or a matter in science, he would go the Rebbe MHM's apartment to ask him. When he would arrive, he would always find him engrossed in learning. "Even between the walls of the University," Rabbi Soloveitchik would say, "in everything he did one could see that he was a tzadik."

A deep insight into the Rebbe MHM can be gained from the following two quotes from Rabbi Soloveitchik. On one occasion he said, "I knew him in the [university] and he was already a Rebbe then." Another time he said, "Even now, when he is revealed as a Rebbe, he is still hidden. Only a little bit is revealed. We can never imagine the depth of his greatness." Even when he was hidden—pursuing his private studies—his greatness was revealed, and even when he was revealed—famous as the Rebbe—his true greatness was hidden. Here, again, we see the combination of the two extremes—majesty and modesty—that characterize Melech HaMoshiach.

That the Rebbe MHM's knowledge of Torah is vast and deep beyond description has been attested to by all the Torah authorities of the world. As Rabbi Soloveitchik once said, "I never saw anyone with such a good memory and such a vast knowledge of Torah." After Rabbi Soloveitchik passed away, his younger brother, Rabbi Aharon Soloveitchik succeeded him as Rosh Yeshiva of Yeshiva University (in addition to continuing as Rosh Yeshiva of Brisk in Chicago) and a leader of Orthodox Jewry. He was one of the great Rabbis who looked to the Rebbe as Melech HaMoshiach.[7]

---

6. My primary reference here is Rabbi Shalom DovBer Wolpe, *Shemen Sasson MeChaverecha* (Israel, 1995). See also Branover and Naveh, *Ultimate Jew,* chap. 7.

7. See Rabbi Soloveitchik's letter published in *The Jewish Press*, June 28, 1996, p. 27.

The great Gaon, Rabbi Pinchus Hirschprung, Chief Rabbi of Montreal, Canada, was famous for knowing the entire Talmud by heart. He was one of the greatest supporters of the Rebbe MHM's campaign to establish the daily study of Rambam. He said that G-d brought the soul of the Rebbe MHM into the world in our generation to enliven "the dry bones," and he called the program of the daily Rambam study "an exalted G-dly enactment."

Rabbi Aharon Soloveitchik and Rabbi Pinchus Hirschprung both wrote letters after the concealment of the 3rd of Tammuz supporting the continued belief that the Rebbe is Melech HaMoshiach and the declaration of *"Yechi Adoneinu Moreinu V'Rabeinu Melech HaMoshiach L'Olam Voed!"* ("Long live our master, teacher and Rebbe, Melech HaMoshiach forever!")

In 1982, the great *gaon* and *posek* (halachic authority) of our generation, Rabbi Moshe Feinstein, wrote a letter honoring the Rebbe MHM on the occasion of his 80th birthday saying that the Rebbe MHM rises above all other Rabbis and leaders of our generation because he has successfully established centers of Torah learning everywhere in the world where there are Jews. He then gives his blessing to the Rebbe MHM for everlasting life—not just long life—since, he says, "we hope that Melech HaMoshiach will come immediately and then the Rebbe *shlita* will merit to serve forever."[8]

The Sephardic Chief Rabbi of Israel, Rabbi Mordechai Eliyahu said of the Rebbe MHM: "There is no Rabbi as great as him in all areas of the Torah in this generation. In the logical understanding of the Torah no secret is hidden from him.... Similarly, in the hidden and Kabbalistic levels of the Torah.... It is absolutely correct to say of him that he is the greatest in the generation. There is no one second to him. But what is most important is his humility.... There is no one like him." Majesty and modesty—the trademark of Moshiach.

The Science of Moshiach

And what about the sciences that the Rebbe MHM has mastered? We mention the experience of the renowned physicist, Professor

8. The letter is reproduced in Rabbi Shalom DovBer Wolpe, *Yechi HaMelech HaMoshiach,* 137.

55 F. D. R. DRIVE
New York, N. Y. 10002

ORegon 7-1222

ר"מ תפארת ירושלים
בנוא יארק

בע"ה

ה' מנחם אב, תשמ"ב

בכללות, בקשר לידידי וחביבי מעלת הוד כבוד קדושת הגאון הצדיק האדמו"ר מנחם מענדל שליט"א שניאורסאהן מליובאוויטש, הנה ידוע בעולם אודות גאונותו בתורה ואיך שהוא באמת מתעסק גדול בתורה בגאון שהוא עובד תמיד בתורה ושהוא באמת בעל תורה.

במשך השנים שאני מכיר האדמו"ר שליט"א שהקשר בינינו הי' בידידות גדולה הי' לי כמה הזדמנויות להשתעשע עמו בענינים מסובכים בין בתורת הנגלה ובין בתורת הנסתר, ותמיד החשבתיו כרום גדולתו וגאונותו.

אבל מאחר שידוע כל זה בעולם אין צורך לכתוב מענינים אלו, רק כדאי להעיר כאן, כמין הקדמה לקונטרס חשוב זה שידידי הרה"ג מו"ה הרב אברהם יצחק שי' עם טוב מפילדלפי' מוציא לאור לכבוד שנת השמונים של ידידי האדמו"ר שליט"א, נקודה אחת אודות גדולתו האישית של האדמו"ר שליט"א והוא,

שקידש עצמו להכנס ראשו ורובו להרביץ תורה בכל העולם, בקטנים ובגדולים בכלל ישראל ביחד ובאמת מתעסק בגאונות בענין של הרבצת תורה, וניכר שכל עבודתו הוא בזה, בין במה שסידר בקיץ מחנות לצעירים ומבוגרים, בנים ובנות, שהם לשם ולתפארת, ומטרתם היא ג"כ מטרה שכולה תורה, ובין בכל השנה בכל פעולותיו המרובים, הציבוריים והאישיים, וזה רציתי להזכיר שזה באמת השבח הכי גדול והוא השבח שהי' לכל גדולי עולם במשך כל השנים שהיו בעיקר בתורה, וגם שבטח אמת שכל הפעולות והענינים שעושים צדיקים גדולים הם דברים גדולים, אבל הענינים של התורה הם באמת היסוד הראשון.

לפיכך אני משתתף בשמחה להענין שנתקבל בהעולם כולו שיש שמחה גדולה שהשי"ת האריך ימיו לשנת השמונים, ואין מספיק סתם לברכו לאריכות ימים ושנים שהיום לא מספיק ברכה זה שמקוים שמלך המשיח יבוא מיד ובקרוב ויזכה אז האדמו"ר שליט"א לעבוד לעולם ועד בתורה וביראת שמים שעבודה זה לא יגמר גם אחר ביאת הגואל וגם אחרי בנין בית המקדש שגם אז זה יהי' היסוד הראשון, ונברך שיזכה ידידי אדמו"ר שליט"א לאריכות זה ושיזכו כל כלל ישראל לגאולה השלימה בקרוב מיד ביחד עם האדמו"ר שליט"א.

בידידות והוקרה

משה פיינשטיין

RABBI M. FEINSTEIN

Letter from the great gaon and posek Rabbi Moshe Feinstein, of blessed memory, in honor of the Rebbe MHM's 80th birthday

Yirmiyahu (Herman) Branover. He was the head of a physics institute in Latvia before leaving the Soviet Union and now heads the MHD (Magnetohydrodynamics) Laboratory at the University of the Negev in Beer Sheva, Israel. He is one of the directors of the Rabbi Yisroel Aryeh Leib Research Institute on Moshiach and the Sciences and some of his lectures at our conferences appear later in this book.

Professor Branover had frequent private meetings with the Rebbe

MHM in which he would present reports on his most recent research. On one such occasion Branover presented his plans for a new invention. The Rebbe MHM told him to do the presentation as he would do it at the university for professional engineers, with all the technical details. Branover proceeded, mentioning the value of the two-phase flow velocity and the value of the efficiency of the system. The Rebbe MHM immediately responded that those two values were inconsistent. Professor Branover was completely taken aback by this statement but was too respectful to ask for an explanation.

When he went back to Beer Sheva, however, Branover and his team reviewed their calculations to try to find the error but found none. It was not until many months later that the error was found. A new doctoral student who had joined their team was studying the report line by line, equation by equation, to learn about the project and in the process he found the error. This was the error that the Rebbe MHM saw instantly.

The Rebbe MHM's mastery of the sciences turned out to be a powerful tool in addressing perceived conflicts between modern science and Torah which often created a barrier between the modern Jew and his commitment to Judaism. Over the years, many Jews would turn to the Rebbe MHM to resolve these issues—from the late Chief Rabbi of Israel, Rabbi Yitzchok Isaac Herzog,[9] himself a great Gaon well versed in the sciences, to professors such as Branover and Paul Rosenbloom, to the working scientist and engineer, and the Rebbe MHM answered these questions, explaining the issues in terms of the sciences themselves.[10] Many Jewish professors, scientists and university students were brought back to Judaism as a result.

But there is a deeper significance to his mastering the sciences for it is a fundamental characteristic of the wisdom of Moshiach that it is derived (in part) from the transformation and refinement (*beirur*) of the wisdom of the world—the sciences—just as the wisdom of King Solomon was derived (in part) from the transformation of the wisdom of ancient Egypt.[11] The revelation of the "wondrous aspects

9. See Rabbi Herzog's letter to the Rebbe MHM in *Igros Kodesh* 13:142–46.

10. We have quoted some of these answers in sec. 3 of this chapter, "Wisdom from Below."

11. See the sicha of *Parshas Emor*, 5751, esp. sec. 6, *Sefer HaSichos 5751*, 527–29.

of the Torah" in the Era of Moshiach, a very deep level of pure Torah intellect, "is brought about by the transformation of the wisdom of the world ... in the same way that *Geula* is brought about through the revelation of the *Alef* in *Gola*[12] and in the same way that the abundant wisdom of Solomon [was derived from] the wisdom of Egypt."[13]

A Great Prophet

Rambam writes that Melech HaMoshiach will be a great prophet, close to the level of Moshe Rabbeinu. The Midrash Tanchuma implies that Moshiach will be a prophet even greater than Moshe Rabbeinu.

Before discussing some of the prophecies of the Rebbe MHM, we should point out that prophecy did not end with the close of the Biblical era of the prophets as is commonly thought. Quite the contrary, the Rebbe MHM has explained in various sichos that it is a fundamental belief of Judaism that prophecy is always available to the Jews. In every generation there is a prophet. Rambam writes that "prophecy will return to the Jews" in an open manner as a prelude to the coming of Moshiach.[14] Specifically, the Baal Shem Tov, the Maggid of Mezeritch, the Alter Rebbe and all the Rebbeim of Chabad were prophets.[15]

The Rebbe MHM discusses this in connection with the Messianic prophecy of Isaiah: "I will return your judges as in the beginning and your advisors as at the start,"[16] explaining that the Previous Rebbe was the "judge of our generation and the advisor of our generation and the prophet of our generation" which has been confirmed by the fulfillment of his blessings and predictions.

The Rebbe MHM then extends this to our generation based on the halacha in Rambam regarding a prophet whose status is verified by another prophet: "A prophet about whom another prophet testifies that

---

12. See the sichos of *Parshas Acharei-Kedoshim* and *Emor,* in *Sefer HaSichos 5751*, for a full explanation of this concept. It is explained briefly in chap. 2, sec. 4 of this book.

13. *Sefer HaSichos* 5751, 525 n. 73.

14. *Igeres Teiman*, chap. 3.

15. *Likutei Sichos* 14:72–74; *Sefer HaSichos* 5751, 2:786–95. These sichos include a full discussion of the issue of prophecy throughout the generations.

16. Yeshaya 1:26.

he [the first one] is a prophet—*as is the case regarding the previous Rebbe and continues in the next generation by his disciples etc.*—has the *chazaka* of being a prophet." The Rebbe MHM is indicating here, in the most humble way, that he is the prophet of this generation.

He then continues with an imperative to publicize this:

> It must be publicized to everyone in the generation that we have merited that G-d has chosen a person whose stature is incomparably higher that the people of his generation, to be the judge, the advisor and the prophet of the generation. He gives instruction and advice regarding the service of G-d of all the Jews and all the people of this generation, in all matters of the Torah and mitzvos and also regarding their daily life and conduct in general, including the attainment of knowledge of G-d through all their actions and applying all their actions to the service of G-d, and including the primary prophecy of "To the Geula immediately" and that immediately—"Behold, Moshiach is coming."[17]

The Rebbe MHM's prophecies regarding world events in general and the Land of Israel in particular are well known.[18] We mention briefly his prediction of great victory in the Six Day War in 1967. Prior to the war, he started the Tefillin Campaign and said that Jews putting on Tefillin gives the Jewish people the power to defeat the enemy. During the summer of 1973 (the year of the Yom Kippur War), when there was no sign of war in Israel, he began a campaign for Jewish children to study Torah saying that this gives us the power to defeat our enemies. He referred to the verse in the Psalms: "From the mouths of young children and babes You (G-d) set a foundation of strength to nullify the enemy."[19] A week before Yom Kippur he declared that there would be a great victory. These predictions were widely publicized in Israeli newspapers.

At the same time he warned against giving any of the liberated land back to the Arabs, saying that even talking about giving them autonomy or land threatens the lives of Jews. Unfortunately, we have seen

---

17. *Sefer HaSichos* 5751, 2:792.

18. See Rafael Heruti, *Nevuat HaGeula* (Kiryat Malachi) for extensive documentation of this.

19. Tehillim 8:3.

this prophecy fulfilled also. Ariel Sharon once admitted in a speech that he gave in Kfar Chabad, Israel (before he became Prime Minister), that if they had listened to everything that the Rebbe MHM told them, they wouldn't be in the predicament that they were in.

When Mikhail Gorbachov rose to power in Russia, the Rebbe MHM called in Professor Branover, who is a Russian émigré, and told him to start building housing and employment projects in Israel for the tens of thousands of Jews who would soon be leaving Russia and coming to Israel. This request surprised Professor Branover and was met with disbelief by his contacts inside Russia who told him that there was no indication of any permission being given for Jews to leave Russia. On the contrary, they told him, from all indications it looked like things were going to get worse under Gorbachov!

Professor Branover went back to the Rebbe MHM to make sure that he understood the message correctly. When the Rebbe MHM confirmed the prediction and repeated the instructions, Branover went to work on housing and employment projects immediately. The rest is history. The Jews came out of Russia en masse—an event which the Rebbe MHM said was the beginning of the fulfillment of the Messianic prophecy of *Kibbutz Goliyos*—the return of the exiled Jews to the Land of Israel in the Era of Moshiach.

Years later, when Gorbachov visited Israel, he met Professor Branover (who headed an Israeli government commission for joint research between Israel and Russia). Branover told him about the Rebbe MHM's prophecy. Gorbachov was astounded. He explained that when he came to power he himself had no such plans regarding Russian Jewry and had no idea that this would eventually happen.

When the Persian Gulf War broke out in 1991, Iraq threatened to attack Israel with chemical weapons. But the Rebbe MHM said that this would not happen and that they did not need the protective gas masks. Furthermore, he said that Israel was the safest place in the world and encouraged people to travel there. While Iraq shot 39 deadly Scud missiles at Israel (which could have killed thousands, G-d forbid) they did not contain chemical weapons and they did not kill anyone. Many people were miraculously saved.[20]

20. See Sima Silman Schechter, *Nisei Milchemet HaMifratz* (Jerusalem: Haktav Institute, 1995) for extensive documentation of the war and the miracles. See especially pp. 187–90 for a collection of the Rebbe MHM's statements about the Persian Gulf War.

Miki Diar survives the destruction of his home by an Iraqi missile

When the war started, the U.S. expected it to be an intense war lasting several months, but the Rebbe MHM indicated that the war would be over by Purim—which is exactly when it ended.

His major prophecy was, of course, *"Higia Zman Geulaschem"*—"the time of your Redemption has arrived," which was made in connection with the Persian Gulf War, as we explained in the Introduction. This book is about the various ways in which we have seen this prophecy fulfilled, especially from the point of view of the scientist.

With The Breath of His Mouth

It is not always possible to distinguish between a prophecy and a miracle. The prophet has the power to predict the future. But a prophet is also a tzadik and a tzadik has the power to *determine* the future. He can cause a miracle to happen. He has the power to make a declaration or a decree and G-d will back it up. In fact, the Talmud says that the tzadik has the power to cancel a decree that G-d Himself has made—and G-d will follow through. So when a prophet says that something miraculous will happen we don't always know if he is predicting something that G-d will do or if he is actually performing the miracle himself, making a declaration that G-d follows through with. For example, when the Rebbe MHM declared during the Gulf War that Israel is the safest place in the world, was he *prophesying* that it would be safe or was he actually *making* it safe by his statement?

But in addition to the prophecies, the Rebbe MHM has performed countless miracles, both for individuals and for the entire world. One of the most dramatic of these was in fact a fulfillment of the prophecy of Isaiah regarding Moshiach: "And with the breath of his mouth he will put the evil man to death." This happened in January, 1953 when the evil Stalin arrested a group of Jewish doctors accusing them of killing some government officials. This was the infamous "Doctors' Plot." It triggered a wave of intense anti-Semitism throughout Russia and the lives of all Russian Jews were in danger. It was reported that Stalin had planned to deport the Jews to a distant northern region where they would be held in concentration camps.

On the night of Purim that year, the Rebbe MHM held a *farbrengen* (Chassidic gathering) in "770." Very late in the night, in the middle of the Purim celebration, the Rebbe MHM's face suddenly became very serious and he began to tell a story. He said that after the fall of the Czar, they held elections in Russia. The Rebbe of Lubavitch at that time, Rabbi Shalom DovBer, instructed the Chassidim to vote in these elections. There was one chosid who was totally removed from worldly matters and had no idea what an election was. But he followed the Rebbe's instruction and went to vote. When he got there, however, he didn't know what to do—how to vote or who to vote for. He met other chassidim there and they told him what to do. After the elections, everyone started shouting "Hoorah! Hoorah!" They were cheering their candidate. But this Chosid thought that they were shouting the Hebrew words *"Hu ra! Hu ra!"* meaning "He is evil! He is evil!" So he began to shout, *"Hu ra! Hu ra!"*

The Chassidim listening to the Rebbe MHM tell this story understood that he was telling them that they should also shout "Hu ra! Hu ra!" So they all stood up and began shouting "Hu ra! Hu ra!"

A few days later, it was reported that Stalin was dead.[21]

---

21. See the Appendix for the full story of the Doctors' Plot and the death of Stalin.

COURTESY RABBI LEVI GOLDSTEIN

The Purim farbrengen, 1953. The Rebbe MHM is in the middle, making a triangular shape with his hands.

Prophecies and Miracles

It would be impossible to give a complete account of all of the Rebbe MHM's prophecies and miracles over the years. (Some of these have been documented in the references at the end of this section.) Yet, I feel that I must mention some stories that have a certain character and illustrate certain points.

I would like to start with a personal story that happened to my family. In January 1978 we moved into an apartment in Minnesota and bought new mezuzos to affix to the door posts. Several months later, when my wife was expecting a baby, we wrote to the Rebbe MHM for a blessing.

In June we received a letter from the Rebbe MHM with a blessing for the pregnancy and birth. At the end of the letter he wrote that the mezuzos should be checked if they haven't been checked in the previous 12 months. We were surprised at this instruction since the new mezuzos which had been put up 5 months earlier had been checked before they were bought. So we took no action.

That summer we moved to Montreal, Canada where I had received a teaching position. The next January my wife gave birth to a baby boy. One Shabbos, when he was a few weeks old, my wife noticed that the baby looked rather weak and was not nursing. After Shabbos we took him to the doctor who sent him to Jewish General Hospital for tests.

At the hospital, when they saw my son's condition they became frantic. Doctors and nurses were running everywhere. They did a spinal tap which came out OK. Then they registered him into the hospital. At first we were not that worried and were confident that everything would turn out all right. However, shortly afterwards, my son was transferred to the intensive care unit (ICU). The doctors had determined that he had a virus called RSV, which is a type of pneumonia. (Later, we found out that there was an epidemic of RSV in Canada that winter from which many babies died.) His condition grew worse and his respiratory rate was very high.

I spoke to the doctor and asked what they could do for him. The doctor said this and that. I asked what they planned to do if that didn't work. The doctor said that they would then try another treatment. Then I asked what they would do if that treatment didn't work. The doctor just shrugged her shoulders indicating that that would "be it," G-d forbid, and walked away.

At that point I became very worried, and began thinking what I should do. Something must be wrong. I had already called the Rebbe MHM's office in 770 to ask for a *bracha* but didn't get an answer. Suddenly, I remembered the answer about checking the mezuzos if they had not been checked in the past 12 months. It was then exactly 12 months since the mezuzos had been bought and checked. I went home, took down the mezuzos and brought them to the *sofer* (scribe). I told him that my son was sick in the hospital and that the mezuzos should be checked immediately. That was Tuesday afternoon.

That evening my son's condition began to improve quickly. They took him out of the ICU and put him in a regular room. By Friday he was well and shortly before Shabbos we brought him home.

The sofer did not know of the Rebbe MHM's instruction to me regarding the mezuzos. But when he returned them to me he said that my mezuzos were written on a thick parchment with a heavy ink so the letters tend to crack. Such mezuzos, he said, should be checked every 6 to 12 months.

Salvation for Slovenia

After the fall of Communism in Russia, other East European countries rejected Communism and declared their independence. Among them was the small country called Slovenia which broke away from Yugoslavia in 1991. This drew a powerful military response from Yugoslavia. The Slovenians, small in number and poorly armed, stood against one of the strongest armies in Eastern Europe. Yugoslavian army tanks advanced toward improvised barricades manned by ill-armed defenders. The Slovenians' struggle was greeted largely with indifference by the rest of the world.

Marian Furlan was a Slovanian Jew who grew up under Communism but rejected it and refused to join the Communist Party. He left Slovenia, moved to Israel and got married there. In the 1980s he and his wife moved to Toronto where they became close to the Chabad community.

When war broke out between Slovenia and Yugoslavia, Slovenia was desperate to get more arms to defend itself. In Toronto, an ambassador of the World Slovenian Congress approached Marian hoping that Marian could help him get arms for Slovenia from Israel through Marian's wife who had served in the Israeli army. Marian explained to him that he could not help him obtain arms, but he mentioned that he was about to travel to New York to visit the Rebbe MHM and could ask for a blessing. The ambassador sent a letter to the Rebbe MHM with Marian on behalf of the Slovenian people.

The next day, Marian also received a letter for the Rebbe MHM from Ivan Plazar, a Catholic priest of Slovenians in Canada. In the letter he wrote:

> Slovenians are hard working, live in peace with everyone (Jews and others).... Now these people are in a very difficult situation. They have been attacked by the communist army and are at war.... Marian Furlan, who is living in Toronto and belongs to Lubavitch is bringing to you our request that you support our tired hands with your prayer and blessing. G-d, Creator of the world, is our only help and He could also bring our nation to victory and liberate them from the slavery of G-dless communism and attain freedom and peace without further bloodshed. We would be very grateful for your prayers, help and advice.

Marian traveled to New York with Rabbi Yisroel Landa, the Chabad Rabbi of the Israeli community in Toronto. When they arrived, they received devastating news: The Yugoslav army was preparing chemical warfare, and the Yugoslav high command had issued orders for a ground assault to break all resistance. The picture was bleak and getting darker. The all-out assault was to begin early the following week.

Marian met with the Rebbe MHM on Sunday, June 7, 1991 at 10:00 a.m. EST and explained the situation in Slovenia. The Rebbe MHM accepted the letters with the requests for blessings and gave his blessing for Slovenia. It was then 4:00 p.m. in Slovenia.

Upon his return to Toronto, Marian learned that shortly after 4:00 p.m., Yugoslavia suddenly requested a cease fire. An air raid which they had launched against Slovenia was recalled just twenty minutes from target. The war was over. Not a single Slovenian life had been lost after 4:00 p.m., the time of MHM's blessing!

A Slovenian government publication that briefly outlines Slovenian history records the events of the war and the Rebbe MHM's blessing. It reads in part: "A Slovenian Jew personally delivered to New York a number of letters from the Slovenian community and individuals, requesting both moral and practical support for embattled Slovenia. These documents were received by world Jewish leader Rabbi Menachem Mendel Schneerson Shlita, and on the very day he blessed the Slovenian nation, hostilities ended."[22]

Hurricane Andrew

Hurricane Andrew of 1992 was the most destructive United States hurricane on record before Hurricane Katrina of 2005. It started modestly as a tropical wave that emerged from the west coast of Africa on August 14. The wave spawned a tropical depression on August 16, which became Tropical Storm Andrew the next day. Further development was slow and the storm almost dissipated on August 20. But by August 21, Andrew had reached the Caribbean and strengthened quickly, reaching hurricane strength on the 22nd and Category 4 status on the 23rd. It blasted its way across south Florida on August 24 with sustained winds measured at 142 mph

22. Reported in the weekly newsletter *L'Chaim!* accessible at http://www.lchaimweekly.org/lchaim/palm/5764/793.htm

and wind gusts as high as 177 mph, and produced a 17 ft storm surge near its landfall point in Florida. It was responsible for 65 deaths and $25 billion in damage in the United States, almost all of it in south Florida.

This is the story of how the Rebbe MHM, after having suffered a stroke earlier in 1992, guided the Lubavitch community in south Florida safely through the hurricane.

The National Hurricane Center had predicted that Miami Beach would be the hardest hit. In fact, they predicted that the ocean would surge on the east, and the bay on the west and Miami Beach would be destroyed. Evacuation orders were issued and throughout the night of Saturday to the morning of Sunday, August 23, the area was evacuated. It was recommended that people go to Kendal where it would be safer, and many people did.

Rabbi Tennenhaus of Chabad of Hallandale called the Rebbe MHM's office to ask if the Chabad community should evacuate. The secretary, Rabbi Leibel Groner, brought the question to the Rebbe MHM in the following manner: He asked if the Chabad community should leave. The Rebbe MHM, who could not speak as a result of the stroke, shook his head to indicate "No." Rabbi Groner then asked the question in the positive—should they stay? The Rebbe MHM nodded his head "Yes." Rabbi Groner asked a third time if they should go and the response was again "No." This was the procedure for asking such questions.

By Sunday, August 23, Hallandale was almost completely evacuated—except for the Lubavitchers. By Sunday evening, Andrew was borderline Category 5. Rabbi Leibel Shapiro, who remained in Miami Beach throughout the storm, described what happened there: On Sunday, as the storm worsened, there was a rumor going around that the Rebbe MHM indicated that people should evacuate. He called Rabbi Groner to check this out. Rabbi Groner went to ask the Rebbe MHM if they should leave since the storm was getting worse. He shook his head "No." Rabbi Groner asked again, "What about the danger?" The Rebbe MHM responded with a strong motion of his head as though brushing the whole thing off and, some people suggest, actually pushing the hurricane away.

Sunday night, at 12:30 a.m. the hurricane hit. At about 1:00 a.m.

all the power went out. They stayed in a room without windows and heard the winds howling the whole night.

When the storm had passed they went out and, to their great surprise, they saw that in their area of Miami Beach, there was very little damage. There was no flooding. There was less water on the ground than from a regular rain storm. The same was true of Miami, North Miami and Hallandale.

Kendal, however, where everyone was supposed to escape to, was the hardest hit. Rabbi Shapiro tells of one man who left Miami Beach for Kendal only to spend the night with his friends standing against the door of the house they were in so that the wind wouldn't blow it open. The roof of that house had already caved in. When he came back to his home in Miami Beach, he found that everything was intact.[23]

Identity as Melech HaMoshiach[24]

In the previous section of this chapter, "A Day in the Life of Melech HaMoshiach," we quoted Rambam's halachic definition of Moshiach in which he describes the emergence of the Jewish leader who will be Moshiach:

> If a king will arise from the House of David who teaches Torah and is occupied in the Mitzvos like David his ancestor in accordance with both the Written Torah and the Oral Torah, induces all the Jews to follow the Torah and to strengthen it where it has been broken and who fights the battles of G-d, he has the assumed status of being Moshiach (*chezkas Moshiach*). If he succeeds further, builds the Beis HaMikdosh in its place and gathers the dispersed of Israel—then he is Moshiach with a certainty (*Moshiach vadai*) and he will direct the entire world to serve G-d together.

As we mentioned above, the Rebbe MHM is a descendant of the Tzemach Tzedek, who was the grandson of Rabbi Schneur Zalman,

---

23. Based on the video, *His Eye on the Storm*, produced by Chabad of Florida. See also "Rabbi's Advice is Simply Divine," *New York Post*, Aug. 25, 1992.

24. See the subsection "In Messianic Times" in the Introduction for a related discussion.

the Alter Rebbe, who was a descendant of the Maharal of Prague who was known to be a descendant of King David. So he is from the House of David.

We have quoted the leading Rabbinical authorities of our generation who have attested to the Rebbe MHM's mastery of the wisdom of the Torah, transcending all other scholars of the generation. "There is no one like him," Rabbi Mordechai Eliyahu has said. His teaching of the Torah and advancing the observance of the mitzvos is known to everyone and in this too he has no equal. This is attested to in the letter of Rabbi Moshe Feinstein discussed above. He has induced all the Jews to follow the Torah and has strengthened it where it has been broken, on a global scale.

He "fights the battles of G-d" primarily in the spiritual sense, as we have described at length above, but also with military advice he has given to Israeli generals and governments officials on many occasions.[25]

Towards the end of 1990, the Rebbe MHM made the historical announcement: "The time of your redemption has arrived." This was the announcement that identified him publicly as Melech HaMoshiach as per the *Yalkut Shimoni* which foretold the Persian Gulf War and that G-d would deliver this message to the Jewish people at that time—"My children, do not be afraid ... the time of your redemption has arrived." Specifically, this announcement would be made by Melech HaMoshiach himself as the Yalkut Shimoni says, "At the time that Melech HaMoshiach comes ... he announces to the Jews, 'O, humble ones, the time of your redemption has arrived!" The Rebbe MHM made it clear that he was making the declaration that identified him as Moshiach and in his talks during 1990-92 the he reiterated this many times, repeatedly making it clear that he is Melech HaMoshiach.

Thus the Rebbe MHM said prophetically that Israel would be safe and encouraged Jews to travel there, and indicated that the war would be over by Purim. Everything happened exactly as he said. Even though 39 Scud missiles fell in Israel, no Jews were killed by them. And the war ended on Purim.[26]

25. This is documented in the video *Faithful and Fortified,* vol. 1, "Defense and Intelligence" (Jewish Educational Media, 2009).

A few weeks before that, in the sicha of Parshas Mishpatim, 5751 (1991), the Rebbe MHM said, "The appointment of *Dovid Malka Meshicha* (i.e. Melech HaMoshiach) has already taken place.... It's just that the people have to accept his sovereignty and the connection between the king and the people has to be completely revealed."[27]

So by Nisan, 5751 it was clear that the Rebbe MHM had the status of *chezkas Moshiach*. In fact, later that year, the Rebbe MHM said that with the miraculous victory in the Persian Gulf War, "the words of the Yalkut Shimoni *have been fulfilled* ... and since then, especially from the month of Nisan, the month of Geula, we are already 'at the time [*sha'ah*—literally *'the hour'*] that Melech HaMoshiach comes.'"[28] And again, "We have already actually seen wonders that confirm that this is *'the year in which Melech HaMoshiach is revealed'* and even *'the hour when Melech HaMoshiach comes'* and announces to the Jews, 'Humble ones, the time of your redemption has arrived' and even the announcement that 'Behold, this is Melech HaMoshiach who came'—*that he already came*."[29]

At the end of Nisan, Rabbinical authorities all over the world began signing a *psak din* (halachic legal ruling) regarding the Rebbe MHM being Moshiach. Initially issued by Chabad Rabbis in Israel on the 30th of Nisan, in the following days it was signed by many more Rabbis from Israel, Europe and the U.S. On the 15th of Iyar, the Rabbis of the Rabbinical Court of Crown Heights (who had already signed it) reissued it as an official *psak din* of the Beis Din of Crown Heights. That same day, the Rebbe MHM publicly encouraged the singing of the declaration *"Yechi Adoneinu Moreinu V'Rabeinu Melech HaMoshiach L'Olam Voed!"* ("Long live our master, teacher and Rebbe King Moshiach forever") following the *davening*.[30]

Then, on the day after Shavuos, Chabad Rabbis from around the world who had come to Crown Heights to be with the Rebbe MHM

---

26. That's not all. A couple of days after Purim, on 16 Adar, 5751 (1991), the Rebbe MHM made a statement in which he said that America would fight a war against Basra, that America is a country of goodness and kindness, and that it would be successful. In 2003, *on exactly that same date*, 16 Adar, the American army, assembled at Basra, attacked Iraq.

27. *Sefer HaSichos* 5751, 1:330.

28. Ibid. 2:595.

29. Ibid., 690.

for the holiday issued a more detailed *psak din.* Referring to Rambam's criteria for determining the identity of Moshiach, it stated in part:

"It is clear in accordance with the absolute law of the Torah that the halachic designation of *chezkas Moshiach* applies to the Rebbe since all the conditions listed in the abovementioned halacha in Rambam [*Hilchos Melachim U'Melech HaMoshiach,* chap. 11, sec. 4] have been fulfilled by the Rebbe."

With the Rebbe MHM's encouragement, this *psak din* was read at

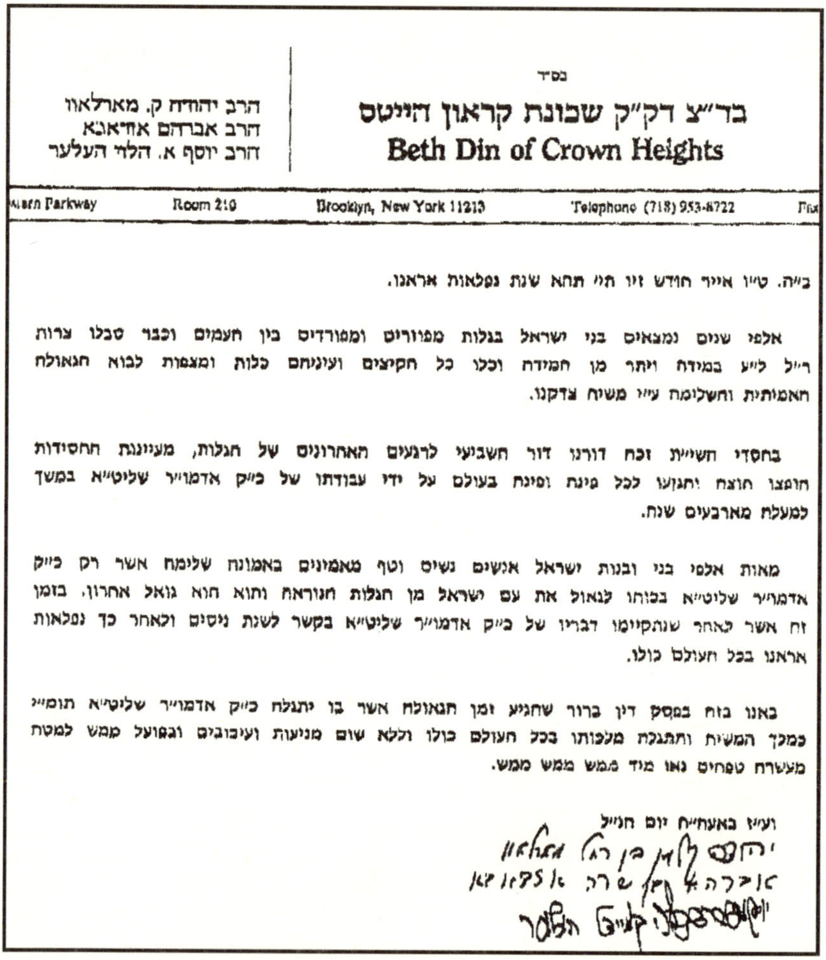

בס״ד

בד״צ דק״ק שכונת קראון הייטס
Beth Din of Crown Heights

הרב יהודה ק. מארלאוו
הרב אברהם אזדאבא
הרב יוסף א. הלוי העלער

...tern Parkway Room 210 Brooklyn, New York 11213 Telephone (718) 953-8722 Fax

ב״ה. ט״ו אייר חודש זיו תו׳ תהא שנת נפלאות אראנו.

אלפי שנים נמצאים בני ישראל בגלות מפוזרים ומפורדים בין העמים וכבר סבלו צרות ר״ל ל״ע במידה ויתר מן המידה וכלו כל הקיצים ועיניהם כלות ומצפות לבוא הגאולה האמיתית והשלימה ע״י משיח צדקנו.

בחסדי השי״ת זכה דורנו דור השביעי לרגעים האחרונים של הגלות, מעיינות החסידות הופצו חוצה יתנוצו לכל פינה ופינה בעולם על ידי עבודתו של כ״ק אדמו״ר שליט״א במשך למעלה מארבעים שנה.

מאות אלפי בני ובנות ישראל אנשים נשים וטף מאמינים באמונה שלימה אשר רק כ״ק אדמו״ר שליט״א בכוחו לגאול את עם ישראל מן הגלות הנוראה והוא הוא גואל אחרון. בזמן זה אשר לאחר שנתקיימו דבריו של כ״ק אדמו״ר שליט״א בקשר לשנת ניסים ולאחר כך נפלאות אראנו בכל העולם כולו.

באנו בזה בפסק דין ברור שהגיע זמן הגאולה אשר בו יתגלה כ״ק אדמו״ר שליט״א תומ״י כמלך המשיח ותתגלה מלכותו בכל העולם כולו וללא שום מניעות ועיכובים ובפועל ממש למטה מעשרה טפחים נאו מיד ממש ממש ממש.

ועז״ב באעה״ח יום הנ״ל

Official ruling of the Beis Din of Crown Heights to the effect that the Rebbe is Moshiach.

the gravesites of the previous Lubavitcher Rebbeim in Russia and at the gravesite of the Ari Zal in Israel.

For the next year the Rebbe MHM continued to make stronger and stronger statements about his being Moshiach and about the Geula being here, and gave strong encouragement to those who were publicizing this.[31] This encouragement continued even after the stroke in 1992. In 1993 he approved the publication of an anthology of selections on Moshiach from his recent sichos,[32] in which he would be explicitly referred to as Melech HaMoshiach.

On 27 Tishrei 5754 (1993), I personally was part of a group that had the privilege to pass before the Rebbe MHM and receive from him a dollar-for-*tzedakah*.[33] As I passed before him, I greeted him saying, "*Shalom Aleichem, Melech HaMoshiach!*" and he nodded to me in acceptance.

At the beginning of 5752 (Nov., 1991), in the sicha of *Parshas VaYera*, the Rebbe MHM said, "In our days all the obstacles have been removed. Therefore, we not only have the *presence* of Moshiach, but also the *revelation* of Moshiach.[34] Now it is necessary only to receive Moshiach in full actuality."[35]

Then, a couple of months later, in the sicha of *Parshas Shemos* 5752 (1992), he made an even stronger statement: "Since we already have the 'king from the House of David who teaches Torah and is occupied in the Mitzvos like David his ancestor ... induces all the Jews to follow the Torah and to strengthen it where it has been broken and who fights the battles of G-d'—thus he has the assumed status of being Moshiach (*chezkas Moshiach*)—he should immediately become Moshiach with a certainty (*Moshiach vadai*)." Just a few paragraphs earlier, he had mentioned the fact that the Rebbeim of Chabad are descended from "the House of David from the tribe of Judah."

---

30. Video recording, 15 Iyar, 5751 (WLCC).

31. Some of this is documented in *And He Will Redeem Us–Moshiach in Our Time* (New York: Mendelsohn Press, 1994), 61–72.

32. *Besoras HaGeula* (New York: Kehot Publication Society, 1993).

33. This has been the only such event since the stroke.

34. In the sicha, he explains that in every generation there is a presence (*metzius*) of Moshiach, in that there is a tzadik worthy of becoming Moshiach who would be revealed as Moshiach if there were no impediments.

In a letter written in 1962, the Rebbe MHM indicated that the one who becomes chezkas Moshiach *will in fact* become Moshiach vadai. He writes, "Quickly in our days we should see the fulfillment of the promise that a king will arise from the House of David who teaches Torah and is occupied in the Mitzvos etc., and will induce all the Jews to follow the Torah and to strengthen it where it has been broken and who fights the battles of G-d [i.e. *chezkas Moshiach*]—he is the one who is Moshiach Tzidkeinu—and he will build the Beis HaMikdosh and gather the dispersed of Israel [i.e. *Moshiach vadai*], as in the *psak din* of Rambam."[36] Thus in 1992, when he says that we already have chezkas Moshiach—which will develop into Moshiach vadai—the only thing to ask for is that it should happen *immediately*.

What does it take to become Moshiach vadai? Rambam states the criteria: build the Beis HaMikdosh and gather the dispersed of Israel. But does he have to *complete* these requirements in order to be Moshiach vadai?

Consider the following quote from the sicha of *Shabbos Chazon*, 5751: "that a Jew believes with complete faith that *Moshiach Tzidkeinu*, 'a king from the House of David who teaches Torah and is occupied in the Mitzvos etc.' *as Moshiach vadai* (by having succeeded, built the Beis HaMikdosh in its place and gathered the dispersed of Israel) enters the shul right now—he should come and redeem us and take us upright to our land—and leads all the Jews to the Holy Land, to Yerushalayim the holy city, to the Temple Mount, to the Third Beis HaMikdosh."

This says that Moshiach, having the status of Moshiach vadai because he already gathered the dispersed of Israel, will enter the shul and take us to the Holy Land. But evidently he has not yet *completed* gathering the dispersed of Israel (because there are still people in the shul who need to be taken to Eretz Yisroel). He has begun to do it, however, and because of that he is Moshiach vadai.

Now, the Rebbe MHM has said that the ingathering of the exiles has already begun with the mass exodus of the Jews from Russia who went to Israel in a pleasant and peaceful manner (i.e. not having to

35. *Sefer HaSichos* 5752, 1:95.
36. *Igros Kodesh* 22:287.

escape, not as refugees).[37] Furthermore, in an amazing sicha published under the title *Kuntres B'Inyan Beis Rabbeinu She-b'Bavel*, he explained that the construction of the Third Beis HaMikdosh has also already begun.[38] Thus the Rebbe MHM may already be considered Moshiach vadai.

In the sicha of *Parshas Mishpatim*, 5752, the Rebbe MHM made a statement similar to the one of Parshas Shemos a few weeks earlier, referring to the *psak din* of the previous year: "The *psak din* of Rabbinical authorities that 'the time of the Geula has arrived'—'a king will arise from the House of David … he has the assumed status of being Moshiach' reaching the status of 'Moshiach with a certainty (*Moshiach vadai*).' This is a *psak din* whose authority originates from Mt. Sinai." This sicha was spoken a month before the stroke.

On the Rebbe MHM's birthday, 11 Nissan, 5752 (1992), a month and a half after the stroke, a *psak din* signed by 200 Rabbis worldwide, was issued stating clearly and emphatically that the Rebbe MHM has the status of *chezkas Moshiach* based on Rambam's halachic definition, and after Gimmel Tammuz, in Tishrei, 5758 (1997) a further *psak din* signed by over 120 Rabbis worldwide declares that he is Melech HaMoshiach. This *psak din* continues to gain more signatures.[39]

The sicha of Mishpatim 5752 itself is devoted primarily to Swords Into Plowshares.[40] In it, the Rebbe MHM discusses a declaration made the previous day by the heads of state of the major world powers at the United Nations announcing a new era of arms reduction, peace and cooperation among the nations of the world. He identified this declaration as the beginning of the fulfillment of the prophecy of Isaiah that in the Era of Moshiach the nations of the world will "beat their swords into plowshares"—transform their military resources into peaceful uses.

The Rebbe MHM proceeds to explain that this declaration came about as a result of the influence of Melech HaMoshiach on the nations of the world. Specifically, he refers to his campaign to pro-

37. Sicha of VaYakhel 5752. See *Sefer HaSichos* 5752, 2:440–41.

38. Printed in *Sefer HaSichos* ibid., 465–75.

39. Photocopies of these *piskei din* are reproduced in Rabbi Raphael Heruti, *Nevuas HaGeulah*, 15–19.

40. See chap. III sec. 2 for a full discussion of the sicha.

mote spread Torah and Judaism to Jews throughout the world, and especially to promote the ideals of goodness, justice and fairness among the nations of the world through the observance of the Seven Noahide Laws, indicating once again that he is Melech HaMoshiach.

At the end of the sicha, he explicitly identifies himself as Moshiach by saying that we are now in the generation of "Moshiach whose name is Menachem."

The Everlasting Life of Melech HaMoshiach

...the Chaim Nitzchi'im*—everlasting life—of the future era, which is attained in our generation without any interruption [of death] in between.*

—the Rebbe MHM[1]

"My beloved is like to a deer."[2]

In this verse in the Song of Songs, Shlomo HaMelech compares Moshiach to a deer. As the Midrash explains, Moshiach "is revealed, then concealed, then revealed again" just like a deer that emerges from the forest, then returns to the forest and runs out again.[3]

On the third day of Tammuz, 5754 (1994), the Rebbe MHM was temporarily concealed from our eyes. As we see from the above Midrash, however, the temporary concealment of Moshiach is itself part of the process of his revelation.

In this section, we will analyze these issues by considering:

- What is known about the passing of *tzadikim* (righteous, holy people) in general from the Talmud, Kabbala and Chassidus;
- How the life of Melech HaMoshiach surpasses the life of all other tzadikim, as he is granted *Chaim Nitzchi'im*—everlasting life—from G-d; and
- How modern science assists in the understanding of the teachings of the Torah on these matters.

1. *Sefer HaSichos 5751*, 496.
2. Shir HaShirim 2:9.
3. *Midrash Shir HaShirim Rabba* 2:22; *Bamidbar Rabbah* 11:2. See also *Likutei Sichos* 9:381.

Everlasting Life

The basic concept of Chaim Nitzchi'im is found in the Torah in connection with the creation of man. Adam was created in such a way that he would live forever. It is only because he ate from the Tree of Knowledge that this everlasting life was not actualized.[4] However, since it was part of the original plan of creation it will be restored in the Era of Moshiach when the Revival of the Dead takes place and all Jews will live forever. The same is true of all the various types of perfection that were built into the original order of creation and were lost. They will be restored in a much greater way in the Era of Moshiach.

It should come as no surprise then, that throughout history some special tzadikim were granted everlasting life, being removed, to varying degrees, from the prevalent order of nature.

The first tzaddik mentioned in the Torah who was granted everlasting life was *Yaakov Avinu* (Jacob) of whom the Talmud says, "Yaakov Avinu did not die."[5] This statement is discussed and analyzed extensively by the commentaries on the Torah and the Talmud. To begin with, we note that Rashi, whose commentary on the Torah is *pshat*—the direct, logical meaning—which is directed to the understanding of even a five year old child who is beginning his study of the Torah, states quite simply: "Death is not mentioned regarding [Yaakov]. Our Rabbis have said that Yaakov Avinu did not die."[6] This is a literal statement intended to be understood even by a child. In his commentary on the Talmud, Rashi states further (explaining why there was a burial etc.), "It appeared to them that he died but he was alive." The Or HaChaim, in his commentary on the Torah, also says that Yakov was actually alive. It was only as if he was sleeping. (That's why Yosef kissed him, which would not have been proper if he had died.)[7]

Talking to Yaakov Avinu

What are the characteristics of the life of Yaakov Avinu? One of the deepest and most analytical commentaries on the Torah, the *Alshich*,

4. *Sefer HaMa'amorim Melukat* 2:277.
5. Ta'anis 5b.
6. Rashi's commentary to Bereishis 49:33.
7. *Or HaChaim* on Bereishis 50:1.

written by Rabbi Moshe Alshich, a student of the *Ari Zal,* provides the technical details. As is well known from the Midrash, Kabbala and Chassidus, there are five parts of the soul which give life to the body. The first three, called *nefesh, ruach* and *neshama*, are contained in the body itself (the *nefesh* is centered in the liver, the *ruach* in the heart and the *neshama* in the brain). The last two, *chaya* and *yechida*, give a life which is higher than the specific life of the first three. They are not contained in any specific part of the body but rather encompass the entire body uniformly.

The Alshich explains that when death occurs, G-d forbid, the three parts of the soul that are contained in the body leave the body. The passing of tzadikim, however, is different. When a tzadik passes away only the ruach and neshama leave the body but the nefesh stays in the body. Thus they are not called dead but rather "sleeping in the dust." Yaakov Avinu, however, transcends even this. The statement "Yaakov Avinu did not die" means that all three parts of the soul—the nefesh, ruach and neshama, together with all the faculties of the soul, all remained in his body. This distinguishes Yaakov Avinu from other tzadikim.[8] The Ramban and Rabbeinu Bachaya, in their commentaries on the Torah, give similar explanations, but in more Kabbalistic terms.[9]

There is a practical application of this on a daily basis. The Midrash Rabba tells the story of Yaakov Avinu blessing his sons. Before he blessed them, he asked them if their belief in Hashem was complete to which they responded, *"Shema Yisroel Hashem Elokeinu Hashem Echod!"* Addressing their father by his title *Yisroel,* they said, "Listen, O Yisroel, Hashem is our G-d, Hashem is one!" Their belief in Hashem was complete. The Midrash continues to say that to this very day when we say *"Shema Yisroel"* twice a day, we are talking to Yaakov Avinu and saying, "Listen, O Yisroel our father, in the Cave of Machpela, we are still faithful to that which you commanded us regarding the unity of Hashem.."[10] Since Yaakov Avinu is alive we can talk to him.

It must be mentioned that there is actually an "advantage" in the

8. This commentary of Alshich is quoted briefly in *Likutei Sichos* 4:1261.
9. Ramban and Rabbeinu Bachaya to Bereishis 49:33.
10. *Midrash Bereishis Rabba* 98:3.

soul of a tzadik leaving his body in that he is freed from the limitations of the body and is able to accomplish more for his followers than he was able to do when his soul was confined to his body.[11] But there is also a state in which the soul of the tzadik is free of the limitations of the body even while it is still in the body. This state was attained by Rabbi Shimon bar Yochai on the day of his passing.[12] So it is possible for some tzadikim to have both advantages—being able to transcend the limitations of the body while the soul is still in the body. The advantage of the soul remaining within the body is, of course, that it is more revealed so it is easier for his followers to communicate with him.

Similarly, it is important to emphasize that the Rebbe MHM is alive so we understand that while he may be concealed he is still our leader and he is very close to us. We can talk to him and declare: *"Yechi Adoneinu Moreinu V'Rabeinu Melech HaMoshiach L'Olam Voed!"*— "Long live our master, teacher and Rebbe Melech HaMoshiach forever." We are loyal to him and anticipate his complete revelation.

"Moshe came back to save us"

The Talmud and Zohar say the same thing about Moshe Rabbeinu (Moses) that they say about Yaakov Avinu: *"Lo mes Moshe"*—Moshe did not die.[13] Rather they compare Moshe's apparent passing to his ascent to Mt. Sinai where he lived miraculously without food or water. The *Mikdash Melech*, a Kabbalistic commentary on the Zohar, explains that the term *VaYomos* ("and he died")[14] which the Torah uses in connection with Moshe is not to be taken literally. He says, "Even though the Torah uses the term 'death,' it may have various meanings: If one does not merit, it is death in the literal sense, but if one merits, it will be concealment in the higher worlds like Eliyahu (Elijah)." This is not death because in this case the soul does not depart from the body. Rather the body ascends with the soul. This is in fact what happened to the prophet Eliyahu,[15] and, according to one interpretation, something similar happened to Moshe. According

11. *Tanya*, "Igeres HaKodesh," chap. 27.
12. *Sefer HaMa'amorim Melukat* 5:272–73.
13. Sota, 13b.
14. Devorim 34:5.
15. This is described in detail in Melochim II, chap. 2.

to other interpretations, he was actually buried.[16] In any case, he is similar to Yakov Avinu whose soul remains in his body.[17]

There are many stories of Eliyahu HaNavi returning to help Jews in difficult times or to teach them the secrets of the Torah. There is also such a story about Moshe Rabbeinu. The Midrash Rabba[18] relates that at the time of the destruction of the first Beis HaMikdosh and the exile, our forefathers came before G-d to plead for the Jewish people, but the decree was not reversed. Then Moshe Rabbeinu came forward and said that he was going to bring the Jews back. He actually went to Babylon to redeem them. The Jews saw him in his physical body at the bank of the Euphrates River and declared, "Look! Moshe Rabbeinu has come up from his grave to redeem us!" However, G-d did not let him follow through with the redemption. Whereupon the Jews cried: "On the rivers of Babylon, we sat and cried."[19]

Dovid HaMelech is Still King of Israel

Similar to the statement that Moshe did not die is the famous statement in the Talmud *"Dovid Melech Yisroel Chai V'Kayam"*—David King of Israel is alive![20] This declaration is part of the *Kiddush Levana* (sanctification of the new moon) service which we recite every month. The Rebbe MHM has clarified the exact meaning of this declaration. He said that it means that *Dovid HaMelech* (King David) is literally alive in this world, for if it meant that he is alive in *Gan Eden* (the afterlife) there would be no point to making the statement, since every Jew is alive in *Gan Eden* after this world. "From the moment he was anointed as King of Israel in Hevron until the end of all generations he continues to live—and continues to live as the King of Israel!"[21]

There are a number of cases on record in the Talmud, Midrash and other seforim of tzadikim who continue to live after their apparent passing; usually it is tzadik who is a Nasi, as we will soon explain.[22]

16. *Sefer HaSichos* 5749, 1:9.
17. *Likutei Sichos* 26:7.
18. *Pesikta D'Eicha Rabbasi.*
19. Tehillim 137:1.
20. Rosh HaShana, 25a.
21. Sicha of 7 MarCheshvan, 5746, in *Hisvaaduyos* 5746, 1:525.
22. In addition, it has been observed that the bodies of certain tzaddikim, who had tobe moved from their graves to other graves many years after their original burial, remained complete and unchanged.

The following story is one of the most dramatic stories on record of a tzadik returning after his passing. First, some background. Rabbi Asher ben Yechiel, (c. 1250–1327) known as the ROSH (literally, "the head;" acronym for "Rabbeinu Asher") was one of the greatest Talmudic and halachic authorities in history. He was one of the leaders of German Jewry until he was forced to flee persecution of the Jews and settled in Spain where he became one of the leaders of Spanish Jewry. His halachic works are one of the foundations of the *Shulchan Aruch*—the Code of Jewish Law. This story is about his father, Rabbi Yechiel, and was recorded in a letter of Rabbi Yehuda, the son of the ROSH i.e., Rabbi Yehuda was writing about his grandfather Rabbi Yechiel.

Rabbi Yehuda writes:

> Since I left Germany when I was a child, I didn't hear much about what happened there, just the little that I heard from my father and my great aunt who told me that my grandfather Rabbi Yechiel was born in the year 4970 (1210). When he was sixteen years old he had a loyal friend in yeshiva named Rabbi Shlomo HaCohen. They made a covenant that each of them would have a part in any mitzva or action that his friend did.
>
> It happened once, on the night of Yom Kippur, that the candle that Rabbi Yechiel lit in the shul went out.... When he saw that his candle was extinguished, he feared for his life. On *Chol HaMoed Sukkos* [a week later] he passed away. It was the custom in Germany that [after the funeral procession] they would place the coffin on a large stone near the cemetery and open it up to check if the body had become twisted during the long procession or if was still straight. At that point, Rabbi Shlomo approached the coffin, crying loudly. He then announced before the entire assembly, 'I am reminding my friend Rabbi Yechiel before all of you that he should remember our covenant.' Rabbi Yechiel then began to laugh inside of his coffin. Everybody saw this.
>
> A few days later, Rabbi Shlomo was sitting and learning in the Beis HaMedrash during the daytime and he noticed Rabbi Yechiel sitting next to him and reading the halachic text along with him. Rabbi Shlomo asked him what's going

on. He answered that he is situated very well in Gan Eden and he has a seat for himself among all the righteous of the generation. Rabbi Shlomo then said to him, "Please, my master, partner of our covenant, I wonder if you have permission to appear before people!"

Rabbi Yechiel responded, "You should know that I have permission to go home just like before, but I don't want to so they don't say that this tzadik is showing off more than all the tzadikim in Gan Eden."

Six months later, on a Friday night, at midnight, he appeared to his wife, who was awake, and said to her, "What are you doing—hurry up and take your sons and daughters and escape quickly from this place because tomorrow they are going to kill all the Jews here." She took my father and his brother…and six sisters.[23]

The story speaks for itself.

The Stature of the Nasi

Yaakov Avinu, Moshe Rabbeinu and Dovid HaMelech are the famous tzadikim, mentioned in the Talmud, Midrash and Zohar, who have everlasting life. In addition, there are other tzadikim who did not pass away and there is a quality with which this is usually associated. We see this in the explanation that the Rebbe MHM gives of the fact that his predecessor, the previous Lubavitcher Rebbe, Rabbi Yosef Yitzchok Schneerson, is still alive: "The *Nasi* (leader) of our generation … who did not die just like Yaakov Avinu."[24] The Rebbe MHM connects this with the fact that the Hebrew letters for the word *"Nasi"* stand for *"Nitzutzo Shel Yaakov Avinu"*—"A spark [of the soul of] Yaakov Avinu." Just like Yaakov Avinu did not die, so too the Nasi does not die. From this we may infer that the same applies (in varying degrees) to the Nasi of every generation.

The uniqueness of the Nasi in this regard can be related to the spiritual stature of the Nasi as explained in Kabbala and Chassidus.

---

23. Quoted by Rabbi Chaim Yosef David Azulay (the Chida), *Shem HaGedolim*, "Ma'areches Gedolim," 11b. Reproduced in Rabbi Shloma Majesky, *Likutei Mekoros*, 2nd ed. (New York: 2008), 1:48.

24. *Sefer HaSichos* 5752, 2:242.

The soul of the Nasi is a "general soul" which includes within it the souls of all the Jewish people and thus the life of all the Jewish people. Just as the Jewish people continue to live, the Nasi continues to live. Furthermore, the Nasi of each generation is the candidate for Moshiach in that generation. This means that if the Geula were to occur in that particular generation, its Nasi would be Moshiach. The innermost part of the soul of every Jew is the spark of Moshiach that is within him. Moshiach (as the Nasi) contains the souls of all the Jews, and with them, all the individual sparks of Moshiach.[25]

The ability of someone, such as the Nasi, to break the barriers of nature is a result of his connection to the Torah, as explained by the great Chassidic Kabbalist, Rabbi Levi Yitzchok Schneerson. The Torah is the blueprint (the "source code") of the universe. G-d used the Torah to create the world. Thus to the degree to which one is connected to the Torah he is above the laws of nature. For example, Rabbi Shimon bar Yochai, who wrote the Zohar and revealed the inner depths of Torah to the world, led a miraculous life. But for *him*—relative to his level in Torah—this was not miraculous. Rather it was *his* nature. For something to be miraculous for him, it would have to be a higher level miracle. One man's miracle is another man's nature.[26]

Of course, Moshiach who reveals the deepest level in Torah is completely above the laws of nature and leads a totally miraculous life even as he exists in a physical body in the physical world, as the Maharal of Prague said, "Moshiach will be ... outside of nature and the way of the world."[27]

The Nasi Does Not Die

Now we are ready to discuss other *Nesi'im*—tzaddikim who were the leaders of their generation—who continue to live after their apparent passing, and who even reappeared in their physical bodies.

We mention first Rabbi Yehuda HaNasi, who compiled the Mishneh. He lived in Eretz Yisroel about 150 years after the destruction of the second Beis HaMikdosh. The Talmud[28] says that after he

25. *Sefer HaSichos* 5751, 1:222; 2:590.
26. *Likutei Levi Yitzchok*, "Igros Kodesh," 264.
27. *Sha'arei Ge'ulah* (Jerusalem: Heichal Menachem, 1992), 2:69 n. 29.
28. Kesubos, 103a.

"passed away" he would return to his home every Friday night in his Shabbos clothes and made *kiddush* for his family![29] He stopped doing this only when it became known publicly, as he felt that continuing this practice would be an insult to other tzadikim who were not able to do this. The Rebbe MHM emphasizes that this means that he came back physically, in his body.[30]

The great Kabbalist, Rabbi Yitzchok Luria, known as the *AriZal* (literally, "the lion"; an acrostic for three Hebrew words that mean "our master Rabbi Yitzchok"), was the Nasi of his generation, in the late 1500s. After he "passed away," during the preparations for his burial (the *tahara*), he immersed himself in the *mikveh* (ritual bath) on his own.[31] The Ari is frequently referred to as the *Ari HaChai*—the Ari who lives.

Many generations later, the *Nesius* passed to the Baal Shem Tov and his successors, the Maggid of Mezeritch, the Alter Rebbe and the *Nesi'im* of Chabad. What happened with the AriZal happened with the Baal Shem Tov[32] and the Maggid[33] also—after they "passed away," at the *taharas*, they immersed themselves in the mikveh on their own.

It is well known that there was great opposition to the Alter Rebbe's extensive teaching of Chassidus. Some of his opponents slandered him to the Czarist government who arrested him under a charge of treason and imprisoned him in the infamous Petropavli prison.

During his imprisonment, the Alter Rebbe was visited in his prison cell by the Baal Shem Tov and the Maggid of Mezeritch—who had "passed away" years before. The Previous Lubavitcher Rebbe relates that he once visited the Petropavli prison and went to see the cell where his ancestor, the Alter Rebbe, was imprisoned. Afterwards, his father, Rabbi Shalom DovBer (the fifth Rebbe of Chabad) asked him if the cell was big enough to hold three people. From this, the Previous Rebbe writes, he understood that when the Baal Shem Tov

29. *Sefer Chasidim*, sec. 1129. Quoted in the notes of Rabbi Akiva Eiger, *Gilyon HaShas*, on Kesubos 103b.

30. *Hisvaaduyos* 5746, 1:205. Reproduced in Majesky, *Likutei Mekoros*, 42.

31. *Igros Kodesh* 20:81.

32. *Toldos Levi Yitzchok*, 329.

33. Ibid.

and the Maggid of Mezeritch came to visit the Alter Rebbe, they came in their physical bodies.[34]

A most remarkable story about the Alter Rebbe's son and successor, Rabbi DovBer of Lubavitch, the *Mittler Rebbe*, was told by Rabbi Nochum, Rabbi DovBer's son. The Mittler Rebbe was buried in the city of Niyezhin in Russia. A shul was built near his gravesite where people would come to pray and Rabbi Nochum attended to the shul. Rabbi Nochum once said that many stories are told of miracles that took place for those who prayed at his father's gravesite "But," he continued, "I only know what I saw with my own eyes." He then related the following:

Once, a fire broke out in the Shul. While everyone ran to bring water he went inside to try to save the *Sefer Torah*. When he entered, he saw that the holy ark containing the Sefer Torah was engulfed in flames and there was no way he could approach it. While he was standing there wondering what to do, his father, the Mittler Rebbe, appeared before him carrying the Sefer Torah. His father approached him and handed it to him. And as soon as he took it, his father disappeared. [35]

The Mittler Rebbe, who had passed away and was buried there, certainly came to him in a physical body since a soul, a spirit, cannot carry a Sefer Torah.

Often, when people hear about the concept that certain tzadikim don't die and their souls remain in their bodies, they ask, "Why should these tzadikim have to remain in their bodies? Why can't they go to Gan Eden like everyone else?" I will comment on this based on my understanding of the commentaries on the Torah[36] and sources in Chassidus.

Firstly, *where* is Gan Eden? Is it a thousand miles away from here? Is it a million miles out in space? No. It is not a physical place so it has no physical location. (We are speaking now of the Gan Eden of the souls, the afterlife, not the lower physical Gan Eden here on

34. Sicha of 19 Kislev, 5711, in *Toras Menachem–Hisvaaduyos 5711*, 113. Reproduced in Majesky, *Likutei Mekoros*, 41. See also the sicha referred to in n. 26 above for a detailed explanation and comparison to Rabbi Yehuda HaNasi.

35. Rabbi Chaim Mordechai Perlow, *Likutei Sippurim* (New York, 1992), 95.

36. Especially the Ramban on Bereishis 49:33.

earth where Adam was placed.) In fact, Chassidus explains that the "air" of Gan Eden surrounds every Jew and in it is registered all the good thoughts, words and actions that he does. As for Yakov Avinu, The Midrash Yalkut Shimoni says that when he entered the tent of his father Yitzchok to receive his blessing,[37] "*Gan Eden entered with him.*"[38] So who says that Yaakov Avinu is not "in Gan Eden" just because his soul remains in his body? He may very well be in Gan Eden.

Furthermore, the body of Yaakov Avinu itself may be higher than Gan Eden. It is explained in Chassidus, that at the time of the revival of the dead, all the Jewish souls will leave Gan Eden and go to a higher level—they will enter their revived physical bodies which will then be very highly refined. This will be a much higher level than Gan Eden as the direct connection of their bodies with Atzmus (G-d's essence) will be revealed.[39] Those certain special tzadikim that we have been discussing, like Yaakov Avinu, may have attained a similar level of refinement in their lifetime and their bodies may have already been higher than Gan Eden.

The Chaim Nitzchi'im of the Rebbe Melech HaMoshiach

All that we have said until now relates to certain very special tzaddikim—the Nesi'im of the Jewish people. But Moshiach is different. Just as Moshiach himself transcends all tzadikim that came before him, so too is the Chaim Nitzchi'im—the everlasting life—that is granted to him qualitatively different from the lives of all other tzadikim. We will now discuss this unique, uninterrupted everlasting life.

There is a verse in *Tehillim* (Psalms) which tells of the request for life that Moshiach makes from G-d, and G-d's response granting him everlasting life: "Life he asked from You. You granted it to him—length of days forever."[40] The Mittler Rebbe discusses this extensively in his Chassidic discourses. He contrasts the life of Melech HaMoshiach with the life of Dovid HaMelech of whom the statement is made: *"Dovid Melech Yisroel Chai V'kayam"* ("David King of

37. Bereishis 27:18–27.
38. *Yalkut Shimoni*, Bereishis, sec. 115. See also *Tanya*, 147a.
39. *Sefer HaMa'amorim Melukat* 3:33–38.
40. Tehillim 21:5. See also Sukka, 52a.

Israel is alive") as we discussed earlier. The Mittler Rebbe writes: "Regarding Moshiach it says 'Life ... You *granted* it to him'. This means a gift which has no interruption. Furthermore, regarding Dovid—his life came only from the ten *sefiros* ... of *Atzilus,*[41] but regarding Moshiach it says '*You* granted it to him' which means from *Atzmus*, the infinite."[42] When G-d is addressed in second person ("You") it refers to *Atzmus* i.e. G-d Himself, His essence.

It is significant that the Rebbe MHM himself spoke of his life source from Atzmus on the eve of Rosh HaShana, 5752, the Rosh HaShana before the stroke. He said, "Regarding Moshiach—in addition to the revelation of the name of Moshiach ... we also have the Atzmus of *Moshiach Tzidkeinu* which is unified with[43] the infinite Atzmus."[44] (He has spoken about the Previous Rebbe in similar terms: "The Rebbe ... is connected with the essence.... Since the Atzmus is down here, the Rebbe is also down here."[45])

Similarly, the Rebbe MHM has spoken of his own everlasting life: "In this generation ... everlasting life without any interruption."[46] He spoke about this often during the years 1990-92.

A life which comes from Atzmus is above all definitions and limitations of natural life and death as we know it. As the Rabbi Shalom DovBer (the fifth Rebbe of Chabad) says, "Relative to Atzmus, life and death are equal."[47] In fact Atzmus transcends all limitations. The miraculous and the natural are equal relative to Atzmus and can coexist.[48] It follows that the life of Moshiach is completely above the laws of nature, even as he lives in the physical world. As the Maharal of Prague writes, "Moshiach will be ... outside of nature and the way of the world." Here too, it is significant that the Rebbe MHM wrote about G-d granting such life in a discourse he published for 3 Tammuz, 5749 (1989): "The life in which He puts our soul is higher than nature."[49]

41. Atzilus is the highest spiritual world. See chap. 1, sec. 2, "Wisdom From Above."
42. *Ma'amorei Admur HaEmtzai*, Devarim, 1:214–15.
43. Lit., "which is one thing with."
44. *Sefer HaSichos* 5752, 6–7.
45. *Toras Menachem–Hisvaaduyos* 5710, 27. Reproduced in Majesky, *Likutei Mekoros,* 236.
46. *Sefer HaSichos* 5751, 496.
47. *Sefer HaMa'amorim* 5670, 100.
48. See chap. VI, sec. 1c, "The Coexistence of Opposites."
49. *Sefer HaMa'amorim Melukat* 3:201.

The Mittler Rebbe further explains the everlasting life of Melech HaMoshiach by drawing a contrast between Moshiach and all other tzadikim. All other tzadikim require some refinement of their bodies and therefore there has to be some sort of *histalkus* (departure or elevation of the soul from the body). Even Eliyahu HaNavi whose body ascended to the heavens and became like an angel, required a transformation of his body from its physical condition on earth requiring food and clothing to its totally spiritual state in heaven.

Moshiach, however, is on a much higher level. "No change or refinement takes place in his body at all. He will have a physical body similar to the bodies of all Jews of that time" which will "eat, drink, wear clothes and sit on a chair," but "Moshiach will remain alive in his body together with the essence of his soul—down here just as it [would be] above, literally." In the case of Moshe Rabbeinu, the Mittler Rebbe continues, "there was no difference at all between his body and his soul … and this will hold true of Moshiach in a much higher way. Therefore … he will not need to ascend above."[50]

The Mittler Rebbe once said, "I accept it upon myself to resolve all the doubts. Wherever I may be—in the infinite levels of Gan Eden—I will provide the Chassidim with the proper understanding in any matter which is causing them difficulty."[51] He has certainly resolved any doubts that anyone may have had about the Chaim Nitzchi'im of the Rebbe MHM.

To See the Rebbe MHM

Once, on *Chol HaMoed Sukkos*, while the Alter Rebbe was a student of the Maggid of Mezeritch, the Maggid called him in to his *sukka*, told him to lock the door behind him and not to get too excited by what he was about to see. A moment later, the Maggid called out, "The Rebbe is coming." The Baal Shem Tov (who had already passed away) then appeared before them and spoke to them.

The Previous Lubavitcher Rebbe explained that the Alter Rebbe merited seeing the Baal Shem Tov because he had the stature of being a *meshores* (literally, an attendant) of the Maggid. The Rebbe MHM analyzes this story and raises a question. The Alter Rebbe

50. *Ma'amorei Admur HaEmtzai*, Drushei Chasunah, 1:131, 155.
51. *Sefer HaSichos 5701*, 52.

stayed with the Maggid in Mezeritch for only a number of months, while being a *meshores* entails continuous long term attending to a Rebbe or a tzadik. This was the status of Joshua who was called the *meshores* of Moshe Rabbeinu. Regarding Joshua the Torah says, "And his *meshores*, Joshua the son of Nun, the young man, would not depart from the tent [of Moshe]."[52]

The Rebbe MHM then explains that since the Alter Rebbe was carrying out the mission of the Maggid even when he was away from the Maggid, he is considered a meshores just like Joshua. The Rebbe MHM concludes that we are carrying out the mission of the Previous Rebbe and therefore we should be able to see him!

From this we can conclude that in our time, since we are carrying out the mission of the Rebbe MHM by preparing ourselves and the world to greet Moshiach and by accepting his sovereignty, we should certainly be able to see him. It must be that one who is a proper *meshores* has this possibility.

Igros Kodesh

Now that we understand that the Rebbe MHM is alive, is with us and that we can talk to him—just as we talk to Yakov Avinu, it is crucial that we have a way to receive a response from him. Since Gimmel Tammuz it has been customary for people to direct their questions to the Rebbe MHM is through *Igros Kodesh*—the volumes of his published letters. One writes a letter addressed to the Rebbe MHM, signs it using his Jewish name and his mother's Jewish name (e.g., DovBer ben Chana) and places it at random in a volume of Igros Kodesh (in Hebrew, English or any other language). The writer of the letter then opens the volume to the place where he put his letter and reads what is printed there. This is his answer. Or, one may open the Igros Kodesh with a question in mind without writing a letter and read the answer.

This method has its basis in the statements of the Rebbe MHM himself, and even further back in Jewish history. After the passing of the Previous Lubavitcher Rebbe, the Rebbe MHM instructed people to continue writing to the Previous Rebbe and said that the

52. Shemos 33:11.

Rebbe would "find a way to answer them." Specifically, he instructed people to place the letters in a printed volume of the writings of the Previous Rebbe. The method of finding an answer to a question that one has by opening a *sefer* (holy book) at random and taking what he sees there as his answer is mentioned by the *Achronim* (latter Rabbinic authorities). There is fascinating story recorded by the Previous Rebbe in which he tells of a young scholar in the 1600s who resolved a doubt that he had by opening a Gemara.[53] The Rebbe MHM explains that the basis for this is the fact that nothing is really "at random." Rather it is by Divine Providence that one opened the book to a particular page.

In fact, this was practiced in our times even before Gimmel Tammuz. A Russian Jew told me that in his hometown of Homil in Russia, during the era of the Communist regime when they were not able to send letters to the Rebbe MHM, they would write the letter and place it at random in the sefer HaYom Yom, the Rebbe MHM's first book. I myself used to do this here in America during the 1980s using the published volumes of the Rebbe MHM's sichos and the Memoirs of the Previous Rebbe.

What is most amazing, in my opinion, is how, after Gimmel Tammuz, Jews all over the world started doing this spontaneously using the Igros Kodesh, which had just recently been published. On the day of Gimmel Tammuz, 5754, the day of the concealment, I was walking around "770" trying to figure out "what was going on" and what I was supposed to do now. I had just been offered a position in the department of Nuclear Engineering at Rensselaer Polytechnic Institute (RPI) in Troy, New York, and wondered if I should still take the position, especially since it involved being away from Crown Heights during the week. I felt that I should be in Crown Heights during this critical period.

Suddenly an idea came into my mind. I decided that I would ask the Rebbe MHM himself. How? Should I take a volume of the sichos like I used to? I didn't do that. Instead I walked over to the shelves where the volumes of Igros Kodesh were and took one at random off the shelf (vol. 15) and opened it. What I read there told me that I should "go to the new place" so I took the position. Why did I

53. *Likutei Diburim*, Hebrew edition (New York: Kehot, 1990), 779.

take Igros Kodesh, which I had never read before, instead of the sichos? To this very day I have no idea. The only thing that I can suggest is that the Rebbe MHM put it into our minds that we should start using the Igros Kodesh to ask him questions.

There are thousands of stories of people who wrote to the Rebbe MHM through Igros Kodesh and received answers and blessings on a wide range of issues, ranging from day to day decisions to urgent medical issues. Miracles have occurred from these blessings. I am personally familiar with the case of someone who was diagnosed with a cancerous tumor and wrote to the Rebbe MHM through Igros Kodesh for a *bracha* (blessing). The answer was a strong *bracha* for good health. At the next medical examination there was no sign of the tumor.

What impresses me most about the answers in Igros Kodesh, however, is their personal nature, i.e. the way in which the answer, originally written to someone else, is often a perfect fit for the question that someone is now asking. This is similar to prophecy, where the recorded words of the prophet are meant for all generations. The Rebbe MHM's letters were written not just to the person who wrote to him and asked for a blessing, but to all those who would seek his instruction and blessing and receive this letter as an answer for years to come. (I have received answers in Igros Kodesh where a certain phrase seemed to me to fit the question that I was asking even more than it fit the letter printed there.)

It is significant to note, in connection with this, what the Rebbe MHM said in a sicha on Rosh Chodesh of II Adar, 5749. He said he was "verifying" receipt of the letters that he had received with requests for blessings etc., and that this certification applied also to letters that were sent to him but not yet received-and even those letters *that would be sent in the future*. This includes the letters that we are writing now.

The Letter That Arrived Twice

Three brothers from Mexico came to Israel to learn in yeshivas. Although they were not Lubavitchers, one of them went to learn in a Lubavitch yeshiva and the other two went to *Litvishe* (Orthodox) yeshivas. In time, one of these two became acquainted with Chabad

Chassidus and transferred to the Lubavitch Yeshiva in Migdal HaEmek. He tried to influence the third brother to do the same but he wasn't interested. Yet, after a while he also became familiar with Chassidus and began to wonder if, perhaps, he should follow his brother to the Lubavitch Yeshiva.

He decided to consult with his *rosh yeshiva* (the dean) on the matter. The rosh yeshiva was sympathetic to the young man's request and told him that, to be honest, he would prefer that the student remain with him. At the same time, he continued, since he had a personal interest in the matter, he could not be completely objective and suggested that the student consult with someone who is totally committed to truthfulness. Thus he recommended that the student consult with the Amshinover Rebbe.

The student went to the Amshinover Rebbe who said that since the young man felt drawn to Lubavitch, he should go there. After a pause he added that this was his own opinion. If he wanted a higher level opinion (which he referred to as *da'as elyon*), the student should write a letter to the Rebbe MHM via the Igros Kodesh.

The student followed through with this advice. He opened Igros Kodesh to a letter that the Rebbe MHM had written to a rosh yeshiva in Germany many years before. The rosh yeshiva was asking the Rebbe MHM to advise him on where he should send his young son to yeshiva. His response was that the boy should learn "where his heart desires."

Now the young man in our story knew that he had his answer—that he should learn in a Lubavitcher yeshiva. He went back to his rosh yeshiva to tell him what happened. The rosh yeshiva was very impressed by the answer his student had received in Igros Kodesh and asked to see the letter. When he saw the letter he was totally astounded. The letter was about him. The rosh yeshiva in Germany was his father who had written to the Rebbe MHM many years before regarding his young son—this rosh yeshiva.[54]

Finding a Way

It is important to remember that the Igros Kodesh is not some

54. "The Letter That Arrived Twice," *Beis Moshiach,* 16 MarCheshvan, 5761, 26.

machine or a computer that has a database of answers. It is the Rebbe MHM himself who is answering our questions. He "finds a way" to answer. Thus we see that there are variations on the way the question is asked and the way that it is answered. Sometimes the answer does not come through the pages of Igros Kodesh. This is because it is the Rebbe MHM who is answering, not the book. Let me give some examples of this.

About ten years ago I had a serious legal problem. I was not particularly worried about it because there were standard procedures for dealing with such a problem and if everything went smoothly I would have the problem solved before the court date. But things did not go smoothly. The assistance that I was expecting did not come through.

As the court date neared things looked very serious and difficult. I realized that it was time to ask the Rebbe MHM for help. I came to "770," went over to the bookcase where the volumes of Igros Kodesh are held, took one at random and opened it. I read the letters on both pages and, to my great disappointment, I didn't see any answer to my question or any statement of reassurance.

As I stood there dumbfounded, wondering what was going to happen to me, I noticed a young man whom I did not know running towards me urgently. As he approached me he called out, "Rabbi Silman, how do you say *'nimna ha'nimnaos'* in English?" When I heard his question, I burst out laughing because his question was my answer. *'Nimna ha'nimnaos'* means that G-d can do the impossible, as I explained to the bewildered young man. Indeed, with G-d's help everything worked out fine for me.

The main thing is to present the question to MHM. He will find a way to answer it.

Finally, I would like to mention that part of the title of this book came from an answer in Igros Kodesh. Initially, the title was "Scientific Thought in Messianic Times." A number of years ago, on the Rebbe MHM's birthday, I wrote a letter to him about my progress on the book. In the letter in Igros Kodesh, the Rebbe MHM had written "Thank you for your book *'She'eris Yisroel.'*" Of course, I understood that the part that said "Thank you for your book" was an answer to me, but the title *"She'eris Yisroel"* did not seem relevant.

Some time later it hit me that *"She'eris Yisroel,"* which means "the remnant of Yisroel," was a very appropriate title for the book since the Rebbe MHM had used a similar phrase in referring to Rabbi Yisroel Aryeh Leib's mathematical paper.[55] When the Rebbe MHM was discussing the manuscript of the paper with Professor Rosenbloom, the mathematician to whom he gave the paper to edit and publish, he said that it was written by his brother who had passed away. "That is the only thing left of him," the Rebbe MHM said.[56]

Hence the title *"She'eris Yisroel."*

Beyond Brain Waves

We have been discussing the concept of Chaim Nitzchi'im—the everlasting lives of certain tzaddikim—especially the Chaim Nitzchi'im of the Rebbe MHM. Our discussion has been based on sources in the Talmud, Kabbala and Chassidus. Can the modern scientific mind grasp all this or must it be accepted totally on faith?

Indeed, one may say that this concept of a tzadik being alive after he apparently passed away is incomprehensible and would have to be accepted on faith. If we reflect on the matter, however, we realize that life itself—a soul in a body—is no less incomprehensible. How is it that a soul gives life to a body? How is it that there is life in the first place? We don't think about this too much because it's part of our everyday experience. We take it for granted.

It is for this very reason that the *Chachomim* enacted that we say blessing on everything that we benefit from in this world—so that we remember G-d and don't take his wondrous creation for granted. The daily blessing *Asher Yotzar* (on the creation and functioning of the body) ends with the words *umafli la'asos*—that G-d acts wondrously. One explanation of this is that it refers to the wondrous way that G-d unites the soul with the body to maintain a person's life.

On the night of Simchas Torah 5711 (1950), the Rebbe MHM instructed the Chassidim going out to the various synagogues for *hakofos* to tell the people there that "Moshe did not die." He was not just quoting the Talmud's statement about Moshe Rabbeinu, whose apparent passing is mentioned in the Torah reading of Simchas

55. Reprinted in the Appendix of this book, with a commentary.
56. See chap. II, sec. 2 for their full dialogue.

Torah. He was delivering a much more contemporary message. Several months earlier, the Previous Lubavitcher Rebbe had "passed away" and the Rebbe MHM was delivering the message that the "Moshe" of our generation i.e. the *Nasi* of the generation—the Previous Rebbe—did not die!

The Rebbe MHM continued to say that if the people hearing this message have difficulty accepting it, the Chassidim should tell them—just like a teacher tells a child—"When you grow up you will understand."[57] From this we see that it's actually possible to understand Chaim Nitzchi'im, not just believe in it.

So again we ask: "How can the modern scientific mind grasp the concept that a tzadik may be alive—body and soul together—while from all external indications it appears that he died?"

We claim that it is particularly the modern scientific mind that is in the best position to comprehend this. In the Introduction we discussed the sicha in which the Rebbe MHM explains that the reason modern science (the "fountains of wisdom from below") was revealed in recent times was in order to provide concepts that would help the mind grasp the revelations of the Era of Moshiach. In recent years a medical procedure was developed which will help us grasp the concept of Chaim Nitzchi'im that we have been discussing.

A woman was found to have a swollen artery (aneurysm) in her brain. Conventional brain surgery could not be performed because if they operated on the artery while the blood was still circulating in the brain the artery could burst at any time and she would die. As the surgeon explained it, it would be like operating on a tense balloon. She was operated on by a medical team at Columbia Presbyterian Hospital, led by cardiac surgeon Dr. Craig Smith and neurosurgeon Dr. Robert Solomon, that has been successful at performing a radically different type of surgery.

In their procedure they circulate the blood of the patient out of the body and through a refrigeration unit, thus lowering the body temperature until first the heart stops, then at about 64°F *the brain waves cease!* At this point the patient is scientifically and legally dead.

The team then proceeds to drain about half of the patient's blood

57. *Toras Menachem–Hisvaaduyos, 5711*, 1:36. Reproduced in Majesky, *Likutei Mekoros*, 241.

out of the body, deflating the artery. The blood is stored until they are ready to transfuse it back into the patient after operating on the aneurysm.

After the operation they warm the blood again as they circulate it through the body, starting the heart again and restoring brain function.[58] When a reporter asked this woman how she felt about the fact that she was "dead" during the procedure, she responded simply, "I'm alive now!"

The patients in these cases are just regular human beings, not tzadikim. If a regular human being, who is subject to the laws of nature, can still be alive for a certain amount of time even after all signs of life—including brain waves—have stopped, it should be easy to accept the statements in the Talmud, Kabbalah and Chassidus that certain tzadikim who are *not subject to the laws of nature* continue to live even though it may appear that they died!

Suddenly He Will Enter His Chamber

On the night of Simchas Torah, 5754 (1993), over a year and a half after the stroke, a chosid[59] approached the Rebbe MHM with the well known photograph of the Rebbe MHM energetically waving both arms. The chosid asked if we would see him again as in the photograph. The Rebbe MHM nodded his head, responding "Yes." After the holiday, to verify the response, the chosid repeated the question in a letter to the Rebbe MHM, asking: "Will the Chassidim again see the Rebbe Shlita doing as in the enclosed picture, <u>with the same physical capabilities (standing, waving both arms, *mitoch simcha*) and even more? We ask the Rebbe Shlita for his blessings that this occur soon.</u>" The Rebbe MHM again responded in the affirmative.

We implore the Al-mighty to bring about the complete revelation of the Rebbe MHM immediately so that "all flesh may see it." We anxiously anticipate the fulfillment of the words of the prophet Malachi regarding Melech HaMoshiach: "Suddenly, the master whom you are seeking will enter his chamber."[60]

58. CBS News, "60 Minutes," April 2, 1995.

59. Mr. Meir Schreiber of Crown Heights. (The underlining in the quote is in his original letter.)

Epilogue: The Lawyers Agree

In 1995 I was working with a group of lawyers to incorporate our organization, the Rabbi Yisroel Aryeh Leib Research Institute on Moshiach and the Sciences, as a non-profit organization. The process was going smoothly and was almost concluded. Then one day my lawyers asked me to submit a paper stating the purpose of the Institute, its background and the concepts on which it was based. I wrote this all out explaining that our purpose is to investigate developments in science and technology that are characteristic of the Messianic Era. I explained that this is based on the statements of the Rebbe MHM, who is still alive, that we are currently in the Messianic Era and that all aspects of the redemption have already begun.

After sending this document to the lawyers, I received a phone call from them. It was a conference call between my two lawyers, myself, and a third lawyer, a friend of mine who had referred me to them. They were all very serious. My lawyers said that they could not submit a document that stated that the Rebbe MHM is alive to the government. I calmly explained to them that there should be no problem, since it is not the business of the government to make decisions regarding people's religious beliefs. But they insisted that they just couldn't do it. This went back and forth for a while.

Thinking that they were just trying to pressure me, I continued to argue my case. Finally, they said that if I insisted on making such a statement in the document, they could no longer represent my organization. They said that it would really cause me no difficulty if they withdrew since the application process was almost complete and I could complete it myself. They would even advise me and tell me how to do it. But they could not officially represent me.

At that point I realized that they really intended to withdraw. Seeing that I had lost my case with them, I decided to make my closing statement. "Listen," I said to them, "if you want to withdraw that's your decision. But you should know that not only is the Rebbe MHM alive, but he has said that his father-in-law, the Previous Lubavitcher Rebbe is also alive."

---

60. Malachi 3:1.

As soon as I said that, they reversed their position completely. They began to apologize. "We're sorry. We didn't mean anything," they began to say. "We can find a way to do it. We will write a cover letter explaining that these are the beliefs of our client. It will be OK." And that's exactly what they did. I had no idea why they reversed their position but everything was approved and the RYAL Institute became registered as a non-profit organization in the State of New York.

Then one evening, several months later, I happened to pick up an old Beis Moshiach magazine and started paging through it. I noticed a reprint of the sicha of 15 Tammuz, 5745 (1985) where the Rebbe MHM was addressing an issue related to the *seforim* (books) of the Previous Rebbe. Someone had made the claim that the Previous Rebbe's seforim should be the inheritance of his heirs. But the Rebbe MHM said that this is not the case at all since the Previous Rebbe is alive and therefore no concept of inheritance applies. He concluded his statement by saying that no Rabbi and *no lawyer* will ever succeed in making a case for the inheritance of these books since the Previous Rebbe is alive!

Suddenly, I understood why my lawyers had reversed their position as soon as I mentioned that the Previous Rebbe is alive. It was because the Rebbe MHM had already decreed that no lawyer will ever succeed in going against the fact that the Previous Rebbe is alive!

The Lawyer's Story

A few years later, on the holiday of Pesach, when Lubavitcher Chassidim, following the Rebbe MHM's directive, go out to various *shuls* (synagogues) to speak, I went to a certain shul in Brooklyn and spoke about the Era of Moshiach and about the Rebbe MHM's statements that we are currently in the Era of Moshiach and that all aspects of the redemption have already begun. Everyone was very interested in what I was saying and very receptive to it, with one exception. There was one man—a lawyer—who took a negative position and challenged some of the things that I said. I returned there on the holiday of Shavuos and spoke again, and once again this lawyer challenged me. We talked privately after each of my speeches and in the end he agreed with me on some things and disagreed on others.

That summer, the rabbi of the shul invited me to lecture at a special program they were making for Gimmel Tammuz. As the participants began to arrive, I saw that the lawyer was among them. This upset me. "Why did he come here," I asked myself, "Just to argue with me?" My lecture was to be about the everlasting life of the Rebbe MHM. I was going to explain to them that the Rebbe MHM is alive. I was concerned about how the listeners—especially this lawyer—would receive this most delicate of all the Messianic concepts.

I began my lecture. I presented to them the same concepts that I wrote about in this section, and I concluded with the story about the incorporation of the RYAL Institute that I mentioned above, explaining that the Rebbe MHM had decreed that no lawyer could ever go against the fact that the Previous Rebbe is alive.

At this point the lawyer spoke up. "Yes, it's all true!" he called out, and he began to tell his story. Some years before, he was working on a case for the government. A certain very important file that he was working on was misplaced and could not be found. He was in great distress. If the file was not found, all his work would be for nothing and it would cause him great difficulty. But the file was nowhere to be found.

As he was involved with Chabad, he decided to go the Ohel of the Previous Rebbe and pray. That night he had a dream in which he saw the Previous Rebbe looking through the file cabinets in a large storage room, searching for the missing file. The storage room, which he recognized from his office building, was full of filing cabinets. Suddenly, the Rebbe pulled a file out of a file drawer and held it up. It was the missing file! The lawyer then woke up.

The next morning, when he arrived at his office, someone walked in to his office and presented him with the missing file. "I believe that the Previous Rebbe is alive," the lawyer concluded his story. "I believe that he came to my office building—maybe even physically—and searched the file room until he found my lost file."

Once again, the lawyers confirmed the fact that the Previous Rebbe is alive.

REFERENCES: CHAPTER I

Wisdom From Above...Wisdom From Below

General

Rabbi Menachem Mendel Schneerson, MHM. *HaYom Yom* (Hebrew-English edition). New York: Sichos in English, 2009.

———. *Igros Kodesh.* New York: Kehot Publication Society.

———. *Likutei Sichos.* New York: Kehot Publication Society.

———. *Sefer HaMa'amorim Melukat,* vols. 1–6. New York: Kehot Publication Society.

———. *Sefer HaSichos.* New York: Kehot Publication Society.

Rabbi Moshe ben Maimon, Rambam. *Mishneh Torah.* "Hilchos Yesodei Hatora" and "Hilchos Melachim U'Milchamoseihem."

———. *Sefer HaMitzvos.*

Rabbi Schneur Zalman of Liadi, Alter Rebbe. *Tanya,* Bilingual Edition. Kefar Chabad, Israel: Kehot Publication Society, 1998. See especially "Translators' Introductions" and "Mystical Concepts in Chassidism," pp. 832-956.

www.otzar770.com. Many of the Rebbe MHM's works—Sichos, Ma'amorim, Igros Kodesh, etc.—are available online at this website.

Sec. 1: Faith and Intellect

Rabbi Levi Yitzchok Schneerson. *Likutei Levi Yitzchok – Igros Kodesh.* New York: Kehot Publication Society.

Rabbi Menachem Mendel of Lubavitch, the Tzemach Tzedek. *Derech Mitzvosecha* (Hebrew). New York: Kehot Publication Society.

———. *Derech Mitzvosecha: A Mystical Perspective on the Commandments* (English). New York: Sichos in English.

Rabbi Menachem Mendel Schneerson, MHM. *Toras Menachem – Hadronim al HaRambam VeShas.* New York: Lahak Hanochos, 1992.

Rabbi Yehuda HaLevi. *Kuzari.*

Rabbi Yosef Yitzchok Schneerson. *Likutei Diburim.* New York: Kehot Publication Society. Original Hebrew-Yiddish edition; all-Hebrew edition; English edition.

Sec. 2: Wisdom From Above

Rabbi Levi Yitzchok Schneerson. *Likutei Levi Yitzchok – Igros Kodesh.* New York: Kehot Publication Society.

Rabbi Menachem Mendel of Lubavitch, the Tzemach Tzedek. *Derech Mitzvosecha* (Hebrew). New York: Kehot Publication Society.

———. *Derech Mitzvosecha: A Mystical Perspective on the Commandments* (English). New York: Sichos in English.

Rabbi Menachem M. Schneerson, MHM. *Inyono Shel Toras HaChasidus*. New York: Kehot Publication Society.

———. *On the Essence of Chassidus*. New York: Kehot Publication Society. English translation of *Inyono Shel Toras HaChasidus*.

Rabbi Shalom DovBer of Lubavitch, the Rashab. *Kuntres UMa'ayon*. New York: Kehot Publication Society. Original Hebrew edition; English edition.

Rabbi Yosef Yitzchok Schneerson. *The Four Worlds*. New York: Kehot Publication Society.

———. *Likutei Diburim*. New York: Kehot Publication Society.

———. *Kuntres Limud Hachasidus*. Original Hebrew edition; English edition. New York: Kehot Publication Society.

———. *Kuntres Toras Hachasidus*. Original Hebrew edition; English edition. New York: Kehot Publication Society.

———. *Lubavitcher Rebbe's Memoirs*, vol. 2. New York: Kehot Publication Society.

Rabbi Schneur Zalman, the Alter Rebbe. *Words of the Living G-d: Selected Discourses from Rabbi Schneur Zalman of Liadi's Torah Or*, vol. 1 (Genesis). New York: Purity Press, 2008. Selections from Torah Or adapted into English by Rabbi Yitzchok Wagshul.

Loewenthal, Naftali. *Communicating the Infinite: The Emergence of the Habad School*. Chicago: University of Chicago Press, 1990. This is a scholarly exposition of Chabad Chassidus. It was referred to by the Rebbe MHM as the author's "scholarly research."

Friedman, Eliyahu. The Great Mission: The Life and Story of Rabbi Yisrael Baal Shem Tov. New York: Kehot Publication Society, 2005.

Schochet, Jacob Immanuel. *Mystical Concepts in Chassidism: An Introduction to Kabbalistic Concepts and Doctrines*. New York: Kehot Publication Society, 1998.

Sec. 3: Wisdom From Below

De Broglie, Louis. *The Revolution in Physics: A Non-Mathematical Survey of Quanta*. New York: The Noonday Press, 1953.

Eves, Howard. *Foundations and Fundamental Concepts of Mathematics*, 3 ed. New York: Dover Publications, 1997.

Feynman, Richard P. *The Feynman Lectures on Physics*. New York: Addison Wesley Longman, 1970.

Gamow, George. *The Great Physicists from Galileo to Einstein*. New York: Dover Publications, 1988.

Kneebone, G. T. *Mathematical Logic and the Foundations of Mathematics.* New York: Dover Publications, 2001.

Livio, Mario. *Is G-d a Mathematician?* New York: Simon & Schuster, 2010.

Reichenbach, Hans. *Philosophic Foundations of Quantum Mechanics.* Berkeley and Los Angeles, 1944.

Smolin, Lee. *The Trouble with Physics.* Boston: Houghton Mifflin, 2006.

Struik, Dirk. *A Concise History of Mathematics,* 4 ed. New York: Dover Publications, 1987.

Wilder, Raymond L. *Introduction to the Foundations of Mathematics,* 2nd ed. New York: Wiley, 1965.

Sec. 4: A Day in the Life of Melech HaMoshiach

Rabbi Menachem Mendel Schneerson, MHM. *Besoras HaGeula.* New York: Kehot Publication Society. Hebrew edition; English edition.

———. *Sha'arei Ge'ulah.* Jerusalem: Heichal Menachem, 1992. A collection of *sichos* on topics of Moshiach and Geula.

Friedman, Avrohom. *MiGola L'Geula.* An anthology of quotes from Chassidus on topics of Moshiach and Geula. English edition translated by Uri Kaploun and published as *Exile to Redemption.* New York: Kehot Publication Society.

Sec. 5: The Emergence of Melech HaMoshiach

And He Will Redeem Us: Moshiach in Our Time. New York: Mendelsohn Press, 1994.

Branover, Herman and Avraham Naveh. *The Ultimate Jew.* Jerusalem: Shamir Publishing House, 2003.

Cohen, Yitzchak, ed. *Wonders & Miracles: Stories of the Lubavitcher Rebbe.* Kfar Chabad: Maareches Ufaratzta, 1993.

Faithful and Fortified, vol. 1, "Defense and Intelligence." New York: Jewish Educational Media video, 2009.

Friedman, Avrohom. *MiGola L'Geula.* (See sec. 4 above.) In English as *Exile to Redemption.* New York: Kehot Publication Society.

Gottlieb, Naftali Tzvi. *Toldos Levi Yitzchok.* New York: Kehot Publication Society.

Heruti, Rafael. *Nevuat HaGeula.* Kiryat Malachi, Israel.

I Will Show You Wonders. New York: Sichos in English. A translation of *sichos* of the Rebbe MHM before and during the Persian Gulf War.

Silman-Schechter, Sima. *Nisei Milchemet HaMifratz.* Jerusalem: Haktav Institute, 1995.

Touger, Eliyahu. *Sound the Great Shofar.* New York: Sichos in English.

———. *I Await His Coming Every Day.* New York: Sichos in English.

Wolpe, Rabbi Shalom DovBer. *Shemen Sasson MeChaverecha.* Israel, 1995.

———. *Yechi HaMelech HaMoshiach.*

Ziegelbaum, Menachem. *Sippurim MeCheder HaRebbe*. Bnei Brak, Israel, 2009. This is a collection of stories about the Rebbe MHM told by his secretaries.

www.otzar770.com.

www.igrot.com. This website is an electronic Igros Kodesh. One may submit a letter to the Rebbe MHM and it will be automatically placed at random in a volume of Igros Kodesh. The pages with the answer will then be displayed.

Sec. 6: The Everlasting Life of Melech HaMoshiach

Rabbi DovBer of Lubavitch, the Mittler Rebbe. *Ma'amorei Admur HaEmtzai, Devarim* vol. 1. New York: Kehot Publication Society.

Rabbi Menachem Mendel Schneerson, MHM. *Sha'arei Ge'ulah*. Jerusalem: Heichal Menachem, 1992.

Majesky, Shloma. *Likutei Mekoros*, 2nd ed. New York, 2008.

CHAPTER II

Rabbi Yisroel Aryeh Leib Schneerson

It is told in the name of Rebbitzen Chana that one time an excited discussion took place in her home in Yekatrinislav between her husband Rabbi Levi Yitzchok, the Rebbe MHM, and Rabbi Yisroel Aryeh Leib on a certain topic in Chassidus known as "Mah uBan." Rebbitzen Chana noted that each of the participants expressed his opinion in his own terminology—Rabbi Levi Yitzchok in the terminology of Kabbala, the Rebbe MHM in the terminology of Chassidus, and Rabbi Yisroel Aryeh Leib in the terminology of Chakira.

Rabbi Levi Yitzchok Schneerson

Rabbitzen Chana Schneerson

Rabbi Yisroel Aryeh Leib Schneerson was the chosid who inspired our Moshiach and science research. Thus our institute is named for him.

In this chapter we present his biography, emphasizing his unique intellectual approach of Chakira. We also explain how he was the inspiration for our work and for the Moshiach and science conferences held every year on his yartzeit. In the process, I describe how he was an inspiration for me personally.

In the Appendix to the book we present his mathematical paper and a mathematical commentary on it.

Man with a Mission

Rabbi Yisroel Aryeh Leib was the chosid whose life and work inspired Moshiach and science research and the establishment of the RYAL Research Institute on Moshiach and the Sciences. In this section we discuss his life and work, emphasizing those aspects which continue to inspire and influence Jewish scientists to this very day.

In 1947, Rabbi Yisroel Aryeh Leib wrote to the Previous Lubavitcher Rebbe regarding a position he was offered in the Department of Theoretical Physics at the University of Liverpool, England. (It has been reported that he went to England to do research in physics at the urging of Ben-Gurion, who wanted him to develop a program in nuclear technology for Israel.) In his letter of response, dated 17 Kislev 5708 (1947), the Previous Rebbe gave the following *shlichus* (mission) to Rabbi Yisroel Aryeh Leib in connection with this position: "What interests me is to bring, with G-d's help, the beauty of *Shem* (Torah and Mitzvos) into the tents of *Yefes* (the universities)[1]—to show the Jewish youth who pride themselves in, and long for, secular knowledge that it is nothing compared to the deep wisdom of the Torah." The Rebbe continued to suggest that Rabbi Yisroel Aryeh Leib meet with the great scientists and language specialists there and work on translating the works of Chabad Chassidus into English, beginning with the Tanya. The Rebbe further suggested translating "the history of the *Baalei-Shem*, the Kabbalists until Rabbi Adam (Baal Shem), Rabbi Adam himself, the Baal Shem Tov and his students, Rabbi Dovber of Mezeritch (The Maggid) and his students, their published works, stories that they told and stories about them."[2]

It was not until 5710 (1950) that Rabbi Yisroel Aryeh Leib was actually able to go to England, and a short two years later, on 13 Iyar, 5712 (1952) he passed away. But the fact that the Previous Rebbe gave him this mission means that he was the man for the job. What were the qualities that he possessed that made him suited for all this, especially for the awesome task of showing the Jewish youth that the sciences are nothing compared to the deep wisdom of the Torah? To begin with, consider the titles with which the Previous Rebbe addresses him in the letter just quoted: "*Vosik* and *Chosid*, a G-d fearing man, a man of exceptionally high ability." From his early childhood Rabbi Yisroel Aryeh Leib exhibited brilliance and diligence in his learning, and this was his trademark throughout his life.

A Biographical Sketch

Rabbi Yisroel Aryeh Leib was born around the beginning of Sivan, 5666 (1906)[3] in Nikolayev, Russia. In 1909, the family moved to Yekatrinislav where his father, *Kvod Kedushas* Rabbi Levi Yitzchok, assumed the position of Rabbi. At the age of three, a visitor to the

1. This is a rephrasing of the blessing that Noach gave to his sons Shem and Yefes. See Bereishis 9:27; Talmud Megilla 9b.

2. *Kovetz Yud Gimel Iyar, Arboim Shonah* (New York: Kehot, 1992), p. 13.

3. This is based on a document which states that his birthday was May 23, 1906, and a statement by his grandfather, Rabbi Boruch Schneur Schneerson, in 1915, that his grandson would be 9 years old "before Shavuos." Another document, however, states that his birthday was May 23, 1909. (Also, at that time, Russia was following the Julian calendar, which was twelve days behind the Gregorian calendar.)

home of Rabbi Levi Yitzchok was amazed to see little Leibel (as Yisroel Aryeh Leib was called) looking into a book with great concentration. Leibel did not take his eyes off the book the whole time the guest was there. Rabbi Levi Yitzchok, noticing the guest's amazement, went over to Leibel, who had been looking into a book of the Mishnah, and asked him what he was learning. Leibel immediately gave a clear explanation of the Mishnah that he had been learning. Rabbi Levi Yitzchok once said of him that he had "the Tzemach Tzedek's head." (The Tzemach Tzedek was the third Rebbe of Chabad, known for his brilliance in all areas of the Torah and especially for his unification of its halachic and kabbalistic aspects.)

His diligence is as legendary as his brilliance. Once, in his youth, during "the Nine Days," he sat through the night learning a *mesechta* (tractate) of the Talmud to make a *siyum* (celebration of the completion of study). His mother, Rebbitzen Chana, wanted to show a visiting chosid, Reb Simcha Gorodetzky, the extent of her son's diligence. She entered his room and called his name, "Leibel!" He was so engrossed in his learning that he did not hear her. So she called his name again, "Leibel!" He still didn't hear her. She then went over to him and pulled on his sleeve and again called "Leibel!" But Leibel was oblivious to the presence of anyone in the room. Rebbitzen Chana then turned to the chosid and, smiling with pride, she said, "Do you see the diligence of my son Leibel?!"[4]

In the summer of 5684 (1924) Rabbi Yisroel Aryeh Leib arrived in Petersburg where he was a member of the household of the Previous Rebbe who cherished him and showed him special attention. The Rebbe MHM and Rabbi Yisroel Aryeh Leib were both present at the *farbrengen* of 19 Kislev, 5687 (1926). The Previous Rebbe said to them, "It's too bad that my father [Rabbi Shalom DovBer, who was the Rebbe before him] didn't see you."

Once, Rabbi Yisroel Aryeh Leib approached the Previous Rebbe with a question on a matter of Chassidus. The Rebbe said he couldn't explain this matter to him. Rabbi Yisroel Aryeh Leib was heartbroken and cried all night. The next day he came to the Rebbe for an entirely different reason and was surprised to hear the Rebbe explain to him the matter which he had asked the previous night. Afterwards,

4. Naftali Tzvi Gotlieb, *Toldos Levi Yitzchok*, 3rd ed. (New York: Kehot, 1986), 169.

the Rebbe explained that since Rabbi Yisroel Aryeh Leib had been heartbroken, he became a *kli* (a proper receptacle) for what the Rebbe had to tell him.

Rabbi Yisroel Aryeh Leib was also very popular among the Chassidim, who would approach him with difficult questions in the Talmud and Chassidus and engage him in lively discussions. At this time he began studying mathematics in the academies in Petersburg where he organized a branch of *Tiferes Bachurim* which held daily Torah classes for the Jewish youth and strengthened their observance of the Mitzvos. His main occupation, however, was the study of Torah which he continued to do with great diligence. He would spend long hours in the Previous Rebbe's library delving into the depths of the books at hand.

One night, while in the library with a friend, Rabbi Yisroel Aryeh Leib became engrossed in his learning. Late into the night, when his friend left for home, Rabbi Yisroel Aryeh Leib was leaning on a lectern, holding a cigarette, lost in deep thought. When the friend returned the next morning he found Rabbi Yisroel Aryeh Leib still leaning on the lectern, standing in exactly the same position that he was in when he left him the night before—still holding the cigarette and still lost in thought.

During the years that the Rebbe MHM and his wife, the Rebbitzen Chaya Mushka, were in Berlin, Rabbi Yisroel Aryeh Leib was there as well. They would receive letters from their father Rabbi Levi Yitzchok addressed to both of them together. For example, a letter dated 9 Adar II, 5692 (1932), a few days before Purim, is addressed to "My dear, cherished and beloved sons, may they live long and good lives." He writes that he is sending the letter "to fulfill the sending of gifts for Purim, each man to his friend (which means [in the Kabbalistic sense] from a father to his sons…)." Since he couldn't send them gifts of food, in accordance with the Mitzva, he says that he is sending them a spiritual gift—a booklet he wrote with Kabbalistic explanations on the Megilla. He writes that he is sending the booklet "to both of you together—the older one and the younger one together" and continues to explain that the essential concept of Purim is the equality of everyone—the great ones and the small ones.[5]

5. *Likutei Levi Yitzchok*, "Igros Kodesh," 267–68.

It is assumed that it was during the Berlin years that Rabbi Yisroel Aryeh Leib wrote his most famous paper on mathematics, "The Location of Eigenvalues and Eigenvectors of Complex Matrices," a topic which has applications to quantum mechanics. (In the next section we will discuss the story of the publication of this paper. The paper itself is reprinted in the Appendix of this book, followed by a mathematical commentary on it by Dr. Neil Miller, a contemporary mathematician.)

In 1933, when the evil Nazis came to power, the Rebbe MHM and Rebbitzen Chaya Mushka moved to Paris. Rabbi Yisroel Aryeh Leib was making every effort to go to Israel but needed certain papers from the German government in order to do so. He approached his brother for help. The Rebbe MHM was ready to go back to Germany to get the papers, but Rebbitzen Chaya Mushka protested that this would be too dangerous for him, and insisted that she go instead. She went, and she succeeded in getting the documents that Rabbi Yisroel Aryeh Leib needed, and he was then able to go to Israel.[6]

While in Israel, Rabbi Yisroel Aryeh Leib got married. The wedding brought joy not only to those present, but also to his parents, Rabbi Levi Yitzchok and Rebbitzen Chana thousands of miles away in the remote Russian village of Chi'illi where Rabbi Levi Yitzchok had been exiled by the Russian government. As Rebbitzen Chana recalls in her memoirs:

> One day the Rav [Rabbi Levi Yitzchok] went to the post office and I stayed home to do the housework and prepare lunch (which had to be prepared *yesh me'ayin*—from nothing!). The Rav came home, his face beaming with joy, and he said, "We have a Mazal Tov!" He showed me a letter which had just arrived … informing us of news … from Israel that our son Leibel had gotten married. This good news was a double *simcha* (rejoicing)—it was the first report we had received telling us that he was still alive, and it contained the happy news of the wedding. It was a ray of light which penetrated the darkness of our exile and a sign of encouragement and hope for the future.[7]

6. *Kovetz Yud Gimel Iyar*, 8.
7. *Toldos Levi Yitzchok*, 170–71.

Rabbi Yisroel Aryeh Leib and his wife lived in Tel Aviv, where he continued his mathematical research while working in a bookstore. She worked as a nurse. Here, too, he would spend long nights studying Chassidus. Frequently, he would study with his *chavrusa* (study partner), Rabbi Nachum Goldschmid, author of the well-known lectures in Tanya, who was his childhood companion from Yekatrinislav. Rabbi Goldschmid's son, who was a little boy at the time, later related the manner in which his father and Rabbi Yisroel Aryeh Leib would study. In the morning, they would sit down together at the table, facing each other, each leaning his head on his hand. They would not talk, but rather would just sit there "thinking to each other." They would sit that way for hours and hours. When the little boy would return later in the afternoon, he would find them still sitting there in the exact same position, just "thinking to each other."

Another close friend of Rabbi Yisroel Aryeh Leib was the chosid, Rabbi Moshe Gurary. His son, Rabbi Nosson Gurary, who knew Rabbi Yisroel Aryeh Leib, described him as "very serious; his external mannerisms, his manner of speech and the appearance of his face were very similar" to those of the Rebbe MHM.

Rabbi Yisroel Aryeh Leib became known for a special style of studying Chassidus. He would start with a concept in Chassidus and explain it in successive stages, each time using less of the technical terminology of Chassidus and more of common everyday speech. On one occasion he remarked that there was a certain concept that he did not yet fully understand, as evidenced by the fact he could not yet explain it entirely in everyday speech.

After the Rebbe MHM came to America, the two brothers kept up a regular correspondence. The Rebbe MHM would send him the most recent publications of Chassidus, and Rabbi Yisroel Aryeh Leib would bring them to the other Chassidim. When he received the Rebbe MHM's book *"HaYom Yom,"* he started to dance with joy and excitement, saying, "My brother sent me his first book!" He remarked that it was too bad that people didn't see the depth hidden in it.

We have talked extensively about Rabbi Yisroel Aryeh Leib's genius and diligence, his love of Chassidus and his work in mathematics, because of their relevance to Moshiach and science research. But we should mention another side of him: On the *Simchas Torah* holiday he was known to dance all night with great exuberance and show tremendous outpouring of *Ahavas Yisroel* (love of one's fellow Jew). Indeed, Ahavas Yisroel and the unity of Jews are basic elements in bringing the Geula, as the Rebbe MHM says in many sichos.

"The Living Shall Take It To Heart"

As we mentioned earlier, in 1950 Rabbi Yisroel Aryeh Leib accepted a position in the Department of Theoretical Physics at the University of Liverpool, England. Two years later, on 13 Iyar, 5712 (1952), he passed away after suffering a heart attack. He is buried in Tzefas, Israel.

At the completion of the *shiva* (seven-day mourning period) for Rabbi Yisroel Aryeh Leib, his brother, the Rebbe MHM, said a short and very deep sicha[8] about comforting the mourners and its connection to the final redemption by Moshiach—the ultimate comfort of the Jewish people. We present a free translation of the sicha:

> The creation of the universe was a progressive descent from above to below with the intention that it be followed by successive ascents from below to above. Since this is the Divine intention, it must actually happen that way since nothing can stand in His way. This is especially so since G-d's plan for the creation precedes the creation itself, and at that point there is obviously no evil in existence.
>
> Thus it must be that one's free will affects only the details but does not affect the creation as a whole, which is in constant ascent since that is the intention of the Creator.
>
> On a deeper level, it must be that the intention of the Creator that there be constant ascent is fulfilled by each individual, as well. After all, why should there be any difference? Man's free will merely sets the pace of the ascent (similar to what the Talmud says, "If they merit, the redemption will happen faster. If not, it will happen according to schedule"). It also determines the manner of the ascent—whether it be by initially leaving room for evil and then completely defeating it, or as a steady ascent with no retreat at all.

8. *Kovetz Yud Gimel Iyar*, 19–20.

From the time that the creative descent was complete—at the end of the six days of creation—the successive ascents began [at the beginning of the first Shabbos]. This can be perceived, however, only by the souls in Gan Eden and by tzadikim in this world, as the Rabbis of the Talmud say: "Tzadikim have no rest—neither in this world nor in the next" for they are in a continuous state of ascent.

Although there are many changes throughout the 6,000 years of history, ascents and descents, the Scriptures say that "a tzadik will fall and arise seven times." The fall is only for the purpose of the ascent which follows it. The fall and the descent are necessary prerequisites for this ascent since any transition between a lower level and a higher level requires that there be *bitul* (nothingness) in between. The same applies to the creative descent: between one level of existence and another there must be nothingness.

This explains the statement of the Baal Shem Tov that although he could have ascended to heaven in a storm of fire like Eliyahu HaNavi, he wanted to fulfill the verse that says "unto the earth you shall return." He meant that the lowest descent, returning to the earth, is a preparation for a level which is higher than ascending to heaven in a storm. Since the descent is a necessary preparation for the ascent to a higher level, it is really not a descent at all, but rather part of the ascent.

This applies, as well, to one who passes away and to the mourners. For the one who passed away, the passing is a return to the earth in preparation for a higher ascent. For the mourners—may they have good life—the mourning which is *katnus hamochin* (a reduced mental state) will also be followed by an ascent to a higher level.

This is also the inner content of the expression used to comfort the mourners—the mourner is included among those who mourn for the destruction of *Yerushalayim*. The level of Moshiach and the Third Beis HaMikdosh is much higher than the level before the destruction. Thus, the Midrash says that Moshiach was born at the moment following the moment of the destruction. This means that immediately following the lowest descent—but no earlier—the Geula began.

But one must work very hard to be on such a level that at the time of the fall and the descent he can actually feel—not just believe or understand—that it is only a necessary preparation for an ascent to a very high level. This must also be the intention of those who comfort the mourners, because only in this way is it possible to be truly comforted—among all those who mourn for *Tzion* and *Yerushalayim*.

An Independent Character

We mentioned above that Rabbi Yisroel Aryeh Leib was known for a unique style of learning. We now attempt to analyze his particular method of learning and research, beginning with a story:

It is told in the name of Rebbitzen Chana that one time an animated discussion took place in her home in Yekatrinislav, involving her husband Rabbi Levi Yitzchok, the Rebbe MHM and Rabbi Yisroel Aryeh Leib, on a certain topic in Chassidus known as *"Mah uBan."* The intense discussion continued for weeks. At every opportunity it was the main topic of discussion. Once, after it had been going on for a few weeks, Rabbi Yisroel Aryeh Leib remarked, "Until now we have just been talking around the issue. What about the essence of *Mah uBan*?!" Rebbitzen Chana noted that each of the participants expressed his opinion in his own terminology—Rabbi Levi Yitzchok in the terminology of Kabbala, the Rebbe MHM in the terminology of Chassidus, and Rabbi Yisroel Aryeh Leib in the terminology of *Chakira* (logical analysis).

It is interesting to note that, on occasion, the Rebbe MHM has referred to a concept as it is explained in all three of these methods of analysis. For example, in the *sicha* of *Vav Tishrei* (the 6th of Tishrei, yartzeit of Rebbitzen Chana), 5742 (1981), he said, "It is explained in Kabbala and in Chassidus and in the books of Chakira that the fact that an object has ten different aspects indicates the completeness of the thing." From this we see, firstly, that each of these three methods has its place in the study of Torah and that, furthermore, the three methods complement each other.

Chakira, Kabbala and Chassidus

We can gain a fuller understanding of what Chakira is and its role in learning by comparing it to and contrasting it with Chassidus. We begin with a very interesting quote from Rabbi Shalom DovBer of Lubavitch, known as the Rebbe Rashab. He once said, "What do we want from Chassidus? If it's the fear of G-d and good character traits, there is the book *Reishis Chochmoh*. If it's Chakira, there are the books of the *Ramak*.[1] But in Chassidus we want G-dliness, and this can be found only in Chassidus." [2]

From the fact that the Rebbe Rashab suggested that what we may want from Chassidus is Chakira, we see that Chakira does in fact play a role in Chassidus. And from the fact that he says that Chakira can be found in the books of the Ramak, which are books of Kabbala, we see that Chakira plays a role in Kabbala. So let's examine the relationship between Chakira and Kabbala, and the relationship between Chakira and Chassidus.

Kabbala was the method of learning of Rabbi Levi Yitzchok. In a letter that he wrote to the Rebbe MHM on 19 Kislev, 5691 (1930), he makes a very fundamental statement about Kabbala: "Kabbala is not comprehensible to the human intellect. It can only be learned in a manner of reception and belief."[3] It has to be received from elsewhere and believed. This explains why it is called *Kabbala*, which means reception. (Through Chassidus, however, it becomes understandable to human intellect, as Rabbi Levi Yitzchok explains later in the letter.)

Chakira is the opposite of this. It starts with assumptions acceptable to human intellect—axioms—and builds on them using the methods of logical reasoning. Rabbi Levi Yitzchok writes about this, too, in another letter. This letter, dated 24 Teves, 5692 (1932) is addressed to both the Rebbe MHM and to Rabbi Yisroel Aryeh Leib. (As we mentioned in the last section, they were both in Berlin together at this time.) Rabbi Levi Yitzchok writes: "In mathematics, and similarly in every matter of intellect, there must be two things:

1. Rabbi Moshe Cordovero, author of the *Pardes*, the great Kabbalist in the generation before the AriZal.
2. *Sefer HaSichos* 5704, 14. Quoted in *Sefer HaToldos Admur Rashab*, 381.
3. *Likutei Levi Yitzchok*, "Igros Kodesh," 223.

1) the initial concepts and the assumptions which remain constant. For example, that the whole is greater than any of its parts… 2) all the theorems and results of mathematics are then built on this."[4] (He continues to explain that this is connected with *Mah uBan*: *Ban* is the axioms and *Mah* is the structure built on the axioms.)

What about Chakira and Chassidus? The difference between them is described in an entry in *HaYom Yom* which identifies three types of wisdom:

1) The teachings of Mussar (ethics), which emphasizes the rejection of physical and material matters.
2) The wisdom of Chakira, which emphasizes the advantage of the inner content and the spiritual element both in intellectual and emotional matters. Chakira teaches one to be attracted to the spiritual rather than to reject the material.
3) The wisdom of Chassidus, which teaches one to refine the material and combine it with the spiritual, forging one unified entity in which the spiritual is in control. In this way, both the material and the spiritual attain their full potential and achieve the purpose for which they were created. Both the material and the spiritual were created by One G-d for one purpose, which is to reveal the light of the holiness of G-d's mighty glory. Only when the material and the spiritual are unified do they fulfill this intention and attain their perfection.[5]

The Light of Chassidus

Now that we have contrasted Chakira with both Kabbala and Chassidus, we should ask how Chakira, Kabbala and Chassidus are all connected. The answer to this can be inferred from what Rabbi Levi Yitzchok writes (in the first letter quoted above) about the relationship of Kabbala and *Nigleh* (the "revealed" part of the Torah—the Talmud, Halacha, etc.) to Chassidus:

> The union of these two lights [Kabbala and Nigleh] is the goal of the revelation of the light of Chassidus—that the hidden part of the Torah and its deepest secrets be understood

4. Ibid., 250.
5. *HaYom Yom*, 7 Kislev.

> in their full abstraction by the intellect just like the comprehension of the revealed part of the Torah. This is in contrast to Kabbala, which is not comprehensible to the intellect. Rather, it is received in a manner of faith. The revelation of the light of Chassidus is a level which is higher than both Kabbala and Nigleh and therefore has the power to combine both of them together.

Similarly, Chassidus is higher than both Kabbala and Chakira and unifies both of them. The Alter Rebbe says in Tanya[6] that the questions raised in Chakira cannot be answered completely unless one uses the concepts of the Kabbala of the Ari Zal. This is what the Alter Rebbe himself did in developing Chabad Chassidus, which unifies Chakira, Kabbala and Nigleh.

In terms of our discussion in chapter I, we may say that Kabbala is the wisdom from above, Chakira is the wisdom from below, and Chassidus is the wisdom that unites them.

There is an advantage to the method of Chakira, even though it is a lower-level method, especially pertaining to the Era of Moshiach. Recall that the Zohar, quoted in the Introduction, speaks of revelations in both the sciences and in Chassidus, which prepare the world for the revelation of the absolute unity of G-d in the Era of Moshiach. Chakira, which is basically a scientific approach, begins with observations and axioms and then builds on them. It is the wisdom from below. It has the advantage of demonstrating unity in the world from within the created universe itself—in its own terms—rather than having it revealed from outside the order of creation, from above. It expresses the G-dliness taught by Chassidus in the world's own terms. As the Rebbe MHM says, "The completeness of G-d's intention of having a *Dirah B'Tachtonim* (a 'dwelling' in the lower worlds) is that the revelation in the world be ... from within the world itself."[7]

"An Independent Character"

Mathematics, which was Rabbi Yisroel Aryeh Leib's profession, has this quality of uncovering the underlying unity in creation, more so

6. *Tanya*, "Igeres HaKodesh," 282.
7. *Sefer HaMa'amorim Melukat* 6:18.

than other sciences. As the famous 19th century mathematician Joseph Fourier said, "Mathematics compares the most diverse phenomena and discovers the secret analogies which unite them."

For Rabbi Yisroel Aryeh Leib, Chakira, his method of learning, was also his method of mathematical research. I mean this in the sense of working "from below"—developing a concept in his own terms, from within himself, as opposed to picking up on someone else's ideas.

This can be seen from what the Rebbe MHM himself said about RYAL in a famous dialogue that took place 23 years after Rabbi Yisroel Aryeh Leib had passed away. The story is as follows:

One day back in 1975, the famous Lubavitcher mathematician, Professor Pesach (Paul) Rosenbloom received a call from Rabbi Leibel Groner, one of the Rebbe MHM's secretaries, regarding a mathematical paper that the Rebbe MHM wanted Rosenbloom to read. When Rosenbloom received the paper and reviewed it, not knowing who the author was, he noticed something highly unusual for a mathematical paper. It had no references to other mathematical works (texts, journals etc.). In fact, it contained mathematical results that had been obtained by some of the great mathematicians of the 20th century—Frobinius and Perron—but in this paper these theorems were proved using entirely different proofs—the author's own proofs. It was evident that he was not aware of the proofs of Frobinius and Perron. (Incidentally, this tells us something about the caliber of mathematician that Rabbi Yisroel Aryeh Leib was.)

This is how Professor Rosenbloom described the event to me:

> One day I got a call from Rabbi Groner asking me who I know that could read a mathematical manuscript in German. I told him that most mathematicians read German, so the question is not whether they can read a mathematical manuscript in German, but rather what specialty someone has to have in order to read it. I said I can't judge without seeing the manuscript and figuring out to whom it should be sent. So he said, "Can I send it to you?" "Yes," I said, "send it to me and I will tell you who I think would be able to read it."
>
> The next day I received the paper via courier. It had no introduction and no conclusion—it looked like it was a part

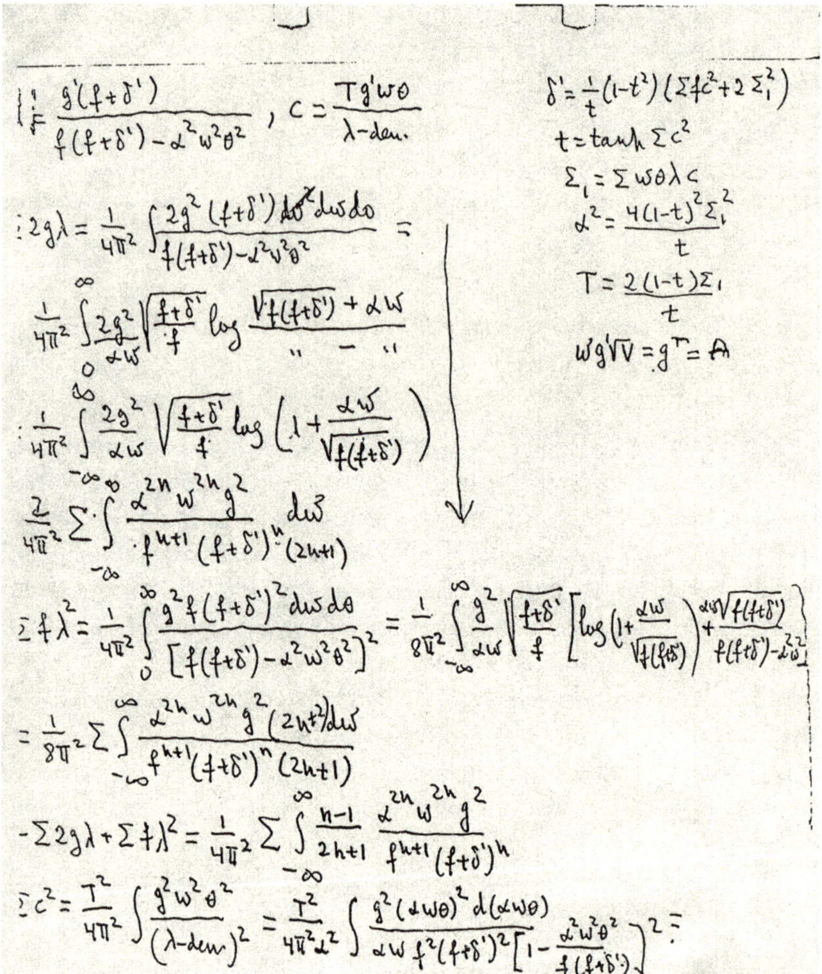

Unpublished mathematical manuscript of Rabbi Yisroel Aryeh Leib

of something greater. There were some elements in it that were well-known, and there were aspects that appeared to be new. It was not in my area of specialty, but I knew enough to know what it was about, and the only expert in this subject whom I knew was a retired professor at the University of North Carolina. So I called Rabbi Groner back and told him

> this, and he asked if I could send a copy to this professor. I said, "Sure I could." But I didn't get an answer from [the professor].
>
> In the meantime, I read the paper and I told Rabbi Groner that I had read it. He asked me whether I could give the Rebbe an answer as to whether it was worth publishing, and I said, "All right, I will write a report for the Rebbe with my opinion on the paper." I figured that this must be someone who has no access to a library, and I thought that it might be some Russian refugee that the Rebbe is interested in helping. I wrote a detailed report supplying it with references and that sort of thing—what had to be done to make it publishable. I brought the report to a farbrengen shortly after that, and I gave it to the Rebbe.

At the farbrengen of 10 Shevat that year, Prof. Rosenbloom approached the Rebbe MHM and they discussed the mathematical paper. The following dialogue then took place:

PROFESSOR ROSENBLOOM: The report—the paper that you sent me .... It needs an introduction. It needs references, and the manuscript is not complete because it mentions some things which are not in the manuscript.

MELECH HAMOSHIACH: But ... the author is deceased. That means that if something can be done by someone else....

PR: Yes, but then someone like myself would have to do it, I suppose....

MHM: I'll be very glad to compensate someone capable of completing it. Maybe you'll supervise his work.... I don't know if you have time enough and ... your time is valuable. But if you can find someone I will gladly pay him. You will only supervise him....

[Rosenbloom understood this to refer to either another professor or a graduate student. But the paper was too advanced for any graduate student, and he didn't think that another professor would be interested in working on someone else's paper.]

PR: Well, this is all you have from him?

MHM: Yes

PR: I see—because he mentions some other things.

MHM: Yes, but that is all. Nothing else. And he's dead ten years ago.

PR: Ten years ago?

MHM: Approximately.

[When Rosenbloom realized that the Rebbe MHM had a special interest in having the paper published, he offered to do it himself:]

PR: Well then, I can do it myself because I've put in some work in this before. I thought it might be of some use to [the author].

MHM: But it will be of use to me. I was not willing to tell you who the author is—not to influence you, but it was written by my brother.... I had a brother. He was younger than I am and he's dead ten years ago. And that is the only thing left of him.

PR: I see. Evidently he had to work independently and he didn't have any access to a library....

MHM: He was an independent character. He was not willing to look at anything, [rather] to write all on his own. Only after that would he look in a reference to see if someone else had explained the same thing. He was very jealous about his independence, and because of that the references are not [there]. But I'll be very happy if it can be printed, and if something must be done to complete it I will be very glad to pay someone for his time if you will supervise him....

PR: Well, I could do it myself.

MHM: You have enough time?

PR: I would put in the time.

MHM [smiling]: May G-d Al-mighty bless you. If so, I'll give you an additional thanks ... *L'Chaim V'Livracha!* (To life and for a blessing)[8]

"And so I worked on it the whole summer," Prof. Rosenbloom concluded his story to me, "and I brought it up to date and simplified it a lot and finally I gave the Rebbe a copy."

Prof. Rosenbloom published it in a mathematical journal called the

8. Prof. Rosenbloom related the dialogue to me in his own words. The text of the dialogue that I used, however, is transcribed from the Jewish Educational Media (JEM) video of the dialogue. The explanatory comments in brackets are based on what Prof. Rosenbloom told me personally.

Journal of Approximation Theory under the title "The Location of Eigenvalues and Eigenvectors of Complex Matrices."[9] The author was identified as Mark Gurari, the name on Rabbi Yisroel Aryeh Leib's passport. (It was common for Jews trying to leave Russia under the Communist regime to use an assumed name.) Prof. Rosenbloom told me that the Rebbe MHM said not to mention that the author was his brother as he wanted the paper to stand on its own merits.

The paper is reprinted in the Appendix of this book, followed by a mathematical commentary on it by a contemporary mathematician, Dr. Neil Miller. (There were a few errors in Rosenbloom's version that were corrected by Dr. Miller.)

Just how important is this quality of deriving results on one's own and not depending on the work of others, to a research scientist? Or, how important is it to be an "independent character"? We can answer this question with a fascinating story from the life of one of the greatest research scientists of our time, Richard P. Feynman.

Feynman was a Jew from Brooklyn who became prominent as a physicist at a young age. He worked on the development of the atomic bomb in Los Alamos during World War II, and afterwards was a professor at Cornell University and then at the California Institute of Technology (Cal Tech). He received a Nobel Prize in physics. He is considered by many to have been the greatest theoretical physicist after Einstein. (See note 5 on page xxvii)

In his autobiography, he talks about the research effort to find the laws of beta decay. He is describing his own involvement in the project. He begins by saying:

"At that particular time I was not really quite up to things … Everybody seemed to be smart and I didn't feel I was keeping up."

Then a discovery was made that the so-called parity law was violated in beta decay and there was a conference on this in Rochester, New York. The physicists Lee and Yang had delivered a paper at the conference describing their discovery that parity had been violated. Feynman attended the conference while staying at his sister's house nearby. He continues:

---

9. Mark Gurari, "The Location of Eigenvalues and Eigenvectors of Complex Matrices," *Journal of Approximation Theory* 22 (1978): 119–49.

"I brought the paper home and said to her, 'I can't understand these things that Lee and Yang are saying. It's all so complicated.' 'No,' she said, 'what you mean is not that you can't understand it but that you didn't invent it. You didn't figure it out your own way, from hearing the clue.'"

She told him to review the paper line by line, making sure that he understands it in his own way. He continues:

"I took her advice and checked through the whole thing, and found it to be very obvious and simple." In fact, in the process he had actually developed the theory a little further. It turned out to be even further than he thought, as follow-up experiments showed that his theory actually explained the remaining unanswered questions about beta decay.

"I had the whole theory!" he continues. "It was the first time, and the only time, in my career that I knew a law of nature that nobody else knew.... It's the only time I ever discovered a new law."

He concludes by saying, "Since then I never pay any attention to anything by 'experts.' I calculate everything myself."[10]

An independent character!

An similar statement of how crucial this quality of independence is to the development of a scientist or mathematician can be found in the memoirs of Alexander Grothendieck, considered by some to be the most "powerful and visionary" mathematician of the latter half of the 20th century. He grew up in refugee camps after World War II, and, in recounting his mathematical development, he writes:

> In those critical years I learned how to be alone.... These three years of work in isolation [1945–48], when I was thrown onto my own resources, following guidelines which I myself had spontaneously invented, instilled in me a strong degree of confidence, unassuming yet enduring, in my ability to do mathematics, which owes nothing to any consensus. ... By this I mean to say: to reach out in my own way to the things I wished to learn, rather than relying on the notions of the consensus.... It is in this gesture of "going beyond," to be something in oneself rather than the pawn of a consensus,

10. Richard Feynman, *Surely You're Joking Mr. Feynman!* (New York: Norton, 1997), 247–50. (See p. xxvii, note 5)

> the refusal to stay within a rigid circle that others have drawn around one — it is in this solitary act that one finds true creativity. All other things follow as a matter of course.[11]

It is not surprising, then, that in the *Sicha* of Shabbos *Parshas BeHa'alosecha*, 5750 (1990), around the time of Rabbi Yisroel Aryeh Leib's birthday, the Rebbe MHM speaks of the importance of accomplishing things through one's own effort: "The foundation of serving G-d is that it be through one's own power.... As is well-known, this is the purpose of the creation of man and having the soul descend into the body ... to work and accomplish with one's own power. Through this he becomes 'a partner to G-d in the act of creation.' [The nature of a person is that he has true pleasure only when he attains a goal through his own work and effort, and not when it's in a manner of 'bread of shame.']"[12]

"Ten Years Ago"

There is another issue in the above dialogue which requires explanation. As mentioned above, Rabbi Yisroel Aryeh Leib passed away in 1952. This dialogue took place in 1975—twenty three years later. Why did the Rebbe MHM say that his brother was "dead ten years ago"? He then qualified this by saying "approximately" but 23 is not approximately 10, so there must be a deeper meaning to this.

In my opinion, this can be understood based on the story in the Torah of Yosef and his brothers. Yosef had been sold into slavery and had risen to become the Viceroy of Egypt. The famine that he had predicted came to pass and his brothers came to Egypt to buy food. He insisted that they bring their youngest brother, Binyamin, with them on their next trip but Yakov did not want to send him because he was afraid that a tragedy befall to him. "Now that he is with me," Yakov said, "I take comfort for his mother [Rochel, who had passed away] and his brother [Yosef, who had disappeared 17 years before]. But if he dies, it will seem to me as if all three of them died on the same day."[13] Binyamin was the last surviving member of the family of Yakov's wife, Rochel. As long as he was alive, Yakov felt that, in a sense, all of them were alive.

---

11. Quoted in Lee Smolin, *The Trouble with Physics* (Boston: Houghton Mifflin, 2006), 327.
12. *Sefer HaSichos* 5750, 505. The brackets are part of the original quote from the Sicha.

Ten years before the dialogue with Professor Rosenbloom, the Rebbe MHM's mother, Rebbitzen Chana Schneerson, passed away. She was the last surviving member of the Rebbe MHM's family, and perhaps as long as she was alive he felt that the whole family was alive. When she passed away in 1964, it was as if his brother, Rabbi Yisroel Aryeh Leib, passed away then as well.

There is another way in which Rebbitzen Chana's passing was also the end of Rabbi Yisroel Aryeh Leib's life. When Rabbi Yisroel Aryeh Leib passed away in 1952, Rebbitzen Chana was in very poor health. The Rebbe MHM went to great lengths to make sure that she did not find out about her son's passing as it would have affected her health adversely. He even arranged to have letters and telegrams sent to Rebbitzen Chana in Rabbi Yisroel Aryeh Leib's name from Europe so she would think that he was still alive. In this way, Rabbi Yisroel Aryeh Leib was "kept alive" until Rebbitzen Chana herself had passed away.

Intellect and Beyond

The completeness of the revelation of G-dliness in the Era of Moshiach (as opposed to other eras in history when there were great revelations of G-dliness such as the time of Shlomo HaMelech and the first Beis HaMikdosh) is that the revelation will be seen from within the world itself, as opposed to being superimposed upon it from above, as it was in the time of Shlomo HaMelech. The Rebbe, MHM emphasizes this in the *ma'amorim* published in Kislev, 5752: "It will be felt in every thing that its true being is G-dliness," i.e. that "G-d's true existence is the real being of everything that exists."[14]

The ultimate purpose of Chakira, which starts from within human intellect and natural law and works from below to above, is to reach a level of understanding G-d where it is known within intellect itself that G-d is above intellect. The Rebbe MHM explains this in a *Hadran* (review) on the Rambam as follows: "The knowledge ... that G-d is not limited to the definitions of intellect—and that this too

13. Rashi on Bereishis 44:29.

14. *Sefer HaMa'amorim Melukat* 6:54, 71. See the "Kuntresim" of 19 Kislev and Chanuka for a detailed comparison between the era of Shlomo HaMelech and the Era of Moshiach.

should be known in an intellectual manner. The intellect itself should know and understand that it is created by G-d, but that G-d is not limited by it. This is expressed in the dictum of Chakira: 'The ultimate attainment of knowledge is that we cannot know You.'"

This will be attained completely in the Era of Moshiach when "the intellect itself will be permeated and filled with the knowledge of the level of G-dliness which is higher than the limitations of intellect."[15]

15. *Toras Menachem–Hadranim al HaRambam v'Shas,* 108–11.

A Continuing Inspiration

When the Previous Rebbe visited Israel in 1929, he visited the graves of several great tzadikim. This, in fact, was the purpose of his visit. When he came to the grave of the Rambam, he stayed an unusually long time. Afterwards he commented on the verse: "And the living shall take to heart," explaining it to mean that one should live his life in such a way that even afterwards he be a *Chai*—one who lives.[1]

In this section we will look at some of the ways in which Rabbi Yisroel Aryeh Leib's life continues to be an inspiration for Jews in general and for Jewish scientists in particular.

Over the years, the Rebbe MHM has said many *sichos* and *ma'amorim* (Chassidic discourses) on the yartzeit of Rabbi Yisroel Aryeh Leib, 13 Iyar. On one such occasion he applied the verse *"echod b'echod yigoshu"* (one to one they approach)[2]— to the yartzeit of Rabbi Yisroel Aryeh Leib. The numerical value of the Hebrew word *echod* is 13. The verse is usually invoked to compare the yartzeit of the Tzemach Tzedek (the 3rd Lubavitcher Rebbe) on 13 Nisan to the yartzeit of his son, Rabbi Shmuel, (the 4th Rebbe) on 13 Tishrei. This shows us the importance of Rabbi Yisroel Aryeh Leib's yartzeit.

In recent years, however, the Rebbe MHM put even more emphasis on the yartzeit by publishing *ma'amorim* for the occasion under the title *"Kuntres Yud Gimmel Iyar"* (the booklet for the 13th of Iyar), a distinction reserved for the most special days on the Chabad calendar.

1. Diary of the Previous Rebbe printed in *Challenge—An Encounter with Lubavitch-Chabad in Israel* (New York: Kehot Publication Society, 1998).
2. Iyov 41:8.

At the Shabbos farbrengen on 13 Iyar, 5751 (1991), Rabbi Yisroel Aryeh Leib's 39th yartzeit, the Rebbe MHM said the famous *"Gola to Geula" sicha*[3] in which he defined Geula as the transformation of Golus (exile), a transformation which is accomplished by "adding the Alef" to Gola.[4] In the same *sicha* he discusses the deeper meaning of the name "Yisroel Aryeh Leib" and the instruction that it holds for every Jew:

The name Yisroel has two opposite meanings: 1) A Kabbalistic meaning interprets the Hebrew letters of *"YISROEL"* as an acrostic for the phrase *"Yesh Shishim Ribu Osiyos LaTorah,"* meaning "There are 600,000 letters in the Torah." This refers to the fact that every Jew has a deep personal connection with the Torah which raises him above the level of the mundane. 2) The original meaning of the name, as it was given to Yaakov after he wrestled with the angel: "You struggled with G-dly forces and with men and you were victorious." This indicates that a Jew must deal with, and sometimes even struggle with, the world.

The message is that a Jew's natural place is in the Torah, which is the source of his life. Thus everything he does must be based on and in accordance with the teachings of the Torah. But his completeness as a Jew is attained only when he takes the teachings of the Torah out into the world and deals directly with the people of the world and the forces of nature, affecting a transformation of the world into a place where G-d is revealed. When a Jew follows this path, he is guaranteed that he will be successful and victorious.

The names "Aryeh" and "Leib" both mean "lion"—Aryeh in Hebrew and Leib in Yiddish. This teaches us that we must approach the task with the courage of a lion. Whether we are dealing with holy matters (as indicated by "Aryeh" in Hebrew, the holy language) or with worldly matters (as indicated by "Leib" in Yiddish, a worldly language) a Jew must overcome all obstacles to accomplish this mission.

Thus Rabbi Yisroel Aryeh Leib's very name continues to be an inspiration.

Two days after Rabbi Yisroel Aryeh Leib's *yartzeit* that year, the Rebbe MHM distributed the famous *Dvar Malchus* booklet contain-

3. *Sefer HaSichos* 5751 2:504-19.
4. See the next section, "The 40th Yartzeit," for a review of this sicha.

ing explanations on the laws of Moshiach in Rambam. He called this booklet itself a *ma'amar*. It was dedicated to the memory of Rabbi Yisroel Aryeh Leib. Another famous event took place that day: After the *davening*, as the Rebbe MHM began walking out of the Beis HaMedrash, everyone started singing *"Yechi Adoneinu Moreinu V'Rabeinu Melech HaMoshiach L'Olam Voed!"* ("Long live our master, teacher and Rebbe, Melech HaMoshiach forever") and he encouraged their singing.[5]

My Story

At this point I feel that it is important for me to relate the effect that Rabbi Yisroel Aryeh Leib has had on me personally which, again, is part of his continuing inspiration.

A Chosid prays at the tziyun of Rabbi Yisroel Aryeh Leib.

I was born into a rabbinical family. My father, a Rav in Minnesota, was a *musmach* of (i.e. received Rabbinical ordination from) HaRav Moshe Feinstein. My grandfather, a Rav in Uzda, Russia—Rav Moshe Feinstein's home town—was a close friend of the Feinstein family. It was in this atmosphere that I was raised. I went to Jewish day schools in Minnesota and Chicago and then to Yeshiva University High School in New York City.

Every yeshiva has its particular *hashkafa* (philosophical outlook), but at Yeshiva University (YU) this was hard to define. Academically, YU maintained a very high standard in both Talmudic and secular studies. Philosophically, however, there was a great tension between these two worlds, and they often collided. In those days, the Rosh Yeshiva, HaRav Joseph B. Soloveitchik, wrote papers and delivered lectures on these issues, but there was a significant gap between what he taught and the policies of administrators and teachers,

5. WLCC, video of 15 Iyar, 5751.

which was often a matter of "jumping back and forth between the two paths" of the holy and the secular.

I remember a discussion that took place when I was 12 or 13 years old, during my freshman year in yeshiva. Two of my friends, Moshe and Chaim, were arguing about what *Torah* is. Moshe said that all knowledge—even the secular studies—is "Torah" since all knowledge is truth (so he claimed) and Torah is truth. Chaim disagreed, claiming that only the Torah itself is Torah while all other knowledge is not necessarily true. I listened to their argument thinking, "This is a very interesting question. I wonder what the real answer is. Anyway, I'm glad they're not asking me...." Suddenly Moshe said to Chaim, "Let's ask Shimon. He's in a higher *shiur* (Talmud class) than both of us. For sure he knows." I was on the spot and I had to think fast. Since I didn't know the real answer, I came up with an answer to satisfy both of them and that put the argument to rest. But I was not satisfied.

After yeshiva, I returned home to Minnesota and enrolled in the University of Minnesota's Institute of Technology with a major in physics. It was clear to me that Torah was *the* source of truth, but I was interested in investigating whether science could also be a source of truth. Here too, at the University of Minnesota, a casual conversation that took place one day became a turning point in this search.

I was sitting at a table in the physics library studying when a fellow student turned to me and asked me to explain to him a Chemistry lab assignment that we were supposed to do. It wasn't the procedure that he didn't understand, but rather the concepts behind it—what was going on among the atoms involved. I began to explain to him the atomic interactions that were taking place in that experiment. As I spoke a feeling of excitement came over me. The intricacies of the structure of the atoms and the complexities of the interactions that I was describing literally filled me with awe and at the end of my explanation I blurted out, "And some people think that all this just evolved!" I did not know what my classmate himself believed or how he would react to my pronouncement, but the smile on his face assured me that he agreed that it was absurd that all this had just evolved.

This casual event was a turning point for me in my search, for it showed me that science itself could be a source of knowledge of G-d. If used properly, it could be a support for the truth of the Torah and there would no longer have to be any conflict between them. A seed was planted that would yet grow.

At this point I had not yet studied Chabad Chassidus. In fact, I had been part of a circle of students at YU who were involved with Lubavitch and would go to Crown Heights occasionally, and a few times I tried to read the *seforim* of Chassidus. But, even though I was well-trained in Talmud and Halacha, I couldn't understand the *seforim* of Chassidus.

One of my trips to Crown Heights was for the holiday of Sukkos 5727 (1966). I was staying at the home of Rabbi Moshe Horowitz *z.l.* (the Bostoner Rebbe) on President Street with some friends from YU. One day, during *Chol HaMoed*, I went for a walk. I was walking absentmindedly down President Street when suddenly I came upon an unusual scene. I saw a man get out of a parked car, open the back door of the car and continue to hold the door open as he stood there, straight as a soldier. For some reason the man was staring straight at me as though he wanted me to get in the car. Since I didn't know who he was and the entire scene appeared very strange, I was scared and froze in my tracks.

I stood there for a few moments trying to decide if I should continue walking or turn around and go the other way. Just as I was about to turn around, the Rebbe MHM walked out of the driveway and, as he passed very close in front of me, looked at me—deeply into my eyes. He continued walking to the car and got into the back seat, and the car drove off. This event made a very strong impression on me which I can still feel forty years later! I was connected.

My first encounter with Chabad Chassidus came about four years later while I was a student at the Technion—Israel Institute of Technology in Haifa. The well-known chosid, Rabbi Reuven Dunin, who lived in the area, used to come to the Technion and talk to the students. Some of us would go to his house for Shabbos, and he would learn a *sicha* with us after the Shabbos dinner. Once he began to doze off as he taught the *sicha* (he worked very hard as a tractor driver all week). I read the next few words out loud to wake him up, but he just said, "Shimon, *tamshich* (continue)," and fell into a deep sleep. Me?! I didn't even know how to learn Chassidus, much less how to teach it. But he had fallen asleep, so I had no choice but to continue. So with G-d's help, I taught. As I read the *sicha* out loud, it became very clear to me. I understood it and was actually able to explain it. The gates of Chassidus had been opened for me.

I continued to learn Chassidus over the coming years and began to identify with it, but all this just increased the tension between my profession and my identity. The more I studied the deep wisdom of Chassidus, the less interested I was in the lower wisdom of mathematics. For me, these were two separate worlds, and they were in collision. Many times I considered giving up mathematics and devoting myself exclusively to learning and teaching Chassidus. Then came my *yechidus* (private audience) with the Rebbe MHM.

A few months after we were married, my wife and I had the wonderful privilege of having *yechidus* with the Rebbe MHM. It was the night after Chanukah, 5734 (1973) at 3:40 a.m. when we entered the *Kodesh HaKodoshim*—the *yechidus* room. I gave our letter to the Rebbe MHM. He read it and then without looking up he asked me, "What is your profession?" Thoughts began to flash through my mind at lightning speed. Firstly, the question took me by surprise. I thought he would ask me questions about what I was learning in Chassidus and things like that. Secondly, I didn't think of myself as actually having a profession at that time. I was a graduate student in mathematics at the University of Minnesota and simultaneously teaching undergraduate courses in mathematics (a position known as a "teaching assistant") but I just thought of that as my studies, not my profession. Thirdly, the assistantship in mathematics did not pay enough to support a family, and I was considering accepting a position I had just been offered to teach Jewish studies part-time at a Jewish school in Minnesota. Maybe that was my profession. (Here, again, the tension between the two worlds appeared.) But I didn't have much time to think and I responded, "I'm a teaching assistant in mathematics at the university."

There is a well-known saying of Rabbi Shmuel of Lubavitch (the fourth Rebbe of Chabad) that the intellectual and emotional struc-

6. *HaYom Yom*, 20 Sivan.

ture of a chosid is established in his first *yechidus*, at which time his personal service of G-d is established in accordance with his true self.[6] Thus it was established in my *yechidus* that my profession would remain teaching mathematics.

But the conflict was not yet resolved. Having identified with Chassidus, I became very critical of mathematics. I began to challenge the use of mathematics as a representation of reality and questioned the very validity of mathematics itself.

(Actually, this critical attack on mathematics proved productive in my profession, as well, as I came to realize the role that the axiomatic method plays in mathematics. Every mathematical theory must be based on certain axioms or assumptions, and those axioms can serve as an interface, or a buffer, between the theory and reality. One can proceed to develop the theory with a clear conscience, knowing that the theory is not about reality itself but rather about the axioms. It's about the assumptions that he made about reality. Based on these assumptions, various conclusions will follow. It is only at the point of stating the axioms that one must be concerned about the extent to which his assumptions accurately represent reality.[7])

It was some years later that I heard about Rabbi Yisroel Aryeh Leib. This happened at a small farbrengen in Minnesota where Rabbi Moshe Feller was talking about a trip to Israel from which he had just returned. He mentioned that he had visited the grave of Rabbi Yisroel Aryeh Leib in Tzefas and continued to speak about him. I was fascinated by this personality who was devoted to the study of Chassidus throughout his life and simultaneously studied and did research in mathematics, having written a major mathematical paper. Later, when I actually saw a copy of the paper, I was overcome by a feeling of joy and exhilaration followed by a feeling of inner peace. The conflict was essentially over. The very fact that such a person had existed, and that he was the son of Rabbi Levi Yitzchok and the brother of the Rebbe MHM, was for me a sign that the two worlds could truly be unified.

The actual unification took place some time later when I saw the

---

7. See chap. I, sec. 3, "Wisdom From Below," where this is discussed at length.

8. *Likutei Sichos* 15:42-48. See the Introduction for a summary and discussion of the sicha.

famous *Sicha of Noach* in which the Rebbe MHM explains that the scientific theories and discoveries of the past 150 years were predicted by the Zohar and they are a prelude to the Era of Moshiach.[8] In the *sicha* the Rebbe MHM explains in detail exactly how the modern sciences prepare the world for Moshiach, the primary way being that they reveal the unity that exists within nature. This prepares the world for the revelation of the ultimate unity—the absolute unity of G-d which is to be revealed in the Era of Moshiach.

Not for Everyone

In case one may have the idea to go college to study the sciences, the Rebbe MHM warns against this. An entire section of the *Sicha* of Noach is devoted to explaining the halachic prohibition and danger in going to college due to the prevalence of ideas studied in college courses that are contrary to the fundamental beliefs of Judaism, and because of the atmosphere of immorality that prevails there.

Elsewhere, the Rebbe MHM explains a more subtle problem in taking college courses together with studying Torah. The concepts, attitudes and ways of thinking which one acquires from the college courses will carry over to his learning of Torah and contaminate it.[9] He will no longer be studying Torah on its own terms, but rather he will be approaching it from a secular viewpoint. This may even result in his interpreting concepts in the Torah in such a way to make them consistent with his secular studies. Unfortunately, I saw many examples of this at Yeshiva University.

Indeed, the Rebbe MHM says that the endeavor of using the sciences for Torah purposes is not for everyone, only for those who are qualified to do it. One must be sure that he is elevating the science rather than having the science pull him down.[10]

9. G-dly intellect and human intellect are fundamentally different. See chap. 1, sec. 1, "Faith and Intellect."
10. *Sefer HaMa'amorim Melukat* 3:60.

The Fortieth Yartzeit

In 1990–91 the Rebbe MHM announced that the time of the Geula had arrived and that everything in the world had to be connected with Moshiach and the Geula. One day, in the fall of 1991, I found a letter of the Rebbe MHM printed in *Likutei Sichos* in which he was responding to someone who had identified himself as a "secular Jew." The entire letter was devoted to explaining that there is no such thing as a secular Jew. There are Jews whose professions deal only with holy matters, and there are Jews whose professions deal with secular matters. But there is no such thing as a *secular Jew*—all Jews are holy.[1] This was exactly my issue. Then I noticed that the letter was dated 13 Iyar, 5732 (1972)—Rabbi Yisroel Aryeh Leib's 20th yartzeit! I realized that in the coming spring, on 13 Iyar, 5752 (1992), we would mark his 40th yartzeit. A wonderful idea then occurred to me to hold a conference honoring his 40th yartzeit on exactly these topics—the holy and the secular in the Era of Moshiach.

When the concept was finally crystallized, the theme of the conference was "The Role of Science in the Era of Moshiach." I worked together with Rabbi Moshe Feller, the director of Merkos-Lubavitch in Minnesota, to organize the conference, which took place on 13–14 Iyar, 5752, in S. Paul, Minnesota.

The guest speakers were Professor Pesach (Paul) Rosenbloom, the famous Lubavitcher mathematics professor who edited Rabbi Yisroel Aryeh Leib's mathematical paper,[2] and Dr. Tsvi Saks, another well-known Lubavitcher mathematician famous for his lectures on mathematics and Chassidus.

1. The letter is now printed in *Igros Kodesh* 27:119–21.
2. See sec. 2 of this chapter.

Holiness From Below...Holiness From Above

The first session of the conference took place on the yartzeit itself, Shabbos afternoon, 13 Iyar, at a *Seudah Shlishis* (the third Shabbos dinner) in the Adath Israel Shul. I reviewed the *ma'amar* that the Rebbe MHM published for Rabbi Yisroel Aryeh Leib's 38th yartzeit, two years earlier, "Kuntres 13 Iyar, 5750."[3]

This *ma'amar* starts with a verse from *Parshas Emor* which speaks about holiness. (In fact, most of the discourses that the Rebbe MHM said on the annual observances of Rabbi Yisroel Aryeh Leib's yartzeit begin with a verse which speaks of holiness.) The revelation of G-d's holiness from below in the heart of the Jew who is yearning for holiness, draws down a higher level of holiness from G-d above. This can be accomplished through the study of Torah which brings holiness into the inner, spiritual dimension of the universe, and through the observance of the Mitzvos which brings holiness into the physical aspects of the universe. Thus there is a triple thread of holiness: 1) the holiness from below; 2) the holiness from above brought down through Torah; and 3) the holiness from above brought down through Mitzvos.

This is symbolized, the *ma'amar* continues, by a parable in the Midrash of a king whose subjects make three crowns for him. He responds by putting one on his own head and the other two on the heads of his sons.

Indeed, this was the point of the Rebbe MHM's letter of 13 Iyar, Rabbi Yisroel Aryeh Leib's yartzeit, 20 years earlier: Every Jew is holy. There is no such thing as a secular Jew.

After I reviewed the *ma'amar*, the Rabbi of the Shul, Rabbi Asher Zeilingold, reviewed the *sicha* that the Rebbe MHM had said the previous year on Rabbi Yisroel Aryeh Leib's 39th yartzeit,[4] focusing especially the deeper meaning of the name Yisroel Aryeh Leib, which we discussed in the previous section.

This was the famous *"Gola to Geula" sicha* in which the Rebbe MHM explained that the Hebrew word for the Redemption—

3. *Sefer HaMa'amorim Melukat* 4:245–48.
4. *Sefer HaSichos 5751* 2:504–19.

Geula—is almost identical to the Hebrew word for exile—*Gola*—except that Geula has the Hebrew letter *alef* (א) added: Geula = Gola + א. The name of this letter—"alef"—means "master," and in this context it refers to the Master of the Universe, G-d. The message here is that Geula is attained by putting the *alef* into Gola i.e. by revealing the presence of G-d and His mastery of the universe while the world is still in a state of exile. This transforms the world and brings it to a state of Geula. Thus Geula does not involve a totally radical innovation in the world. It is simply a matter of bringing out the truth in the world as we know it.

At the end of the *sicha* the Rebbe MHM talked about how the U.S. Air Force was used to bring food and medicine to victims of disasters in countries far away from the U.S. that have no direct connection with this country. The sole motivation for this assistance was goodness and kindness. This is an example of how the nations of the world have begun to change direction in international policy from militaristic to peaceful. The next year, the Rebbe MHM said that this process had advanced much further and that the prophecy of Isaiah that the nations of the world will "beat their swords into plowshares" had begun to be fulfilled. We discuss this at great length in chapter III.

Professor Rosenbloom then spoke about the connection between the yartzeit of Rabbi Yisroel Aryeh Leib and the Torah reading of that week, *Parshas Emor*, along with the weekly chapter of *Pirkei Avos*, chapter 3. This chapter of *Pirkei Avos* ends with a statement of Rabbi Eliezer Chisma contrasting mathematical calculations which are used for Torah with the actual Halachic decisions of the Torah.

Dr. Saks then reviewed the "Kuntres of 22 Shevat," 5749 (1989), a *ma'amar* in which the Rebbe MHM discusses the transformation of science into Torah. This can happen in two ways—on two levels. The lower level is where one uses the scientific knowledge which he has in order to increase his knowledge of Torah. In this case the scientific knowledge becomes an intermediary to aid in his understanding of the Torah. This extends the influence of Torah into the world and brings the Geula closer. There is an even higher level where the science actually becomes part of Torah, such as when Rambam uses Greek mathematics and astronomy to explain issues in the Laws of *Kiddush HaChodesh* (determining the new months and years). Rambam uses the actual calculations of the Greek mathematicians to decide matters of Torah law. He thereby refined these Greek sciences, elevated them and transformed them into Torah. The mathematics is no longer an intermediary through which our understanding of Torah is enhanced but actually becomes a part of Torah as they are now part of Rambam's book on Jewish Law, Mishneh Torah.[5]

Thus there is mathematics which is used for Torah, and mathematics which becomes part of Torah. Generally speaking, these are the two levels that Rabbi Eliezer Chisma speaks of in the mishneh from *Pirkei Avos* that Prof. Rosenbloom discussed.

The second session of the conference was a *Melaveh Malka* (dinner following the conclusion of Shabbos) at the home of Gershon Gitter, a Lubavitcher engineer. At a Melaveh Malka, which is a dinner connected with Moshiach, it is customary to tell Chassidic stories, especially of tzadikim. Dr. Saks told some chassidic stories, and I reviewed Rabbi Yisroel Aryeh Leib's biography and told stories about him, his father, Rabbi Levi Yitzchok, and his mother, Rebbitzen Chana. Some of these stories appear in the earlier biographical sections.

On the next day, Sunday, 14 Iyar, the third session took place at the local Lubavitch House. Professor Rosenbloom discussed Rabbi Yisroel Aryeh Leib's mathematical work and related the story of the manuscript of Rabbi Yisroel Aryeh Leib's mathematical paper that the Rebbe MHM had sent him to prepare for publication. Dr. Saks delivered a lecture on the concept of Infinity in Mathematics and Chassidus. The content of his lecture is contained in his paper in chapter V of this book. Rabbi Moshe Feller spoke about Prof. Rosenbloom's special relationship with the Rebbe MHM and Rabbi Manis Friedman reviewed a ma'amar of the Mittler Rebbe on the refinement of the sciences.

A *Siyum HaRambam* was also held in memory of Rabbi Yisroel Aryeh Leib. Throughout the proceedings the connection with Moshiach and Geula was emphasized, and there was an overall atmosphere of joy and elation.

The RYAL Institute

That fall, my family and I moved to Crown Heights, and the next

5. *Sefer HaMa'amorim Melukat* 2:58–60.

Spring I contemplated convening another conference for Rabbi Yisroel Aryeh Leib's 41st yartzeit. After thinking it through, I decided against it since I was new in Crown Heights and not so familiar with the people and places, etc. Also, I did not feel the same sense of urgency as I did at the 40th yartzeit—a major milestone.

Just to be sure, however, I wrote a letter to the Rebbe MHM asking: 1) Should I make a Moshiach and Science conference in New York similar to the one I had made in Minnesota the year before? and 2) Should I publish the proceedings of the conference of the previous year? To both questions he answered "Yes."

To carry out the Rebbe MHM's first instruction, I organized a second Moshiach and Science conference which took place on 11–13 Iyar, 5753 (1993) in Crown Heights. To carry out the second instruction, I am writing this book.[6]

The main speaker at the 1993 conference was Rabbi Dr. Naftali Berg, a Lubavitcher chosid and director of research of the Advanced Technology Office of the Pentagon's Army Research Laboratory. He spoke about "Swords Into Plowshares at the Pentagon" which appears in chapter III.

On the morning of 11 Iyar that year, before I went to the conference, it occurred to me that there should be an institute to organize such conferences every year and to carry out other Moshiach and science research. I wrote a letter to the Rebbe MHM that morning asking if we should establish such an institute to be called the "Rabbi Yisroel Aryeh Leib Research Institute on Moshiach and the Sciences." Later in the afternoon, towards the end of that day's session, I received a message that that the Rebbe MHM gave his consent and blessing.

Thus the RYAL Institute was established, and with G-d's help we have held a Moshiach and Science conference every year since, in addition to various lectures, publications, etc.

Some of the main highlights of our conferences over the years have been:

- In 1993 and 1994, Dr. Berg delivered major lectures on

6. Many sections of the book have been published separately over the years in *Beis Moshiach* magazine and in the *Voice of Moshiach* newsletter.
7. See chap. 5, secs. 4–5.

Quantum Mechanics and String Theory.[7]

- In 1994 Dr. Alvin Radkowsky, designer of the U.S. nuclear navy, spoke about his design of a new non-proliferative nuclear reactor—"A Nuclear Reactor for the Era of Moshiach."[8]
- Professor Yirmiyahu Branover, a member of our board of directors, spoke at several conferences. In 2003 we held our conference at Touro College and Prof. Branover delivered a major lecture covering his entire philosophy of Moshiach, Torah and science as well as his own personal story.[9]
- In 1997 we held our conference at New York University (NYU). It focused on Rabbi Yisroel Aryeh Leib's mathematical paper which was presented by Dr. Neil Miller who had written a mathematical commentary on it. (RYAL's paper and Dr. Miller's commentary are reprinted in the Appendix.)
- In 1996, the RYAL Institute together with the Committee for the Geula directed by Rabbi Zimroni Tzik of Bat Yam, Israel, organized a major conference on Geula and science at Tel Aviv University. It was attended by over 250 Israeli scientists.

We have seen some of the ways in which Rabbi Yisroel Aryeh Leib

COURTESY DR. LARRY LASHER

Professor Silman putting on Tefillin with Dr. Lawrence Lasher, Pioneer Project Manager at NASA Ames Research Center, at the Moshiach and Science Conference of 2002.

---

8. See chap. III, sec. 7.
9. See chap. VI, sec. 2.

continues to inspire Jewish scientists to this very day and, as the revival of the dead is a part of the Era of Moshiach, we look forward to his immediate return. He should come back and join us and direct the work of the Institute that bears his name.

Timeline

1906—Rabbi Yisroel Aryeh Leib is born in Nikolayev, Russia.

1924—Joins the household of the Previous Lubavitcher Rebbe in Petersburg, Russia; begins serious study of mathematics and organizes Torah classes for Jewish youth.

Early 1930s—He is with the Rebbe MHM in Berlin; presumably writes his paper "Location of Eigenvalues and Eigenvectors of Complex Matrices" at this time.

Late 1930s—Goes to Israel and settles in Tel Aviv. Works in a bookstore while studying Chassidus with Rabbi Nochum Goldschmid.

1947—Receives offer to accept a position in the Department of Theoretical Physics at the University of Liverpool; receives the blessing of the Previous Rebbe for it.

1950—Moves to Liverpool, England and assumes his position at the University.

1952—Passes away on 13 Iyar and is buried in Tzefas, Israel.

1953—His paper "Self Energy of Slow Electrons in Polar Materials" is edited and published by his colleague H. Frohlich.

1975—Rebbe MHM gives manuscript of RYAL's "Location of Eigenvalues" paper to Prof. Rosenbloom to publish.

1990—Rebbe MHM publishes a *kuntres* (ma'amar) for RYAL's yartzeit.

1991—Rebbe MHM again publishes a *kuntres* for RYAL's yartzeit and publishes booklet on Rambam's Laws of Moshiach in memory of RYAL.

1992—First Moshiach and science conference is organized commemorating RYAL's 40th yartzeit.

1993—Rabbi Yisroel Aryeh Leib Research Institute on Moshiach and the Sciences is established; continues to hold annual conferences on Moshiach and science.

2010—The book "Scientific Thought in Messianic Times" is published to document proceedings of Moshiach and science conferences.

REFERENCES: CHAPTER II

Rabbi Yisroel Aryeh Leib Schneerson

Sec. 1: Man with a Mission

Gotlieb, Naftali Tzvi. *Toldos Levi Yitzchok,* 3rd ed. New York: Kehot Publication Society, 1986.

Kovetz Yud Gimel Iyar Arboim Shonah. New York: Kehot Publication Society, 1992.

Sec. 2: An Independent Character

Rabbi Levi Yitzchok Schneerson. *Likutei Levi Yitzchok - Igros Kodesh.* New York: Kehot Publication Society.

Rabbi Menachem Mendel Schneerson, MHM. *Toras Menachem Hadronim al HaRambam VeShas.* New York: Lahak Hanochos, 1992.

Gurari, Mark. "The Location of Eigenvalues and Eigenvectors of Complex Matrices," *Journal of Approximation Theory* 22 (1978): 119–49.

Smolin, Lee. *The Trouble with Physics.* Boston: Houghton Mifflin, 2006.

Sec. 3: A Continuing Inspiration

Rabbi Menachem Mendel Schneerson, MHM. *Sefer HaSichos, 5751.* New York: Kehot Publication Society.

Challenge: An Encounter with Lubavitch-Chabad in Israel. New York: Kehot Publication Society.

CHAPTER III

Swords into Plowshares: *The Trend of History*

The main innovation in the revelation of Moshiach is in that he will fix the entire world and its nations. No facet of the world will remain outside the realm of "the sovereignty shall belong to G-d."

—the Rebbe MHM

At that time there will not be any of the "iron which shortens the life of man," as it says, "They will beat their swords into plowshares." Even more than that—death itself will be annulled.... There will be everlasting life literally—the soul in the body—since the "iron which shortens the life of man" will have been annulled.

—the Rebbe MHM

And it shall be at the end of days
that the mountain of the House of G-d
will be firmly established at the head of the mountains
and will be exalted above all other hills
and all the nations shall flow to it.

And many nations shall go [to confer with each other]
and they shall say one to another,
"Let's go up to the mountain of G-d
to the house of the G-d of Yakov
and [Melech HaMoshiach] will teach us of G-d's ways
and we will follow His paths."
For out of Tziyon shall go forth Torah,
and the word of G-d from Jerusalem.

And [Melech HaMoshiach] will judge between the nations
and admonish many peoples,
And they shall beat their swords into plowshares
and their spears into pruning hooks;
nation shall not raise a sword against nation
nor shall they learn any more war.

—Yeshaya[1]

It's happening now...

An Agenda for Peace

On January 31, 1992, the heads of state of the major world powers met at the United Nations in New York City. This was the first Security Council Summit in history, i.e., the first meeting held by the Security Council at the level of heads of state.

At this meeting they issued a joint statement announcing their intention to reduce arms production and to redirect the resources to improve economic conditions throughout the world.

1. Yeshaya 2:2–4.

The next day, at the Shabbos farbrengen of *Parshas Mishpatim* in "770," the Rebbe Melech HaMoshiach announced that the statement issued by the heads of state at the United Nations was the beginning of the fulfillment of the prophecy of Isaiah, thousands of years ago, that in the Era of Moshiach the nations of the world will "beat their swords into plowshares." He explained that this declaration of intent by the world leaders was the direct result of the influence of Melech HaMoshiach himself on the nations of the world. He continued by describing the details of this influence over several decades, especially the promotion of the ideals of goodness, fairness and justice through the observance of the *Seven Noahide Commandments*, throughout the world. This brought about a refinement of the nations of the world, the climax of which was the collapse of the atheistic Communist regime in Russia and its replacement with a government committed to justice, fairness and peace based on the belief in G-d.[2] (In the next section we will present a translation of excerpts of this sicha.)

This new heightened sense of morality was described by the Secretary General of the United Nations in his report on the January 31 Security Council Summit. In this report, titled "An Agenda for Peace,"[3] he wrote: "It is possible to discern an increasingly common moral perception that spans the world's nations and peoples."

Another aspect of the Geula has begun to unfold before our eyes. Melech HaMoshiach brought the nations of the world to the point where they made a resolution and a declaration regarding world peace. This was followed by Melech HaMoshiach's own announcement that this declaration was in fact the beginning of the fulfillment of Isaiah's "Swords Into Plowshares" prophecy.

At the RYAL Institute we have investigated extensively three general areas of Swords Into Plowshares (SIP) transformations. They are: 1) Countries that have transformed significant amounts of military resources into peaceful uses; 2) Nuclear technology that has been transformed to peaceful uses; and 3) Satellite technology that has been transformed into peaceful uses. Every year, at the annual conference of the Institute, we report on "What's new in Swords Into Plowshares."

2. *Sefer HaSichos* 5752 2:362–65.
3. "An Agenda for Peace," www.un.org/Docs/SG/agpeace.html

What is "Swords Into Plowshares"?

Before beginning our reports on these areas of SIP transformation, it would be instructive to get a good definition of exactly what "Swords Into Plowshares" is. To do this, we quote the Radak (Rabbi David Kimchi), one of the best-known classical commentaries on the Torah and Prophets. In his commentary on Isaiah's prophecy, he explains that Melech HaMoshiach will be the master of all the nations; he will admonish them and tell them to "get in line." He will bring peace among the nations so they won't need their weapons any more. So they will beat them—transform them—to make them into instruments for working the land.

We see from this a few components of a SIP transformation:

1) It is orchestrated by Melech HaMoshiach himself. His influence on the nations of the world brings them to this state of affairs.

2) There is a reduced threat perception. Nations do not feel threatened by each other so they feel that they do not need so many weapons. As a result of this, there is a surplus of weapons and military resources.

3) There is noticeable arms reduction and redirection. Weapons, specifically, are removed from military use and transformed into instruments to produce food, and, more generally, military resources are redirected towards economic goals.

A Global Phenomenon

We want to emphasize that this is an ongoing process involving all technologically advanced nations of the world. We will describe many examples of this in the coming sections. This Swords Into Plowshares phenomenon has been so extensive over the past decade and a half, however, that it is impossible to document it completely. There is an institute in Bonn, Germany that has been documenting the conversion of military resources to peaceful uses extensively since the process began. The institute, called the Bonn International Center for Conversion (BICC), is an international think tank which conducts research, makes policy suggestions and provides consulting services to governments and private organizations. It collects data and information on a wide variety of SIP transformation issues and publishes various reports, most notably its annual "Conversion Survey," a book-length report on Swords Into Plowshares transformations throughout the world, as well as periodic shorter reports on specific countries and specific technologies. The institute describes itself as "an independent, non-profit organization dedicated to promoting peace and development through the efficient and effective transformation of military-related structures, assets, functions and processes." Its members "support and promote the processes by which people, skills, technology, equipment, financial and economic resources are shifted from the military or defense sector towards alternative civilian purposes."[4] The BICC reports are among the sources for this chapter of our book.

Another major work documenting the global Swords Into Plowshares transformation is the book *From Defense to Development*. Commenting on this global phenomenon, the author writes:

> If there is a single turn-of-the-century trend that gives us hope…it is the phenomenon of conflict resolution and its companions—cooperation, education and investment—in place of war, destruction and privation. The 1990s was a period of remarkable progress in peaceful transition in the world, from South Africa to the former Soviet Union.… Spending on the military plummeted almost 40% worldwide. Although tensions and a worrisome rise in militarism are still percolating in the world, our knowledge and practice of conflict resolution has expanded enormously. Almost all people, even our military leaders, would agree that investments such as clean water, agricultural productivity and, above all, educatioare far more powerful in achieving human welfare than spending on war and defense.[5]

One may find examples of military-to-peaceful-use transformations described in the many science and technology journals, newspapers and on the internet. Frequently, the authors say explicitly that the development that they are describing is an example of Swords Into Plowshares!

---

4. See www.bicc.de
5. See www.epsusa.org/publications/newsletter/2003/nov2003/markusen.htm

In addition to technologies, the armed forces of many countries have been transformed and redirected, and their role has been considerably redefined. In countries where the prospect of external aggression has been greatly diminished or is now non-existent, the traditional emphasis on self-defense as the primary function of the military has been reconsidered. Greater prominence is attached to other functions such as international peacekeeping or peace-enforcement operations, humanitarian assistance and emergency aid in cases of natural disasters, etc.

The Rebbe MHM has emphasized how the U.S. Air Force has been used to bring food and medicine to victims of such disasters in countries far away from the U.S. that have no direct connection with this country. The sole motivation for this assistance was goodness and kindness.[6]

Furthermore, the narrow concept of *national* security, embodied above all in the armed forces, has been surpassed by that of *global* security and, in tandem with this, the realization that global security cannot be achieved without seriously reducing poverty, improving health care and extending good governance throughout the world. In other words, there is a growing understanding that *plowshares* may be more powerful than *swords* in achieving global security.

Even some of the most militaristic and violent countries, such as Russia and China and the like, have come to realize that Swords Into Plowshares is the most desirable path to take. After the defeat of the regime of Sadam Hussein, a member of the new Iraqi Governing Council was interviewed by an American news network. He described how they were attempting to transform Iraq into a democratic country. In the course of his statement he listed several components that a society should manifest in to be considered democratic. Among the usual qualities such as freedom of speech, representative government, etc., he mentioned "the transformation of power for peaceful purposes."

The Swords Into Plowshares transformation is now so pervasive throughout the world that a group of physicists from the China Academy of Engineering Physics, in a paper entitled "The Practice And Progress Of Defense Conversion in China," were prompted to write, "The peaceful use of military technology represents the trend of history."

---

6. *Sefer HaSichos 5751* 2:517.

The day after the heads of state of all the nations on the U.N. Security Council made their declaration regarding a new era of peace and cooperation in which military resources would be transformed into peaceful uses, the Rebbe MHM said a sicha in which he explained that this declaration marked the beginning of the fulfillment of the prophecy of Isaiah that in the Era of Moshiach the nations of the world would "beat their swords into plowshares." This was the famous sicha of Shabbos Parshas Mishpatim, 5752. In this section we present excerpts from this sicha[1] together with comments and some discussion. (The comments are in italics.)

It should be pointed out that while Swords Into Plowshares is a fundamental part of the Geula in its own right, it is also connected with the Everlasting Life of the Era of Moshiach as the Rebbe MHM writes, "At that time there will not be any of the 'iron which shortens the life of man,' as it says, 'They will beat their swords into plowshares.' Even more than that—death itself will be annulled.[2]... There will be everlasting life literally—the soul in the body—since the 'iron which shortens the life of man' will have been annulled."[3]

The Swords Into Plowshares Sicha

The Sicha

It is well known that everything that happens in the world is through Divine Providence, especially major events connected with countries of great importance whose influence and leadership affect the whole world....

1. Our translation of the *sicha,* as well as all the translations in this book, are so-called "free translations." While we try to preserve the original language as much as possible, words or phrases may be left out and we sometimes paraphrase in order to convey the content when a direct translation may not sound as clear in English as it is in Hebrew. The reader is encouraged to look up the original sicha.

2. *Sefer HaSichos* 5752, 2:235 n. 96.

3. Ibid., 239, n. 139.

First of all, there was the event which occurred on the eve of this Shabbos—a conference of the heads of state of the major world powers, led by the leaders of the two superpowers. At that meeting they decided upon and announced a new era in international relations—an era in which warfare would be eliminated, weapons would be reduced or destroyed, and peace and unity, cooperation and mutual aid would prevail among the nations of the world—for the good of all mankind.

This event was preceded by the speech of the President of this country to the entire nation—the State of the Union Message—in which he announced a plan of action for the reduction and elimination of weapons and the reallocation of military funding toward the improvement of the economic conditions of the citizens of the country. This announcement was then approved by the Congress, the body which establishes the laws of the land (with the authority of the Torah, in accordance with the rule "the law of the government is the law")....

One of the prophecies of the Geula by Moshiach regarding the conduct of the nations of the world is: "They shall beat their swords into plowshares and their spears into pruning hooks. Nation shall not raise a sword against nation nor shall they learn any more war."

This will be brought about through the actions of Melech HaMoshiach himself as it says at the beginning of that verse, "And he will judge between the nations and admonish many peoples." As the Radak explains, the "judge" is Melech HaMoshiach who will be the master of all the nations. He will admonish them and tell anyone who is doing an injustice to straighten out the inequity. Because of this there will be no war between nations for he will make peace between them. Therefore they won't need weapons and they will refashion them to make tools to work the ground.

The Rebbe MHM then quotes the beginning of the prophecy with the commentary of the Radak:

"And it shall be at the end of days that the mountain of the House of G-d will be firmly established...and all the nations shall flow to it. And many nations shall go and they shall say one to another, 'Let's go up to the mountain of G-d to the house of the G-d of Yakov, and he will teach us of G-d's ways and we will follow His paths.' For out of Tziyon shall go forth Torah and the word of G-d from Jerusalem."

The Radak explains that the "teacher" will be Melech HaMoshiach and it is regarding him that the verse says, "He will judge."

Since these days the heads of state of the nations of the world are making decisions and announcements regarding the reduction and elimination of weapons and the increase in things required for the maintenance of the economies of their individual countries and the world as a whole, which is the content of the prophecy "They shall beat their swords into plowshares"—dismantling the instruments of war to construct from them instruments for working the ground, "the earth from whence comes forth bread"—this is a clear sign of the beginning of the fulfillment of this prophecy as part of the true and complete Geula by Moshiach."

In the last section we presented a 3-point definition of Swords Into Plowshares. The 3rd point was:

Arms reduction and redirection. Weapons, specifically, are removed from military use and transformed into instruments to produce food, and, more generally, military resources are redirected towards economic goals.

A careful reading of this sicha and consideration of the examples given indicates, I believe, that the 3rd criterion itself can occur on three levels:

1) *"Reduction and elimination of weapons and the increase in elements required for the maintenance of the economies of their individual countries and the world as a whole"—i.e., the nations initiate dual (two-fold) programs of arms reduction and, simultaneously, an increase in advancing economic growth. This was the case in the UN declaration.*

2) *Arms reduction where the money saved would be redirected to advance economic growth as in the case of the U.S. president's State of the Union Message.*

3) *"Dismantling the instruments of war to construct from them instruments for working the ground" i.e., the instruments of war themselves—the weapons and their technology—are transformed into instruments for economic benefit.*

From what we have seen so far in the sicha, we note that the Swords Into Plowshares process is the result of the influence of Melech

HaMoshiach on the world and that this process has already begun. The Rebbe MHM is clearly indicating that he is Melech HaMoshiach. This becomes more explicit as the sicha continues:

To explain further: Since we are now at the climax of the coming of Moshiach, we see the beginning of the effect of Melech HaMoshiach on the nations[4] ("And he will judge between the nations and admonish many peoples, and they shall beat their swords into plowshares") by the fact that G-d put the idea into the hearts of the kings of the nations[5] to come together and make a decision and an announcement regarding a new state of affairs—Swords Into Plowshares....

The Rebbe MHM now gives the background and the details of the influence of Melech HaMoshiach on the nations of the world:

When the Previous Rebbe came to this part of the world, America became the point of origin for the project of spreading Torah and Chassidus. From here it was spread to the ends of the earth by the *shluchim* (representatives) that he sent throughout the world, reaching even to the most far-flung corners of the world, with the goal of spreading Torah and Judaism among all Jews and, of primary significance, to spread the concepts of goodness, fairness and justice among the nations of the world by their observing the Seven Noachide Laws. Extreme emphasis has been put on this in recent years as we have been getting closer to the time of the coming of Moshiach when, G-d says, "I will transform all the nations that they make a clear statement, all calling in the name of G-d and serving Him uniformly."

One of the most significant ways in which "the concepts of goodness, fairness and justice among the nations of the world by their observing the Seven Noahide Laws" has been spread among the nations is by presidential proclamations establishing "Education Day U.S.A." For example, in 1982 President Ronald Reagan wrote, in part, "The Lubavitcher Rebbe ...has provided a vivid example of the eternal validity of the Seven Noahide Laws, a moral code for all of us regardless of religious faith."

This is documented extensively in a special section in the Appendix of this book.

4. The italics are in the original sicha.

5. The Rebbe MHM refers here to the saying "The hearts of kings and officers are in the hand of G-d," based on Proverbs 21:1.

Recently this project reached a climax with the opening of the borders of Russia, a country which, in its time, "engraved on its banner" a declaration of war on the spreading of Torah and Judaism, and even a war on the belief in G-d, Creator and Master of the Universe. This included the arrest of the Previous Rebbe, resulting in his leaving Russia. Nevertheless, the activities of his *shluchim* in spreading Torah and Judaism continued secretly with great self-sacrifice until recently when this dictate of the government was cancelled and it became possible to continue the activities of spreading Torah and Judaism openly, with even greater strength. Finally the previous regime fell and a new regime was established which proclaimed fairness, justice and peace, based on belief in the Creator and Master of the Universe.

The conference which took place yesterday between the new leader of Russia and the president of this country with the participation of the heads of state of several more powerful nations was a continuation of this process. And the content of their proclamation was "Swords Into Plowshares."

Now, the fact that the conference and the proclamation occurred in the city of the Rebbe indicates that all this is the result of the activities of spreading Torah and Judaism, fairness and justice throughout the world, which has been done—and continues to be done—by the Rebbe, the Moshiach of the generation. Through the completion of this work in our time, the world has become all set for the commencement of the action of Melech HaMoshiach that "He will judge between the nations and admonish many peoples, and they shall beat their swords into plowshares."

It is noteworthy that the United Nations, whose aim is to bring unity and peace among the nations, was established in America shortly after the *Nasi* came to America. The decision to set it up was first made during the war around 5702 (1942),[6] then, more significantly, at the end of the war, around 5705 (1945).[7] Its location was

6. The "Declaration by United Nations" was made on January 1, 1942, when representatives of 26 nations pledged their governments to continue fighting together against the Axis powers.

7. In 1945, representatives of 50 countries met in San Francisco at the United Nations Conference on International Organization to draw up the United Nations Charter. It was signed on June 26, 1945.

set to be in the city of the *Nasi* [New York] and its building was constructed around 5711 (1951)—emphasizing that the unity and peace among nations is the result of the activities of the *Nasi* in improving the world by spreading Torah and Judaism, fairness and justice throughout the world. [*1951 marked the beginning of the Nesius of the Rebbe MHM.*] The more that the world is improved through the *Nasi*, the more established this organization becomes.[8] The completion of this process is attained with the fulfillment of the prophecy "they shall beat their swords into plowshares" which is the foundation of this organization. This verse is even engraved on a wall of its building [*the "Isaiah Wall"*]....

The Isaiah Wall at the United Nations

Even in the timing of the declaration we see the connection to the *Nasi*. It took place on the eve of the last Shabbos of the month of Shevat, the 10th day of which is the day of the passing of the Previous Rebbe, and the 11th day of which marked a renewal and continuation of his *Nesius* with even more power [*by the Rebbe MHM whose Nesius began on that day*] for 42 years, reaching the completion of the work of our generation—the last generation of the Golus which is also the first generation of the Geula.

8. But there is also darkness and falsehood at the U.N. Prime Minister Benjamin Netanyahu reported that in 1984, when he came to the U.N. as the ambassador from Israel, the Rebbe MHM told him that he was going to a "house of darkness"—a place where there is darkness and falsehood, and that Netanyahu should light a "candle of truth" there.

The Logic of Peace

This event took place in the week of Parshas Mishpatim and there is a connection between them....

Mishpatim are laws between man and man that can be understood by human intellect [*not to kill, not to steal etc.*]....The reason these laws are mentioned in the Torah right after the story of the revelation at Mt. Sinai is to tell us that even though these laws are perfectly logical, they must be fulfilled not based on their rationality but because they are commanded by G-d—they are "from Sinai."

On a deeper level, the reason these laws are rational in the first place is "from Sinai." The intellectual imperative that they carry is a result of the fact that, for these laws, G-d's supra-rational will was expressed also through His wisdom in the Torah, and since the Torah is the "blueprint" the creation of the universe,[9] these laws became a necessity according to the definitions of the created universe and thus an imperative in human intellect [*which is part of the order of creation*].[10] Thus the human mind is inclined toward ideas that are expressed through G-d's wisdom. Had the Mishpatim not been given in this manner from Sinai, they would not make sense to the human mind....

This applies to the nations of the world also with respect to the fulfillment of the Seven Noachide Laws, as Rambam writes that even though the human mind is inclined toward accepting these laws, "One must accept them and fulfill them because G-d commanded them in the Torah. If, however, one fulfills them based only on an intellectual decision, he is not considered to be of the righteous of the nations of the world nor even of their scholars."

The ultimate in this will be achieved in the Era of Moshiach when, as Rambam writes, "The entire world (including the non-Jewish nations of the world) will be concerned only with the knowledge of G-d, as it says, 'the world will be filled with the knowledge of G-d,'" meaning that the knowledge of G-d (even the abstract Kabbalistic concepts) will pervade human intellect....

"Beating swords into plowshares" is something that is understood

9. The Midrash *Bereishis Rabba* starts by saying that G-d used the Torah as his plan for creating the universe, just like a builder uses a blueprint to construct a building.

10. See chap. I, sec. 1, where this is discussed in the context of Faith and Intellect.

and an imperative even according to human intellect (*Mishpatim*) because man's intellect dictates that that the world be run based on fairness and justice, by rejecting war which brings destruction and desolation, G-d forbid. This should bring about the dismantlement of weapons of war ("beat their swords") and their transformation into instruments that benefit civilization ("plowshares"). Nevertheless, throughout all the generations many wars were waged among the nations of the world, causing destruction and desolation—in direct contradiction of the dictates of human intellect!

So it must be that the real reason that recently we can discern a drive to put an end to the era of wars and to begin a new era of running the world based on fairness and justice, peace and unity—as emphasized in the recent declaration by the heads of state—is not just because of the intellectual imperative, which was always there, but primarily because we are nearing the time of "swords into plowshares," as declared by the Torah ("from Sinai").

Based on this, we can say that the decision and declaration of the heads of state on the eve of Shabbos Parshas Mishpatim regarding the state of affairs of "Swords Into Plowshares" is the result of the decision and declaration of the Rabbinical authorities (*"Malchei Rabonan"*[11]) that "Behold, here comes Melech HaMoshiach."[12] This began with the *psak din* of the Previous Rebbe that for some time already our work in Golus has been complete and we are already prepared to receive Moshiach, as well as the [recent] *psak din* of Rabbinical authorities that the time of the Geula has arrived[13]—that, as Rambam writes, "A king will arise from the House of Dovid…he has the assumed status of being Moshiach (*chezkas Moshiach*), reaching the status of being Moshiach with a certainty (*Moshiach vadai*)—a *psak din* "from Sinai" that has pervaded the defining characteristics of the world to the extent that the heads of state are deciding on their own and making a declaration in their legal chambers (which have legal standing following the Halachic principle of "the law of the government is the law") regarding the state of affairs of Swords Into Plowshares….

May it be the will of G-d that the true and complete Geula by Moshiach should come in its full actuality, immediately….So be it

11. Literally, "Rabbis who are kings."
12. Shir HaShirim 2:8 as explained by the Midrash *Shir HaShirim Rabba*.

for us in actuality *teikef u'miyad mamosh* (= immediately, literally), with all the connotations of the word *miyad*. This includes its being an acrostic—MYD representing three tzadikim throughout the generations connected with Moshiach: Moshe Rabbeinu, Yisroel Baal Shem Tov and Dovid Malka Mishicha.[14] More specifically, with respect to our own generation, MYD alludes to the three eras related to the Previous Rebbe, starting with the one closest to us: Moshiach, whose name is Menachem (*the Rebbe MHM, successor of the Previous Rebbe*); Yosef Yitzchok (*the Previous Rebbe himself*); DovBer (*the middle name of Rabbi Shalom DovBer, the father and predecessor of the Previous Rebbe*).

Similarly, it should be so with all the connotations of the word *mamosh*,[15] first and foremost being that it should happen immediately, literally in the plain sense of these words.

In summary, it appears that there were four factors in the Swords Into Plowshares declaration:

1) *The spreading of Torah and Chassidus started by the Previous Rebbe and reaching its climax with the work of the Rebbe MHM which made the world ready for it.*
2) *The direct effect of the Rebbe MHM on the nations in providing judgment and instruction to them, especially by promoting the concepts of goodness, fairness, justice and the Seven Noachide Laws.*
3) *Once the Rebbe MHM brought the world to a high state of refinement, G-d Himself put it in the minds of the heads of state to make the declaration.*
4) *The* piskei din *of the Rabbinic authorities had the effect of turning the ideal of Swords Into Plowshares into a logical, sensible reality.*

---

13. This declaration was initially made by the Rebbe MHM himself, then followed up by *piskei din* of Rabbinic authorities worldwide. See chap. I, sec. 5 for a full documentation of this including an image of one of these *piskei din*.

14. *Dovid Malka Meshicha* is a title which refers to both King David and to Melech HaMoshiach.

15. The Rebbe MHM gives no detail here. One meaning generally understood is that it is an acrostic for the Rebbe MHM's own initials: Menachem Mendel Schneerson.

Nations

The Swords of Russia

Russia and China are lands that the Rebbe MHM has mentioned explicitly as examples of countries where major political change has taken place in a peaceful, nonviolent manner. Russia in particular stands out because of its transformation from one extreme to another. The oppressive Communist dictatorship which fought against the Torah, Judaism and even the belief in the Creator, dissipated like a cloud and was replaced by a government based on justice, fairness and belief in the Creator. Russia's Swords Into Plowshares (SIP) transformation was a follow up to this. Chabad chassidim were once exiled and executed for maintaining Jewish life in Russia. Now a new generation of Chabad Rabbis—and some of the old ones who survived—has returned to Russia to spread Judaism. And a kosher kitchen has been installed in the Kremlin!

We begin with a Swords Into Plowshares story from the former Soviet Union which illustrates not only the transformation from "military" to "peaceful," but also the spirit of mutual cooperation between countries, and their combined efforts, that the Rebbe MHM speaks of in the *sicha* of Parshas Mishpatim.

From Silos to Sunflowers

Visitors to the farmland around Pervomaysk in Ukraine today would see fields of sunflowers—an important cash crop in this rich agricultural land. But not so many years ago vegetation coexisted with missile silos, which held more than 700 nuclear warheads aimed at the United States and Europe.

The story begins with the breakup of the Soviet Union. As a legacy from the Soviet war machine, four new nations inherited nuclear weapons. Strategically located, Ukraine's territory held the third largest nuclear arsenal in the world; with 2,000 nuclear warheads, this arsenal outranked those of China, France or the United Kingdom.

The United States knew that Ukraine needed financial and technical help in order to dismantle this lethal legacy. Politically, they knew that any efforts had to involve the Russians as well. In January, 1994 the three nations signed a Trilateral Agreement, confirming Ukraine's commitment to return the nuclear warheads to Russia in exchange for power reactor fuel from Russia and security guarantees from the United States. Senators Richard Lugar and Sam Nunn succeeded in getting the Congress to approve legislation to give Ukraine, Kazakhstan, Belarus and Russia support for reducing the nuclear and other dangerous weapons on their territories. This support extended beyond removing the missiles and dismantling the silos. The United States also helped build housing for the military personnel and their families who no longer had to guard and launch these missiles.

Then U.S. Secretary of Defense William Perry was directly involved in the transformation of the silos around Pervomaysk over a period of three years. In 1994 he visited the operating silos and saw the map on the wall that had a glowing bulb for every U.S. and European city targeted by a live warhead. A year later he came back to monitor progress and saw the silo being dismantled.

The following winter Perry returned. Perry, accompanied by Russian Defense Minister Pavel Grachev and Ukraine Defense Minister Valeriy Shmarov, simultaneously turned the keys that only two years earlier would have launched nuclear missiles at the United States. This time the keys ignited the explosives that blew up the empty silo.

When Perry next visited the site in June, the snow and ice of winter had given way to summer. The concrete and barbed wire of the military enclave were also gone, and in its place were furrowed fields. The three men—once sworn enemies—together dug into the fertile earth and planted sunflowers.[1]

1. http://www.nti.org/j_features/j1_sunflower.html

U.S. Secretary of Defense William J. Perry (right) Ukraine Minister of Defense Valeriy Shmarov (center) and Russian Federation Minister of Defense General of the Army Pavel Grachev (left) celebrate the completed dismantlement of Silo 110 by planting sunflowers in the field where the missile silo used to be near Pervomaysk, Ukraine, June 4, 1996.

The ISTC

In Russia, which was the most militant of all the military nations after World War II, there is a unique development in the area of Swords Into Plowshares. Just a few days after the Rebbe MHM's announcement that Isaiah's prophecy that they shall beat their swords into plowshares had begun to be fulfilled, President Boris Yeltsin and President George Bush sat down together at Camp David and made a plan to establish the International Science and Technology Center (ISTC) in Moscow. This was to be an institute that would employ former Russian military scientists to work on projects transforming the technology and science that they had developed for military purposes into peaceful uses. That was to be the entire purpose of the Center. Over the next two years this became a reality. By 1994 there was a functioning institute of science and technology in Moscow funded by four countries: the United States, Russia, Japan and the European Union. These governments invested hundreds of millions of dollars into the institute. The first executive director of the institute was an American named Glenn Schweitzer who wrote the book *Moscow DMZ* in which he documents the ISTC's accomplishments during its first years.[2]

The first point that should be made about the ISTC is that it was a pure Swords Into Plowshares transformation. This was in sharp contrast to the "dual use technology" policy of the United States at that time, whereby military technology was considered simultaneously for both military and peaceful uses—which was not a complete SIP transformation. For the ISTC, however, "*dual* use technology" would not have been good enough. The United States, Europe and Japan would not be interested in pumping millions of dollars into Russia to finance scientific projects which could be used for military purposes as well as peaceful purposes. They would be interested only if the projects had uniquely peaceful applications with no possibility of being used for any military purposes, and these were the only kind of projects that the ISTC considered.

ISTC Projects

We now turn our attention to some specific examples of the projects, and the scientists involved, that the ISTC sponsored. We will present these examples, as well as other Swords Into Plowshares examples in Russia, in considerable detail to demonstrate that SIP is not just an ideal but is well established and has been widely developed in the real world. As we follow this trend of transformation of military technology, military research institutes and their personnel, military airbases, army bases and naval bases, etc. into peaceful uses, and the redirection of the funds previously used to finance these operations to peaceful purposes, we see that Swords Into Plowshares is indeed "the trend of history."

In *Moscow DMZ*, Schweitzer estimates that when the Russian military machine was functioning fully, there were approximately one million scientists, engineers and technicians working for the Russian military. It was estimated that after the fall of the Communist government about 60,000 of these staff members had to be targeted for international science and technology projects. These were the people who were most volatile and dangerous. They were scientists and engineers who, if approached (as many of them were) by Middle

2. Glenn E. Schweitzer, *Moscow DMZ: The Story of the International Effort to Convert the Russian Weapon Science to Peaceful Purposes* (New York: Sharpe, 1996).

Eastern countries to work in those countries to develop their nuclear technology programs, they would be the ones most useful to those countries. If a low-level technician took such a job, he could not cause much harm. But if a high-level engineer or scientist took such a job, his participation would be very beneficial to their projects, and thus a threat to world peace. So out of one million scientists there were 60,000 who had to be reemployed.

In fact, Russian scientists and engineers needed employment very badly. Once their military contracts were cancelled, a lot of them had no source of income. It was found that many top Russian scientists were earning lower salaries than bus drivers. Thus they were an easy target for Middle Eastern countries that would offer them lucrative salaries to come join their nuclear projects.

Of the 60,000 scientists, it was estimated that 30,000 were in aeronautical science; 20,000 in nuclear science and 10,000 in biological science. A third of them (about 20,000) remained employed by the Russian military. Another 15,000 got jobs in the commercial sector. The remaining 25,000 were the ones that the ISTC took on.

Some of the projects that they worked on were the following: Nuclear scientists who specialized in the design of nuclear weapons were put to work on designing safer nuclear reactors. Russia was notorious for having poorly designed reactors. They disregarded nuclear safety. The accident at Chernobyl was the most dramatic of all of Russia's nuclear accidents, but it was not the only one. There were many, many incidents of radiation leaks. Parts of Russia are contaminated by radiation. Nuclear waste was not dealt with in any safe manner. Often it was simply dumped into rivers. Many ISTC projects involved putting these nuclear scientists to work designing safer, better and more efficient nuclear reactors, and to clean up the contamination in those areas of Russia that had been contaminated by nuclear accidents and by careless disposal of nuclear wastes.

Chemical and biological specialists who were trained in biological and chemical warfare were given projects involving pollution cleanup. While in the United States pollution is already pretty much under control, in Russia it has always been disregarded. The communist government never concerned themselves with it. Now the chemical and biological specialists are working on chemical and biological pollution problems.

Aerospace specialists have been working on projects ranging from designing special environmental observation aircraft to the research of nuclear propulsion technology for exploration of Mars.

The ISTC has expanded considerably since its establishment in 1994 and is currently involved in many more projects.[3]

In addition to ISTC projects, there are many other SIP transformations taking place in Russia. A particularly noteworthy example is the case of a biological weapons installation called Vector in Novosibirsk which was transformed into a pharmaceutical research organization now called the State Research Center of Virology and Biotechnology. Whereas previously the top Russian biological scientists worked on devising biological weapons to kill people, now they are doing research on viruses, collaborating with the United States Centers for Disease Control, contracting internationally to produce pharmaceutical drugs. The same scientists and the same technology which was once used to produce biological weapons are now being used to produce drugs and to research cures for diseases!

The Nunn-Lugar Program

Another way in which the U.S. has helped Russia transform its military resources to peaceful uses has been through the Cooperative Threat Reduction (CTR) Program.[4] It assists the states of the former Soviet Union in controlling and protecting their nuclear weapons, weapons-usable materials, and delivery systems. To aid in the implementation of arms control agreements, CTR also contributes to the dismantling and destruction of a number of nuclear weapons and their associated delivery systems.

The Cooperative Threat Reduction Program, also known as the "Nunn-Lugar Program" (named for sponsoring Senators Sam Nunn and Richard Lugar), began in 1991 as a piece of U.S. legislation entitled "The Soviet Nuclear Threat Reduction Act of 1991," which stemmed from Soviet President Mikhail Gorbachev's request for assistance in dismantling Soviet nuclear weapons, and US President George Bush's subsequent proposal to assist in the disposition, dismantling, and destruction of nuclear weapons in the Soviet Union.

---

3. See their website www.istc.ru
4. See www.nti.org/db/nisprofs/russia/forasst/nunn_lug/overview.htm

In the Soviet Nuclear Threat Reduction Act, Congress noted that aid to the Soviet Union, and its potential successor states, would be in the national security interest of the United States as a means to address the threat of nuclear weapons proliferation. Specifically, the concern was that substandard methods for protecting, controlling and accounting for nuclear weapons and materials might result in the dangerous transfer of nuclear weapons, components or weapons-related knowledge.

With these threats in mind, the Act delineates a two-fold objective: 1) to facilitate on a priority basis the transportation, storage, safeguarding, and destruction of nuclear and other weapons in the Soviet Union, its republics, and any successor states; and 2) to assist in the prevention of weapons proliferation. For Fiscal Year 1992, the Act allotted $400 million of transferred Department of Defense funds for this purpose. In October 1992, an additional $400 million was allocated to establish the Safe and Secure Dismantlement (SSD) Talks under the Former Soviet Union Demilitarization Act.

Allocation of all funds was contingent upon a U.S. presidential certification that the recipient country is: "making a substantial investment of its resources for dismantling or destroying such weapons; forgoing any military modernization program that exceeds legitimate defense requirements and forgoing the replacement of destroyed weapons of mass destruction; forgoing any use of fissionable and other components of destroyed nuclear weapons in new nuclear weapons; facilitating United States' verification of weapons destruction; complying with all relevant arms control agreements; and observing internationally recognized human rights, including the protection of minorities."

In October, 1993, the Soviet Nuclear Threat Reduction Act evolved into the "Cooperative Threat Reduction Act of 1993" and in June, 1999, the United States and Russia signed an agreement to extend the legal framework for the CTR program through June 2006.

The CTR Program had a human side, too. In a meeting with US Senator Richard Lugar in December, 2000, Russian Strategic Rocket Forces Commander Vladimir Yakovlev stated that the U.S. CTR Program ought to address the social needs of military servicemen. He proposed adjusting the program to provide housing for servicemen involved in Russia's nuclear weapons program.

Megatons to Megawatts

One of the outcomes of the cooperative agreements between the U.S. and Russia to transform Russian nuclear weapons into peaceful uses has been—surprisingly—the fueling of nuclear reactors in the United States to produce electricity. Here's the surprise: 45% of the fuel in American nuclear reactors currently (2009) comes from dismantled Russian atomic bombs. Together with another 5% that comes from American bombs, this fuel produces about 10% of the electricity in the U.S.

The fuel from missiles that may have once been aimed at your home may now be lighting it!

By comparison, hydropower currently generates about 6% and solar, biomass, wind and geothermal power together account for 3%.

Thousands of decommissioned warheads have been converted into civilian reactor fuel for America's 104 nuclear reactors since the treaties went into effect. This program, known informally as Megatons to Megawatts,[5] is scheduled to expire in 2013, but the U.S. and Russia are negotiating another agreement to continue diluting[6] Russia's highly enriched uranium after the expiration of this program. The proposed new program is known in the nuclear industry as HEU-2 (High Enriched Uranium-2).

In addition to uranium, an agreement between the U.S. and Russia from the year 2000 to convert plutonium into reactor fuel[7] has been reaffirmed by Russia's president, Dmitri Medvedev.

The primary practical benefit of using the Russian nuclear material has been its abundance, which has drastically reduced the price of the fuel, resulting in a lower cost of the electricity produced.

Smaller-scale programs are transforming the nuclear material from American bombs for use in nuclear reactors. For example, since 2005, the American company Nuclear Fuel Services began diluting

5. The power of atomic bombs is measured in megatons and electric power is measured in megawatts.

6. Weapons grade uranium is enriched to about 90% U-235 so it must be diluted to the reactor level which is less than 5% U-235.

7. See sec. 7 of this chapter on the Radkowsky reactor which uses plutonium as a fuel.

uranium from the 217 tons the government declared surplus. It is used at the Tennessee Valley Authority plant.

At a factory being built at the Energy Department's Savannah River site in South Carolina to dismantle warheads from the American arsenal, technicians are transforming plutonium into reactor fuel. The 34 tons to be recycled there are expected to generate enough electricity for a million American homes for 50 years.[8]

Military Institutes

Another example of a Swords Into Plowshares transformation in Russia is the restructuring of a major Russian military research institute, the Central Aerohydrodynamic Research Institute (TsAGI) from a military research center into a research center for peaceful applications.[9]

TsAGI was established in Moscow in 1918 and grew to become the largest aerospace test center in the world. Its facilities were moved to Zhukovsky outside Moscow in 1939. By World War II, TsAGI had already branched out into several divisions. Zhukovsky is now a city of over 100,000 people whose economy revolves around the aviation industry.

TsAGI has been involved in the design and testing of every major Soviet military and commercial airplane, helicopter, missile and spacecraft. It also engages in flight simulator design and training. Its facilities include wind tunnels for subsonic, transonic, and supersonic aerodynamic tests, as well as chambers for thermal strength testing. Some of these facilities are the most advanced in the world. TsAGI has extensive software development and computational capabilities as well as prototype manufacturing facilities that are used for producing test instrumentation and experimental models.

After the Swords Into Plowshares declaration, TsAGI began to reorganize itself in a way that would make maximum use of its technology and facilities to develop new products and provide services of economic benefit while retaining its key technical personnel and providing challenging scientific work for them. (While many of its

8. http://www.nytimes.com/2009/11/10/business/energy-environment/10nukes.html
9. See BICC Report 3, "Conversion of the Defense Industry in Russia and Eastern Europe," at www.bicc.de . See also www.tsagi.ru

employees left to enter commercial pursuits, this itself is considered by many analysts to be a "passive conversion" from military to peaceful activity.)

To achieve these goals, TsAGI formally consolidated its core aerodynamic, thermal and mechanical testing facilities into the TsAGI State Science Center (TSSC) in 1994. The Science Center itself has three categories of subsidiaries. Category I comprises companies that are most closely linked to the Science Center's core capabilities and activities. They all involve high technology and use key scientific personnel from what was TsAGI before the establishment of the Science Center. The main difference between these Category I subsidiaries and the Science Center itself is that while TSSC provides technology services for the state (it's similar to government-owned national laboratories in the U.S.), the Category I subsidiaries are involved in providing these services for private customers, both inside and outside Russia. In addition to carrying out commercial businesses, these subsidiaries engage in advanced applied research that may have further commercial potential.

Subsidiaries in this category include a helicopter plant; facilities for producing measurement instruments, supplying testing equipment and testing services; and a software firm. Aerospace technology has been reapplied by enterprises in this category to produce automobile testing equipment and medical equipment.

The second category of subsidiaries also involves technology and skilled personnel from TsAGI, but the technology is used to develop new commercial products and services that don't involve the basic testing services of the Science Center as the main element of the product or service. These subsidiaries are basically free to do business on their own. Much of the required capital comes from private investors.

As an example, one of these Category II subsidiaries specializes in solar heaters. The enterprise grew out of a research group at TsAGI that was working on commercializing technical processes developed at the institute. They used technology developed at TsAGI for the production of flexible, thin metal sheets to develop piping for the panel that was cheaper and more efficient than the traditional copper models. TsAGI's aerospace technology in surface coverings was also applied to the solar panel to increase its absorptive capacity and dampen energy release. While the solar collector project was initially oriented toward developing a solar water heater, in 1994 TsAGI researchers began developing applications in air conditioning, desalinization, drying equipment and other areas. The firm has received orders from the Moscow Regional Government, and even from some foreign companies.

Other enterprises in this category include a firm that provides repair and certification work for foreign aircraft, an enterprise that uses a large vacuum chamber for high-altitude thermal strength testing, a producer of machine tools for wood processing, and a flight simulator.

Category III comprises subsidiaries that were established to provide employment for excess production workers, to utilize idle space and equipment, and to generate revenue without monetary investment. Some of these subsidiaries stress employment while others stress income. It is important to emphasize the significance of this step. One of the aspects of Marxist philosophy that the Rebbe MHM criticized was the idea that the individual good must be sacrificed "for the good of society." Judaism, in contrast, emphasizes the significance of the individual and says that a single act or even a single thought of an individual can bring about a change for the good of the entire world.[10] In the formation of these Category III subsidiaries, we see a complete reversal of the Marxist philosophy: An institute is reorganized and a subsidiary is set up whose primary purpose is to provide employment for the former employees of the original institute. (How often do we see such a step taken in the reorganization of U.S. corporations?)

The Category III subsidiaries are low-technology product ventures such as production of furniture and shoes, or services such as a cafeteria and a management training program, but they have little or nothing to do with the science center's base activities. From a Swords Into Plowshares perspective, however, these subsidiaries are especially important since they were established primarily to provide employment, and their products are consumer goods.

Of particular interest are Category III subsidiaries that are involved in joint manufacturing projects with Israel. The first one is a sub-

10. Rambam, Hilchos Teshuva, 3:4.

sidiary that manufactures various instruments. TsAGI provides the building, equipment, and personnel; an Israeli firm provides technology and some equipment; a US firm provides marketing and sales; and a Swiss firm provides financing. Each partner has 25 percent ownership. The second Category III subsidiary involved in joint manufacturing projects with Israel is an industrial diamond manufacturer. This is a joint venture with 55 percent ownership by TsAGI and participation by firms from Israel, the United States and Sweden.

Here we have seen how a major military research institute has been subdivided into divisions in which the military technology itself is used for peaceful uses, as well as divisions which directly provide consumer goods.

The Transformation of Military Bases

We reported above on various Russian military technologies and military institutes that have been transformed for peaceful uses. But a military machine involves other resources such as military bases. The Soviet Union abandoned thousands of military sites as a result of its withdrawal from Central and Eastern Europe. What happened to all these military bases used by Russia's vast army for many decades? Have they also undergone a Swords Into Plowshares transformation?

Military bases throughout Russia and Eastern Europe, as well as military research institutes, have been transformed and are being used for peaceful purposes. We will look at two Russian military installations in Poland and see how they have been transformed.[11]

Following World War II, the Soviet Union took over the former German barracks facility in Poland's northwestern port city of Szczecin. Situated directly between Berlin and Gdansk, this city of 420,000 is the logistic and population center of the region. Shortly after the Russian soldiers withdrew from Szczecin, the barracks complex was converted into an educational complex consisting of three schools: 1) The Institute of Computer Science of the Szczecin Technical University, 2) A campus of the University of Szczecin and

11. See BICC Report 11, "Base Closure and Redevelopment in Central and Eastern Europe," at www.bicc.de

3) The Maritime University of Szczecin. The barracks hospital was converted into a civilian hospital owned and operated by the city. The current level of activity in this complex greatly exceeds the former, military base level.

The city of Legnica is located in southwestern Poland and is the capital of Legnica Province. During World War II it was the Eastern Forces Headquarters of the German army. As the Russian troops advanced, the Germans abandoned it. The Russian army took it over and in 1955 it became the headquarters for the entire Soviet Northern Group Forces.

During their presence in Legnica the Soviet troops occupied 1200 buildings—840 that had been built by the Germans and 360 that the Russians themselves built. To assure security and separation from the residents of the city, the Russians built walls around various sections of the city which contained these military buildings. One such walled section, called the Russian Square, a five-block area in the heart of Legnica, housed the Russian elite—officers and high ranking party officials. It consists of massive houses, which are at least a hundred years old, with large yards.

After the Russians left, these buildings were used for housing for local residents. The larger buildings were bought by developers to transform into apartment buildings and the smaller buildings were sold to individuals for private homes.

One building located just outside Russian Square had been built by the Germans in the 1800s as a villa for visiting athletes. After World War II, the Soviet military took over the residence and used it as a guest house for visiting officers. After the Russians left, the city government converted the building into a retirement home under the management of the Legnica City Office of Social Assistance. After a lengthy and expensive renovation, it began functioning as a home for seniors in 1996.

Finally, several military airbases have been transformed into civilian airports. This process, however, has taken much longer because of the tremendous expense involved in removing the pollution and contamination left by the military, rebuilding and expanding the buildings and the runways and, in general, meeting the safety requirements for a civilian airport.

Disarmament and Conversion

A full Swords Into Plowshares transformation would consist of two components: 1) disarmament, and 2) conversion. Disarmament is the *reduction* in military resource use (i.e., reduced armaments, demobilization and demilitarization). Disarmament by itself, however, does not necessarily involve the transformation of these arms into peaceful uses. Conversion, on the other hand, is the *reuse* of military resources for peaceful purposes. Examples are the reuse of defense expenditures for civilian government expenditures, and the transformation of production capacities for tanks to production capacities for trucks, soldiers becoming farmers, etc. But it may happen that while conversion is taking place in one area of military resources, there may be a buildup in another area and the military as a whole may be expanding. In other words, in such a situation disarmament is not taking place. Thus a comprehensive Swords Into Plowshares transformation involves both disarmament and conversion.

The main disarmament treaty for conventional weapons was the Treaty on Conventional Armed Forces in Europe (CFE), which was signed on Nov. 22, 1990 in Paris. The main principle of the CFE Treaty is parity at the lowest possible level in heavy attack weapons between what were once the two military blocs of NATO and the WTO (Warsaw Treaty Organization—Russia and its allies). The basic objective was to radically curtail the possibility of launching massive-scale surprise attacks on the continent of Europe. Emphasis was therefore laid on heavy weapons, excluding ships.

In line with the original purpose of the treaty to completely eliminate weapons from arsenals, destruction of the weapons was foreseen as—and in fact turned out to be—the predominant method of arms reduction. As a result of the treaty, more than 50,000 heavy weapons had been destroyed in what was the longest peacetime conventional disarmament operation in the history of the world!

Conversion of heavy armored vehicles

However, destruction of the weapons was not the only method of arms elimination provided for by the treaty. Conversion to nonmilitary purposes was also provided for. This combination of disarmament and conversion, which came to actualization in the mid 1990s,

A dramatic example of disarmament

was a significant Swords Into Plowshares event. Specifically the treaty provided for certain types of tanks and ACV's (armored combat vehicles) to be converted into vehicles for nonmilitary use.[12]

12. See sec. 10 of this chapter, "Into Plowshares—Literally," for a follow up of this.

Transforming the Swords of China

After giving various examples of Swords Into Plowshares transformations in Russia in previous section, we now turn our attention to China.

When discussing Swords Into Plowshares (SIP) in China, as is the case with all matters of China, the issues are unique and complex. China has always isolated itself and done things its own way. It's hard to say "isolated" when the country has a population of over a billion people, but on the other hand, China's large population had enabled it to develop in its own way for decades with minimal international interaction. This applies as well to its nuclear weapons development which, following a short period of collaboration with Russia, it did mostly on its own.

Thus the analysis of China vis-à-vis the Swords Into Plowshares conversion is itself complex and requires some background. We will first give a brief review of China's nuclear development. Our report is based on a paper of Prof. Yitzhak Shichor of the Hebrew University entitled "Peaceful Fallout: The Conversion of China's Military-Nuclear Complex to Civilian Use."[1]

China's Nuclear Technology Development: Plowshares Into Swords

The initial stages of China's nuclear weapons development began in about 1952. China had recently established the Research Institute

1. BICC Brief 10 at www.bicc.de

of Modern Physics, drawing on Chinese physicists at home and abroad. Those who returned from abroad brought with them personal connections, books, equipment and materials which provided the foundation for China's nuclear research.

In January, 1955, China and Russia began collaborating on nuclear research. In March 1956, China, Russia and 10 other socialist countries formed the Joint Atomic and Nuclear Research Institute in Dubna (near Moscow). The next month Mao Zedong (Tse-tung) declared that China "would not only have more aircraft and large guns, but also atomic bombs." Russia continued its nuclear science and technology assistance to China and on October 15, 1957 the two governments signed an "Agreement on Producing New Weapons and Military Technical Equipment and Building Comprehensive Nuclear Industry in China."

Early in 1958, however, Russia secretly decided to back off from this deal and, as Sino-Soviet relations began to deteriorate, China decided to start from the beginning and develop an atomic bomb on its own. In the spring of 1960, scores of scientist were mobilized to work on nuclear weapons and missiles projects for the Nuclear Weapons Research Institute. Despite a major setback later that year when Russia suddenly stopped the supply of equipment and recalled hundreds of nuclear experts who took blueprints and information back with them, China proceeded with its plan to develop an atomic bomb, more determined than ever.

At this point a *plowshares into swords* conversion took place in China, a process called a *grand strategy* by political scientists, in which all of a nation's resources, including its civilian resources, are directed to a military purpose. Much effort was exerted to ensure the supply of all the people, materials and equipment needed by the nuclear industry. By the end of 1963, over 400 factories, institutes, colleges and universities all over China contributed to the effort. Throughout the 1960s and 70s, China's nuclear weapons development proceeded at a steady pace. As its nuclear program gathered momentum, China's threat perception increased. Mao claimed, "If we are not to be bullied in the present-day world, we cannot do without the bomb."

Toward *Swords Into Plowshares*

In the late 1970s, after Mao's death, China put forward a policy that its surplus military manpower and equipment be used for civilian purposes. This was not yet a Swords Into Plowshares policy since China's threat perception was still very high, considering Russia as the most immediate threat to its security. The policy was put forth primarily for economic reasons. China realized that much of the nuclear and conventional defense industry potential, used in the past almost exclusively for military production, had a surplus capacity which could be used for civilian purposes. It was a matter of putting this surplus to practical use.

This policy was described by the Chinese as "preserving military needs while converting to civilian production." It was a sort of *dual use technology* policy—assuring military production while developing civilian goods.

In the case of China's nuclear technology, there were four main obstacles to such conversion:

1) China's nuclear industry had been developed in total secrecy and under tight security measures, totally secluded from market forces.

2) The technologies it used were dangerous and the nuclear industry was so specialized for military purposes that the existing production facilities and equipment could not be directly converted to civilian production. Furthermore, its management system, operational methods and work styles were incompatible with the requirements of civilian production.

3) Two thirds of China's nuclear complex were located in deserts and mountain regions, far away from urban commercial centers.

4) The nuclear industry had no experience in what was to become its principal civilian application—nuclear power plants.

It was necessary to reduce centralized state control and increase flexibility, initiative and competition. One of the organizational measures taken in this direction in the 1980s was to allow each of the defense industries to establish its own trading corporations. Several corporations were set up to deal with specific civilian aspects of China's nuclear industry, such as the China Isotope Co., China Nuclear Instruments and Equipment Corp. and the Nuclear Industry Development Research Center. At the same time, a gradual process of changing the names and the nature of the defense-industrial government ministries began. For example, in 1988 the Ministry of Nuclear Industry was replaced by the China Nuclear Industry General Corporation (CNIGC) under a newly created Ministry of Energy Resources.

Swords Into Plowshares at Last

With the Swords Into Plowshares declaration in 1992, China's threat perception was greatly reduced and a genuine Swords Into Plowshares process began in China. Russia was no longer the threat that it formerly had been, and China found that it no longer needed the vast nuclear weapons machine that it had developed, nor the vast army that it had organized.

Reduced threat perception is a defining characteristic of a Swords Into Plowshares transformation, as we mentioned in the beginning of this chapter. Regarding this aspect of SIP, the Rebbe MHM quotes the commentary of the Radak, who writes, "There will be no war between one nation and another.... They will not need instruments of war and they will beat them to make from them instruments for working the ground." We might say that the highest form of Swords Into Plowshares occurs when peaceful conditions bring about a change of attitude on the part of the countries of the world. They will not feel threatened by each other; they will feel secure so they will not need instruments of war.

One example of the transformation of nuclear technology was the Modern Physics Research Institute (founded in 1950) which played a key role in China's nuclear weapons program. It has been renamed China Institute of Atomic Energy (CIAE). It has 50 research offices, an overseas engineering projects office, and various research and testing centers. Its services include application of nuclear energy and radiation, development of energy saving products, environmental science including environmental protection, biological engineering, etc. It produces high polymer materials, new medicines, food additives, fine chemicals, isotope instruments and meters, electronic components and devices, heavy water research reactors, cyclotrons, sterile generators, etc.

Another Chinese institution which emerged from the SIP nuclear transformation was the China Institute for Radiation Protection (CIRP), which has eleven research offices dedicated to labor hygiene and occupational disease, radiation medicine, radioactive waste disposal, radioactive organisms, nuclear safety, application of nuclear techniques, etc. Its services include providing comprehensive environment evaluation, technologies for harnessing radioactivity and industrial waste, biochemical preparations, bio-engineering and tissue culture techniques, immuno-diagnosis medical kits, techniques for radiation sterilization and processing, etc. It produces continuous environment monitors, aerosol sampling machines, rock stratum stress gauges for earthquake and landslide forecasts, radon education meters, nucleon scales, and various medicines and medical kits.

What the Chinese Say

How do the Chinese themselves view the Swords Into Plowshares transformations going on in China as well as in the rest of the world? In the introduction to their article titled "The Practice and Progress of Defense Conversion in China," three professors from the China Academy of Engineering Physics write:

> The peaceful use of military technology represents the trend of history. The fundamental element of the conversion of military technologies to civilian ones for each country is to reconfigure the factors of production in peacetime, which consists of land and natural resources; capital and labor; science, technology and information; etc. The purpose of conversion is to make an efficient contribution of postwar economic recovery and construction which is determined by national defense, economy, strategy of science and technology development, industry policy and labor distribution. Defense conversion is a fundamental issue that is related to the world peace, disarmament, international security, and economy development. The influence of the conversion has been more significant after the cold war; and a lot of coun-

> tries have paid their attention to it. Many scientists and outstanding people have been working hard on promoting the defense conversion.
>
> The most important aspect of peaceful use of defense science and technology is its development and applications. It is a common demand of the world['s] people and wishes of most scientists and engineers to convert the advanced military technologies to productive forces, to promote national economy development and bring benefits to human beings. The national condition of China determines that the objective of Chinese conversion be to develop productive forces and meet the increasing demand for material and civilization [*sic*]. Development and applications of advanced defense technology in civilian area will promote significantly the economy, technology and society, and bring new social culture and material prosperity. Science and technology are the first important productive forces. In order to be converted to productive forces, science and technology must be effectively converted to products that meet the market requirements.[2]

This concludes our discussion of Russia and China. Since these two countries were the most militant countries of the post World War II era, their SIP transformation clearly emphasizes that "Swords Into Plowshares is the trend of history."

---

2. I downloaded this article from a Chinese website (of the CAEP?) in the mid 1990s. It no longer appears to be available.

Swords into Plowshares at the Pentagon

by Rabbi Dr. Naftali Berg, z.l.

Rabbi Dr. Naftali Berg, z.l.

Imagine that you are a research scientist and you want to find out what kind of scientific research is going on at the Pentagon. For that purpose, you take a trip to Washington D.C. and visit the Pentagon. There you are directed to the Army Research Laboratory (ARL), and when you tell them that you are interested in knowing what kind of new technologies are being developed there, they direct you to the Advanced Technology Office. There you ask to speak to the man in charge, and they set up an appointment for you with the Director of Research. Now you are about to meet the Director of Research of the Advanced Technology Office of the Pentagon's Army Research Laboratory. Before continuing, stop for a moment and try to imagine what you expect this man to look like.... Now look at the picture above.

The man you are looking at is Rabbi Dr. Naftali (Norman) Berg, lecturing on quantum mechanics at the Moshiach and Science conference of 1993. He was the Director of Research of the Advanced Technology Office until he passed away in the summer of 1994. He was a Lubavitcher chosid who pursued a career in physics after completing his yeshiva studies and smicha. *After receiving a Ph.D. in physics, he was offered a position in defense research and a higher-paying position in industry. The Rebbe MHM told him to take the position in defense research as he would be able to help the Jewish people that way. Indeed, many times Dr. Berg was instrumental in obtaining U.S. military aid for Israel.*

When I organized the 2nd Moshiach & Science conference in 1993 (commemorating Rabbi Yisroel Aryeh Leib Schneerson's 41st yartzeit,) I invited Dr. Berg to be the keynote speaker. He spoke brilliantly on two themes: 1) "Swords Into Plowshares at the Pentagon," the topic of this section, and 2) "Physics and Technology in the Era of Moshiach," in which he discussed certain discoveries in modern science that support the concept of creation. This will appear in a later section of the book.[1]

When I first asked him to come and lecture on how military technology at the Pentagon was being transformed for peaceful uses, he asked me, "How do you know that military technology at the Pentagon is being transformed for peaceful uses?" (I guess people aren't supposed to know what's going on inside the Pentagon.) I replied that it must be so since a year before the Rebbe MHM had said that Isaiah's prophecy that the nations of the world will "beat their swords into plowshares" in the Era of Moshiach has begun to be fulfilled. So it must be happening at the Pentagon, too. Dr. Berg laughed and said that, indeed, there was a lot of Swords Into Plowshares activity at the Pentagon.

In this lecture he draws on his first hand knowledge and experience at the forefront of military research to show how "they shall beat their swords into plowshares" is an ongoing process of the transformation of military technology into technology for the benefit of people—another aspect of the Geula that is unfolding right before our eyes.

Dr. Berg's Lecture

I've been asked to talk about how I as a research scientist—a director of research for the U.S. Army—see this concept of *"v'chitzu charvosom l'itim"*—"they shall beat their swords into plowshares." I want to give a few examples of this.

There appeared an article in Scientific American in January 1991 entitled, "The Future of Space Reconnaissance." I'll read you the introduction and then I'll explain it: "As the superpowers continue to launch spy satellites, many nations are planning to orbit their own.

1. See chap. V, sec. 5.

Such expensive proliferation will complicate international politics into the next century." Basically what this article talks about is the fact that the U.S. and other countries have developed very, very capable techniques to be able to see and probe the earth's surface from satellites which can be hundreds, even thousands, of miles away using either optical techniques—shall we say, better cameras, or infra red, looking at thermal emissions, or using radar. And these techniques have allowed these countries, the U.S. and other countries, including Eretz Yisroel which is mentioned here, to be able to see what their enemies are doing. This is, of course, very laudable. You always want to see what your enemies are doing. This is a defensive utilization. However, sometimes these satellites are used for offensive purposes as well, since you can plan your offensive strategy based on what they are doing.

But as a result of this space reconnaissance technology, we have developed, for commercial application, satellites which are used to scan the earth's surface to see areas which can be better used for agriculture. There are three current programs in this area. One is called Landsat, which means "land satellite"; another one is called Land TM, which means "land thematic mapper." It sees the heat emitted from the earth's surface, and from there one gets the idea of what is under the earth's surface.[2]

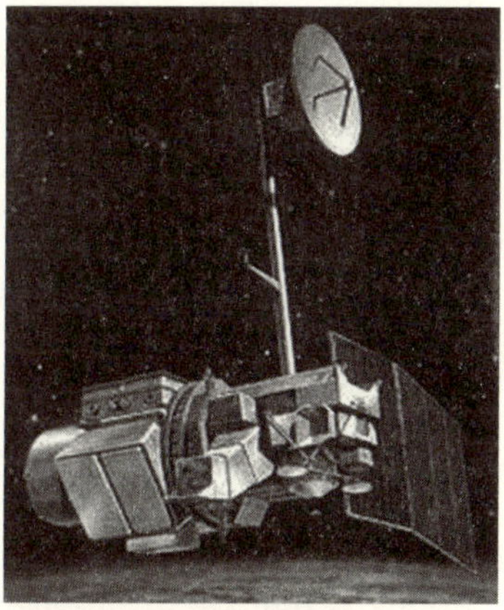

Landsat 5 Satellite
Courtesy ASA/JPL–Caltech.

There is also a series of satellites called "earth resource satellites." One program, called ERS1, uses a radar system to scan the earth's surface. From that we can understand the quality of the soil, the covering of the earth's surface and water resources underneath, and also chemical content, minerals and oil which are under the earth's surface.

There is also a French program in this field called SPOT. It stands for four words in French which I don't speak so can't tell you what it

2. The TM sensor has seven spectral bands: six acquire earth reflectance data, and one acquires earth temperature data.

is.[3] The SPOT system is a satellite which uses both visual and thermal imagery to scan the earth's surface. These satellites are actually used to monitor agriculture to see how effective plowing and other agricultural activities are. The goal is to develop increased and enhanced uses of resources for agriculture, for food production, for monitoring forestry and also, as I said before, to find minerals and petroleum or other types of resources, including water. This is certainly a direct application of military technology for peaceful and actual food production uses.

Fluidics, etc.

The second example is something which I am personally connected to. At my laboratory, which used to be called the Harry Diamond Laboratory, the only army laboratory named after a Jew—Harry Diamond, who was an inventor—we discovered a number of years ago a phenomenon called "fluidics." This basically means to replace electronics, a transistor, by a fluid, a compressed gas flowing through some kind of structure. In other words, you can control hydraulic surfaces by the flow of small amounts of gas rather then using electronics. The purpose of this for military technology was to be able to use these in an environment which would be too harsh for electronics, such as nuclear environments. The relevant issue from the military point of view is that the military hardware should survive this mission. So this fluidic technology was developed. As an offshoot of it, we developed a fluidic amplifier, a loudspeaker, which does not use electricity at all. It's based purely on the flow of gas.

In 5738 (1977), after what occurred at the beginning of that year,[4] I thought that this would be a very useful thing to have in "770" to be able to amplify the voice of the Rebbe MHM so that we could hear it without much difficulty. There is a man who owns a company who has actually developed this commercially, but in order to use it in a shul or a *beis hamedrash* (study hall), you have to have a *psak halacha*—a *heter* (halachic permit, especially for use on Shabbos or Yom Tov) for this. For the last eight years, we have been trying, unfortunately without much success, to find a *posek* (halachic authority),

3. *Satellite Pour l'Observation de la Terre* or *Systeme Pour l'Observation de la Terre.*
4. The Rebbe MHM had a heart attack on Shmini Atzeres of that year.

who would put his name on the line, so to speak, and give a *psak* that we should be able to use it. But, G-d willing, perhaps this will happen eventually, especially now that we're in the Era of Moshiach.

A third example which I want to bring is a program which has been started by the new administration, the Clinton administration, called "Dual Use Technology." This program has set aside for the first year five hundred million dollars. A half a billion dollars of the defense research budget is to be used by commercial industries to take military technology and use it for commercial application. If any of you have a company, we can talk after the lecture. If a company puts up 50% of the money, the other 50% comes from this fund, and the military then would give its technology to this company to develop a commercial peacetime application.

This program has literally just started so I can't give any example of things which have actually been accomplished. But some of the ideas which have been put forth from our laboratory are, for example, the following: We have developed a radar called an "ultra wide band" radar which is able to actually see deep into the earth. Its purpose for military applications is to find mines and other things of a military nature that are buried deep inside the earth. The peaceful application is to find toxic waste materials in order to be able to dig these up and remove them so they will not harm human beings.

Another technology which we hope to be able to use is very small radars which we have built for bombs, and put them on cars for collision avoidance. If you have radar, you can see where a car is and stop your car before you hit it. A third technology, which is in one of the areas in which I work, is also an acoustics technology, uses sound—very sensitive microphones—to pick up enemy vehicles, like tanks and airplanes, and is able to follow them around on the battle field. The application of this for peaceful purposes is to monitor traffic or to put them at small airports that can't afford radar.

Anyway, the main point is that we really see the revelation of Moshiach now that the administration is taking a half of a billion dollars per year and putting it into these "Dual Use Technologies." It is certainly a fulfillment of the Rebbe MHM's *sicha* of a year ago that *"v'chitzu charvosom l'itim"*—"they shall beat their swords into plowshares"—that the nations of the earth will take their military budgets and convert them into food production and other peacetime uses.

Finally I would like to give you two very simple examples, from areas which I'm connected with, of the general application of technology in the Era of Moshiach. You know that we are developing now what is called interactive video. It is, to pardon the expression, television. Using fiber optics, cables and switching techniques, not only can you see what is going on in another place, but you can actually talk back and forth. In other words, many people can be involved in a conversation or a discussion or in a class simultaneously using interactive video. Certainly this interactive video will be developed for the Rebbe MHM to be able to teach hundreds of thousands, even millions of people, the Torah.

Another example which is in the category of Swords Into Plowshares: We've developed, in the last number of years, very powerful laser beams which can cut through metal, through steel and through rock. I propose that when the third Beis HaMikdosh is built, if there will be any human involvement at all, if it won't come completely built from Hashem, we will use lasers to cut the rocks for the Beis HaMikdosh since we don't have the *shamir*, the special worm [that cut the stones in the first Beis HaMikdosh], nowadays. But the main thing is that Hashem should open our eyes to see the Rebbe Melech HaMoshiach complete the Geula and take us to Yerushalayim and the Beis HaMikdosh and show us all the miracles and wonders of the Era of Moshiach now!

Technologies

The Transformation of Satellite Technology

Satellites have played a major role in modern military technology since they were first launched in the 1950s. The spy satellite—the eye in the sky—was a trademark of the Cold War. In this section, after some historical background,[1] *we examine how this technology has been transformed to peaceful uses.*

Satellites as Swords

It all started back in the dark days of the Cold War. Scientists from the United States, the Soviet Union and several other countries discussed the possibility of launching a satellite as part of the research effort known as the International Geophysical Year (IGY). But IGY was just a cover. Both the U.S. and Russia were working on developing satellites for military purposes—to spy on each other. The U.S. had been secretly working on this since 1955.

Dr. John Hagen, a senior scientist with the Naval Research Laboratory, headed the American effort to launch a satellite for the IGY, code named Project Vanguard. The launching was behind schedule and over budget.

On the night of October 4, 1957, Hagen arrived early at a reception at the Soviet Union's Embassy in Washington, D.C. to compare notes with a few Soviet scientists, whom he considered personal friends, and learn the status of the Russian satellite effort. Suddenly,

1. Based on Roger D. Launius, "Sputnik and the Origins of the Space Age," available from http://history.nasa.gov/sputnik/sputorig.html. Launius is a NASA historian.

a reporter from the New York Times at the reception received a phone call informing him that the Russians had just launched Sputnik 1, the world's first Earth-orbiting satellite.

Sputnik 1 was a 22-inch aluminum sphere with four spring-loaded whip antennae trailing behind. It weighed 183 pounds and traveled an elliptical orbit that took it around the earth every 96 minutes. In contrast, the Project Vanguard satellite was to weigh only 3.5 pounds.

The reaction of the American public to the successful launch of Sputnik 1 was shock and panic. The event had a "Pearl Harbor" effect on American public opinion. It created the appearance of a serious technological gap between the U.S. and the Russians that had menacing implications. As Senator Lyndon B. Johnson (later to be President Johnson) put it, "I remember the profound shock of realizing that it might be possible for another nation to achieve technological superiority over this great country of ours." One of his aides spelled it out even more clearly, saying, "The simple fact is that we can no longer consider the Russians to be behind us in technology. It took them four years to catch up to our atomic bomb and nine months to catch up to our hydrogen bomb. Now we are trying to catch up to their satellite." Johnson is also reported to have said that soon the Russians would be dropping bombs on us like kids drop stones on passing cars going under an overpass.

The Space Race

Even before the effects of Sputnik 1 had worn off, the Soviet Union struck again. On November 3, 1957, less than a month later, it launched Sputnik 2. While the first satellite had weighed less than 200 pounds, this spacecraft weighed 1,120 pounds and stayed in orbit for almost 200 days. But America's Project Vanguard was a disaster. In a test launch on December 6, 1957 the rocket rose about three feet above the platform, shook briefly, and disintegrated in flames. In another attempt, on February 5, 1958, the Vanguard launch vehicle reached an altitude of four miles and then exploded. Another team of U.S. (formerly German) rocket scientists, however, initiated Project Explorer and successfully launched Explorer 1 into orbit on January 31, 1958.

Thus began a competition between the U.S. and Russia—the race for space—that continued for decades. It was clear that whoever dominated in the space race had a distinct military advantage over the other. Spy satellites—the eye in the sky—became the routine technique for each country to determine the military capability of the other. The U.S. spared no effort to gain the upper hand in the space race. Science and mathematics education were revised in the U.S., and technological training was enhanced. The National Aeronautics and Space Administration (NASA) was formed from the National Advisory Committee for Aeronautics, a civilian research agency of highly qualified scientists that had a very close working relationship with the military.

This discussion of Satellites as Swords is a historical introduction to our discussion of how these swords have been transformed into peaceful uses (plowshares), as will be discussed in the coming subsections. But just to see how far we have come, consider what happened when the space shuttle Columbia exploded in the year 2003 and the entire U.S. space shuttle fleet was grounded. At the time, there were U.S. astronauts in the international space station in outer space who had no way of returning to earth since they would usually travel between the space station and earth in a space shuttle. Who brought them back to earth? Our old space rival, the Russians.

This is the Era of Moshiach.

Stopping the Spread of Disease

We now begin a detailed discussion of how satellites are being used today for peaceful purposes. Our starting point will be the famous lecture of Dr. Naftali Berg *z.l.* at the 1993 Moshiach and Science Conference of the RYAL Institute, presented in the previous section of this chapter. He spoke about how satellites are now being used to locate and identify land that is suitable for agriculture and to find resources deep beneath the surface of the earth which can't be located by conventional means. He discussed the Landsat, Land TM and ERS1 satellites specifically.

I would now like to update Dr. Berg's report.[2]

---

2. This update on the various satellite systems and their peaceful uses is based on information I collected from various sources, especially Scientific American, Science News, NASA websites, and various other websites, over the years since 1993.

The Landsat 7 satellite, a much more technologically-advanced version of the Landsat satellite discussed by Dr. Berg, was launched on April 15, 1999. Landsat 7's sensor has been called "the most stable, best characterized Earth observation instrument ever placed in orbit."[3]

The Landsat satellite program drew on technologies developed under U.S. reconnaissance satellite programs like Corona (which we will discuss later). Commercial high-resolution systems currently prepared by US consortia are essentially civilian copies of earlier U.S. military systems, usually developed by the same companies that developed the military systems.

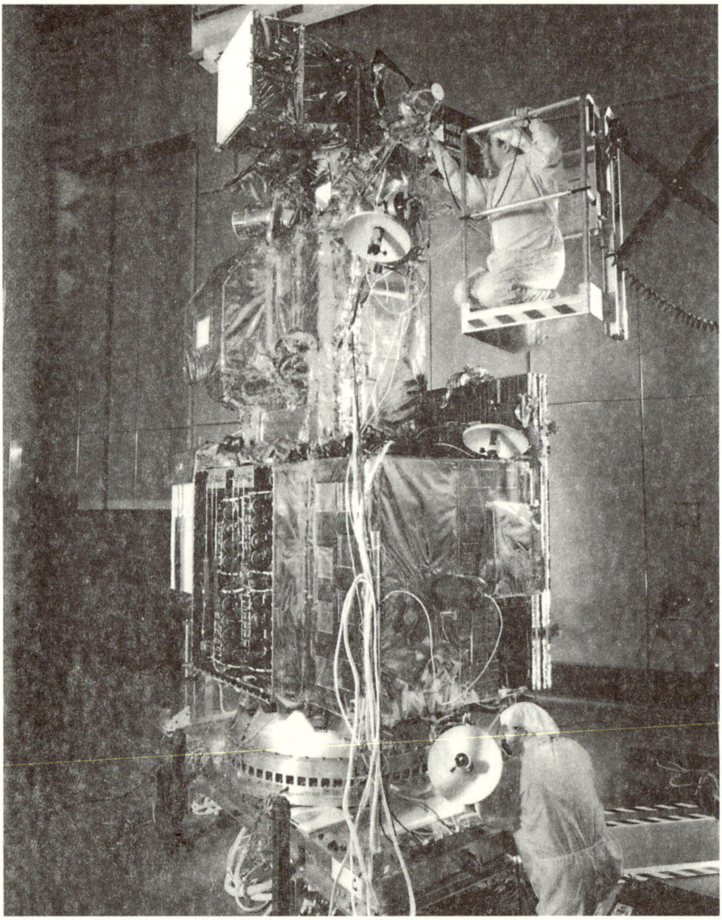

Landsat 7 being prepared for launch (Courtesy NASA)

3. http://landsat.gsfc.nasa.gov/about/landsat7.html.

Landsat and other similar satellites are now being used by NASA's Center for Health Applications of Aerospace Related Technology (CHAART) to track the spread of diseases, especially those whose spread is extremely unpredictable. As reported in the medical journal Emerging Infectious Diseases, "Perhaps the broadest use of Landsat and SPOT data has been to identify and map vegetation or crop types. This factor is important because the distribution of vegetation types integrates the combined impact of rainfall, temperature, humidity, topographic effects, soil, water availability, and human activities. Nearly all vector-borne diseases are linked to the vegetated environment during some aspect of their transmission cycle; in many cases, this vegetation can be sensed remotely from space."

One of the topics at the forefront of scientific research is to try to learn what is causing certain diseases to spread. Take, for example, cholera. Certain parts of the world are especially prone to outbreaks of cholera, such as India and other countries in that region. But scientists have observed that since the early 1800s there have been seven worldwide epidemics of cholera, leaving scientists mystified: how did the disease spread throughout the world?

Landsat has been used to identify oceanic conditions contributing to the reproduction of the bacteria that cause cholera. It was found that the bacteria itself actually grow in the oceans, and can therefore be transmitted by ocean currents. The growth and spread of the cholera bacteria is a result of the warming of the ocean in certain areas, the sea level, and certain green plants in the ocean on which this bacteria feed. Landsat is being used to identify those locations in the ocean which are conducive to the spread of the bacteria and then to forewarn those countries to which it is likely that the epidemic would spread.

In Brazil, a certain disease was spreading in urban areas. The disease was known to be transmitted by infected flies, but scientists did not know why it would hit certain areas of the cities whereas other areas were not affected. Landsat was used to identify how the flies were spreading the disease. It appeared that forest animals such as foxes were emerging from the forest, carrying the flies infected with the disease, and entering certain areas of the city which border on the forest, thereby spreading the disease to the city. The disease can now be stopped at the point that it enters the city before it spreads.

Closer to home, Landsat has achieved the same kind of success right here in New York State with respect to Lyme Disease. Lyme Disease is spread by ticks carried by forest animals. Usually deer carry the ticks, and if a person is bitten by the tick, he may contract the disease. Landsat has identified certain areas in Westchester County, for example, which are particularly conducive to the spread of this disease because of their proximity to the forest (so that forest animals coming out of the forest make contact with people nearby), as well as the abundance of shade and moisture in those areas which also contributes to the spread of the disease.

Tracking the spread of diseases is one example of the prevalent of the use of satellites for peaceful purposes. The very fact that NASA has established a Center for Health Applications of Aerospace Related Technology shows the extent to which this transformation has occurred.

The Terra Satellite

"Terra" (meaning land) is the name of the Earth Observing System (EOS) flagship satellite, launched on December 18, 1999. The five sensors aboard Terra are comprehensively measuring the earth's climate system. Its goal is to observe and measure how the earth's atmosphere, cryosphere, lands, oceans, and life all interact. Data from this mission are used in many research and commercial applications.

Flooding is a way of life along the lower Mekong River in Vietnam and Cambodia. Every year between August and November, monsoon rains fill the rivers of Southeast Asia, and the Mekong River Delta broadens well past its dry season levels.

New composite surface images from the MODIS (MODerate-resolution Imaging Spectroradiometer) instrument on board Terra have proved to be a huge help to researchers monitoring the flooding in recent years in Southeast Asia. For example, in the year 2000 Southeast Asia experienced the worst flooding in 40 years. Hundreds of thousands of people living along the Mekong River in Cambodia were forced to flee. But using Terra, the Dartmouth Flood Observatory was able to chart the flooded areas and determine the areas hardest hit long before this information was available on the ground. The maps were so good that they were able to guide relief workers as they delivered emergency food supplies to the effected areas. Terra can image most of the world every day, and it views a very wide area.

In 1996 the Dartmouth Flood Observatory began work to develop a global flood monitoring system with the idea of converting satellite imagery into flood alerts distributed to relief agencies online. One of their goals is "to know how big a flood event is and keep a global record of such events in order to determine if climate change is accompanied by changes in the frequency and magnitude of big floods."

In addition to Terra, they now use images from National Oceanic and Atmospheric Administration (NOAA) satellites, the Canadian Radarsat and the Landsat 7 satellite.

The United Nations World Food Program has been making good use of the monitoring system. In areas like Cambodia and Laos where there are weak data collection systems on the ground, the MODIS data enables them to identify the areas that need assistance.[4]

The Rebbe MHM said that Isaiah's prophecy of "Swords Into Plowshares" began to be fulfilled with the declaration in the U.N., at the end of January, 1992, regarding the transformation of military resources to peaceful uses. The Terra satellite was a genuine product of this new era. Landsat 7 was also a product of this era but the Landsat program had started earlier. Nevertheless, the satellite technology for these and similar satellites was a direct transformation of military technology. It was derived directly from the U.S. spy satellite program known as Corona.

As we will now explain, it was not just the Corona technology that was transformed to peaceful uses, but—in this new era—the Corona spy photographs themselves were declassified and used for peaceful purposes.

The Corona Satellite Program

The technological challenges facing the developers of the Corona spy satellite were almost insurmountable. In addition to the satellite itself, a new type of camera, as well as film, had to be developed. And

4. http://www.scientificamerican.com/article.cfm?id=satellites-help-find-floo.

after the film was used, it would be dropped to earth by the satellite and had to be caught in midair. The project involved a joint effort of rocket scientists, optical scientists from Kodak and Polaroid, and MIT.

In fact, the attempt to launch the satellite failed 12 times! The 13th launch was successful, but that flight did not carry a camera. Thus it wasn't until the 14th mission, on August 18, 1960, that Corona was successful. In order to deliver the film to the U.S., the satellite dropped a capsule with the film toward Earth. Then, two small parachutes on the capsule opened to slow its speed. Finally, an Air Force C-119 aircraft swooped in to snatch the capsule in midair and returned it to Earth, where CIA and Air Force technicians processed the film.

In a single day, Corona would yield more images of the Soviet Union than did the entire U-2 spy plane program. In all, the Corona program involved more than 100 satellite flights from 1960 to 1972 at a cost of about $850 million. The imagery allowed the U.S. intelligence community to catalog Soviet air defense and antiballistic missile sites, nuclear weapons-related facilities and submarine bases, along with military installations in China and Eastern Europe. Corona also provided pictures of the Six-Day War in Israel in 1967, and monitored Soviet arms control compliance.

The world at large did not learn of Corona until February 1995, when President Clinton ordered it declassified. More than 866,000 images have been transferred to the National Archives and Records Administration in College Park, Maryland, outside Washington, D.C.

It is interesting to note that even during the Cold war years when Corona was functioning as a spy satellite, it brought the world closer to peace than to war. This happened because the satellite proved conclusively that the Soviets' missile force in 1960 did not number in the hundreds, as the public had feared, but rather amounted to somewhere between 25 and 50. (This fact, however, that was hidden from the American public for years.) As physicist Richard Garwin, one of the founders of Corona, put it, "It kept the Cold War from developing to a hot war."

Finally, we note that the satellite technology introduced into modern warfare by Russia has come full circle and is now being used to help Russia reduce its nuclear arsenal. In the 1960s Corona satellites photographed secret Russian nuclear weapons facilities. After the Swords Into Plowshares declaration, Russia attempted to reduce its nuclear arsenal and safely store the warheads it kept. Adequate secure storage space was needed.

Thousands of Corona photographs, analyzed by scientists such as Joshua Handler of Princeton University, showed that Russia had enough secure storage space to enable thousands more nuclear warheads than previously thought possible to be removed from missiles and stored safely under the Strategic Arms Reduction Treaty.

Precision Farming

The general meaning of Isaiah's prophecy that in the Era of Moshiach "They shall beat their swords into plowshares" is that military resources will be redirected towards peaceful uses. We have discussed several ways in which satellites, originally designed for military purposes, have been transformed into peaceful instruments for purposes beneficial to mankind. But now we discuss a transformation of satellite technology "into plowshares" in a very literal sense — converting it into an actual instrument to aid in farming. The satellites being used are those of the French SPOT satellite system, and the farmers they are helping are the farmers of South Africa.

The SPOT *(Systeme Pour l'Observation de la Terre)* satellite system was developed by the French *Centre National d'Etudes Spatiales* (CNES). The first SPOT satellite was launched in 1986, the second in 1990, and the third in 1993. Using its "off-nadir" viewing capability, i.e., its ability to view points on the earth that are off to an angle rather than directly beneath its detectors, an area on the ground can be viewed as often as every three days. This off-nadir viewing can be programmed from a ground control station and is quite useful for collecting data in a region not directly in the path of the satellite's scanner where timeliness of data acquisition is crucial.

In the traditional method of farming, fields are managed as homogeneous units by applying seed, fertilizer, pesticides, etc. uniformly throughout the field ("whole-field management"). But this is very inefficient since not all parts of a field need the same quantities of these materials. Moreover, farmers, and the agricultural sector in general, are increasingly subjected to financial pressure due to factors such as increasing input costs, unpredictable climatic conditions, unstable economies and fluctuating market prices. A more effective approach to farming is needed.

A new scientific approach to farming called "precision farming" aims to characterize the inherent variation that exists within individual fields, to determine the reasons for this variation, and to manage it by applying only the right quantity of the right input where and when it is needed ("site-specific management"). The end results are lower inputs and cost, along with increased production and profit, as well as less pollution of ground and water resources.

A primary requirement for implementing precision farming is *spatial information* to map the variation that exists on each field. Although aerial photographs and on-ground surveys can be used, they are not cost-effective when applied over large areas. The high costs involved therefore put the acquisition of critical information out of reach of most farmers. Another requirement for precision farming is the availability of timely and repetitive information to monitor the development of crops and to detect changes in a timely fashion. Neither aerial nor on-ground surveys are suitable to satisfy these operational requirements since they cannot repeatedly cover a large farm area.

Better Than Plowshares

Satellites to the rescue! The only reliable and fully operational source of the information needed for precision farming is satellite data. Satellites are ideally suited to satisfy the spatial requirements of precision farming because they image large parts of the earth's surface at regular intervals. Data collected by different sensors contain important information regarding on-ground processes and, since the data is recorded in digital format, it can be processed rapidly.

The Satellite Applications Center in Pretoria, South Africa has developed an innovative product called AGRi-I (AGRicultural Information) to make the satellite data accessible to and easy to use for the farmer. Its goal is to provide affordable, understandable and useful agricultural spatial information at regular and timely intervals to individual farmers that can be used as input to their precision farming system. It was introduced to farmers in South Africa during the 1999–2000 growing season.

AGRi-I is a time-series of Vegetation Index (VI) maps that are delivered in near real time via the internet to the farmers. The VI map is a sensitive measure of biomass and chlorophyll activity of plants that enables a farmer to determine and monitor the growth potential of various agricultural crops such as maize, wheat, sunflower and sugarcane accurately. VI values indicate the condition of specific crops. Land areas under young crops with sparse leaf coverage have low values due to the low biomass or chlorophyll activity, as well as the relatively high incidence of bare soil per unit area, as sensed by the satellite. On the other hand, land areas where there is denser vegetation, such as mature maize fields, will have high VI values. The VI values are grouped into intervals which are then color coded and conveniently placed on a map.

This information enables the farmer to map the variation on his fields, to monitor crop development, to detect crop stress in time to ameliorate the problem, and to do early yield estimations. The SPOT satellites used for this project capture regular images over vast areas of the earth's surface at a spatial resolution of 20 meters. This means that individual blocks of information are captured that correspond to areas of 20m x 20m on the ground. The farmer receives up to 5

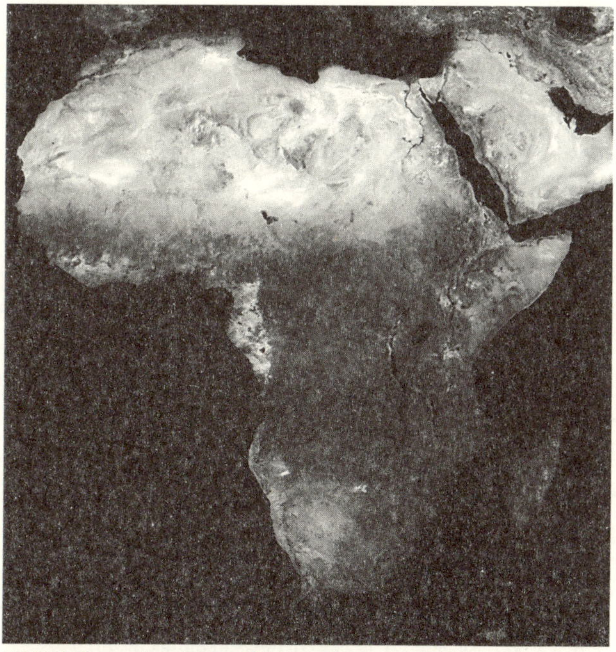

SPOT Vegetation Map of Africa (www.cnes.fr)

Vegetation Index maps of his area of interest over the growing season. Delivery is done via the internet on a monthly basis to ensure that information is timely, relevant and useful when reaching the farmer.

This redirection of military satellite technology to farming—plowshares—is one more demonstration of the fact that... *This is the Era of Moshiach!*

Communication Satellites[5]

In the fall of 1945 a Royal Air Force electronics officer named Arthur C. Clarke, later to become famous as a science fiction writer, wrote a short article in Wireless World that described the use of satellites in 24-hour orbits high above the earth to distribute television programs. This concept, however, did not "take off" until 1955 when John R. Pierce of AT&T's Bell Telephone Laboratories wrote an article analyzing the utility of satellites for communication. He compared the communications capacity of a satellite, which he estimated at 1,000 simultaneous telephone calls, to the capacity of the first trans-Atlantic telephone cable, which could carry only 36 simultaneous telephone calls (and cost 30–50 million dollars.)

After the 1957 launch of Sputnik I, the space race was on and many people considered the benefits, profits, and prestige associated with satellite communications. The first communication satellite, "Signal Communication by Orbital Relay" (SCORE), was launched on December 18, 1958. It orbited the earth for 12 days until the batteries failed.

NASA began to develop a "passive" satellite communication program and on August 12, 1960, the United States launched Echo 1, a passive reflector satellite ("mirror") with no amplification possibilities—it could only reflect the radio waves back to earth. But the Department of Defense pursued the development of "repeater" or "active" satellites which amplify the signal received at the satellite—providing much higher quality communications. The other technology was soon abandoned.

Bell Laboratories had assisted in the Echo 1 project and then went on to develop Telstar, an experimental satellite that relayed television

5. See http://www.hq.nasa.gov/office/pao/History/satcomhistory.html.

signals. It was launched into medium-earth orbit (4,000 miles high) in 1962. In the same year the Communications Satellite Corporation (COMSAT) was formed as a result of the Communications Satellite Act, and in 1965, COMSAT's first satellite, "Early Bird," was launched from Cape Canaveral. Global satellite communications had begun.

Intelsat

On August 20, 1964, agreements were signed which created the International Telecommunications Satellite Organization (INTELSAT), a new international organization which would ultimately assume ownership of the communication satellites and responsibility for management of the global system. (The Early Bird satellite was known as Intelsat 1.)

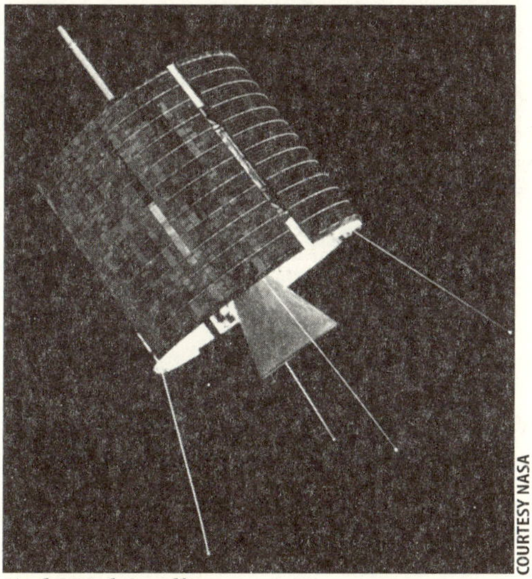

Early Bird Satellite

From a few hundred telephone circuits and a handful of member nations in 1965, INTELSAT has grown to a present-day system with more members than the United Nations and the capability of providing hundreds of thousands of telephone circuits.

Communication satellite technology reached a new level of sophistication with the launch of the Advanced Communications Technology Satellite (ACTS) in 1993. It was the first all-digital communications satellite.

Cost to carriers per circuit has gone from almost $100,000 to a few thousand dollars.

Cost to consumers has gone from over $10 per minute to a fraction of a dollar per minute, and if the effects of inflation are considered, the reduction in cost is even greater.

COURTESY NASA

Intelsat 8

INTELSAT provides services to the entire globe, not just the industrialized nations.

The extremely popular cellular telephones also make use of communication satellites for their long-distance connections. Cell phone technology is not expensive, and it is possible to place a cellular system in a developing country at a very reasonable price.

The unity among the nations of the world, as well as individuals, that communication satellites have accomplished is itself a phenomenon of the Era of Moshiach.

Communications satellites play a unique role in the Era of Moshiach. They play a fundamental role in two major aspects of the Messianic Era: 1) bringing unity to mankind, as described above, and 2) filling the world with the knowledge of G-d.

In explaining the statement of the Zohar that the science of our times (after 1840) will be a preparation for the Era of Moshiach, the Rebbe MHM said that one of the ways that this is accomplished is by using modern technology to spread Torah—the knowledge of G-d—throughout the world. For many years the Rebbe MHM's own public addresses (*farbrengens*) were broadcast live throughout the world via satellite. Even the internet, now the most advanced technological method of teaching Torah worldwide, is being greatly enhanced through the use of communication satellites.

Since 1989, Chabad-Lubavitch has been broadcasting the lighting of the Chanuka menora in "770" throughout the world via satellite and, in recent years, over the internet. But it's not just a broadcast. On many occasions it has actually been a simultaneous satellite hookup of many public menora lightings at various locations around the world. Those assembled at each location can view the menora lightings at all the other locations around the world that are connected to the satellite. Thus the satellite hookup unifies Jews throughout the world in their observance of this mitzva. It also unifies people of various nationalities, as many non-Jews also view these menora lightings and participate in the publicizing of the miracle of Chanuka.

It has also been the custom of the Rebbe MHM to say a *sicha* at these gatherings. The *sicha* is broadcast throughout the world, a foretaste the Messianic prophecy of Isaiah that the earth will be filled with the knowledge of G-d. And the Jews listening to the *sicha* are united in the study of Torah—the Torah of Moshiach.

We now present excerpts of the Rebbe MHM's *sicha* at the "Chanuka Live" menora lighting of 5752 (1991). Note, in particular, his reference to satellites and their significance in unifying the Jewish people in the observance of the mitzva of lighting the menora, and in unifying mankind in general.

The Purpose of Creation

> The Torah begins with a statement which serves as the foundation of the life of each individual Jew and of the entire Jewish people: "In the beginning G-d created the heavens and the earth." The entire universe—from the high heavens to the far flung corners of the earth—and everything contained therein, was created by G-d.
>
> The purpose of this creation—as alluded to by the Hebrew word "*b'reishis*" ("in the beginning")—is for two things that

are referred to as *"reishis"*: The Torah and the Jewish people. This means that the universe was created so that every Jew would conduct himself in accordance with the instructions of the Torah regarding everything in the world.... The very knowledge that everything was created by G-d brings one to act in accordance with the will of the Creator as expressed in the Torah.

The Talmud says, "Everything that G-d created in His world, was created only for the sake of His glory." This applies also to all the scientific discoveries of recent years. [They are also part of "everything that G-d created" and] their purpose is to increase the glory of G-d by using them for purposes of holiness, Torah and mitzvos.

In fact, it is of fundamental importance to bring out into the open this fact of creation by G-d. For example, when one says the blessing *"Shehakol Nihya Bidvoro"* ("everything came into being by His word") on a glass of water etc., he reveals the glory of G-d in the world....

Chanuka Candles and Communication Satellites

This concept is especially apparent in the act of lighting a candle, as in the mitzvos of lighting Shabbos candles or Chanuka candles, which brings physical light into the world in addition to revealing the glory of G-d. The same applies to the observance of the Torah and its mitzvos in general as it says, "A mitzva is a candle and the Torah is light."

Chanuka Candles express this idea in an even deeper way for a few reasons. Firstly, their purpose is to publicize the miracle of Chanuka. Also, an additional candle is lit every day, which brings about a continuous increase in light. And, most importantly, the Chanuka candles are lit in a such way that they will shine outside one's house—in public places where both Jews and non-Jews pass by. This brings spiritual light—goodness, fairness and justice—to the people around him and ultimately to the whole world.

In recent years there has been a dramatic development in this area. The custom was established to broadcast the lighting of the Chanuka candles on a global scale by using satellites. The satellite itself is thus being used for the glory of G-d. Through it the act of lighting the Chanuka candles that's taking place at one location can be seen at various places around the world simultaneously. Anyone who wants, whether Jew or non-Jew, can activate the equipment required to see the lighting of the Chanuka candles.

The Significance of Satellites

There are several significant aspects to publicizing the miracle of Chanuka via satellite hookup. Firstly, as it relates to each individual Jew, the satellite hookup demonstrates that each individual Jew—even a small child—has the ability to do an act that will bring light to the entire world. Just now we saw an individual Jew light a Chanuka candle, and instantly this light was seen throughout the world.

Regarding the Jewish people as a whole, through the satellite hookup unity was attained among Jewish people throughout the entire world. Even when they are dispersed among the nations to the ends of the earth, they are one people who are connected to the One G-d through one Torah. Just now we saw the fulfillment of the mitzva of lighting the Chanuka candles at various locations throughout the world. All those participating were united by this since at every location they saw the menora lightings at all the other locations.

Another significant aspect to this global satellite hookup was that the Chanuka miracle was publicized throughout the world in a way that included the nations of the world. They saw that through an act performed by the Jews there was an increase in a light that reveals the glory of G-d throughout the entire world. In this way the whole world becomes enlightened with the light of truth. Justice, fairness and peace are thereby attained throughout the world.

A Lesson for Life

This event also teaches us an important lesson that relates to the fundamental principle-of-life for every Jew—that the whole world was created for the Jews and for the Torah:

A Jew may think that an act that he performs in his own private space has no relevance to the rest of the world, especially to parts of the world that are far away from him. And thus, he would think, it would certainly have no relevance to the heavens above. [But this is not so.]

In order to make it easier for a Jew to understand that an act that he does wherever he is has an effect on the whole world—in the heavens and the earth—G-d revealed certain secrets of nature through which we can connect the heavens with the earth, and parts of the earth that are very far apart. Through the satellite that hovers in space and receives signals that are sent to it, in order to communicate them to the ends of the earth, a person can remain in his own place and be connected with someone at the other end of the world. He can talk to him and see him, to inform him of assistance that he will give him or to give him some good advice, etc.

We see this right now. From here in New York we see and speak via satellite with Jews in various places throughout the world—Israel, Moscow, Paris, Japan, India (Calcutta), Australia etc.—regarding matters of Judaism, Torah and mitzvos. We encourage them to increase in the observance of Torah and mitzvos, starting with the mitzva of "Love your fellow Jew as yourself" by giving *tzedaka* (charity) both in the material and spiritual sense, especially in matters related to the observance of Chanuka. This increase in the observance of Judaism will also bring about an improvement in material matters—good health and sustenance, etc.—a healthy and enlightened Jewish life.

We can take this a step further. In addition to using the satellite to inform someone at the other end of the world about assistance that will be sent to him, the money can actually be transferred to him instantly via satellite. It can be deposited in his bank account or his account at a store to buy food, household items, etc.

This, in fact, is one of the main purposes of satellites: To bring more unity to mankind; to enable one person to help another even when they are far apart, whether it be material help or spiritual help. Thus satellites help people to increase in matters of justice and fairness, peace and unity throughout the world.

What we said above about the light of the Chanuka candle being spread instantly throughout the world by the satellite has even more significance in our time when the faith in and the anticipation of the complete revelation of Moshiach is stronger than ever. We will actually see the true and complete redemption by Moshiach—when the whole world will be enlightened with the true light of the revelation of the glory of G-d in its ultimate perfection.

The GPS

The most well known satellite system is the Global Positioning System (GPS), developed by the Department of Defense for military navigational purposes. Following the "Swords Into Plowshares" declaration, however, it was transformed from a military instrument to a peaceful one. This "Swords Into Plowshares" transformation was even more significant than that of the other satellites that we have been discussing for two reasons: 1) The other satellites that we considered were military only in the sense that satellite technology in general had military origins. In the case of the GPS, however, the satellites themselves were developed as military instruments. 2) On the other hand, the transformation itself was a genuine conversion of military technology to peaceful uses since, for the most part, the GPS satellites have left the military domain and are now used primarily for peaceful purposes. While the technology is still in use by the military, civilian GPS receivers vastly outnumber military ones.

We will trace the development of the GPS as a military tool and its transformation to a peaceful one, and examine some of its prominent peaceful uses.

Where Am I?

Throughout history, navigation was a matter of concern for travelers. How is one to determine his location if he is in a caravan traveling through the desert or on a ship sailing across the sea? From the earliest times, mariners relied on angular measurements to celestial bodies like the sun and stars to calculate their location.

In the 1920s, "radio navigation," a technique based on radio waves, was introduced. This technique allowed navigators to locate the direction of shore based radio transmitters. Satellites were first used in position-finding in the 1960s in a simple but reliable two dimensional system called "Transit." This laid the groundwork for a system that would later revolutionize navigation forever—the Global Positioning System.

The Military Development of the GPS

GPS navigation and position determination is based on measuring the distance from the user position to the precise locations of the GPS satellites as they orbit. By measuring the distance to four GPS satellites, it is possible to establish three coordinates of a user's position (latitude, longitude, and altitude) as well as GPS time.

The Defense Department's primary purposes in developing GPS were to use it in precision weapon delivery and to reduce the need to develop specialized navigation systems that supported only the requirements of specific missions. Beginning in the early 1960s, the U.S. Department of Defense began pursuing the idea of developing a global, all-weather, continuously available, highly accurate positioning and navigation system.

The Transit system, sponsored by the U.S. Navy, was the first operational satellite based navigation system. Developed by the Johns Hopkins Applied Physics Laboratory under Dr. Richard Kirschner in the 1960s, Transit consisted of seven low- altitude, polar-orbiting satellites that broadcast very stable radio signals; several ground based monitor stations to track the satellites; and facilities to update satellite orbital parameters. Transit users determined their position on earth by measuring the Doppler shift of signals transmitted by the satellites.

The concept for Transit evolved from observations of the Russian satellite Sputnik in 1957. Researchers at the Applied Physics Laboratory (APL) discovered that measurements of the Doppler shift as the satellite passed by were adequate to determine the entire satellite orbit. They noted that conversely, if the satellite orbit were known, position on the earth could be determined using these same Doppler measurements. Transit was designed to meet the Navy's requirement for locating ballistic missile submarines and other ships at the ocean's surface.

Transit was important for the later development of GPS because it resulted in a number of technologies that were extremely useful to GPS, and it demonstrated that a space system could offer excellent reliability. "Timation," a second forerunner of GPS, was a space based navigation system technology program the Navy began work on in 1964.

In the meantime, the Air Force was working on a similar technology program that resulted in a design concept called System 621B. It provided three-dimensional (latitude, longitude, and altitude) navigation with continuous service.

Finally, in April 1973, the Deputy Secretary of Defense designated the Air Force as the lead agency to consolidate the various satellite navigation concepts into a single comprehensive Department of Defense system to be known as the Defense Navigation Satellite System (DNSS).

By September 1973, a compromise system was evolving which combined the best features of earlier Navy and Air Force programs. The system concept that emerged is what is known today as the NAVSTAR Global Positioning System.

From the beginning, the GPS was a military tool developed by the military for the military. Its most prominent military success was its use during the Persian Gulf War of 1991. But after the Gulf War, as we entered the "Swords Into Plowshares" era, the GPS was quickly transformed from a military instrument to one that is primarily peaceful.

We will now discuss the structure of the GPS and its use in the Persian Gulf War. Then we will describe its transformation to peaceful uses.

The Structure of the GPS

The first developmental GPS satellites were known as Block Is. Between 1978 and 1985, a total of eleven Block I satellites were launched. The military actually had a dual role planned for the GPS satellites. In addition to carrying navigation and timing equipment,

GPS satellites would carry nuclear detonation (NUDET) sensors designed to detect nuclear weapon explosions, assess nuclear attack, and help in evaluating strike damage. The system would also contribute to monitoring compliance with the nuclear test ban treaty. The first GPS satellite to carry a nuclear explosion detection sensor was the sixth Block I satellite, launched in April 1980. The next series of GPS satellites were known as Block IIs. The first one was launched in February 1989 after a two year delay caused by the Challenger Space Shuttle accident.

The GPS consists of three segments: space, control, and user:

COURTESY NASA

GPS1 Satellite

- The Space Segment consists of operational satellites in six circular orbits 20,200 km above the earth at an inclination angle of 55 degrees with a 12-hour orbital period. The satellites are spaced in orbit so that at any time a minimum of six satellites will be in view to users anywhere in the world. The satellites continuously broadcast position and time data to users throughout the world.
- The Control Segment consists of a master control station in Colorado Springs, Colorado, as well as five monitor stations and three ground antennas located throughout the world. The monitor stations track all GPS satellites in view and collect

ranging information from the satellite broadcasts. The monitor stations send the information they collect from each of the satellites back to the master control station, which computes extremely precise satellite orbits. The information is then formatted into updated navigation messages for each satellite. The updated information is transmitted to each satellite via the ground antennas, which also transmit and receive satellite control and monitoring signals.

- The User Segment consists of the receivers, processors, and antennas that allow land, sea, or airborne operators to receive the GPS satellite broadcasts and compute their precise position, velocity and time.

GPS and the Persian Gulf War

The 1990–91 Persian Gulf War was the first major test of GPS in a combat situation and it proved beyond a doubt the importance and utility of the NAVSTAR. GPS revolutionized combat operations on the ground and in the air during Operation Desert Storm and was, as one Allied commander noted, one of two particular pieces of equipment that were "war winners" (the other was night vision devices). GPS navigation proved to be a crucial technique for desert warfare. GPS satellites enabled coalition forces to navigate, maneuver, and fire with unprecedented accuracy in the vast desert terrain almost 24 hours a day despite difficult conditions, frequent sandstorms, few paved roads, no vegetative cover, and few natural land marks.

Although on average, each U.S. Army maneuver company (e.g., tank, mechanized infantry, or armored cavalry) had at least one GPS receiver, the demand for receivers was so great that 10,000 more units were hastily ordered during the crisis so that more coalition forces could benefit from the system. Operations such as precision bombing, artillery fire support, the precise positioning of maneuvering troop formations, and Special Forces operations such as combat search and rescue missions were all enabled or greatly enhanced by the GPS. In addition to being carried by foot soldiers, GPS receivers were attached, in some cases with tape, to vehicles and helicopter instrument panels and were also used in F-16 fighters, KC-135 tankers, and B-52 bombers.

Peaceful Uses

It was the military use of the GPS in the Persian Gulf War, just prior to the Swords Into Plowshares declaration, that gave the commercial GPS market its biggest boost. The success of GPS in Operation Desert Storm sparked a surge in a market that had barely existed just a few years prior to the war. In addition to precipitating a rise in demand for GPS commercial receivers, the war provided GPS technology and the suppliers of GPS receivers broad exposure. News coverage of the conflict served as free publicity for the two main wartime suppliers.

Since the Persian Gulf War, the United States has employed GPS in various aid operations. During Operation Restore Hope in 1993, GPS was used to air drop food and supplies to remote areas of Somalia because of lack of accurate maps and ground based navigation facilities. During the Balkan crisis, GPS assisted in delivery of aid to the Bosnians by guiding U.S. Air Force transport planes at night to their drop zones where food and medicine was then parachuted close to towns and villages.

The launch of the 24th Block II satellite in March 1994 completed the GPS constellation. "Initial Operational Capability" (IOC) was formally declared December 8, 1993 in a joint announcement by the Department of Defense and the Department of Transportation. The IOC notification means that the NAVSTAR GPS is capable of sustaining the Standard Positioning Service (SPS), the 100-meter positioning accuracy available to civilian users of the system on a continuous, worldwide basis. This IOC announcement for GPS had purely civilian connotations.

In 1991, the Federal Aviation Administration's (FAA) promised that GPS would be available free of charge to the international community beginning in 1993 on a continuous, worldwide basis for at least 10 years. This offer was extended the following year when the United States offered SPS to the world for the foreseeable future and pledged to provide at least six years' notice prior to termination of GPS operations or elimination of the GPS SPS. Thus the GPS became available to all nations of the world for peaceful uses.

One of the earliest civilian uses of the GPS was surveying, since this could be done even before all the GPS satellites were operational. Surveyors' applications could accept the limited availability of satellite signals. They did not need to use their data in real time, so they could make observations whenever sufficient satellite signals were available. GPS surveying offered greater productivity and cost savings over traditional survey methods. Tasks that normally required several weeks or months to finish could now be completed in a fraction of the time using GPS, at one-fifth to one-tenth of the cost of conventional surveying.

Aircraft navigation was another early application. In fact, the first U.S. announcement regarding civil use of GPS came in 1983 following the downing of Korean Airlines Flight 007 after it strayed over territory belonging to the Soviet Union. At that time, President Reagan announced that the Global Positioning System would be made available for international civil use once the system became operational.

The Global Positioning System has proved to be one of the most pervasive technologies in our society. Currently, many automobiles are equipped with GPS navigation aids. One particularly sophisticated system called OnStar performs a variety of functions. For example, in a crash, built-in vehicle sensors automatically alert an OnStar Advisor and relay critical details of the incident. An Emergency Advisor is immediately connected to the vehicle to see if help is needed. Even if the driver is hurt and can't respond, the Advisor knows his exact location through GPS technology and can direct emergency responders to the scene. The Advisor also provides critical crash data to the responders so they are better prepared to administer treatment when they arrive at the scene. Automatic Crash Response enables OnStar to send help even if the victim can't ask for it—and save lives.[6]

Another application, called SafetyNet, can identify bad neighborhoods for a driver so he can avoid them. If he happens to end up in such an area, his mobile phone will switch into watchdog mode. If trouble arises, all he has to do is shake the handset and the phone will set off an alarm, start sending his location to friends and family every 30 seconds via Google Talk, take a picture, turn on its speaker, and dial 911 or another emergency number. The application

6. See http://www.onstar.com/us_english/jsp/plans/safe_sound.jsp.

announces each of these steps out loud so a potential attacker can hear what's happening and run away. At the very least, he won't want to steal a cell phone that's shrieking about calling the cops.[7]

GPS Guided Vehicles

It's a cold night in Minnesota. The snow begins to fall. It falls and falls and falls.... They send out the snow plows to clear the roads—the streets, the highways and the long, winding and treacherous narrow country roads. But the driver of the snow plow can hardly see two feet in front of him, so how does he avoid hitting a passenger car that may be on the road? What if there's an oncoming car?

It's a dark night in the San Joaquin Valley in California. A heavy fog sets in. People are trying to get home from work but they can hardly see a few feet ahead on the road. Even if they drive at normal speeds, a car may appear suddenly in front of them. Will they be able to stop in time? And what if someone is speeding? Dozens of accidents involving multiple car pileups occur every year in this area during the "fog season." For example, on November 1, 2007 a 108 car pile-up occurred on Highway 99 in Fresno, killing three people and injuring over 100. It was blamed on a drunk driver going too fast in the morning fog, causing a chain reaction.

Back to Minnesota. As in every big city, traffic congestion is a problem in the Twin Cities of Minneapolis and S. Paul. Public transportation buses often suffer long delays in slow-moving traffic. To alleviate this, a 260 mile network of special bus-only lanes has been designated on the shoulders of various roads. But these narrow lanes have problems of their own. How does a nine-foot-wide bus swing into one of these ten foot wide lanes without error? And driving in these lanes is no picnic. With stalled traffic on the left and guardrails or barriers on the right, drivers say, they feel as though the lane is closing in on them. When snow masks the pavement markings, bus drivers don't see the shoulder's barriers. Bus drivers have to deal with other motorists crowding the shoulder and cars merging on and off freeways.

I, myself, lived in Minnesota for many years. Often, I would drive

7. http://www.wired.com/gadgets/wireless/magazine/17-02/lp_10coolapps?currentPage=9.

to New York with my family for a *farbrengen* such as *Chof Av* (the 20th of Av—the yartzeit of Rabbi Levi Yitzchok Schneerson, father of the Rebbe MHM) or one of the Jewish holidays. The first time I made such a trip someone said to me, "You're going to drive all the way to New York on the freeways—with all those trucks?!" I didn't see the problem. It's a good thing that I didn't know the statistics that I saw years later (while doing the research for this book) regarding car-truck collisions. In 1997, 5,355 people were killed in car-truck collisions, and it has been the subject of Congressional hearings.

Could a system of satellites developed by the military to help navigate battleships and fighter jets be the solutions to all these problems and save thousands of lives every year? The answer is yes! Professors Max Donath and Craig Shankwitz of the Intelligent Transportation Systems Institute at the University of Minnesota, have developed a system that uses the Global Positioning System (GPS) to maintain snow plows, buses and trucks on the proper course, steer them back on course if they waver, and alert them to nearby vehicles and other objects that they may not see.

The TechnoBus

Take, for example, the TechnoBus developed by the Intelligent Vehicles Lab of the Intelligent Transportation Systems Institute. This is how it works: Two GPS receiver units mounted on the top of the bus determine its location to within centimeters. An onboard digital map database of bus-only shoulder lanes contains locations of the shoulder lane striping, also to centimeter accuracy. Using the GPS-provided position, the map database is queried for information about the bus shoulder. The information is projected onto a "head-up" (overhead) display. This type of display was developed by the Air Force for fighter aircraft—another Swords Into Plowshares transformation. The display allows the driver to see a computer representation of the road. The GPS based computer representation enhances the driver's visibility of the shoulder lane boundaries—even when snow and fog obscure the actual view.[8]

8. See http://www.its.umn.edu/Research/ProjectDetail.html?id=2001046 and http://www.its.umn.edu/Publications/AnnualReport/2006/LabsFacilities/intelligentvehicles.html for the latest on this.

Similar technology has been installed on several snow plows in Minnesota and has been successful.[9] And such technology has been developed for trucks to help avoid collisions with nearby cars and to maintain vehicles on the road in the event that the driver may have lost control of the vehicle or fallen asleep (the SafeTruck program).[10] It is also being considered for California Highway Patrol vehicles that lead convoys of cars through the fog—a system they call "pacing."

The location of missing children and anti-theft devices that track auto thieves are other important uses of the GPS. Now that new wireless devices such as handheld computers and smart telephones are popular, GPS technology plays an even greater role in our daily lives.

The TechnoBus

Bringing Aid to Disaster Areas

In 1998, Hurricane Mitch devastated Central America, killing more than 9,000 people, leaving two million homeless and causing more than $5 billion in damage. It was the worst storm in the Western Hemisphere in 200 years.

Satellites to the rescue! Satellite images of the areas hit—including images from Russian satellites— were used by relief workers and reconstruction crews to locate landslides and retrace roads, utilities and shifting waterways to deliver needed aid quickly, and to spot potential danger areas before they were hit. The effort was coordinat-

9. See http://www.its.umn.edu/Research/FeaturedStudies/driverassistive.html.
10. http://www.its.umn.edu/Research/ProjectDetail.html?id=1997033.

ed by the Center for Integration of Natural Disaster Information of the U.S. Geological Survey. The information has also facilitated the rebuilding of the devastated areas. These methods can be used to coordinate efforts to respond with aid to areas hit by a natural disaster even in a remote part of the world.

Satellite image of Hurricane Mitch **9/26/98**

The Indonesian Tsunami

On December 26, 2004, an earthquake measuring 9.0 on the Richter scale hit the Indian Ocean 150 miles northwest of Indonesia's island of Sumatra. Thirty minutes later waves 100 feet high wiped dozens of villages off the map in Sumatra's Aceh province. Tsunamis also traveled across the Indian Ocean, striking Thailand, Sri Lanka, India, Malaysia, and several other countries.

In the following days and weeks, the death toll due to the tsunami rose by the thousands and tens of thousands, reaching over 230,000.

By the beginning of January 2005, the World Health Organization had estimated that half a million people had also been injured and 5 million people made homeless by the tsunami and warned of a health disaster if sanitary drinking water was not made available. They feared that the number of deaths from the spread of disease could match those due to the tsunami itself.

The earthquake was the most powerful one to occur in the world in the past 40 years, and the death and devastation that it brought made it the worst natural disaster in recent history. And the nations of the world responded with the largest humanitarian aid operation ever.

In the *sicha* of 13 Iyar, 5751 (1991), the Rebbe MHM described the commitment to goodness and kindness by the nations of the world in connection with aid delivered by the U.S. military to countries devastated by war:

> There are ways in which we see how the nations of the world are doing acts of goodness and kindness, which further reveals that there is a Master of the universe. This is a preparation for the fulfillment of the Messianic prophecy that "sovereignty will be G-d's alone."
>
> This is in contrast to the way it used to be—that only the Jewish people, even though they are a minority, would be fulfilling G-d's intention of making this world a "dwelling place" for Him, while some of the other nations would act only in their own self-interest, being cruel to each other, conquering other countries, etc. Now we see that the nations of the world are preparing themselves for the Geula by doing acts of kindness....
>
> It is well known that in recent days this country (the U.S.), which is founded on charity and kindness, used its military power to provide aid and rescue people in a far-off place in the world even though this brought no direct benefit to the citizens of this country. The U.S. sent many soldiers, bringing food, clothing and medicines, to help refugees in those places. Instead of using its air force for conquest and instead

> of using the food and clothing for its own citizens, it used them to save these destitute people, especially children, preventing loss of life.
>
> Thus we see how merciful this nation is. When they saw that people are suffering they moved quickly to help them, even though they were not their relatives and they never even had any connection with them. This is consistent with the charitable policies of this country that they help people throughout the world without waiting to be asked.[11]

This is a theme which we have been discussing since the beginning of this chapter—the recent refinement and redirection of the nations of the world towards goodness and kindness, culminating in the Swords Into Plowshares declaration.

Largest Humanitarian Aid Operation Ever

In the case of the tsunami, the aid provided by the U.S. and other nations was the largest humanitarian aid operation ever undertaken. By January 1, international aid pledges had reached two billion dollars. During the last week of 2004 alone, the U.N. reported receiving 1.5 billion dollars. By the middle of January public and private pledges to the devastated countries were over 10 billion dollars. Billions more followed over the following months.

But money alone was not enough. Food, water and medical supplies had to be delivered to the survivors, some of them in remote areas. Once again, the U.S. used its military resources for peaceful purposes. An aircraft carrier battle group was sent to Indonesia and began launching helicopters heavy with supplies. A flotilla carrying Marines and water-purifying equipment headed for Sri Lanka, and a former staging base for B-52 bombers in Thailand roared with the takeoffs and landings of giant cargo planes. More than 20 vessels with thousands of sailors and Marines were dispatched within a week of the disaster, along with some 1,000 land-based troops. Millions of pounds of food were brought in.

There was another amazing element in the U.S. aid for Indonesia. Indonesia is not a friend of the U.S. at all. On the contrary, it is quite

11. *Sefer HaSichos 5751*, 2:517–18.

unfriendly to this country and did not want U.S. Navy ships cruising its territorial waters! Yet, the U.S. continued to help the Indonesian people.

Satellites to the Rescue

The relief effort was seriously hampered by the fact that many roads and bridges were destroyed and many areas were flooded. It was difficult to know exactly where to bring the aid and which areas were inaccessible. This became a major problem.

Here too, the U.S. Department of Defense used its military satellites to help workers manage their relief efforts. The National Geospatial Intelligence Agency (NGA), the intelligence agency that tasks spy satellites to monitor military maneuvers and find secret weapons plants, and provides spy satellite imagery and analysis to the U.S. government, was pressed into service to help in the relief effort following the earthquake and the resulting tsunami.

The agency used its satellite imagery to measure the scope of the tsunami and to funnel information, including damage assessments of roads, bridges, ports and airfields, to the US agencies handling disaster relief. Those updates were used to guide where workers and life support supplies should be sent. Maps created using the satellite imagery were delivered instantly over the internet. Supplies were then delivered by helicopter to the inaccessible areas.

Several months later, NGA satellites were used once again to aid in the relief effort for the victims of Hurricane Katrina in the U.S.

And it's all thanks to military technology being transformed to peaceful uses—and "a little help from Above."

The Hand of G-d

One of the first thoughts that comes to mind when hearing of this awesome disaster is the central prayer *Unesaneh Tokef* ("We declare the power of the holiness of the day"), recited on Rosh HaShana and Yom Kippur, which states: "On Rosh HaShana it is written and on Yom Kippur it is sealed: who will live and who will die…who by fire and who by water." Even in the midst of the mighty tsunami wave when people are being washed away by the tens of thousands, one could clearly see the hand of G-d, the Divine providence, applied to each individual. Here are two such stories.

A man on an island off India was washed away with the rest of the residents of his island when the first tsunami wave struck. But the second, more powerful wave dumped him back on the shore. He was the only survivor of the island, and he lived off coconuts for 25 days until he was rescued.

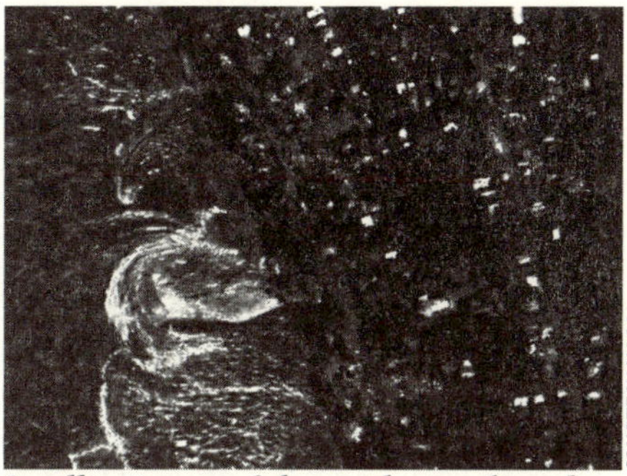

COURTESY NASA

Satellite image of the swirl preceding the tsunami.

A ten year old school girl from England was on the beach in Thailand with her family when she noticed the ocean water begin to swirl and foam and then recede. She had just learned in her geography class two weeks earlier that these were the signs of an imminent tsunami wave. She told her parents that a tsunami was coming, and they warned everyone in the area. They all escaped, heading for the high ground. About 100 lives were saved as a result of her warning. There are many similar stories.

The International Charter

Military satellites from European and other countries are also being used in disaster relief. In July 1999, following the UNISPACE III conference held in Vienna, Austria, the European and French space agencies (ESA and CNES) initiated the "Charter On Cooperation To Achieve The Coordinated Use Of Space Facilities In The Event Of Natural Or Technological Disasters" (often called "Space and Major Disasters" for short). It was later signed by Canada, the U.S., Argentina, Japan and China.

The International Charter aims at providing a unified system of space data acquisition and delivery to those affected by natural or man-made disasters. A user can request the mobilization of satellites such as ERS, SPOT, LANDSAT, NOAA satellites and others to obtain data and information on a disaster occurrence.[12]

Satellites 2010

The process that we have been describing—the use of military satellite technology for peaceful purposes and even to save lives—continues as we go to press. We will report on two such uses within the past year.

We discussed how satellites have been used to track the spread of diseases thereby enabling the people in the path of the disease vector to take precautionary measures. In 2009, Dr. Timothy Ford, Vice President for Research and Dean of Graduate Studies at the University of New England, published a paper in the *Journal of Emerging Infectious Diseases* in which he described the use of earth-monitoring satellite data to help public health officials predict and minimize the effect of infectious disease outbreaks. This information may be used to predict outbreaks of cholera (as we described above), Ebola virus, hantavirus, and plague.

In a follow-up report on Dr. Ford's paper, the Journal of the American Medical Association (JAMA) notes that, "Even a few weeks of warning could help target the efforts of public health officials to those areas that are at greatest risk and allow them to deliver public health messages that can substantially reduce the burden of disease. For example, Ford noted that advance warning of a potential cholera outbreak could allow public health officials to stress the importance of basic interventions such as filtering drinking water through sari cloth, which can remove plankton from the water and reduce mortality by 50%." [13]

We also discussed how satellites have been used to help victims of natural disasters such as hurricanes and earthquakes. In May, 2009,

---

12. See http://www.disasterscharter.org/about_e.html. See also http://www.disasterscharter.org/about_e.html.

13. http://my.une.edu/news/vpforresearchtimothyfordinjamaarticleonbenefitsof satellitedataforpredictinginfectiousdisease-1. [Yes, this is a real URL!]

Honduras was hit by an earthquake. Disaster officials knew exactly where to send help since a state-of-the-art Earth observation system called SERVIR directed them to the hardest hit areas. The satellite system uses imagery to zero in on places where a flood, fire, hurricane, or earthquake has left destruction in its wake. Team members combine satellite data with ground observations, and display a near real-time map of crisis points. At a glance, decision-makers can see the locations of most severe damage so they can send help in a hurry. SERVIR is a joint effort of NASA and several other agencies.

Another aspect of the relief effort was the international cooperation characteristic of the Swords Into Plowshares era, as the Rebbe MHM explained in the sicha of Parshas Mishpatim. Shortly after the earthquake struck, the Taiwanese imaged the hardest hit areas with their Formosat-2 satellite. Dr. Cheng-Chien Liu of the National Cheng-Kung University of Taiwan emailed Dan Irwin, SERVIR Project Director (at NASA's Marshall Space Flight Center) who was in Berlin to tell him that they had images ready to send, and stayed up until the early hours of the morning to send them.[14]

In the recent earthquake in Haiti, satellites were put to work immediately in two ways. Firstly, satellite imagery showed the exact areas devastated by the earthquake. Images also indicated the areas where large numbers of people went for refuge. This information told rescue workers where to send food, water and medical supplies.

Here, too, it was a joint effort of many nations, whose satellites collected this data, including the European Space Agency's ERS-2 and France's SPOT-5 satellites which we discussed earlier. These satellites can use radar imagery to monitor a region even when it's covered with clouds, and even detect the surface deformations that may signal aftershocks or landslides to come.

Secondly, satellites were used for communication. When the rescuers arrive, one of the most immediate needs beyond food, water and medicine is the ability to communicate so they can coordinate their efforts and bring in the necessary aid. Here again, satellites came to the rescue. Telecoms Sans Frontieres, a French-based international relief organization, was among the first on the scene with BGAN (Broadband Global Area Network) terminals. The

14. http://www.nasa.gov/mission_pages/servir/earthquake.html.

BGAN devices are about as big as a netbook or laptop computer, and provide a mobile hookup for phone or data communication (wireless or wired) through the Inmarsat satellite network. Rescue teams from around the world converged on Haiti bringing dozens of BGAN terminals with them.[15]

15. http://cosmiclog.msnbc.msn.com/archive/2010/01/14/2174324.aspx.

A Nuclear Reactor for the Era of Moshiach

Interview with Professor Alvin Radkowsky

There is something awesome in contemplating nuclear energy—the release of the innermost power of matter known to man...

In fact, the Rebbe MHM said that there is a fundamental lesson that we must learn from nuclear energy regarding the power of the individual. No individual should feel that he is too small or too insignificant to have a positive effect on the world. For we have seen that a small amount of nuclear material can bring destruction to an entire city. But the power of good is much stronger than the power of destruction and every person, no matter how small he may feel, has a tremendous reserve of spiritual power within himself so he has the power to do an incredible amount of good for the world.

In discussing Swords Into Plowshares—the transformation of military technology to peaceful uses—we could certainly say that the greatest "sword" of all time has been nuclear weapons. Realizing the danger that a nuclear war poses to the world, those nations that have nuclear technology have made agreements and treaties not to provide this technology to other nations. This restriction must also apply to nuclear reactors designed to generate electricity, since a byproduct of the nuclear reaction (when uranium is used as the fuel) is plutonium, which can be used to make bombs.

Unfortunately, this prevents many poor nations of the world from having access to the cheap electricity provided by nuclear reactors and puts them in a position where they must take drastic measures to obtain energy—such as cutting down their forests for firewood—or else do without it.

Wouldn't it be nice if someone would design a nuclear reactor that could not be used for making bombs, but could only be used for peaceful purposes—like providing cheap electricity for poor countries? What a Swords Into Plowshares development that would be!

In this section we tell the story of just such a development.

The Ad in the New York Times

In the summer of 1993, Rabbi Yitzchok Springer of Crown Heights, Brooklyn, New York placed a full-page ad in the New York Times presenting the statements and prophecies of the Rebbe MHM concerning the fact that we are now in the Era of Moshiach and that all the Messianic prophecies have begun to be fulfilled. Among them he mentioned the Swords Into Plowshares prophecy about military resources being transformed into peaceful uses.

Not long afterward, he received a phone call from a Manhattan lawyer named Joel Grae who explained that he was the president of a corporation[1] that was designing a nuclear reactor that could only be used for peaceful purposes (like we described above). Mr. Grae told Rabbi Springer that he saw this new nuclear reactor as a clear example of the Swords Into Plowshares prophecy that he read about in the ad in the New York Times. The reactor was designed by a religious Jewish nuclear engineer named Professor Alvin Radkowsky.

Being familiar with the work of the RYAL Institute, Rabbi Springer contacted me and showed me the material on the nuclear reactor that he had received from Joel Grae. A close relationship then developed between Rabbi Springer, Joel Grae and myself, which continues to this day. Mr. Grae has spoken a few times at our annual Moshiach and Science conferences on Rabbi Yisroel Aryeh Leib's yartzeit (13 Iyar). Prof. Radkowsky himself accepted our invitation to participate in our conference of 1994, but his schedule required that he be in Europe at that time. Instead, he made himself available for an interview earlier in the year when he was in New York, on January 19. The video of this interview was subsequently viewed by the Rebbe Melech HaMoshiach himself, and was shown at the Moshiach and Science conference that year.

1. The Radkowsky Thorium Power Corp., now called Lightbridge. See www.ltbridge.com.

Prof. Alvin Radkowsky

Radkowsky, a student of Edward Teller, was the original Chief Scientist for the US Naval Nuclear Propulsion Division directed by the late Admiral Hyman Rickover. Radkowsky headed the design teams for the US Navy's nuclear submarines, aircraft carriers, and other surface ships. He also headed the design team that built the first full-scale civilian nuclear power plant at Shippingport, Pa. The technology for this electricity-generating plant was, to a large extent, based on the technology used on the nuclear ships.

In 2001, Dr. Radkowsky was awarded the Alvin M. Weinberg Medal for outstanding international technical and policy leadership in nuclear science and technology. It was given to him as recognition for his "seminal contributions and innovations in the engineering development of nuclear power." He passed away in 2002.

In 1972 he moved to Israel, where he was appointed a professor at Tel Aviv University. Before leaving the U.S., he was encouraged by his friend (and ours) Dr. Naftali Berg to meet with the Rebbe MHM. The meeting was arranged and lasted for 45 minutes. They discussed many matters of Torah and science and he received the blessing of the Rebbe MHM, who also encouraged him to continue writing on topics of Torah and science. Subsequently in Israel, Radkowsky began the design of his new nonproliferative nuclear reactor which could supply cheap nuclear energy to any nation of the world without fear of the materials being used to make nuclear weapons.

In the first part of the interview we discuss the Era of Moshiach in general and some topics of modern physics as they relate to Judaism. Afterwards we discuss the nuclear reactor and its connection with the Era of Moshiach.

The Limits of Knowledge

PROF. SHIMON SILMAN: Professor Radkowsky, welcome to our conference. I would like to begin our discussion by explaining the theme of our conference, "Seeing Moshiach through the Eyes of Science." A couple of years ago the Rebbe MHM announced, "*Higia Zman Geulaschem*—the time of the redemption has come." Now everything in the world has to be connected with Moshiach and prepared for the redemption. That is why we organized this series of conferences. The first one took place two years ago on the 40th yartzeit of Rabbi Yisroel Aryeh Leib, the youngest brother of the Rebbe MHM. He was a great mathematician who worked in areas of theoretical physics, especially quantum mechanics, in the 1930s and 1940s. He passed away in 1952 while he was working at the Department of Theoretical Physics at the University of Liverpool.

The purpose of this conference is to investigate how we, as professional scientists, can make our scientific professions ready for Moshiach. We explore the role that science will play in the Era of Moshiach and how we can bring the revelation of Moshiach into our scientific work, our study of science and the discipline of science itself. In short, we delve into the question of how to bring Moshiach into the world of science.

Many years ago, the Rebbe MHM spoke about the statement in the Zohar that says that in the 600th year of the 6th millennium (1840) the windows of the heaven above will open up and the fountains of wisdom below will break open, and this will prepare the world for the coming of Moshiach. He explains that the opening of the windows of the heaven above refers to the first publication of the works of Chabad Chassidic Philosophy developed by Rabbi Schneur Zalman of Liadi, the *Alter Rebbe* (the first Rebbe of Chabad). Chabad Chassidus talks of the unity of G-d as it is revealed from above. The fountains of wisdom below refers to new developments in the sciences which took place at the same time, which prepare the world for the coming of Moshiach by showing unity within nature. These two developments together prepare the world for the revelation of the absolute unity of G-d which will be revealed and evident to everyone in the time of Moshiach.

Can you explain for us how the developments in the past century or so in quantum mechanics, atomic theory, and nuclear physics show a greater degree of unity in nature then was ever before conceivable?

PROF. ALVIN RADKOWSKY: There is a great deal to be said on this subject. I will confine myself to one aspect which I think is extremely revealing, and that is the role of quantum mechanics. As most of you know, the recent articles in Physics Today and Scientific America have shown that there seem to be contradictory experiments when it comes to the role of particle physics. This is based on the fact that ultimately it is not possible to observe the momentum and the position of a particle at the same time. If we want to have full knowledge of a position, we cannot have full knowledge of the momentum, and vice-versa. It seems that there is a limit to our understanding. We can never have a complete understanding. This was reconciled by Neils Bohr with the idea of complementarity. If you have a lens, from one point of observation it seems to be convex and from the other point—the opposite point—it seems to be concave, and both things are true yet at first glance they seem to be contradictory. In actuality, there is a unity between these two viewpoints which seem to be absolutely contradictory.

There is an ancient problem in our understanding of G-d's rule of the universe, which was especially mentioned by the Rambam: How can we reconcile free will with the fact of G-d's knowledge of the future? The Rambam says that the answer to this is as deep as the sea and as wide as the earth. In other words, it's impossible to have an understanding of it. But actually if we use the idea of complementarity, we do see that it is possible to have both. There is no contradiction between the two because just as you cannot simultaneously observe the convexity of a lens at the same time that you observe its concavity, in the same way you cannot observe the prior knowledge of G-d with the person's exercise of his free will. So even though this seems to be contradictory, it's in line with the basic principle of complementarity which we find exists all throughout the science of the universe. I think that is all I can say in a short time.

[*In fact, the question is much deeper than just "observing" G-d's knowledge and free will. For an in depth discussion of this, see the Rebbe MHM's "Hadran on the Rambam—5746."* [2]]

SS: Rambam writes in *Hilchos Yesodei HaTorah* (Laws of the Foundations of the Torah—at the beginning of his halachic work Mishneh Torah) about the knowledge of G-d and free will. The halacha with which Rambam begins these laws is: "The foundation of all foundations and the pillar of all wisdom is to know there is a First Being Who causes everything else to exist, and everything that exists in the heaven and earth and that which is between them exists only by virtue of the true existence of the Creator, G-d." This applies

2. *Likutei Sichos* 27:251–52.

not just to objects like a table and a chair or a video camera, but to more abstract entities like a concept or a scientific discipline, as well. How can one see within the study of physics or science in general that the true existence of this entity—the scientific discipline—is not as an independent entity, but rather that its true existence is the existence of the Creator?

AR: I think it's really what I said before, that this fundamental paradox we observe in quantum theory, while it's first brought to our attention in quantum theory, it involves the entire universe. So the whole understanding of the fundamental way in which G-d interacts with nature is understood through this fundamental principle of complementarity.

Another important thing is the fact that there are limitations on our knowledge which are imposed by, you might say, relativity and quantum mechanics. While our knowledge has grown very great—we can go into outer space, and we have great computing powers—at the same time we realize our limitations because mankind has to be great and humble at the same time. The very existence of the fact that there is a limit on the speed of light makes it impossible for us ever to yell very far into the universe because human life is finite and the universe is so great. The same principle limits our power to calculate and compute because there is a fundamental limit on computing. The "distance we can go" is limited by the speed of light because after a certain point we cannot increase faster then the signals will travel between different components of a computer. So in both cases our knowledge has advanced so much that we finally realize our limitations.

[*The Rebbe MHM objects to the notion that the speed of light is an* absolute *limit on motion.*[3]]

SS: So then you would say that we can see within science itself that it has reached its limitation and points to something beyond itself?

AR: Yes.

SS: Something higher then itself.

AR: Yes.

3. See chap. VII, sec. 1.

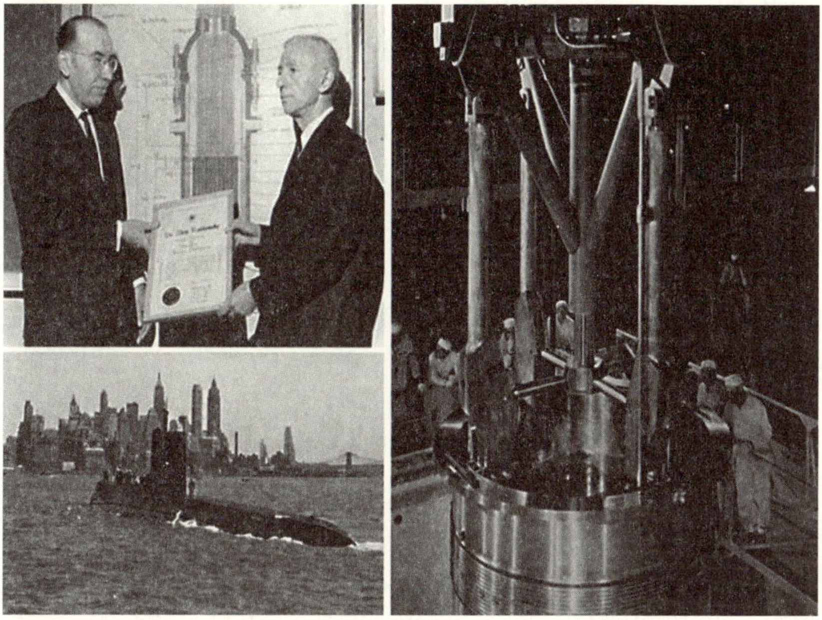

COURTESY BGU

Radkowsky & Rickover / Nautilus / Shippingport

SS: Would you say that this is a way in which we can actually apply the first halacha in Rambam to our scientific professions?

AR: ...We can go very far—much further then we thought possible before—but as we go further and further we realize our limitations.

SS: Then hopefully we see beyond that and become aware of the existence of the Creator Who sets up and maintains the whole system of the universe and the scientific disciplines that study it.

AR: Yes.

The Nuclear Reactor

SS: You yourself have attained international fame as a physicist because of your work in nuclear reactor design which goes back to the time you worked with Admiral Rickover and designed the nuclear submarines for the U.S. Navy. Can you review some of that history for us and say something about your work with Admiral Rickover and the developments which lead up to the design of your most recent reactor?

AR: Well, my work was not only on nuclear submarines; it was on nuclear aircraft carriers, destroyers, guided missile ships—all sorts of nuclear ships.

SS: Can you review some of that history and what you did?

AR: Yes. Our initial efforts were based on military aspects because the ability of nuclear propulsion was even more important for nuclear aircraft carriers then for nuclear submarines because a conventional aircraft carrier is dead in the water after about two or three days of full power sailing, while a nuclear-driven aircraft carrier can go for ten years without refueling.

The most important thing is that this led to the possibility of peace time reactors—using nuclear reactors for peace. Unfortunately, the conventional nuclear reactors generate an immense amount of plutonium—tons and tons of it, which can be used for weapons. So while we created an unlimited source of energy, we created at the same time a potential destroyer of mankind. This has led to a great deal of opposition to nuclear energy because of the potential danger that it might lead to nuclear war. Now I have developed a means of nuclear energy which cannot possibly lead to the production of weapons. I believe this will lead to a universal source of energy which can be applied throughout the world.

The International Atomic Energy Commission has said that the application of nuclear energy in third world countries is limited by the possibility of nuclear proliferation. These countries are inherently unstable and, although they are greatly in need of energy, the conventional nuclear stations create the possibility of these governments' using the reactors to produce weapons. But this new type of reactor can be used all over the world. Many of these third world countries have no sources of energy—they have no oil or coal, and they cut down most of their forests for fire wood. This is going to lead them to impoverishment and ultimately will lay the seeds for a split between "have" nations and "have-not" nations. By means of this nuclear energy invention, there will be unlimited amounts of energy made available to these third world nations without any danger of weapons being produced. This will lead to a potentially great rise in prosperity and universal peace.

SS: This reminds one of the conference two years ago [1992] of the heads of state of the major world powers—the United States, Russia and others. They met at the U.N. and made a joint declaration saying that they were going to cut back on the production of weapons, and the money and resources saved would be used to increase food production. At the time the Rebbe MHM said that this was the beginning of the fulfillment of the prophecy of Isaiah that in the Era of Moshiach "They will beat their swords into plowshares."

AR: This is sort of an extension of that idea. Instead of making nuclear reactors which can be used for weapons, we now make them so that they can be used only for peace and to give these impoverished nations a cheap and safe source of energy so that they can apply it and also grow wealthy.

SS: So you see your new reactor as—

AR: —a major step toward world peace.

SS: And the Geula?

AR: Yes.

SS: It is certainly an exciting development to witness. I recall that before you moved to Israel, you had the opportunity to have a private audience with the Rebbe MHM.

AR: Yes.

SS: Dr. Berg told me that you discussed matters of science and technology and that the Rebbe MHM encouraged you to continue writing on matters of Judaism with respect to the sciences. What was the outcome of that audience?

AR: Well, I think the outcome was many of the new ideas which I have been working on in Israel regarding the use of nuclear energy for peacetime, and the more profound ideas of the implication of scientific knowledge to the unity of G-d and His supreme powers which man can never hope to match....

For me it's been a series of miracles in my work. I hope that it will be successful, *im yirtze Hashem* (G-d willing), and will really lead to something which will benefit mankind.

SS: Thank you very much.

Nuclear Renaissance

Much has happened in the development of the Radkowsky reactor since Prof. Radkowsky and I first spoke. There is a renewed interest in nuclear energy now because of concerns over carbon emissions from other fuels and because of the increase in the demand for energy overall.

Since the 1986 Chernobyl disaster there had been a steady decline in the number of new nuclear reactors built. The last time planning permission was granted for a nuclear reactor in the UK was 1987. No new reactor has been built in the US since 1979, when an accident at the Three Mile Island power plant caused the reactor core to melt. During the 1990s, early developers of nuclear power such as Italy and Germany promised to phase out their nuclear energy altogether.

But global electricity demand grew by more than 60 per cent from 1980 to 2004, and is forecast to double by 2030. So nuclear power has emerged, once again, as the energy of the future because of its ability to deliver massive quantities of energy (without carbon emissions).

In Europe, Finland is scheduled to open its first new plant since 1982 in 2011, and Britain is considering building 10 new reactors. The US Nuclear Regulatory Commission is expecting as many as 32 applications for new reactors by 2010, after a 20-year pause.

In Asia, India is working towards having five times its current nuclear capacity by 2020, and China, which has 11 reactors, wants 10 times that number.

There are also countries hoping to go nuclear for the first time. According to the World Nuclear Association, these include Chile, Nigeria, Vietnam, Ireland, Turkey and Indonesia. In the past few years, nearly thirty governments have announced their intentions to launch peaceful nuclear programs, and most of them are being taken seriously. Between February 2006 and January 2007, thirteen governments in the Middle East announced a need for nuclear reactors.

As of 2009, there are nearly three dozen reactors under construction worldwide, and at least 50 more are in the planning stages.

This revival of interest in nuclear energy is being called "the global nuclear renaissance" in nuclear technology circles.

But the threat to global security posed by the proliferation of nuclear technology is awesome—both from conventional nuclear weapons and from a possible "dirty bomb"—which can be produced from the of the nuclear reactions in reactors designed to produce electricity.

Furthermore, even if a country has absolutely no intention of producing a nuclear weapon and has adequate safeguards to insure that no one else gets hold of its nuclear materials, it may be in danger from its neighbors who *think* that it may intend to produce a nuclear weapon.

And this is where the Radkowsky reactor comes into the picture. The Radkowsky Thorium Fuel (RTF) cycle, described in full technical detail by Prof. Radkowsky himself in the next section, does not produce any nuclear materials that could be used effectively to make a weapon. It is emphasized also that Radkowsky's system is not an entirely new reactor design but rather a "fuel cycle" that can be fitted into any lightwater reactor so that any existing lightwater reactor can be transformed into a Radkowsky reactor.

Radkowsky's company, formerly known as Thorium Power, Ltd. But now called Lightbridge, has been working with Russian researchers since the mid-1990s when it was funded in part by the U.S. Department of Energy as a way of keeping former Soviet scientists employed.[4] In recent years, it has been testing its fuel design in a research reactor at the Kurchatov Institute in Moscow.

4. See sec. 3 of this chapter on "Swords Into Plowshares in Russia."

Using Thorium in a Commercial Nuclear Fuel Cycle

by Professor Alvin Radkowsky

Background

Nuclear power plants (NPPs) produce energy by "burning" nuclear fuel. The nuclear fuel presents a complex engineering product and is a subject of extensive research and development studies. A separate discipline within the nuclear engineering domain, Nuclear Fuel Management, is devoted to analysis of nuclear fuel cycle activities.

The stages of the fuel cycle, typical for a Light Water Reactor (LWR) Plant of current technology, are depicted in the figure. Naturally occurring uranium (U) element is mined and processed to produce U_3O_8 concentrate (yellow cake). At the next stage, the yellow cake concentrate is converted into gaseous form and is enriched to a U-235 content level required by the core design and operational requirements. Enriched uranium is then reduced to UO_2 powder, which is used to fabricate fuel elements and finally fuel assemblies. The fuel assemblies are loaded into a reactor core and are "burned" during the power production period. The loading pattern of fuel assemblies and reshuffling strategies constitute a subdomain of Nuclear Fuel Management, namely in-core Fuel Management. All stages preceding the loading into a reactor core are designated as a front-end of the fuel cycle. Nuclear fuel is produced from uranium or thorium (Th) which are naturally occurring elements. Natural uranium contains a fissile component, U-235, and a fertile element, U-238, while natural thorium contains only a single fertile isotope Th-232. Therefore, uranium, enriched in U-235, is a standard fuel for the Light Water Reactors (LWR) of current technology. Additional fissile isotopes generated by transmutation of fertile isotopes are Pu-239 and Pu-241 (uranium chain) and U-233 (thorium chain). These isotopes play an important role in the process of generating fission energy during the fuel burnup process; these artificial fissile isotopes are created and "burned" in situ, contributing much of the energy generated by a nuclear fuel.

The aim is to show that current designs of LWRs can use a thorium fuel with improved performance and economics, while the world gains in reduced risk of further nuclear proliferation.

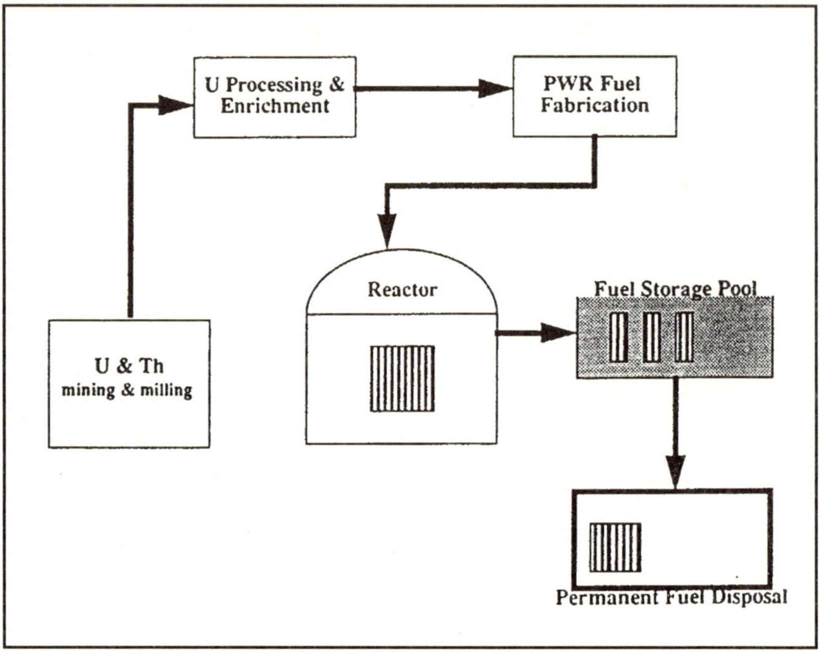

The core of a power reactor using the thorium-based fuel known as RTF (Radkowsky Thorium Fuel) fits into exactly the same space as the cores of current standard uranium reactors, thus requiring only minor changes in the plant. The importance of this feature is evident—ensuring virtually no risk to the huge investments already made in the present LWR plants throughout the world.

The RTF has the following advantages:

- RTF spent fuel is completely nonproliferative for all practical purposes. In contrast, standard uranium LWRs (1 GWe) discharge between 250 and 300 kg of plutonium. Only 5 to 7 kg of such plutonium, with the services of only 3–4 people, is sufficient to fabricate a bomb.

- There is a dramatic reduction in radwaste in quantity, toxicity, radioactivity and heat emission.
- Safety is enhanced due to the elimination of soluble boron control during operation. Otherwise, kinetic behavior is similar to that of standard LWRs, and the core meets accepted thermal specifications.
- Natural uranium ore requirements are reduced by about 20%.
- Fuel cycle costs are reduced by 20% to 30%.
- There is no need for fuel reprocessing; the once-through cycle is optimal.
- The RTF is especially suited for burning and disposing of either weapons-grade or reactor discharge plutonium.
- RTF is especially suited for burning and disposing of HEU from dismantled nuclear weapons.

Radkowsky Thorium Power Corporation (RTPC) obtained initial funding from Raytheon Nuclear Inc and has arranged for a Department of Energy-USIC grant through Brookhaven National Laboratory to fund the Kurchatov Institute in Moscow. The plan is to construct lead test assemblies and test them in a VVER-1000 by the year 2002.

Ever since the Manhattan Project there has been interest in utilizing thorium for nuclear power. Thorium is much more plentiful and cheaper than uranium; U-233, which is formed by neutron absorption in thorium, has the highest value of of all fissile nuclei in the spectra of LWRs. Thorium oxide, being stoichiometric, can withstand burnups as high as 100 GWd/t, much higher than uranium oxide. The main difficulty is that natural thorium lacks any fissile component. Previous attempts to utilize thorium in LWRs have therefore required adding a fissile fuel homogeneously to the Th, the fissile fuels being either plutonium or highly enriched uranium (HEU), which "invest" neutrons into the Th in order to build up its reactivity.

However, such fuels are expensive and highly proliferative. The

RTF provides a practical and attractive solution to the utilization of thorium by making use of the seed-blanket concept invented by Radkowsky and successfully demonstrated in the series of cores used in the Shippingport reactor. Indeed, had the nuclear cores in India been equipped with RTF, they could not have pursued their nuclear testing programme.

The RTF core consists of seed-blanket units (SBUs), equal in number and outer dimensions to the fuel assemblies of a standard uranium LWR. The basic idea of an SBU is to form two separate regions: the mainly thorium section, the blanket, which is the outer region of the unit, and in which the U-233 is burned in place; and the inner region, the seed, which consists of partially enriched uranium, the U-235 content being below 20%, a percentage generally accepted as being nonproliferative. The purpose of the seed is to supply neutrons to the blanket in the most efficient way by thermalizing the seed spectrum (high water content) so as to minimize plutonium formation. This necessitates the use of metallic fuel, U/Zr, alloy in the seed, as in the first two Shippingport cores. The blanket fuel consists of oxide rods of Th plus a small amount of uranium enriched to less than 20%. The uranium is added for two reasons:

- Natural thorium has no fissile content; therefore enriched uranium is required to provide adequate power in the blanket during the period of U-233 buildup.
- The U-238 and other nonfissile isotopes denature the discharged blanket residual U-233 so that it cannot be used for weapons.

The RTF core design concept can in principle be used for all LWRs. Our initial work has concentrated on PWRs, of which there are two kinds: the Russian VVERs, which employ a hexagonal arrangement, and the Western PWRs having a square arrangement. We have completed a preliminary detailed design for the VVERT-1000 (3000 MWth output), the T standing for thorium, and are now working on a PWRT design.

Optimization of the seed and blanket lattice parameters was aimed at efficient generation and in-situ burning of U-233 (in blanket). This optimization resulted in moderator-to-fuel volume rations (Vm /Vf) of 3.2 and 1.9 in seed and blanket regions, respectively. The relatively high Vm /Vf value in the seed was chosen in order to reduce the epithermal absorption of U-238 and consequently the buildup rate of Pu isotopes. The blanket Vm /Vf value was chosen as an optimal intersection point of two processes: U 233 buildup (low Vm /Vf) and efficient fission (high Vm /Vf). The resulting value of 1.9 was found almost identical to that of a standard PWR core lattice.

In-Core Fuel Management

One of the main features of the RTF is its in-core fuel management scheme. The standard multibatch fuel management is replaced by a more complicated scheme, based on two separate fuel flow routes: blanket route and seed route. The residence time in the core of the SBU blankets is quite long (about 10 years) in order to achieve a very large accumulated burnup for the Th fuel of about 100 GWd/t. Basically, the in-core management for the seeds is a 3-batch scheme, similar to that for standard uranium fuel assemblies. One-third of the seeds is fresh (F), one-third once burned (O), and one-third twice burned (T). After each annual (or 18-month) cycle the T seeds are replaced by F seeds. The F seeds together with the remaining partially depleted seeds are reshuffled into "empty" blankets so as to obtain a "low-leakage" configuration with optimum power distributions. The power density distribution across the core and the power density variation throughout a cycle are typical of that for a standard uranium PWR. Thorium oxide also has better thermal conductivity than uranium oxide and will be used in RTF blankets, where the power density is less than in standard uranium cores. According to Kurchatov Institute metallurgists, the endurance of zirconium-clad thorium oxide fuel can be enhanced by adding a small amount of niobium to the thorium.

Reactivity Control System

The RTF reactivity control system is based on the use of control rods (CR) and burnable poisons (BP) without the utilization of soluble boron control. This is not unusual, as boiling water reactors are controlled in a similar manner. The RTF cannot utilize soluble boron control during normal power operation because it would depress the power in the blankets too much. The blanket power is proportional to 1/(1 Kb), where Kb is the blanket multiplication factor and should be as close as possible to unity in order to maximize the thorium power output. Thus, it is necessary to control the RTF by BP and CRs in the seeds. Two types of BP rods are used in the current design in order to compensate for a major part of the burnup-related excess reactivity. The WABA type (Westinghouse Advanced Burnable Absorber) and Gadolinium loaded fuel rods. It was discovered that due to the close interaction of our SBUs, the standard PWR control rods, which are only about one-third the number of assemblies, are adequate to control the RTF. In all other seed-blanket core designs, a separate moving control means was required for each seed.

The WABA, presenting a standard PWR technology, is introduced to compensate the long-term burnup-related excess reactivity (from BOL to about 2/3 of cycle), and the Gd bearing rods to compensate for the excess reactivity of the first 30–50 days of the cycle. Our calculations indicate that the reactivity control requirements are almost identical for all cycles and that the total burn-up reactivity shift is about 5% p, about half of that for a typical uranium-fuelled PWR. Soluble boron will be used only for refueling and for cold shutdown. The NRC and EPRI have long been desirous of eliminating soluble boron control during hot pressurized operation in order to avoid small boron leaks, which may corrode high strength steel parts and impede operation of vital valves. There is also the possibility of a reactivity incident if the emergency cooling water is not borated.

A complete 3-D calculation was performed for the reactor core during the power production cycle and following fuel discharge. The amount of natural U is about 20% less than for a standard PWR and the number of fuel rods about 25% less. These factors, together with a major reduction in back-end fuel storage expenses, result in an overall fuel cycle cost reduction of 20% to 30%.

Proliferation resistance of Radkowsky Thorium Fuel

The average annual U-235 discharge from the Radkowsky Thorium Fuel (RTF) is 128 kg, compared with approximately 750 kg of U-235 loaded annually into the core. Thus the seed U-235 is about 83% depleted and that of the blanket nearly 100%. The high burnup of the seed is due to the fact that most of the seed energy comes from burning U-235. As a result of the higher burn-up, the RTF plutonium has far greater Pu-238 and Pu-242 content than that of a PWR.

This effect is crucial in rendering the RTF nonproliferative for all practical purposes.

There are four factors involved in determining the proliferation resistance of the RTF plutonium:

- The amount of plutonium discharged per year is about 20% of that of a standard uranium-fueled PWR.
- The critical mass of RTF Pu is greater than for PWR Pu.
- Because of the far greater spontaneous fission source in RTF Pu, the nominal yield from a weapon is almost negligible and the chance of a fizzle yield is relatively high.
- Because of the much higher Pu-238 content of the RTF Pu than of PWR Pu, the heating in a weapon made of RTF Pu storage can be sufficient to cause a phase change in the Pu metal and disintegration of the explosive, resulting in extreme difficulty in fabrication and storage of the explosive device.

The total spontaneous fission source for a critical mass of PWR grade Pu is 7 times larger than that of a weapon-grade Pu. The RTF seed and blanket Pu is 13 and 22 times larger, respectively, than weapon-grade Pu. The nominal and fizzle yields are estimated for each of the considered Pu compositions and are presented in the table below.

Probability of an Indicated Yield

Yield	Super Grade (Trinity)	Weapons Pu grade	PWR Pu grade	RTF-seed Pu grade	RTF-blanket Pu grade
Nominal	0.88	0.68	0.07	0.006	0.002
Fizzle	0.02	0.06	0.35	0.55	0.74

The probability that an explosion device, constructed from the RTF Pu, will deliver a nominal yield is small (seed) to negligible (blanket), and a probability of a fizzle yield is relatively high. Thus, it is shown that the RTF Pu will produce an unreliable weapon.

U-233 diversion potential

The RTF cycle is based on extensive utilization of thorium, which produces through a nuclear reaction the fissile isotope U-233. U-233 has been determined to be superior to U-235 and at least as efficient as Pu-239 as a weapon material. Therefore, a special effort was invested in the RTF design to create effective barriers to diversion of U-233.

The total amount of U-233 in the spent blanket fuel, discharged once in 10 years, is about 630 kg (including Pu-233). The annual equivalent of 630 kg may represent a major proliferation potential. In order to eliminate this potential the U-233 created in the blanket is denatured by the initial addition of about 20% enriched uranium. The amount of U added for a dilution of fissile components was carefully chosen to reduce the overall content of fissile uranium isotopes in a discharged blanket fuel well below 17%, which is roughly equivalent to 20% of U-235 enrichment. Additional uranium isotopes created during the long in-core residence time of the blanket are U-232, U-234, U-235, and U-236.

Isotopic separation of the U-233 will be far more difficult and costly than the separation of U-235 from natural uranium. This is because of the several nonfissile isotopes mixed with the U-233 and contamination by the hard gamma emitter T1-208, which will necessitate remote operation. Thus there is no practical means to use the blanket discharge fuel for weapons.

From Nuclear Weapons to Nuclear Medicine

Radiation and radioactive materials have been used in medical procedures for decades, but treatments have often been dangerous, and results, unsatisfactory. In one letter in Igros Kodesh many years ago, the Rebbe MHM advises someone to have surgery rather than radiation treatment because of the unreliability of the radiation treatment.

Now, in the Era of Moshiach, however, with Swords Into Plowshares in progress, great advances in the use of radiation in medical procedures have come about based on the application of military technology. In this section, we first consider an example where mathematical methods and computer programs developed for the design and use of weapons have been applied to make the use of radiation in medicine more accurate. Then we describe an example where nuclear weapons themselves have been the source of very pure radioactive materials for use in medicine.

A Bag of Water

Back in the 1990s, scientists discovered a very disturbing statistic regarding people who received radiation treatment to destroy cancerous tumors: "Each year, more than 100,000 cancer patients who are treated with radiation in hopes of a cure die with active tumors at the primary cancer site." Not only was the radiation treatment not successful—the patient died—but the tumors remained intact; the radiation did not destroy them.

Research into the cause for this showed that a serious miscalculation had been made in determining how the radiation should be delivered. Scientists had modeled the human body simply as a "bag of water" and had directed the radiation toward a tumor at a certain location inside this bag of water. What this calculation ignored is the fact that radiation travels in different ways through different media such as muscle, bone, blood and air cavities inside the body. Furthermore, every time the radiation travels through an interface between two such media, it changes direction—just like light (electromagnetic radiation), traveling through the air, changes direction when it enters water. Thus, in many cases, the radiation had missed the tumor and left it intact.

The Monte Carlo Method

Where were medical scientists going to get accurate calculations that could take into consideration the different media through which the radiation had to travel? In fact, the physicists at Lawrence Livermore National Laboratory (LLNL) had actually made such calculations in connection with their development of atomic bombs. These calculations were based on a mathematical theory called the "Monte Carlo Method," a branch of probability theory. The Monte Carlo Method was developed by the atomic scientists and mathematicians working on the Manhattan Project to build the first atomic bomb.

In 1993, a research team at the Lawrence Livermore Laboratory applied the Monte Carlo Method to directing radiation to attack tumors. Their new method, called the "Peregrine System," simulates the radiation treatment "particle-interaction by particle-interaction." During treatment, a patient receives trillions of photons or particles of radiation. The Monte Carlo Method reconstructs the treatment by selecting a random sample of photon particles and tracking them through a computer model of the radiation delivery device and a model of the tumorous region, based on a computed tomography (CT) scan of the patient. Everything that happens to the photons after they leave the x-ray machine—colliding with an electron in the skin, ionizing a hydrogen atom in the blood, perhaps being absorbed by calcium in the bone—is calculated in the model.[1]

1. See www.llnl.gov/str/Moses.html

We think that one can hardly find a better example of "Swords Into Plowshares" than this—nuclear weapons transformed into nuclear medicine. Bill Richardson, then Secretary of Energy, thought so as well. In a statement announcing the production and marketing of the Peregrine System, he said:

"Peregrine could change the way cancer is treated in America. This technology was developed through advances resulting from nuclear weapons research and with the multidisciplinary scientific expertise of a Department of Energy national laboratory. This is an excellent example of turning swords into plowshares."

A Cure for Cancer

Scientists at one of the country's most contaminated nuclear waste sites are collaborating with medical researchers to turn nuclear waste into a therapy for cancer. Patients with Hodgkin's disease and brain, ovarian, and breast cancers may be able to receive the new radiation-based treatment.

The Hanford nuclear site in Richland, Washington at one time produced plutonium for nuclear weapons for the U.S. Department of Defense. Plutonium production has been halted, but large quantities of plutonium remain. Pacific Northwest National Laboratory (PNNL) and Westinghouse Hanford Company have entered a joint venture to investigate radioisotope production from the waste. Radioisotopes are unstable elements that emit particles of radiation. Until recently, about 90% of all medical radioisotopes used in the United States were imported from countries such as Russia and Canada. But recently, scientists at the Hanford site found a way to chemically extract a pure form of the radioisotope yttrium-90 from strontium-90, a by-product of plutonium production.

The technique to extract pure yttrium-90 from strontium-90 was discovered by Lane Bray, a Hanford chemist. This technique produces the purest form of yttrium-90 in the world. The advantage of yttrium-90 over other radioisotopes is its short half-life. Its radioactive emissions are halved every 64 hours, and within a few days, the amount of radioactivity it generates is too small to be measured. The use of this isotope for treatment of patients with recurring Hodgkin's disease began in the early 1990s.

Smart Bullets

The problem is that radioactivity destroys cancer cells by emitting particles that physically damage *any* surrounding cell. Therefore, for successful treatment with minimal side effects, the radioisotope must be targeted specifically to the tumor cells. This is accomplished by the use of proteins called "monoclonal antibodies," also referred to as "smart bullets." These proteins circulate in the body and lock onto tumor cells, allowing the attached radioisotope to act primarily on the diseased tissue.

The treatment consists of attaching the isotope to a specially engineered antibody that seeks out cancer cells within the body. Once it is attached, beta particles emitted from the isotope destroy the cancerous cells while sparing normal surrounding tissue. "Clinical trial results to date have been very promising," said Tom Tenforde, Manager of the Radioisotopes Program at Pacific Northwest National Laboratory. "At M.D. Anderson Cancer Center in Houston, Texas, a positive response was seen in 80 percent of the approximately 130 lymphoma patients whom all conventional therapies had failed to cure. And, the therapy can be administered on an outpatient basis and doesn't cause severe side effects, such as hair loss and nausea, associated with conventional therapies."

In another test at the M.D. Anderson Cancer Center, yttrium-90 treatment produced positive results in about 85% of the 80 Hodgkin's disease patients that were treated. Hodgkin's disease is a cancer of the lymph nodes that is often successfully treated with the traditional methods of chemotherapy and radiation. Patients who were not helped by traditional treatments were enrolled in a clinical trial of yttrium-90. Huibert Vriesendorp, a radiation oncologist at M.D. Anderson who conducted the trial, reported that "The patients in the trial had failed all other conventional cancer treatments such as chemotherapy and radiation. These patients have previously received very aggressive treatments and are relieved to find that the yttrium-90 treatment has relatively few side effects and is patient-friendly." Another advantage of this type of outpatient therapy is that it is highly cost-effective. According to Vriesendorp, "Use of radiolabeled antibodies is much cheaper than traditional cancer therapies. Insurance companies are willing to pay for these procedures."

"Another exciting application of yttrium-90 therapy is for rheumatoid arthritis," noted Vriesendorp. "The stabilized isotope is injected directly into the joint to attack the inflammatory cells that are at the root of the disease."[2]

2. http://www.ncbi.nlm.nih.gov/pmc/articles/PMC1522183/pdf/envhper00356-0016-color.pdf

Persian Gulf War Technology to Plowshares

Low Vision Enhancement Systems

"Opening the Eyes"—Literally

Some three million Americans suffer chronic visual impairments known to ophthalmologists as "low vision," which means that they are not totally blind but retain some degree of useful vision. These disabling impairments cannot be corrected medically, surgically or with conventional eyeglasses. But now, technology used in the Persian Gulf War has been transformed into a high technology visual aid to help these people see.

Pilots in the Persian Gulf War wore head-mounted units which displayed before their eyes the region on the ground containing their target. By merely focusing their eyes on the image of the target on the display before their eyes, and then pressing a button, a missile would be shot at the target on the ground below.

This technology is the basis for the Low Vision Enhancement System (LVES), a video headset that offers people with low vision a view of their surroundings equivalent to the image on a five-foot television screen four feet distant from the viewer.

The LVES was developed by scientists at the Johns Hopkins University Wilmer Eye Institute in collaboration with NASA's Stennis Space Center and the Veterans Administration.

The device is designed to enhance and compensate for low vision in people whose eyesight with conventional eyeglasses is worse than 20/100 in their better eye, but better than 20/800.

"LVES does not fix vision or restore vision. Instead, it alters images to make them easier for people to see with the vision they still have," says Dr. Robert W. Massof, professor of ophthalmology at Hopkins and lead inventor. He also directs the Lions Vision Research and Rehabilitation Center at Wilmer.

The lightweight headset, worn like goggles, is fitted with three miniature, black-and-white video cameras. Two of the cameras, one over each eye, provide a normal, 3-D view to observe what's happening in their environment. A third, more complex camera is used for seeing facial features, fine details of objects, distant objects, or for close-up, detailed work and reading. A control unit allows the user to select and control the cameras, and to adjust contrast and image polarity to suit the user's needs. The cameras feed the images to a computer that corrects for the particular vision problem of the user, and then sends the images to the video display in the goggles.

Visionics Corporation of Minnesota, the manufactures of LVES, gives the following description of the technology: "The LVES does for video images what headphones do for sound. It presents the enlarged image of a video screen to each eye in a way that creates a sense of being immersed in a video scene." Brad Blankenship, president of Visionics, says, "Our engineers are experienced in building vision systems for the military, like the helmet mounted displays that helicopter pilots used during Operation Desert Storm. To use such high technology for this uniquely peaceful, beneficial application is very satisfying." Swords Into Plowshares!

Controls built into a battery pack worn on the belt let the wearer adjust contrast and magnify images from 1.5 to 10 times. The unit automatically compensates for changes in lighting to reduce glare in bright light. In addition to displaying images from the built-in video cameras, a cable connection lets the LVES become a personal large screen display for input from televisions, videocassette recorders or computers.

The system is used by people with a variety of low vision conditions, particularly those who have experienced loss of central vision, the part of vision normally used for reading. These patients may have macular degeneration associated with aging or diabetic retinopathy, in which diabetes causes swelling and leakage of fluid in the center of the retina, the macula. The system also has benefited people who have lost peripheral or side field vision, a problem associated with glaucoma, an increase of fluid pressure inside the eye that damages the retina and optic nerve, and people suffering from retinitis pigmentosa, a progressive degeneration of the retina that results in tunnel vision and extreme sensitivity to light. Persons with optic nerve disease and congenital damage to the retina also use the LVES successfully.

LVES was introduced to the commercial marketplace in 1994. In the year following more than 200 of these video-based systems were successfully dispensed by some 30 clinics across North America. The LVES has been accepted by the medical and user communities as an effective and credible device. Many users enjoy the improved quality of life the LVES has provided in their daily activities.[1]

[Some people with low vision were first made aware of the existence of LVES by reading the reports of the RYAL Institute on it. They contacted us and we directed them to medical offices in the New York area that provide it.]

JORDY

The technology was advanced by Enhanced Vision Systems of Huntington Beach, California which produced a new and improved low-vision headset called Joint Optical Reflective Display (JORDY) in 1999.

Worn like a pair of glasses, JORDY magnifies objects up to 30x; when placed in its docking stand, it magnifies up to 50x. The user can also adjust the contrast, brightness, and display modes, depending on what works best for their condition.

Another enhancement incorporated an even smaller, high-resolution camera that fits into discreet, lightweight glasses. Increased miniaturization means people can wear the device more comfortably for longer periods of time.[2]

COURTESY NASA

JORDY™

1. See www.hopkinsmedicine.org/press/1994/June/199421.htm
2. See www.nasa.gov/vision/earth/technologies/jordy.html

"And All Flesh Will See..."

One of the Messianic prophecies of Isaiah is that in the Era of Moshiach "the glory of G-d will be revealed and all flesh will see it." [3] The Rebbe MHM emphasizes that this does not just mean that all people will see it but that the actual flesh – not just the eyes – will see it. How is it possible for flesh to see? This process too has already begun.

While the low vision enhancement systems LVES and JORDY enable people with low vision to see better, in an even more dramatic development, people who are actually blind are now able to see through their flesh; specifically, through their tongue.

COURTESY NIH

BrainPort

A device called BrainPort, developed by Wicab, Inc., translates information from a digital video camera to the tongue through gentle electrical stimulation. It consists of a postage-stamp-size electrode array for the top surface of the tongue, a base unit, a digital video camera, and a hand-held controller for zoom and contrast inversion. Visual information is collected from the user-adjustable head-mounted camera (with a field of vision range of 3–90 degrees) and sent to the BrainPort base unit. The base unit translates the visual information into a stimulation pattern that is "displayed" on the tongue. The image is created by presenting white pixels from the camera as strong stimulation, black pixels as no stimulation, and gray levels as medium levels of stimulation. Users often report the sensation as pictures that are painted on the tongue with Champagne bubbles.

The BrainPort vision device is currently an investigational prototype and is not yet for sale but Wicab plans to develop a device that will be available commercially in the near future. With the current system (arrays containing 100 to 600+ electrodes), study participants have been able to recognize high-contrast objects, their location, movement, and some aspects of perspective and depth.[4]

3. *Yeshaya* 40:5
4. See http://vision.wicab.com/technology

Brains of Cruise Missles Into a Cure for the Brain

An alternative to surgery for the removal of a brain tumor is the use of radiation to destroy the tumor. Radiation has the obvious advantage of eliminating the need to open the patient's skull to operate on his brain. But there have also been some problems associated with it. As we mentioned in the section on nuclear medicine,[1] the calculations used to model the human body have often been inaccurate. This problem was solved by adapting mathematical methods used in the development of atomic bombs to model the human body more accurately—a Swords Into Plowshares development.

But there is another problem. Assuming the body has been modeled accurately, the location of the tumor has been identified and the correct dose of x-ray radiation is then directed at the tumor, what if the patient moves after the radiation is sent? Small movements of the body are normal and cannot be avoided, while even the slightest movement will cause the radiation to miss its target and destroy normal tissue instead.

The traditional solution to this problem in the procedure known as "Stereotaxic Radiosurgery" has been to screw the patient's skull into a frame to prevent movement, a process that necessitates anesthesia and post surgical hospitalization while the bolt wounds heal.

Surgeons are not the only ones who face this problem. In a war, military personnel may need to destroy an enemy ship by shooting a Cruise missile at it. But again, after the location of the target is identified and the missile is shot, the target may move, causing the missile to miss it. To solve this problem the Cruise missile is designed with special software that continuously tracks the target. If the tar-

1. Sec. 8 of this chapter.

get moves, the software recalculates the path of the missile and redirects it to the target's new position.

The Persian Gulf War was the first combat test of the Cruise missile system, and it was used with great success. Within the first few minutes of Operation Desert Storm, Tomahawk Cruise missiles struck with accuracy at Iraqi command centers and radar installations. In the war, Tomahawks were used to destroy surface-to-air missile sites, command and control centers, electrical power facilities and Iraq's presidential palace.[2]

Let us return to the operating room example. In an awesome Swords Into Plowshares development, Professor John Adler of the Stanford University Medical Center has adapted Cruise system military software to control the x-ray beams being directed at tumors.

Dr. Adler's method, called Computer Mediated Stereotaxic Radiosurgery (CMSR), works as follows: First, a CT scan is taken of the patient. The CT scan identifies the location of the tumor in relation to the bones in the body. This information is then fed into a computer which stores an accurate map of the surgical target and locks the radiation beam on the tumor. Then, a computer controlled robotic arm, known as a Cyber Knife, moves around the patient, shooting beams of radiation from several different angles. The cumulative effect of the beams is a high dose of radiation at the target point.

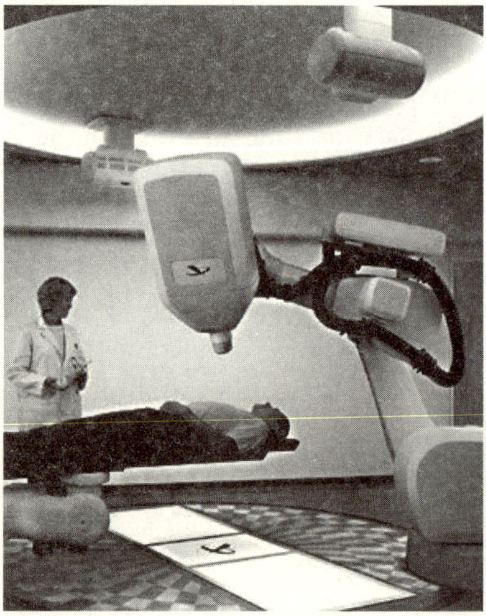
COURTESY MEDGADGET.COM

The Cyberknife

The tracking software guiding the robot arm matches what it sees with the computer-stored map and reconfigures the angles of the beam and redirects the robotic arm if the patient moves slightly during treatment.

Unlike most modern procedures in radiation oncology,

2. See http://www.pbs.org/wgbh/pages/frontline/gulf/weapons/tomahawk.html.

CMSR relies on powerful but precisely aimed doses that can kill malignant tumors with a few—and often only one—treatment sessions, Prof. Adler explained. In traditional radiation therapy, physicians administer small doses of radiation over weeks or months in order to allow healthy tissue surrounding the tumor to recover from the onslaught. But CMSR attacks the tumor with short pulses of very precisely directed, high-energy radiation.

"Because we crossfire in many different directions, we can minimize the radiation dose reaching healthy tissue and prevent injury," Prof. Adler said. "The robot arm is extremely flexible and can position the beam at any of a large variety of angles to treat a small site."

CMSR can be applied to tumors that could not be treated by ordinary Stereotaxic Radiosurgery. The spine, for example, cannot be screwed into a frame to prevent it from moving; thus, traditional radiosurgery would not be attempted. In addition, regular surgery on the spine poses a risk of damaging nerves and causing paralysis. But with CMSR, the problem of movement of the patient is obviated since the tracking system redirects the beam to follow the patient's movements.[3]

Other conditions can also be treated with CMSR. An arteriovenous malformation (AVM) is an abnormal connection between an artery and a vein. When an AVM occurs, blood from arteries, where the blood pressure is higher, flows through the abnormal connections into the vein where the pressure is lower. This causes weakened blood vessels to form in the area and, because of the higher pressure of blood from the artery, these weakened blood vessels are prone to rupture.

Usually AVM's are treated with surgery, but if they are located in areas that are inaccessible—90% of them occur in the brain—or in areas where surgery is too risky, such as the spinal cord, CMSR can be applied.

3. "Stealth Surgery on Brain Tissue," *Science News* (Aug. 26, 1995).

Into Plowshares—Literally

Tanks *for the Plows*

The large number of tanks and other heavy armored vehicles which have been made surplus in Eastern Europe and the former Soviet Union has spurred ideas for the civilian use of such vehicles. A considerable number of vehicles have been converted on a trial basis to a variety of civilian uses in such countries as the former German Democratic Republic, Belarus and Russia. Civilian uses which are envisioned include:

- Bulldozers
- Firefighting vehicles
- Cranes
- Power-unit vehicles
- Mineral fine-crushing vehicles
- Quarry vehicles
- Rescue vehicles
- Casualty evacuation vehicles
- Transportation vehicles
- Oil-rig vehicles
- Oil and chemical product
- Spill-cleaning vehicles
- Tracked ice-breaking prime movers
- Environmental vehicles
- And, last but not least—plowshares, literally!

In fact, a great deal of military equipment is easily adaptable for civilian use because the equipment item itself is not lethal, although it is part of the military arsenal of the armed forces. In addition, most actual weapons systems can also be converted because they are comprised of a piece of transport equipment (such as vehicle chassis, an aircraft or a ship) and a lethal element (such as a gun or missile) mounted upon the transport part. In some cases, such as towed artillery, the transport equipment part is minimal while in others, such as lightly armed landing ships, the weapon itself is of minor importance. It is therefore not surprising that many of the transformations of weapons for civilian use were accomplished simply by removing the weapon element and using the transport equipment element.

A converted Russian tank serves as a plow.

In addition to armored vehicles, other types of military equipment that have undergone a Swords Into Plowshares transformation include:

- Helicopters for police missions, search-and-rescue and transport. In the United States alone, hundreds of surplus military OH-58 and Bell Jet-Ranger helicopters have been converted for law enforcement and public safety-related use.
- Military training aircraft have been reassigned to civilian flight training, sometimes after the removal of specific avionic equipment.
- Small landing ships are now used as ferries, and small transport ships and tenders, for civilian transport purposes.

Bombs—Away!

Dr. Moshe Alamaro, an Israeli engineer working at MIT, has devised a revolutionary method for transforming air force bombers ("swords") into tree-planting aircraft ("plowshares"). Equipment installed in the huge C-130 transport aircraft used by the military for laying carpets of landmines across combat zones has been adapted to deposit the trees in remote areas. Alamaro founded a company, Aerial Forestation Inc., to market the concept. Lockheed Martin Inc., the manufacturer of the C-130 aircraft, is excited about the fact that in this project their military aircraft will be used for peaceful purposes.

A spokesman for Lockheed Martin said, "Equipment we developed for precision planting of fields of landmines can be adapted easily for planting trees. There are 2,500 C-130 transport aircraft in 70 countries, so the delivery system for planting forests is widely available—mostly mothballed in military hangers waiting for someone to hire them. The possibilities are amazing. We can fly at 1,000 ft. at 130 knots, planting more than 3,000 cones a minute in a pattern across the landscape—just as we did with landmines, but in this case, each cone contains a sapling. That's 125,000 trees for each sortie and 900,000 trees in a day." (See image below)

COURTESY M. ALAMARO

Artist's conception of bombers planting trees.

Millions of acres of forest have been destroyed in the last century due to the lumber industry, forest fires, etc. In many poor countries where fuel is not available, people have cut down their forests for firewood. Traditional reforestation methods, using machinery on the ground, is tedious and time-consuming and can replace only a tiny percentage of these trees.

Alamaro's system is incredibly efficient. He designed open-topped conical canisters of a starchy biodegradable material, each of which contains a seedling packed in soil and nutrients. The tree cones are pointed and designed to bury themselves in the ground at the same depth as if they had been planted by hand. The canisters are dropped from a low-flying plane so that they hit the ground at 200 m.p.h. and imbed themselves in the soil. Then the canisters decompose and the young trees take root. They contain fertilizer and a material that soaks up surrounding moisture, watering the roots of the tree.

Alamaro uses a combination of ballistics and navigation technology to place the saplings accurately. His canisters are strong enough to withstand the impact but still decompose quickly. Moreover, his system is overseen by an airborne surveillance system, which guarantees safety and also monitors the early growth of the trees. It uses military technologies developed for the Star Wars program and Operation Desert Storm—the Persian Gulf War.

The system will work in any area where trees once grew. Even in desert areas, the cones can be adapted for planting suitable desert shrubs. Some of the areas considered for reforestation are the Scottish mountains which used to be forested; large areas of the Black Forest that were cut down during the cold war to provide "line of sight" across the Iron Curtain between East and West Germany; areas in North Africa; and the tundra of Canada, Australia and the US. A project is also planned for the Sinai Desert.[1]

The Benefits

A major benefit of reforestation is the prevention of soil erosion. Trees hold the soil intact, and areas that have lost their trees risk losing valuable soil to wind and especially to rain. The issue is, in fact,

1. See http://www.guardian.co.uk/uk/1999/sep/02/paulbrown.

far more serious than the loss of soil. Landslides caused by heavy rains on such areas often result in loss of lives. In a country such as Honduras, for example, which has many hills and mountains and is frequently hit by hurricanes, thousands of lives are lost to landslides.

Dr. Alamaro visited us at the RYAL Institute in 2001. When it was explained to him that his invention was part of the Swords Into Plowshares transformation fulfilling the Messianic prophecy of Isaiah, he said that he was very happy to be a part of it.

Global Security

In the previous sections we described in great detail some of the Swords Into Plowshares transformations taking place throughout the world. Indeed, we see how the Swords Into Plowshares prophecy has begun to be fulfilled.

But now we address a basic question which is often raised regarding the current state of world affairs:

How do we reconcile the presence of war and terrorism in the world with the prophecy of Swords Into Plowshares which is associated with the cessation of conflict?

We will answer this in two parts. In the first section we present the facts—that is, the exact current state of affairs. We present the "Human Security Report," an ongoing project begun in 2002 to investigate the state of global security. You will be surprised at the findings—as were the authors of the report themselves!

Following that, we analyze the issue itself—war in the Era of Moshiach.

The Human Security Report

We have been discussing the progression of the Swords Into Plowshares process, which began in 1992. Now we can look back over the years and ask ourselves, "Do we see a different world, a better world, a more peaceful world?" If we read the newspapers and listen to the news, we hear a lot about war and terrorism. But the Rebbe MHM told us to open our eyes to see the Geula. This may involve a deeper, more analytical look than just reading the newspapers.

On the other hand, the Swords Into Plowshares phenomenon, as described in the *sicha* of Parshas Mishpatim, focuses on events that were well-known and reported on in the newspapers. Indeed, all the SIP phenomena that we have been describing in this chapter can be found in various scientific journals, on websites and in published reports. One need only collect and analyze the information. This is exactly what happened in the case of the *Human Security Report.*

On October 17, 2005, the Human Security Center issued the *Human Security Report 2005: War and Peace in the 21st Century*."[1] Funded by five governments[2] and three years in the making, the *Report* tracks and analyzes trends in political violence around the world. "Its findings are sharply at odds with conventional wisdom. It shows that most forms of political violence have declined significantly since the end of the Cold War—and finds that the best explanation for this decline is the huge upsurge of conflict prevention, resolution and peace-building activities that were spearheaded by the United Nations in the aftermath of the Cold War," according to the Human Security Center's own description. And this is exactly how the Rebbe MHM described the Swords Into Plowshares phenomenon in the *sicha*.

I must admit that I, too, was worried about the wars and their ramifications for the Geula, but when I read the Human Security Report I felt a sense of exhilaration. When the Rebbe MHM tells us to open our eyes, we may think it means something mystical. Sometimes all it means is to look more carefully and get all the information. Read the right reports. The following is a summary and analysis of the *Human Security Report*:

The post World War II era witnessed an extraordinary increase in the number of wars, most of them civil wars. This was followed by a *steep decline* in warfare after the end of the Cold War, around the time of the Swords Into Plowshares declaration.

There are two key questions to be asked: 1) How do we explain the decline in the use of force in relations *between* states and 2) What brought about the decline in wars *within* states? Strangely, neither of these important trends has been the subject of much scholarly investigation. Scholars have generally been more interested in explaining the causes of war than the causes of peace. This is particularly true with respect to the remarkable post-Cold War decline in civil wars.

---

1. *Human Security Report 2005: War and Peace in the 21st Century* (New York: Oxford University Press, 2006).

2. The Canadian Department of Foreign Affairs and International Trade; the Norwegian Royal Ministry of Foreign Affairs; the Swedish International Development Cooperation Agency; the Swiss Agency for Development Cooperation; the Swiss Federal Department of Foreign Affairs; and the British Department for International Development.

Much of the research on the causes of international peace over the past 20 years has used large datasets and statistical inference models to examine the "correlates of war"—the economic, political, and sociological factors associated with interstate war—and peace. This research points to a number of long-term global trends that are associated with reduced risks of international conflict:

- A dramatic increase in the number of democracies.
- An increase in economic interdependence.
- A decline in the economic utility of war.
- Government participation in international organizations that encourage the peaceful settlement of disputes.

While these factors are associated with reduced conflict, further analysis is required to determine if they are the causes of peace or the results of the peace. Which is the cause and which is the effect?

The report suggests some possible reasons for the dramatic decrease in wars in this era. One of them is that the end of the Cold War removed a major cause of armed conflict from the international system. Approximately one-third of all wars in the post-World War II period had been driven wholly, or in part, by the geopolitics of the Cold War. The end of the political confrontation between East and West in the late 1980s not only removed the only real threat of war between the major powers, but also meant that Washington and Moscow stopped supporting proxy wars in the developing world. Denied the external assistance that had long sustained them, many of these conflicts simply petered out, or were ended by negotiated settlements.

Furthermore, the end of the Cold War was followed by a sudden increase of international activism directed toward stopping ongoing wars and preventing wars that had ended from restarting. The greatly increased involvement by governments in international organizations, which play an important role in building global norms that encourage the peaceful settlement of disputes, has helped reduce the incidence of conflict. This little-analyzed but critically important development, the report says, appears to offer the most compelling explanation for the steep decline in warfare, starting in 1992.

But then the report goes on to give a different explanation, which is along the lines of the explanation given by the Rebbe MHM in the sicha.

The Power of Ideas: A War-Averse World

A quite different explanation for the decline in interstate war stresses neither the role of liberal economic and political institutions, nor military deterrence, but a gradual normative shift against the use of violence in human relationships. Among the key indicators of this general shift in attitude that the report mentions are the outlawing of war crimes, crimes against humanity and genocide. These antiviolence norms play an important role in constraining behavior. They also inform the creation of laws and institutions, which in turn can provide the monitoring and enforcement mechanisms to help encourage compliance.

Nowhere is this normative shift more evident than in changing public attitudes toward war. Prior to the 20th century, warfare was a normal part of human existence. For governments, war was simply an instrument of statecraft. Today the forcible acquisition of territory is universally perceived as a blatant transgression of international law, and resort to force against another country is only permissible in self-defense, or with the sanction of the UN Security Council.

There has been a similar change in attitudes to colonialism. While colonial subjugation is now universally abhorred, such conquests were once accepted as a normal part of empire and were often depicted as morally justified, in that they brought the benefits of civilization to the colonized. Ideologies that glorify violence and see war as a noble and virtuous endeavor are today mostly absent. Insofar as similar ideologies still exist, they are mainly found not in governments but in small, fanatical terrorist organizations, such as those associated with al-Qaeda. In addition, the sort of hypernationalism that drove Nazi German and Imperial Japanese aggression in the 1930s and 1940s is now extremely rare. Some scholars argue that the rise of war-averse sentiment in the industrialized countries has been the critical factor in the worldwide decline in international war. The reason that liberal democracies live in peace, according to this view, is not because they have democratic modes of government, but because their leaders and peoples have become more averse to war.

From this perspective, interdependence and the growth of membership in international organizations are the *result* of the peace achieved by increased war-aversion, not the cause of the peace.

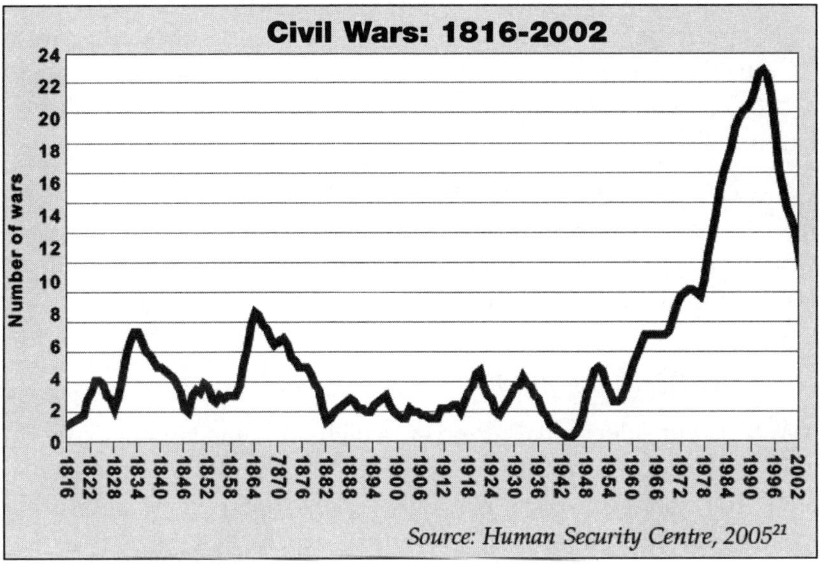

Civil Wars, 1816–2002

The Rise and Fall of Civil War

Until now we have been discussing international conflicts. But the decline in civil wars immediately after the Swords Into Plowshares declaration was even more sudden and more dramatic. The number of civil wars taking place around the world initially increased after World War II—from 2 in 1946 to 25 in 1991. Prior to World War II the maximum number of civil wars in the world in any one year had never exceeded 10. As the graph shows, the escalation in the number of civil wars from 1946 to 1991 was the largest in the entire 1816 to 2002 period. This increase was due mainly to the rise in Cold War-related conflicts and to struggles for control over the new states created by the end of colonialism.

But the decline in the number of civil wars that began after 1992 was steeper than the increase from 1946 to 1992. In just 10 years, the number of civil wars fell by 80%!

The Human Security Brief

The Human Security Report has been updated annually since it was originally released, by the Human Security Briefs of 2006 and 2007. Again and again the compilers were surprised that the shift from militarism towards peaceful relations—Swords Into Plowshares—continues year after year. While this may not be obvious to one who follows the news, it is absolutely clear once all the data is collected.

The 2007 Human Security Brief (published in 2008) focuses on three areas: 1) Global terrorism, 2) Armed conflicts in sub-Saharan Africa and 3) Armed conflicts globally. In each of these areas there has—surprisingly—been a decrease.[3]

The Decline in Terrorism

We first consider global terrorism which, of course, is the issue of greatest concern. Careful analysis of the data shows that the global death toll from terrorism has fallen. This decrease is surprising because the consensus among Western experts has been that the threat of terrorism, particularly Islamist terrorism, has been increasing since the September 11 al-Qaeda attack on the U.S. This was the view held by the 2006 and 2007 US National Intelligence Estimates, by a 2007 survey of 100 foreign policy and security experts published in the US journal Foreign Policy, by a 2007 report on the terrorist threat to Europe from the director of the UK's Security Service, and by a 2008 report from the official US National Counterterrorism Center (NCTC).

But the reason these organizations consider terrorism to have increased is because they include in their data attacks against civilians by armed groups in Iraq. The Human security Brief points out, however, that this is incorrect since these attacks in Iraq are part of the civil war going on there and thus are not part of what is generally considered "terrorism." Once the attacks in the civil war in Iraq are removed, the data shows a net decline of more than 40% in deaths from terrorism since 2001.

---

3. The content of this section is based on, and quoted from, the Overview of the Human Security Brief 2007, www.humansecuritybrief.info/access.html.

Also, the civil war in Iraq has greatly subsided since 2007, so even if that data is included, reports issued by the NCTC at the end of 2007 and in 2008 show a 40% decrease in deaths from terrorism between July and September of 2007, and a further decrease of 20% between October and December of that year.

Another report, published by The US based Intelcenter think tank in mid-2007, examined the 63 "most significant" attacks launched by al-Qaeda and its affiliates over a period of nearly 10 years.

The study included the major attacks most associated with Islamist terrorism — those in Bali, London, Madrid, Amman, and Jakarta, as well as the September 11 attack in the U.S. It did not include civil wars. They found that by mid-2007 the *number* of Islamist attacks around the world had declined by 65% from the high point in 2004, and that *fatalities* from such attacks had declined by more than 90%.

In other words, the fall in Islamist terrorist violence has been remarkable. But can this decline be expected to continue? According to the Human Security Brief, not only has Islamic terrorism decreased, but also the *threat* of terrorism has diminished. They give three reasons for this:

- Counterterrorism efforts are more widespread, more coordinated, and more effective today than they were prior to 9/11.
- Their global network has broken down and there is infighting within and defections from the terrorist groups.
- There has been an extraordinary drop in support for Islamist terror organizations in the Muslim world over the past five years.

The last factor is of particular interest. There is overwhelming evidence to indicate that the more the Islamists attempt to impose their extremist ideologies and their harshly repressive policies on other Muslims, and the more violence they perpetrate against their coreligionists, the more they lose support in the Muslim world. By late 2007, in Afghanistan, just 1% of Afghans "strongly supported" the presence of the Taliban and foreign jihadi fighters in their country. In Pakistan, which is considered the world's "crossroads of terror," support for Islamist political parties has collapsed—dropping dramatically between the 2002 and 2008 national elections. And in the

Northwest Frontier Province where al-Qaeda has its strongest presence in Pakistan, support for Osama bin Laden dropped from 70% in August 2007 to 4% in January 2008.

Yet we still feel the pain of the terrorist murder of the Chabad *shluchim*, Rabbi Gavriel and Rivka Holtzberg in Mumbai, India in November, 2008. The attack originated in Pakistan.

Historical evidence suggests that terrorist campaigns that lose public support will, sooner or later, be either abandoned or defeated. As Muslim populations increasingly reject Islamist policies and terror tactics, they are more likely to cooperate with official counterterror campaigns. This is precisely what happened in Iraq, where Sunni insurgents became so alienated from their former al-Qaeda-in-Iraq allies that they joined with the US in an anti-Islamist alliance to defeat them.

Sub-Saharan Africa

After the Swords Into Plowshares declaration of 1992, when military conflict in much of the rest of the world subsided, sub-Saharan Africa remained the world's most war-afflicted region, with a battle death toll that was greater than the rest of the world's combined.[4] As late as 1999 the *Economist* reported, "From north to south, east to west, large swathes of the continent are at war."[5]

But Swords Into Plowshares is an ongoing process. After 1999, sub-Saharan Africa's security landscape was dramatically transformed. The Human Security Brief of 2007 documents the decline in wars during the years 1999–2006. The number of armed conflicts being fought in the region fell by more than half. The number of people being killed dropped even more steeply. By 2006 the annual battle death toll was just 2% of that of 1999.

This was not the only positive change. Conflicts are conventionally defined as armed confrontations between one government and another government, or between a government and an insurgent group. But this definition—one used by almost all of the major con-

4. American news organizations don't report very much about conflicts in Africa, except for the major ones, such as the conflict in Darfur. The BBC, however, frequently reports on them.

5. "Africa's Wars," *The Economist*, January 14, 1999.

flict datasets—completely ignores communal and other "non-state" conflicts, those in which a government is not one of the warring parties. To address this omission, the Uppsala Conflict Data Program (UCDP) created a new non-state conflict dataset for the Human Security Report Project (HSRP) that has revealed just how important this previously uncounted category of conflict is, particularly in sub-Saharan Africa.

In 2002 there were 26 of these non-state conflicts in sub-Saharan Africa—twice as many as those in which a government was one of the warring parties. But in this category, too, there has been a sharp decline in both conflict numbers and death tolls since 2002. Between 2002 and 2006 non-state conflict numbers had dropped by more than half across the region, and their death tolls had fallen by some 70%.

"One-sided violence" i.e. violent campaigns waged against defenseless civilians by governments or non-state armed groups (e.g. rebel groups) constitute a third type of political violence. Once again, we find a positive change taking place in recent years. Between 2002 and 2006 the number of campaigns of "one-sided violence" against civilians fell by two-thirds, and their death tolls fell by more than 80%.

Sub-Sahara African governments have also been highly vulnerable to coups d'état in the past. Indeed, the region has suffered almost half of the world's coups since 1946. But here, too, there has been a positive change. The average number of coups per year since 2000 has been some 40% lower than in the 1980s, the peak decade for coups in the region.

So what happened to bring about such a dramatic reduction in wars in this region? The Human Security Brief analyzes this, first attempting a conventional explanation:

> The most robust finding on the causes of war by researchers is that the higher the per capita income a country enjoys, the lower its risk of armed conflict. This is why most wars take place in very poor countries. The evidence for this war-poverty association is overwhelming.
>
> But while economic growth is clearly associated with reductions in the incidence of both conflicts and coups over

> the long term, it cannot explain the steep recent decline in the number of armed conflicts in sub-Saharan Africa. Since 2000, income levels have not risen high enough, or over a long enough period, to explain the reduction in new outbreaks of conflict. This is also true of all the other "structural" explanations that researchers have focused on.... The explanation for the radical improvement in sub-Saharan Africa's security climate in the new millennium must lie elsewhere.

Then what is the real explanation?

> The HSRP's research suggests that the drivers of this remarkable decline in armed conflicts in sub-Saharan Africa are to be found not in long-term structural change, but in the post–Cold War surge of policy initiatives designed to stop wars ("peacemaking") or prevent them from starting again ("postconflict peacebuilding").... These peacemaking and postconflict peacebuilding initiatives—which include third party mediation efforts to end ongoing conflicts, humanitarian missions, and peacekeeping and peacebuilding operations—have all greatly increased in number since 1990.

In other words, it was the new drive for "peace and unity, joint efforts and mutual assistance among the nations of the world, for the good of all mankind," as described by the Rebbe MHM in the *sicha* on the Swords Into Plowshares declaration.

Conclusion—the State of Human Security

The Human Security Report 2005 found that armed conflicts involving a government as one of the warring parties had declined by more than 40% around the world from the end of the Cold War to 2003. The most severe conflicts—and the number of genocides—had declined by some 80%. Coup d'état numbers were down by 60 % from the high point in 1963.

The Report also found that the number of battle deaths in state-based armed conflicts had declined even more steeply than the conflict numbers, though over a much longer period. The average number of battle deaths per conflict per year—the best measure of the deadliness of warfare—had fallen from 38,000 in 1950, to just 600 in 2002. By 2006 the estimated global battle death toll had declined even further.

The HSRP now has five years of data on "non-state" conflicts—those fought between communal or rebel groups or warlords, but in which the government is not a warring party. Five years is long enough to detect trends, and they now report that worldwide, non-state conflict numbers have undergone a marked and consistent decline since data were first collected in 2002. In fact, they declined by a third—from 36 to 24—between 2002 and 2006. Reported battle deaths from these conflicts declined by 60% over the same period. Much of this global reduction in the number of non-state conflicts and associated fatalities has been driven by the improvements in sub-Saharan Africa.

When we look at the combined global total of state-based and non-state conflicts, we find that there has been an 18% decline—from 68 in 2002, to 56 in 2006.

Campaigns of one-sided violence often take place during civil wars, with sub-Saharan Africa having the lion's share. These conflicts, like armed conflicts, have to result in at least 25 fatalities a year to be counted. In 1989 there were 19 such campaigns, but their number grew unevenly throughout the 1990s, indicating that targeting civilians had become an increasingly prevalent post–Cold War security issue. In 2004, the peak year, 38 campaigns of one-sided violence were being perpetrated around the world, but since then there has been a sharp drop. In 2006 there were just 26—a net decline of 32% from 2004.

The Lion and the Terrorist

In studying the Human Security Report, we discovered that wars have steadily decreased and have become less deadly since Swords Into Plowshares began, and even terrorism has declined in recent years. Yet the question is often asked: How do we reconcile the presence of war in the world with the prophecy of Swords Into Plowshares which is associated with the cessation of conflict? How could there be any war in the Era of Moshiach, especially after the Swords Into Plowshares prophecy has begun to be fulfilled?

Swords and Plowshares

Firstly, we must point out that even the wars waged since the SIP transformation began (at least those fought by the U.S—a country of goodness and kindness) have a new aspect, which we may call "Swords and Plowshares." In the war in Afghanistan, for example, on the first day of the war, U.S. military airplanes dropped food packages for hungry civilians at the same time that they dropped bombs on military targets. In the war in Iraq, the very precise bombing of Baghdad enabled by advanced satellite technology saved the lives of many civilians. Also, military ships and helicopters were used to demine the harbor of Umm Qasr so that food, drinking water and medicine could be brought in by ships to civilians in Basra and surrounding areas who were experiencing severe shortages. The point is that in these wars, humanitarian aid is an integral part of the strategy right from the start.

Now, regarding the occurrence of wars in the era of Swords Into Plowshares, the essential point is that the Rebbe MHM said that the prophecy of Swords Into Plowshares—like all the other aspects of the redemption—had *begun* to be fulfilled. But they have not yet reached their completion. Indeed, the Midrash speaks of wars of terror that the Ishmaelites (the Arabs) will wage *in the Messianic Era*, as we will discuss shortly. But the Swords Into Plowshares transformations continue throughout the world, as we have been documenting in this chapter.

I believe that a careful reading of the Swords Into Plowshares prophecy itself will clarify the issue. It reads:

And they shall beat their swords into plowshares
and their spears into pruning hooks
nation shall not raise a sword against nation
nor shall they learn any more war.

We notice something very curious here. First the prophet Yeshayahu (Isaiah) says that the nations will beat their swords into plowshares because, as the Radak explains, they won't need them any more as a result of Melech HaMoshiach's influence on the nations of the world. Then he says, "Nation shall not raise a sword against nation." It would seem that the order should be reversed. First it should say that the nations will not raise swords against each other; then, since they won't be fighting each other they won't need their weapons any more so they will beat them into plowshares.

From the fact that beating the swords into plowshares is stated first, we see that this process will begin even before all wars cease, as the world becomes more peaceful and the nations don't feel threatened by each other. Then, as a further development, "nation shall not raise a sword against nation," i.e., there will be no more war. And finally, "nor shall they *learn* any more war." They will not even maintain a readiness for war. There will be no military exercises, no military training or even any computer simulations of conflicts.

The Swords Into Plowshares prophecy is famous as a prophecy of Isaiah. This prophecy is, in fact, also a prophecy of the prophet Micha.[1] In a *sicha*,[2] the Rebbe MHM discusses another prophecy of Micha which indicates that there will be wars during the beginning of the Geula. So the two are not mutually exclusive. The war and ter-

1. Micha 4:1–3.
2. *Likutei Sichos* 18:293.

rorism that we have experienced is not inconsistent with the fact that we are in the Era of Moshiach. As we mentioned above, there are statements in the Midrash predicting the occurrence of these events in the Messianic Era. For example, the Midrash Yalkut Shimoni makes the following prediction:

> In the Era of Moshiach (*Achris HaYamim*) the Ishmaelites will wage three wars of terror—one upon the sea because of an arched bow, one upon the land because of a raised sword, and one against a great city which will be more severe than the other two…and from there the son of Dovid (Moshiach) will sprout forth and show us the destruction of the others. From there he will go to the Land of Israel.[3]

Evidently, the terror attack against a great city refers to the September 11th terror attack against New York City. The attack upon the sea may refer to the attack on the U.S.S. Cole in the Gulf of Yemen in 2000, and the attack upon the land, to the terror attack on the U.S. embassies in East Africa in 1998.

There is a law of action and reaction in operation here. In Chassidic philosophy it is explained that it is part of the natural order that whenever a new level of good is introduced into the world, there is a resistance to it and thus a corresponding evil appears to oppose it.[4] For example, the declaration of "The time of your redemption has arrived" coincided with the Persian Gulf War and Saddam's attack on Israel.

A New World Order

Following the Persian Gulf War came the Swords Into Plowshares phenomenon—the transformation of military technology to peaceful uses, a decrease in war and a tremendous increase in international cooperation and mutual assistance which continues to this very day. Consider, for example, the speech of Gordon Brown, the Prime Minister of England, at the G20 economic summit in 2009, convened to discuss the global financial crisis:

---

3. *Yalkut Shimoni,* Yeshayahu, sec. 421. The *Zohar* on *Parshas* Balak makes a similar prediction.

4. See, e.g., *Sefer HaSichos* 5751 1:73.

> A few years ago, meetings such as this could not have happened with so many different countries from diverse continents involved. Far less could there have been an agreement amongst them. But today the largest countries of the world have agreed to a global plan for recovery and reform.... For the first time we have a common approach....
>
> We have maintained our commitment to help the world's poorest....
>
> This is collective action—people working together at their best. I think a new world order is emerging, and with it, the foundations of a new and progressive era of international cooperation.

This is exactly the new attitude of the nations of the world described by the Rebbe MHM in the *sicha* of Parshas Mishpatim—and expressed by the Secretary General of the United Nations in his report "An Agenda for Peace"[5] on the summit meeting at the UN in 1992 where the Swords Into Plowshares declaration was made.

But along with these developments there appeared a new evil—a culture in which evil people were willing to destroy themselves in order to kill multitudes of innocent and defenseless men, women and children—terrorism.

It is in the nature of this action-reaction process that, as the good continues to blossom, the evil falls away as the outer shell is removed from a fruit and discarded. Thus, as the Messianic prophecies continue to unfold, war and terrorism continue to decline. They are being phased out. Soon the completion of the Swords Into Plowshares prophecy will be achieved and "nation shall not raise a sword against nation."

"And a calf and a lion...shall lie down together"

In contrast to the remnants of war, back on the peaceful side, we have begun to see a new dimension in the fulfillment of the prophecies of peacefulness, for they have not only pervaded humanity, but have begun to penetrate the animal kingdom as well.

5. See sec. 1 of this chapter.

The prophecy that the nations of the world will beat their swords into plowshares in chapter 2 of Isaiah is closely related to another Messianic prophecy in chapter 11 of Isaiah: "A calf and a lion...shall lie down together."[6] According to the Rambam this prophecy is an allegory referring to the peaceful relations that will exist between previously belligerent nations in the Era of Moshiach[7]—along the same lines as the "Swords Into Plowshares" prophecy. Many commentaries, however, explain this prophecy literally[8] to refer to a change in the nature of wild animals and the peaceful relations in the animal kingdom.

The Rebbe MHM resolves this controversy by explaining that Rambam's allegorical explanation applies only to the first phase of the Messianic Era when, according to Rambam, the world will still follow its natural order. In the second phase of the Era of Moshiach, however, a miraculous order will be in operation alongside the natural order. Under this miraculous order, the prophecies of chapter 11 will be fulfilled in the literal sense.[9]

These two phases, however, are distinct only conceptually, not chronologically. The Rebbe MHM emphasizes that we must demand from Hashem, and we can expect to get from Him, the fulfillment of all the promises of both Messianic eras immediately. This applies even to the revival of the dead which, chronologically, is the last on the list of Messianic events.

Thus it should not come as a surprise that several years ago the world witnessed an event of "The calf and the lion...shall lie down together"—literally. This happened in a nature reserve in Kenya, Africa. The following is based on reports from Reuters and the BBC:

A grown lioness baffled experts in Kenya by adopting a baby oryx, a kind of antelope that lions like to *eat*. The lioness came across the oryx shortly after it was born on January 5, 2002 in Samburu National Park, finding it lying in wait for its mother who had gone to search for food. Defying nature, the lioness quickly adopted the oryx, giving it affection and protection from other predators as if it were

---

6. Yeshayahu 11:6.
7. Hilchos Melachim U'Melech HaMoshiach chap. 12, 1.
8. Rabbi Avraham ben David (the Ra'abad) in his *halachic* comments on the Rambam, as well as other commentaries.
9. *Likutei Sichos* 27:191–206 in Yiddish. Hebrew translation in *Sha'arei Geula* 2, sec. 31.

her own cub. Once, a leopard approached and tried to kill the oryx, but the lioness protected the calf.

The Lion and the Calf

Extraordinarily, the lioness still allowed the mother oryx occasionally to come and feed her calf. The baby oryx was very close to the lioness. The lioness would lie down to rest in the afternoon and the oryx would curl up beside her.

After two weeks of taking care of her adopted baby, the lioness, tired and weakened, led the oryx to the river to drink and fell asleep, failing to notice a hungry male lion in the area. While the lioness was taking a nap, the baby oryx was playing around and it was caught by the male lion. The baby was killed. The lioness was very angry. She went around the lion about ten times roaring, and then the lioness disappeared. The lioness has not been seen since.

"And the cow and even the bear shall graze"

We have seen the prophecy of the calf and the lion fulfilled. Now, what about the next verse that says: "And the cow and even the bear shall graze." Well, we may have seen that too—right here in America in the town of Hygiene, Colorado. The story, as reported by the

Associated Press, reads as follows:

> A curious cow and a young wild bear touched noses and hung out together in a pasture before the bear dashed off with the cow in pursuit, neighbors said. But whether the cow was chasing off the bear or just wanted to play is a matter of debate.
>
> Residents of this rural town about 30 miles northwest of Denver say the encounter took place Sunday after the bear climbed into an apple tree in the cow's pasture and then climbed down.
>
> The cow is named Apple because she loves eating the fruit of that very tree.... Apple ambled over for a look, said Penny Cox of Hygiene, who witnessed the get-together. Eventually the bear took off with Apple running after it.

The Baal Shem Tov said that once a miracle occurs it becomes part of nature. It is no longer miraculous and can happen again and again.

REFERENCES: CHAPTER III

Swords Into Plowshares

<u>*General*</u>

Rabbi Menachm Mendel Schneerson, MHM.
Sefer HaSichos, 5752, *vol.* 2. New York: Kehot Publication Society

BICC Yearbooks, Conversion Survey 1996 – 2005. This series of yearbooks is published annually by Bonn International Center for Conversion and has one of the most detailed documentations of the Swords Into Plowshares transformation available. See also their website: www.bicc.de

Markusen, Ann, Sean DiGiovanna and Michael C. Leary, *Defense to Development*. Routledge, 2003. This is a thorough report on the military to peaceful transformation worldwide.

<u>*Sec. 3: Russia and Eastern Europe*</u>

BICC Report 3, *Conversion of the Defense Industry in Russia and Eastern Europe*, available at www.bicc.de.

BICC Report 11, *Base Closure and Redevelopment in Central and Eastern Europe*, available at www.bicc.de

Schweitzer, Glenn E. *Moscow DMZ: The Story of the International Effort to Convert the Russian Weapon Science to Peaceful Purposes*. New York: M.E. Sharpe, 1996.

www.istc.ru. This is the website of the International Science and Technology Center.

www.tsagi.ru. This is the website of TsAGI.

<u>*Sec. 4: Transforming the Swords of China*</u>

BICC Brief 10 *Peaceful Fallout: The Conversion of China's Military-Nuclear Complex to Civilian Use*.," available at www.bicc.de.

<u>*Sec 6: The Transformation of Satellite Technology*</u>

Rabbi Menachm Mendel Schneerson, MHM.
Sefer HaSichos, 5752. New York: Kehot Publication Society

NASA websites under www.nasa.gov.

<u>*Sec. 7-8: A Nuclear Reactor for the Era of Moshiach/Using Thorium*</u>

http://web.bgu.ac.il/engn/nuclear/research/NucEnergy/RTF.htm

www.tau.ac.il/taunews/97fall/nuclear.html

www.ltbridge.com.

Sec. 9: Persian Gulf War Technology to Plowshares
http://www.hopkinsmedicine.org/press/1994/JUNE/199421.HTM
http://www.nasa.gov/vision/earth/technologies/jordy.html

Sec. 10: Into Plowshares - Literally
"Bloodless Spine RadioSurgery," available at http://www.thefreelibrary.com/Bloodless+Spine+RadioSurgery%3A+CyberKnife+Center+of+Miami+First+With...-a0132112604
"Computerized robot arm aids in surgery," USA Today (Society for the Advancement of Education), Feb, 1996, available at http://findarticles.com/p/articles/mi_m1272/is_n2609_v124/ai_18748011/
"Stealth surgery on brain tissue - computer-mediated stereotaxic radio surgery uses missile-tracking software to treat certain kinds of cancer." Science News, Aug. 26, 1995, available at http://findarticles.com/p/articles/mi_m1200/is_n9_v148/ai_17288651/

Sec 11: The Human Security Report
Human Security Report 2005: War and Peace in the 21st Century. New York: Oxford University Press, 2006.
"Overview of the Human Security Brief 2007," available at www.humansecuritybrief.info/access.html.
www.humansecurityreport.info. This is the website of Human Security Report.

CHAPTER IV

The Convergence of Science to Torah:

The Trend of Science

"True science whose purpose is only to seek the truth—its conclusions cannot contradict the Torah which is the Torah of truth. Rather, on the contrary, the more that [the scientists] will deepen their research, the more they will increase their recognition of the truth of both the principles and the details of our religion, the religion of Israel."

—the Rebbe MHM

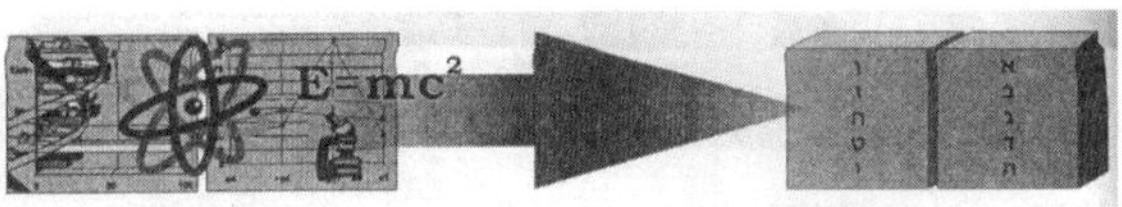

A belief in science and religion is consistent....There are very many scientists who do believe in G-d....Their belief in G-d and their action in science is thoroughly consistent.

—Richard Feynman

Deeper Research

Science was once used—more accurately, misused—as the basis for all challenges to the belief in the existence of G-d. Now, however, after the revelation of the Wisdom from Above and the breaking open of the fountains of Wisdom from Below, science is moving closer to the famous declaration of Dovid HaMelech, "The heavens relate the glory of G-d."

In this chapter we tell the story of how the practice of science has been subjected to critical analysis by the scientists themselves, resulting in old theories being discredited and paving the way for "true science whose purpose is only to seek the truth" to lead to an awareness of the Creator—recognition of the existence of G-d.

The Branover Interview

Prof. Brunover at SUNY Stoney Brook, 2004

Probably the foremost authority in the world on matters of Torah and science is Professor Yirmiyahu (Herman) Branover of the University of the Negev in Beer Sheva, Israel. Born in Russia, Professor Branover became famous for his pioneering work in Magnetohydrodynamics (MHD) while he was director of the Physics Institute of Latvia. Most recently, he directed the MHD Laboratory at the University of the Negev in Israel.

As a member of the Rabbi Yisroel Aryeh Leib Research Institute on Moshiach and the Sciences, Professor Branover has spoken at our annual Moshiach & science conferences. We also interviewed him to learn his views on a wide range of issues relating to Moshiach and science.

The Disappearance of Evolution

PROF. SILMAN: The Rebbe Melech HaMoshiach explains in a sicha—based on a statement in the Zohar—that the scientific discoveries of the past 150 years, together with the revelation of the teachings of Chassidus, were revealed in order to prepare the world for Moshiach. Can you comment specifically on scientific theories and developments in the past 150 years that are characteristic of the Era of Moshiach?

PROF. BRANOVER: Without a doubt, the inclination of modern science is certainly converging with the world outlook of Torah. The Rebbe Melech HaMoshiach always stresses that we have to learn from everything we see, and we see very much in this time and age all kinds of expressions in science and scientific literature, and in practical results of science and different technologies, evidence that supports this convergence. I will give a few examples of how modern sciences are converging toward Torah views, and how so many of those once famous arguments between science and Torah have been shown to be no longer valid.

What happened was that all of the so-called conflicts between science and Torah just fell apart. Nobody came along and officially destroyed them, but they disappeared on their own by virtue of science's very progress.

Let's use the theory of evolution as an example. First of all, the theory itself was transformed several times, so it's already no longer Darwinism. Darwinism was later changed, because of the development of genetics, into modern synthesis by Huxley and others. But no matter what the version of the theory, it became absolutely invalid because they could never find the missing links between certain species among the fossils, so they came to the complete inversion of the theory. And that is ridiculous because Stephen Gould and others punctuated the equilibrium and they say, "No, there are no missing links between the fossils because there haven't been missing links. One species was transformed into another in paleontological frame in zero time. It happened, so to speak, overnight." Therefore, for those who look to fossils to present a case for the evolutionary process—there hasn't *been* an evolutionary process. There has been a jump from one situation to another. The whole reasoning process simply doesn't make sense. Darwin at least had a nice explanation, but now those who accept this punctuated equilibrium can only say, "Now we will investigate what the mechanism was. Why did this jump?" They foolishly stick to evolution because it is important for ideological reasons, but they don't even have a theory as poor as Darwin's. At least Darwin's theory looked elegant externally. Now they have nothing.

Evolution is Not a Theory

Modern philosophers of science, e.g. Carl Popper, say evolution is simply not a scientific theory; it does not meet the criteria. What it is, in fact, is a metaphysical system of ideas. In other words, it's nothing. It has nothing to do with science. Without going into detail, this is important because many people still make a big deal out of evolution without knowing that evolution is dead. Nobody who is serious speaks about evolution. It is only those who have ideological motives or who make their living from evolution that stick to it.

Now the same situation exists regarding geological evolution which has been very much replaced by a "catastrophic" approach. The scientists involved claim that the changes have been catastrophic rather than evolutionary. According to this theory, geological phenomena could change in a very short time. There are many other examples. But what is important is not only that those ideas which claimed to contradict or refute Torah disappeared, collapsed, or were annihilated; but that the newest developments in science are in accordance with what the Zohar says and what the Rebbe MHM is always quoting. He stresses that in our days the material world is prepared for the coming of Moshiach, and nothing can reflect the material world more than scientific thoughts, scientific inquiries, and scientific explanation. Science was always supposed to be, and has to be by definition, rational, cold and removed. But suddenly all these new theories are coming, with accelerating speed, back to classical fundamental Torah concepts.

The Center of the Universe

In this segment, Prof. Branover explains how Einstein's theory of Relativity removes all questions raised regarding the Torah's statement that the earth is the center of the Universe.

I mentioned before that all of the so-called conflicts between science and Torah eventually fell apart because of the progress of science. I will give a few more examples.

First, the oldest argument of all is probably the structure of the solar system. When Copernicus wrote his book 450 years ago, he changed the whole idea of its structure. He replaced the Ptolemeyan geocentric system by a heliocentric system. For the sake of his calculations, he exchanged the places of the sun and the earth. It's not because he believed in the inverse structure, but the calculations are easier if you assume the sun is standing still and the earth is revolving. But this caused a whole revolution and a lot of arguments. In the beginning, the whole debate was completely unknown to Jews—they didn't have to deal with this conflict. But the *Haskala* movement transplanted it into Jewish communities, into Jewish schools, and that is how this so-called conflict between Torah and science started.

Now, as the Rebbe MHM always stresses, we live in the era of modern science, and modern science has several theories as its foundation: Einstein's theory of relativity, quantum theory and others. Einstein's theory says clearly and unequivocally that there is no absolute space and no absolute moment; therefore the whole question (of the placement of the sun and the earth) is scientifically irrelevant. We can assume either way: a geocentric or a heliocentric system.

An interesting book was published back in 1926, just 10 years after Einstein published his general theory of relativity. The book, *The Philosophy of Space and Time,* was written in German by one of Einstein's disciples, Hans Reichenbach, and is now available in English.[1] In it he writes very clearly and very convincingly that we can be modern and adhere to modern physics and Einstein's theory and accept the idea that the earth is standing still at the center of the universe.

Remarkably and regrettably, people in universities are still not aware of this. Because of their narrow-mindedness, even some people who are working in relativity have never come to this point. They learned in elementary school that the earth revolves around the sun, and they never progressed beyond that point. Working in relativity theory, they know their equations, but they work in their specific narrow area and never have taken time to perceive it on a global scale.

Here in Chabad, we even held a special seminar on this question: how to understand the Rebbe MHM's words on these matters and how to accept without any fear or apologetics the idea that the earth is standing still. We couldn't find the sources so we needed to go to the Rebbe MHM himself and ask his advice. Then we found this book by Reichenbach and we understood. Even we professional scientists who belong to Chabad couldn't find the answers. We needed direct guidance from the Rebbe MHM.

---

1. Hans Reichenbach, *The Philosophy of Space and Time* (New York: Dover Publications, 1957). See, especially, p. 217.

Quantum Mechanics and Free Will

In this segment of the interview, Professor Branover explains how the theory of quantum mechanics overturned the Laplacian theory of determinism and incorporated Free Will into modern science.

The most fundamental contradiction (I should say, *apparent* contradiction—it wasn't a real contradiction) which dominated the area of Torah and science for several centuries since the inception of classical science by Newton, Laplace, and others was the issue of free will. Since Newtonian science, classical science assumed that everything in nature is deterministic—that there are chains of causes and effects and that the present state of a physical system determines the future state and that science is omniscient—that it can explain not only the human body, but also the human mind, thought, and soul.

Thus they came to the conclusion that everything is predictable using the laws of physics and mathematical computations, and since, in their perception, man is some kind of machine—a robot—acting on the basis of physical and chemical impulses, they reached the conclusion that there is no free will because we can calculate how a certain man a hundred years from now who wasn't yet born will behave and what he will do in the year so-and-so and on a certain date. This, if it would be true, is a terrible contradiction to free will, which is one of the fundamental principles of Torah. To make the choice whether or not to fulfill G-d's 613 commandments, we need free will. Hopefully, we will make the right choice, although we have the possibility of doing the opposite.

A Nondeterministic Universe

Now with quantum physics, nondeterminism was reinstated because, as anybody who has even an elementary knowledge of quantum physics knows, particles, the bricks of the universe, known as electrons, protons, photons (particles of light), etc., all have dual structures. They are simultaneously particles and waves. That in itself is difficult to understand, which demonstrates how sophisticated the world is and how simplistic, primitive, and wrong we had been in the past. Second, because of this wave character of the universe, it turns out we can only deal with probabilities regarding elementary particles. But great physicists like Richard Feynman, a Nobel Prize winner in quantum physics, along with others, argued that it applies not only to microscopic, but also to macroscopic (big) objects. Feynman says an atomic bomb can be triggered by a single electron hitting or not hitting a target. An electron's hitting or not hitting a target depends on probability and cannot be predicted.

So the whole world is becoming nondeterministic—which brings us back to free will. Even more so, as the Copenhagen school of quantum physics has stated (I will not go into the difference now between the Einstein school on the one hand, and Niels Bohr and the Copenhagen school on the other), if there wouldn't be an observer, the whole world would scientifically cease to exist. It would just be a collection of probability waves. Or, said another way, the world could exist simultaneously in an infinite number of states. But that is impossible because the real world is one, so there couldn't be two, ten, hundreds, or millions of infinite worlds. So something has to bring about the collapse of probability waves into reality.

A Creator and Free Will

After having checked all the possibilities, they come to two possible conclusions: One is that there has to be a Creator, a Supreme Being that is not dependent on nature, because, by simple logic, to establish nature you need something *higher* than nature. So it must be a Creator, a Supreme Being. But being indoctrinated as they are, and being atheists, many scientists reject this idea. They say, "No. We are scientists and cannot accept this idea of G-d." But this, of course, has nothing to do with science; it has to do with their faulty education.

The other possibility is they *do* say man possesses free will. One of the greatest scientists in quantum theory, Eugene Wigner, said, "I know nothing, no other entity which is independent of nature, except for free will which man possesses." Therefore the free will of a human being is bringing about the collapse of probability waves and, in this way, reality "evolves"—which is remarkable! So science leads us directly toward the acknowledgement of free will and the acknowledgement of G-d. Also in modern mathematics, in mathematical logic for example, you can find many examples of how this happens.[2] The newest developments lead quickly, irrevocably, and convincingly back to the Torah world outlook.

That is exactly what the Rebbe Melech HaMoshiach is saying: You only have to open your eyes, and everything—even the coldest, most rational study that science is supposed to be—is leading you directly to Torah concepts.

Revival of the Dead

Professor Branover continues to discuss how modern science opens our minds to understand various aspects of the Era of Moshiach. In this section he discusses Techiyas HaMeisim—*the revival of the dead.*

SS: Professor Branover, it is a basic concept of Chabad Chassidus that even things that are miraculous and beyond comprehension should, as much as possible, be brought within the grasp of the mind. Now, *Techiyas HaMeisim*—the revival of the dead—is the most miraculous event of the Messianic Era. Are there any theories in science that can help the mind grasp such a concept—that those who are dead are going to come back to life?

YB: I can answer on two levels. First of all, the sciences are coming back with ever-growing speed to the Torah world, converging to the Torah world outlook and Torah teachings. A hundred years ago it may have been difficult to comprehend, difficult to have "peace of mind" intellectually, with some supernatural ideas which are deeply rooted in Torah such as miracles, revival of the dead, and the final redemption—the *Geula*. These supernatural events can appear to be frightening to people who stick to atheistic materialistic realms, especially if they compare them with the conclusions of classical science. But classical science today is obsolete and looks very simplistic and primitive.

Now, through modern science we have learned how sophisticated, how complicated, this world is, and how we have only scratched the surface. The more we go into scientific explanation, we realize how much more we don't know, how phenomena discovered by science—through mathematics and through empirical observation—lead to

2. Prof. Branover is referring here to three-valued logic (and, more generally, many-valued logic). This is discussed in chap. V, sec. 2.

conclusions which our limited minds are unable to perceive and accept and comprehend.

Accepting the Incomprehensible

I already mentioned the dual structure of the particle. Whether it is light or an electron or any other particle—even if it is very small and cannot be seen by the naked eye—one can still imagine that it is something one can touch, something one can perceive. Simultaneously, it is a wave. A normal human mind within its natural limitations cannot visualize or rationalize that. Nevertheless, we accept it.

So science is teaching us that we have to accept things which may be the counter-intuitive conclusion of our own investigation, or may have been passed over to us from a higher authority—from the Creator Himself—and from our great teachers: from Moshe Rabbeinu and the Rebbe Melech HaMoshiach. Despite the fact that we are limited, and therefore we can't understand or visualize every event, we have to accept these things because they are real—more real than things which we assume we are viewing, seeing or listening to in the process of our exploration of nature.

Modern science is showing us clearly that we need to get rid of simplistic approaches, get rid of obsolete perceptions. We need to make peace with the fact that sometimes, if you deal with difficult concepts, you just have to accept the incomprehensible either because they are logical and scientific conclusions, or else they are revelations and teachings of the great, great teachers we have.

In this context, I mention the many miracles of the Geula such as revival of the dead—*Techiyas HaMeisim*. Also, amongst the miracles of the Geula spoken about by the Rebbe MHM is the revelation of Moshiach himself. Now, Moshiach is already here—we just have to open our eyes to see it. But the actual *seeing* of Moshiach (after *Gimmel Tamuz*) will be miraculous. The fact that our nature-oriented mind doesn't understand it completely doesn't matter; we have to learn and to have full trust and confidence, and then our understanding will also improve. After all, we pray three times every day for things which we probably can never understand rationally. But if our prayers are said with real *kavana* (sincerity), then when we talk about miracles such as revival of the dead and complete Geula, we can have full confidence and belief in them despite the fact that we may not be able to visualize them happening. This is one of the results of the evolvement of modern science. Another is the acceptance by some scientists of the idea of revival of the dead....

A Secular Scientist's View

Prof. Branover now discusses a book written by a prominent scientist on "revival of the dead."

YB: It is remarkable that certain scientists are coming directly to the idea of *Techiyas HaMeisim*. An interesting book, published some years ago, *The Physics of Immortality, b*y Frank Tipler is a very good example. A number of years earlier he also wrote, together with John Barrow, about the Cosmological Anthropic Principle. In their book, they wrote that man is central to the universe and the earth is central to the universe; and that in the entire universe, with all its galaxies, the earth has the only possible conditions under which life—specifically man—can exist. Then, with his new book, he went a step further. As an atheist, he isn't relying on spiritual revelation; but through mathematical and physical explanations he comes to the conclusion of resurrection of the dead, not only as a physical possibility but, as I understand him, from his point of view, it will inevitably, necessarily occur. He also speaks about the necessity and existence of free will. Here you have a splendid example of how completely the external sciences and a man who was very external to Torah are coming to the same conclusions because his profession leads him in this direction.

SS: Can you comment specifically on his theories?

YB: As religious Jews, we should have a problem with the details of his theory, because he speaks about certain time frames for the processes which he anticipates. He even brings a long argument whether he should use the term "G-d" or some other term such as "omega point." Of course, coming from a scientist who is external to Torah and probably never exposed to Torah, this is not surprising. Regarding his time frames, he comes to the ideas of resurrection and free will based on the accumulation in supercomputers of vast amounts of information about human beings, about the structure of their body and their spiritual activity, and so on. It doesn't fit the Torah frame on those issues. So we can argue about the details—and I think we should argue because he should learn about Moshiach from classical Torah sources.

But the fact that Tipler even comes to the same conclusions is most admirable and remarkable. That would have been unthinkable to me, who spent more than 40 years of my life in the scientific and university world. It's so unexpected, so miraculous. Why is it that nowadays scientists and serious physicists are coming to these conclusions which just 20-30 years ago had been completely taboo? Who in the scientific community would have even dared to mention them?! I am sure that because of the materialistic indoctrination and atheism of many scientists, they will object strongly to what Tipler is writing in his book, but nevertheless he decided to publish it and to make it known to the general public.

Furthermore, the fact that he came to these conclusions from scientific investigation only, and decided to publish his conclusions, and the book itself was actually published and is widely publicized and distributed—this whole thing is so remarkable to me, having spent my young years as an atheist in the Communist Soviet Union. We had been taught from kindergarten the slogans that religion is an opiate for the masses and that science refutes and denies religion. Now suddenly look what happens: Scientists are coming directly to Torah concepts!

The Purpose of Science

SS: There is an interesting statement in the Talmud Sanhedrin that says, *"Lo Ivrei Alma Ela L'Moshiach"*—that the whole world was created only for Moshiach. Now, of course, the sciences are also part of the created universe. Would you say then, based on what we have been talking about, that science has now reached the point where it is playing its proper role in the order of creation now that it is converging to the truth of the Torah and serving the purposes of Moshiach?

YB: The Rebbe MHM's point of view is that everything that was created, including science, was created for a holy purpose. Everything depends on us and our free will. How do we use it? We can use science for refuting religion, for struggling against religion, as we did over the course of several generations. Many people made a new form of idol worship out of science by worshiping science and scientists. That's the wrong approach. Many people may have been well-meaning but they were misled; they have been blind. In the era of classical science it was difficult to find the right way to use science. You needed a giant such as the Rebbe MHM to explain all these matters. But he didn't just start speaking about all those things—that science is leading to Torah and Moshiach—only recently. He spoke about it *fifty years ago*, when all those books I mentioned and all those new approaches were unknown. Fifty years ago the public hardly knew about quantum physics and relativity and so on. Certainly we now see how the whole universe and its reflections in science are really serving and leading directly to Moshiach.

The Inner Reality

We learn from Chabad Chassidus, from the Baal Shem Tov and the Alter Rebbe, that the ultimate goal and purpose of the Chassidic movement was to accelerate the coming of Moshiach. Now that we are in the seventh generation, the generation of Moshiach, we are so fortunate that we are so close, that we live in this time. So now it depends on us. The Rebbe MHM says that Moshiach is here and we just have to open our eyes to see it. It is very important that Jews work in this area. We need at least ten people, but preferably many more, to make it the foundation of their lives to do something to bring Moshiach immediately now through strengthening Mitzvos, through helping people, through unity, through kindness—to bring about this world of kindness for which we are longing so deeply.

SS: It is interesting that you mention that this process goes back to the Alter Rebbe since one of the things that you spoke about a few minutes ago—that we have to change our perception of reality and not take the world or even our scientific analysis of the world at first glance—is a point discussed by the Alter Rebbe at length in chapter 42 of Tanya. He says a Jew must train himself not to look at the world superficially and think that what he sees is the true reality. Rather he must realize that the whole physical world that he sees is only an *external manifestation* of G-d, and the Jew has to train himself to look at the *inner reality*—to realize that the true reality of the world is the inner life, which is the life that G-d gives to the world.

A Tribute to Dr. Berg

SS: Our colleague, the late Dr. Naftoli Berg *z.l.* [Norman Berg, Director of Research of the Advanced Technology Office of the Army Research Laboratory of the Pentagon], spoke at the Moshiach and Science conferences of the RYAL Institute in 1993 and 1994 about some of the developments in scientific areas that again show that we are in the time of Moshiach. Could you comment on him or on any of his ideas?

YB: Our dear friend Naftoli Berg, *olav hashalom* (of blessed memory) lectured about Torah and science and about convergence of sciences with Torah in the Era of Moshiach and about the Rebbe Melech HaMoshiach himself. But beyond just lecturing, his whole life was a personal example. He was a manifestation of a Jew who is bringing Moshiach closer because he worked in one of the most important, most famous laboratories of the Pentagon and was involved in military research. I think he won all the awards that were ever issued by the Pentagon. I visited him many times in his laboratory. On their "wall of fame," where they exhibited their awards, nobody was mentioned as many times as Naftoli Berg, and there he was—a Jew with a full beard and *tzizis* (ritual fringes) with a big black yarmulka! In his laboratory, he had all the necessary accoutrements for keeping mitzvos—including a *kvort* (special container) to wash his hands for eating bread, etc. He used to arrange a *minyan* (prayer group) in the Pentagon itself, and that in itself was remarkable. People knew about it all over the world. We heard a conversation on the streets of London where one Jewish youngster told another, in great astonishment, that he heard there was a man in the Pentagon who is walking around with a beard and *tzizis*. The boys were overwhelmed because they couldn't imagine that to be possible. The *Kiddush Hashem* (sanctification of G-d's name) was enormous.

The Everlasting Life of Melech HaMoshiach

The Talmud and the works of Kabbala say that such great tzadikim as Yakov Avinu, Moshe Rabbeinu, Dovid HaMelech and Melech HaMoshiach have everlasting life. Even though they may appear to have passed away, they are actually alive. The Rebbe Melech HaMoshiach himself has often said that this is true of his father-in-law, the Previous Lubavitcher Rebbe, as well.[3]

In the concluding section of the interview, Professor Branover discusses this from the point of view of a scientist.

SS: Using the concepts we've been discussing, and using ideas in science to help us understand events of the Era of Moshiach that are currently unfolding, I'd like to ask you to comment specifically on the event of *Gimmel Tammuz* (3 Tammuz, 5754 (1994), the day of the concealment of the Rebbe MHM). Since *Gimmel Tammuz* we have not been able to see the Rebbe MHM with our physical eyes. Yet we know from his most recent talks, particularly those from 1991-92, that the Rebbe MHM has *Chaim Nitzchi'im*—uninterrupted everlasting life. This is similar to what the Talmud and Zohar say about Yakov Avinu—*"Yakov Avinu Lo Mes"*—our father Jacob never passed away. It is also similar to what the Talmud and Zohar say about Moshe Rabbeinu—that he did not die—and about Dovid HaMelech (King David)—*"Dovid Melech Yisroel Chai V'Kayom"*—King David continues to live. The Rebbe MHM has explained that this does not mean that King David continues to live in *Gan Eden* (the world of the souls) because that is true of every Jew. Rather, it means that he continues to live in this world and continues to be the king of the Jewish people—starting from when he was anointed in Hevron thousands of years ago and continuing through the rest of time.

The Talmud and works of Kabbala and Chassidus also say similar things about Melech HaMoshiach. So we know that the Rebbe MHM continues to live an everlasting life. In connection to this, we continue to announce publicly, *"Yechi Adoneinu Moreinu V'Rabeinu Melech HaMoshiach L'Olam Voed!"* (Long live our master, teacher and Rebbe Melech HaMoshiach forever)—a declaration that the Rebbe MHM has explained gives life to the king: When the subjects of the king, the nation, make an announcement, *"Yechi HaMelech"*—"Long live the king"—this gives life to the king. Can you integrate this with our discussion now, about how scientific theories give the Jew the ability to see the world differently and prepare him for the events of Moshiach, particularly that the Rebbe MHM will reappear before us? We understand that it won't be a resurrection, because he

3. See chap. 1, sec. 6, for an in-depth discussion of this.

is still alive—it will be a reappearance. Can you give us some structure?

YB: First of all, from what was discussed before, I think the bottom line from all the discussion here, as you mentioned before, is also mentioned in Chassidus—in Tanya. What our eyes of flesh are seeing is not reality. Reality is much more complicated, much more sophisticated. Therefore, the whole issue of certain great personalities in Jewish history living eternally, beginning from Yakov Avinu and others whom you mentioned, is a statement which cannot be removed from the Torah. As long as you accept the Torah, you have to accept all of it. You cannot divide Torah into parts. Even if you change one letter, and certainly even more so, one concept, it is not the Torah any more. This would create a falsification of the Torah. I believe that all we have discussed about the sciences—how science is teaching us that we have to accept concepts even when we don't see them, visualize them or comprehend them—all that is a good preparation for accepting and believing with full confidence this concept of *"Yakov Avinu Lo Mes."* Certainly it is applicable to the Rebbe Melech HaMoshiach. It is this aspect and *not* resurrection of the dead which we rely on.

In one sicha[4] where the Rebbe MHM discusses this in detail, he says that, yes, from a very earthly viewpoint, it appears as if Yakov Avinu died. The Egyptians saw how people dealt with his body and all the behavior surrounding the event, and yes, if you are only capable of seeing on the surface, then it looks like Yakov Avinu died. But the Torah makes a statement that this was not the case. In reality he did not die.

The Rebbe MHM stresses that we shouldn't be entrapped by the temptation to explain it in a spiritual sense, that he is living spiritually—that is obvious; that is elementary. The point is that *he is living also physically and bodily* and we have to rely on the teachings of the Torah to educate ourselves. There have been many things that we couldn't understand but now science itself recognizes them. At today's level, science may not be able to give direct explanations for this very lofty concept that Yakov Avinu is alive physically, but we should learn from our experiences regarding other questions. This is certainly something which should be a foundation of our faith.

4. Sicha of 20 Menachem Av, 5731.

This brings me to the question of *"Yechi."* I believe that it is very important that we say *"Yechi Adoneinu Moreinu V'Rabeinu Melech HaMoshiach L'Olam Voed."* As you said, it gives *chayus*—life—to the Rebbe MHM. It gives life to all of us because Moshiach, as the Talmud states and as logic says, is the foundation, the ultimate goal of the entire creation. So by reciting it—if we recite it from the depths of our hearts and the depths of our souls—it adds life, energy, so to speak, if I may use a technological term, to ourselves and to the entire universe. But the necessity is not only to recite it, which of course is also very important, but also to make it publicly known, to involve other people. But it is even more important to sincerely believe in it and to live with this concept that Melech HaMoshiach—the Rebbe—is with us. He lives. He is not only guiding us and supervising everything that is happening here in this world on a spiritual level, but he is also physically with us, bodily, despite the fact that temporarily we cannot see and cannot hear him. That should become the foundation of our life. It may not be easy. These are difficult concepts, but the important ones are the most difficult, and the most difficult are the most important. You have to study, you have to train your imagination, your perception, and more than anything else you have to have *bitochon* (trust) and *bittul* (humility). You have to nullify yourself. We should not rely on our cleverness—that is a fundamental issue of Chabad Chassidus. If a person has an exaggerated self-esteem, he is not leaving space for G-d. This person views himself as the center of the universe.

So first of all there should be *bittul*, and on the grounds of this *bittul* there should be an unlimited trust and confidence. Then it will not only make us happier and more fulfilled, but it certainly will accelerate enormously, and make absolutely imminent and immediate, the coming of Moshiach—the Rebbe Melech HaMoshiach.

SS: Based on what we have said, I think it is appropriate to conclude the interview by declaring—*"Yechi Adoneinu Moreinu V'Rabeinu Melech HaMoshiach L'Olam Voed!"*

[Prof. Branover and Prof. Silman said the declaration of *"Yechi"* together.]

The Logic of Intelligent Design

"By learning and publicizing the concept of continuous creation, which is one of the fundamental precepts of Chassidus, we bring the Geula closer and make it happen faster."[1]

—the Rebbe MHM

Where did the world come from? Was it always here? Did it start with a bang? How does one even begin to answer these questions?

The first statement in the Torah is, "In the beginning, G-d created the heavens and the earth." We Jews have held fast to this belief throughout history while the philosophers of other nations maintained that the world was always here. It was not just a matter of curiosity. If G-d created the world—and us—then our existence has a purpose and we have a responsibility to our Creator to fulfill that purpose. We are not free to live as we please.

This is why the mitzva of Shabbos is so important. It is an observance that expresses our awareness that G-d created the world in six days and rested on the seventh. He is our Creator and we were created to serve him. We therefore have an awesome responsibility to live our lives in accordance with His will and we experience a tremendous joy in having the merit to serve Him.

The creation of the world by G-d further implies that G-d Himself can do what He wants with it and give it to whomever He wants. He gave the Land of Israel to the Jews as an inheritance and it is ours forever.

1. *Sefer HaMa'amorim Melukat* 3:181.

Intelligent Design

Our belief in the Creator did not remain a matter of belief only. Jewish sages and philosophers throughout the ages presented logical arguments that prove that the world had a beginning. Rabbi Akiva, the great sage of the Talmud, once explained to a non-believer that just as the existence of an intricately crafted object proves that there was a craftsman who crafted it, so too the universe itself proves that there is a Master Craftsman who created and crafted it.

The greatest Jewish philosophers such as Rav Saadia Gaon (in *Emunot V'Deot*), Rabbi Yehuda HaLevi (in the *Kuzari*) and Rambam (in the *Guide for the Perplexed*) gave extensive philosophical proofs that the universe was created. Some of these arguments are quoted and summarized by the Tzemach Tzedek, the third Rebbe of Lubavitch, who also explains their Kabbalistic basis (in *Sefer HaChakira*).

Throughout history, however, this remained a matter of controversy between those who believed in the Torah and those who didn't. One of the most important revelations of the Era of Moshiach will be the G-dly force which maintains the existence of the world. Just as the physical aspects of the universe are plainly visible now and no one needs to be convinced of them, in the Messianic Era, the G-dly creative force will be visible to all and it will be clear and obvious that "G-d created the heavens and the earth."

One of the early Jewish philosophers, Rabbeinu Bachaya, in his book *Chovos HaLevavos* (Obligations of the Heart), proves the existence of the Creator as follows:

> If someone brought before us an orderly script which could only have been produced with a pen, and he would say that ink fell on the paper and the shapes of the words were formed on it by themselves, we would immediately reject his statement and say that this script could only have been the result of an intentional act.
>
> If we agree that written words could not have been formed by themselves, how could one say regarding [the universe], whose structure is even more refined and whose design is infinitely deeper, that it was done without the deliberate intention of One of great wisdom and great ability?[2]

2. *Chovos HaLevavos*, "Gate of Unity," chap. 6.

This brings us to a discussion of the theory which has become known as "intelligent design." The concept of intelligent design postulates that the complex structure of the universe, and everything that is in it, attests to the existence of an intelligent being that designed and created it. While this is very logical, it has encountered stubborn resistance from the scientific community. At the same time, intelligent design has growing support among many scientists and educators. Some school boards have introduced it into the public school curricula in their communities.

When the school board of Dover, Pennsylvania introduced intelligent design into the school curriculum in 2005, its members were taken to court by a group of unhappy parents. During the proceedings, a scientist on the witness stand who was defending intelligent design was approached by the prosecutor who placed a pile of books on evolution in front of him and asked the scientist how it is that he doesn't accept what is presented in all those books. The scientist replied simply that the arguments presented there are not conclusive and have not been proven.

I submit that we could have given a much better answer, based on what we explained above. I submit that those very books on evolution prove that evolution is wrong and that the intelligent design concept is correct. For, would any normal person think that all those books with all those words and paragraphs and ideas—whatever the ideas may be—were the result of an explosion in a printing shop? No! They were obviously designed by an intelligent author. Likewise, the entire universe which is certainly much more structured and complex than a pile of books. Case closed.

Using a Supercomputer to Prove Creation

In a letter responding to a group of students who had asked for a proof of the existence of a Creator, the Rebbe MHM gave the following proof based on the analysis of an organized system:

> When we see an organized system consisting of several components which have no control over each other, but that relate to each other with a precise coordination, we conclude with absolute certainty that there is an external force that connects and unifies all the components. And the very fact that it unifies and connects the parts of the system proves

> that it is greater and stronger than them and controls them.
>
> For example, if we enter a factory which is running fully automatically and we see no one there, we would have no doubt that somewhere there must be a big engineer who knows about all the machines and their parts and controls them in such a way that they are unified with each other and with some central unit. Furthermore, the freer the factory is from human involvement and the more automatically it runs, the more extraordinary the engineer must be.
>
> Now, if this is so regarding a factory, where we are speaking of hundreds, thousands or even tens of thousands of parts, how much more so does this apply when we consider our world—the inanimate or plant and animal life, and certainly the structure of the human body (as it says in the verse, "From my flesh I see G-d"). According to modern science, every object is composed of billions of atoms, and each atom contains many smaller particles, so it seems that there should be tremendous chaos and disorder among them. Nevertheless we see a wondrous order and an amazing coordination between the small parts, the big ones and the very large ones, and coordination between the constituent parts of the microcosm and the macrocosm, etc. It is thus clear without any shadow of a doubt that there is an 'Engineer' that is responsible for all this.

In the mid-1980s, I was working as an applications consultant for a supercomputer company (ETA Systems, a subsidiary of Control Data Corporation). I worked in a consulting office at the University of Minnesota, advising scientists and engineers on the use of the Cyber 205 supercomputer. There were two other consultants in the office, one black and one Chinese. The black consultant would often ask me questions about Judaism. One day he said to me, "I know that it says in those books over there [referring to the *Chitas* (*Chumash, Tehillim* and *Tanya* in one volume) and *Siddur* that I had on the shelf] that G-d created the universe...." He went on to say that a physicist friend of his told him that the universe had evolved. He wanted to know how one can prove that the Torah is right. (He was not trying to argue against it; he was genuinely interested.)

"Look," I said to him, "the universe has such a complex structure.

How could it have just evolved? Someone must have designed it and set it up." He agreed that that made sense but wanted to know how we could actually *prove* it. "Take, for example, the Cyber 205," I responded. "Do you agree that it is a very complicated computer?" The entire purpose of our office was to explain the intricacies and complexities of the Cyber 205 to the users. "Of course!" he exclaimed. I then continued, "What if someone told you that the Cyber 205 had evolved? No one designed it. No one built it. It just got there by itself—from an accident in some engineering laboratory. What would you say?" To this he did not respond. He just burst out laughing at the absurdity of such a notion. "So how much more so," I concluded, "the entire universe—of which the Cyber 205 is just a small part!"

A scientist recently wrote in the Harvard Magazine, "Any researcher who can prove the existence of intelligent design within the accepted framework of science will make history and achieve eternal fame. He will prove at last that science and religious dogma are compatible. Even a combined Nobel Prize and Templeton Prize (the latter designed to encourage search for such harmony) will fall short as proper recognition."[3]

It is obvious that we have proved it and, I guess, we have achieved eternal fame.

But I want to emphasize that intelligent design is more than just a valid argument. For the underlying concept—that when you see a complex design you understand that some intelligence designed it—lies at the foundation of all scientific thinking. Otherwise, why look for structure and design in the universe in the first place? Who says that there is any logical structure? And even if we have seen structure in so many things, why assume that the next thing one finds has any structure, and why exhaust oneself to try to find it? If the universe and life on earth came into being merely by a sequence of random events, why assume that everything in the universe is structured and has a logical explanation? Some things may, by chance, happen to be logically structured, but the composition of most things would merely be random. Yet we see that scientists are driven to find logical structure in everything, including underlying theories and even unifying theories.

3. Edward O. Wilson, "Intelligent Evolution," *Harvard*, Nov–Dec 2005, 31. (This was brought to my attention by Yehudah Freeman.)

In fact, the early scientific researchers, going back to the time of Isaac Newton, were actually motivated by a belief in G-d. Since the universe was created by G-d, they reasoned, He must have created it in an orderly fashion. They then set out to find that order. As Newton wrote at the conclusion of his Principia, "The most beautiful system of the sun, planets and comets could only proceed from the counsel and dominion of an intelligent and powerful Being."

A century later, the great mathematician Leonhard Euler incorporated the concept of the Creator into his development of mathematics. He wrote, "Because the shape of the whole universe is most perfect and, in fact, designed by the wisest Creator, nothing in all of the world will occur in which no maximum or minimum rule is somehow shining forth."[4]

In fact, the very existence of intelligence and logic itself indicates intelligent design. Our intelligence is logically structured; it is not random. The design of intelligence proves intelligent design.

Furthermore, even in those phenomena where we see randomness, mathematical analysis often shows that an overall structure can be identified, patterns can be discerned and conclusions drawn. There is intelligence in randomness, too. To my mind, this is a most amazing aspect of G-d's creation.

The Anthropic Principle

Many scientists have taken note of the fact that if the properties of certain forces of nature or particles of matter or the values of certain physical constants were slightly different, life on earth would be impossible. The seemingly arbitrary and unrelated constants in physics have one unique thing in common —they are exactly the values required if there is to be a universe capable of producing life.

Thus they speak of an "Anthropic Principle," which states that it is implicit in the structure of the universe that it was designed for man.[5] As theoretical physicist Michio Kaku expressed it, "The anthropic principle…makes us realize that a miraculous set of "accidents"

4. Quoted in Jerrold E. Marsden and Anthony Tromba, *Vector Calculus,* 5th ed. (New York: W.H. Freeman, 2003), 181.

5. There are a few different versions of the Anthropic Principle. See Michio Kaku, *Parallel Worlds* (New York: Doubleday, 2005), 247–248.

makes consciousness possible in this three-dimensional universe of ours. There is a ridiculously narrow band of parameters that makes intelligent life a reality, and we happen to thrive in this band. The stability of the proton, the size of the stars, the existence of higher elements and so on, all seem to be finely tuned to allow for complex forms of life and consciousness...Countless worlds exist in deep space devoid of life, much less of intelligence. It should make us appreciate how delicate life is, and what a miracle it is that it flourishes on Earth."[6]

The stability of the proton refers to the fact that if the proton were just 1% heavier than it is, protons would decay into neutrons and all nuclei would become unstable and disintegrate. Atoms would just fall apart and, of course, life would be impossible.

The earth is fine-tuned for the existence of life. The improbability of this "just happening" would be comparable to "a Boeing 747 aircraft being completely assembled as a result of a tornado striking a junk yard," as astronomer Hugh Ross put it. The great theoretical physicist Stephen Hawking went even further. In contemplating "the odds against a universe like ours emerging" he concluded, "I think there are clearly religious implications."

Now, if we go further and consider man himself, we come to the awareness of the Creator in an even more awesome way. Begin by considering the structure of the atoms and molecules that make up the human body. This alone is mind boggling—one needs all of quantum mechanics to understand atoms and molecules. Now these particles make up biological cells and tissues which have their own amazing structure. And these make up organs and systems of organs within the body which work together and are coordinated by a brain —truly an awesome structure!

But it doesn't end there. All that we have said until now applies equally to an animal as to a human. Above and beyond all this is the human who has—intelligence, the ability to think and speak. Where does intelligence come from if not from the Omniscient *Intelligence*? Where does all this structure come from if not from the Master *Designer*?

Thus, Rabbi DovBer of Lubavitch, the Mittler Rebbe, wrote about

6. Kaku, *Parallel Worlds,* 348–349.

"the wondrous wisdom that is evident in the anatomy of the human body. All the scholars of medicine have exhausted themselves trying to understand it...because it is a G-dly wisdom. Similarly among the animals and the plants and the earth and everything contained in it, there is G-dly wisdom through which all this was made."[7]

To better appreciate this, I think we need some perspective. For if something is too awesomely mind boggling, it will simply go right over our heads and we will have no appreciation of it at all. Consider, for example, a student untrained in science or mathematics who looks at a book on quantum mechanics. It will have absolutely no effect on him, and if you tell him that there is deep wisdom there, he may take your word for it but he will not be impressed. Now, if he studies calculus, differential equations, classical mechanics and electromagnetic theory for a few years and then looks at the book on quantum mechanics, he will certainly be impressed, perhaps awestruck.

That is why my colleague at ETA Systems was not affected by my first point regarding how complex the universe is. But when I mentioned the Cyber 205, which was very complex yet something that he was very familiar with, and then pointed out that the entire universe is so much more complex, the point struck home.

So let's go back to computers for a moment. Recently, the BBC aired an interview[8] with a computer scientist who was one of the pioneers of the internet and is still at the leading edge of internet research, regarding the next generation of the internet. After discussing some of the amazing advances being developed, the interviewer asked him if he thinks that some day the computer will replace the human brain. His response was an emphatic "No!" He said that we don't even begin to understand the human brain: "We don't even understand the brain of a fly. We don't even understand the brain of a worm!"

So if we understand the computer and we know that it's incomparable even to the brain of a worm,[9] how much more awesomely complex is the structure of the brain of a human.

7. Quoted in *Sefer HaMa'amorim Melukat* 5:120–121.

8. BBC World Service, Nov. 28, 2008.

9. For those who are interested, see Nivedita Chatterjee and Sitabhra Sinha, "Understanding the mind of a worm: Hierarchical network structure underlying nervous system function," *in Progress in Brain Research: Models of Brain and Mind—Physical, Computational and Psychological Approaches,* vol. 168 (Amsterdam: Elsevier, 2007), 145–153.

We would think that everyone would be convinced of these points. But remember that there is a catch: Once you accept the Intelligent Designer, the Creator, G-d, you are forced to accept the instruction manual that comes along with the creation—the Torah and its 613 Mitzvos for Jews and Seven Mitzvos for non-Jews. There are implications as to how we are going to live our lives. This of course accounts for the irrational resistance to the concept of intelligent design. In a documentary video called *Expelled*, Ben Stein documents this stubborn resistance, as well as the persecution of the proponents of intelligent design by the scientific community. One prominent scientist whom Stein interviewed, Richard Dawkins, said that he could accept that life on earth was designed by aliens from outer space, but not by G-d. (But then who designed the aliens?)

Professor Anthony Flew of England was not convinced of these points. He was an atheist. In fact, he was the world's leading authority on atheism for half a century, lecturing, writing books on it, etc. But then the genetic code was broken and this was too much even for him. On Channuka of 2004, he announced that he believes in G-d. The structure of DNA, he explained, was so awesomely complex that it could not have just evolved. It must have been designed and created by—G-d. This radical declaration shook up the whole international atheistic community which was relying on him for guidance for half a century. When asked how he could make such a statement after all those years of atheism, he replied simply, "I am 81 years old and I can say whatever I want!"

Meanwhile, the scientist who led the team that cracked the human genome, Francis Collins, director of the U.S. National Human Genome Research Institute, in his book *The Language of G-d*, explains his belief in the Creator this way:

"When you have for the first time in front of you this 3.1 billion-letter instruction book that conveys all kinds of information and all kinds of mystery about humankind, you can't survey that going through page after page without a sense of awe. I can't help but look at those pages and have a vague sense that this is giving me a glimpse of G-d's mind."[10]

---

10. Francis Collins, *The Language of G-d: A Scientist Presents Evidence for Belief* (Free Press, 2006).

Irreducible Complexity

Recently scientists have identified certain biological structures that are not only very complex, but the structure is "irreducible" in the sense that if any of its components were missing it could not function at all. They explain that such structures could not possibly have evolved by natural selection from a simpler form to a more complex form because any simpler form would be useless. Biochemistry Professor Michael Behe of Lehigh University uses the term "irreducible complexity" to describe this.

Professor Behe defines irreducible complexity as follows: "By *irreducibly complex* I mean a single system composed of several well-matched, interacting parts that contribute to the basic function, wherein the removal of any one of the parts causes the system to effectively cease functioning. An irreducibly complex system cannot be produced directly (that is, by continuously improving the initial function, which continues to work by the same mechanism) by slight, successive modifications of a precursor system, because any precursor to an irreducibly complex system that is missing a part is by definition nonfunctional."[11]

Irreducible complexity has blown a huge hole through evolution and for many scientists it has sunk the ship—they have abandoned evolution because of it and have come to believe in G-d, the Creator.

Einstein and Creation

The Rebbe MHM was often asked how one could prove the existence of G-d. In response, he frequently referred to Rabbeinu Bachaya's proof mentioned above. For example, in one letter he wrote:

> When we see a printed book with hundreds of pages, it is obvious to the observer that there is a printing house where they arranged all the letters in such a way that they explain intellectual concepts etc., not that there was an ink spill and by chance the drops of ink became arranged upon hundreds of pages in the form of letters explaining various things. This applies much more so when we see a piece of wood or stone

11. Michael J. Behe, *Darwin's Black Box: The Biochemical Challenge to Evolution,* 2nd ed. (Free Press, 2006), 39.

> etc. and then we find out about the arrangement of its atoms which number not merely in the hundreds or thousands. Rather there are billions of billions of them and they are all arranged in a wondrous manner and follow extremely precise fixed laws. This is explained at length in the *Chovos HaLevavos* and in the *Kuzari* etc. See also the *Sefer HaChakira* of the Tzemach Tzedek.

He called this "the most obvious proof."

Similar thoughts were once presented by one of the greatest scientists of all time—Albert Einstein. He was once asked how he understood G-d and in response he explained:

> We are in the position of a little child entering a huge library filled with books in many languages. The child knows someone must have written those books. It does not know how. It does not understand the languages in which they are written. The child dimly suspects a mysterious order in the arrangement of the books but doesn't know what it is. That, it seems to me, is the attitude of even the most intelligent human being toward G-d. We see the universe marvelously arranged and obeying certain laws but only dimly understand these laws.

Indeed, it was one of Einstein's own theories that brought the scientific world to the awareness that the universe had a beginning. The story of its discovery was told by Einstein's wife, Elsa:

It began like any other day. Elsa Einstein was preparing breakfast for her husband, Albert. As she later recalled: "The Doctor came down in his dressing gown as usual for breakfast but he hardly touched a thing. I thought something was wrong, so I asked what was troubling him. "I have a wonderful idea," he said, and after drinking his coffee, he went to the piano and started playing. Now and again he would stop, making a few notes then repeat: 'I've got a wonderful idea, a marvelous idea!' I said: 'Then for goodness' sake tell me what it is, don't keep me in suspense.' He said: 'It's difficult, I still have to work it out.' "

He continued playing the piano and making notes for about half an hour, then went upstairs to his study, telling her that he did not wish to be disturbed, and he remained there for two weeks. "Each day I sent him up his meals," she said, "and in the evening he would walk a little for exercise, then return to his work again. Eventually, he came down from his study looking very pale. 'That's it,' he told me, wearily putting two sheets of paper on the table. And that was his theory of relativity."

It was Einstein's theory of General Relativity. It was a new way of looking at the universe based on the non-Euclidean geometry of Riemann and it had some startling implications that even Einstein was not prepared for. Among other things, it implied that the universe had a beginning—that it was not always here. Einstein himself had been of the opinion that the universe was a "steady state" universe i.e. that it was always here.

As a result of further theoretical and experimental research, it is now an accepted part of modern science that the universe had a beginning. One contemporary authority on cosmology praises the Jews for holding on to the belief that "G-d created the heavens and the earth" throughout history while the rest of the world believed otherwise. Such admiration for the Jews was foretold in the Torah in the last verse of the Song of *Ha'azinu:* "The nations will praise His (G-d's) people."[12] Rashi explains this to mean that in the times of Moshiach "the nations will praise the Jews saying, 'Look how praiseworthy this nation is! They remained attached to G-d in spite of everything that happened to them. They didn't leave him. They knew His goodness and greatness.'"

The true significance of the fact that now science also understands—in its own terms—that there was a beginning, is that it is a sign of the times—the Messianic times—when, as we say in the *Rosh HaShana* prayers, "Every creation will know that You created it and...will understand that You formed it and every living thing will declare 'The G-d of Israel is King and His domain extends over everything.'"

Math, Music and Moshiach

Einstein's General Relativity brought to the attention of the scientific world the fact that the universe is a "new universe" i.e. it had a

12. Devorim 32:48.

beginning and it was not always here. And science brought Einstein closer to an awareness of the Creator of the universe.

There is an interesting story about another way in which Einstein came to an awareness of G-d. On April 30, 1930, Yehudi Menuhin, the famous Jewish boy wonder violinist, a descendant of Rabbi Schneur Zalman of Liadi, the Alter Rebbe, gave a magnificent performance in Berlin. After the concert, several prominent members of the audience, Albert Einstein among them, were allowed to go backstage to meet Yehudi. Einstein rushed over to him, embraced him and said, "Now I know there is a G-d in heaven."

So Einstein knew that "There is a G-d in heaven," and that "Someone must have written those books" but what about connecting with Him by observing G-d's mitzvos? In fact, as a young boy, he had observed some mitzvos, such as keeping kosher, for a number of years, and while he was not observant in his adult life, he maintained an interest in Judaism. He read Maimonides (Rambam) in translation and said that he wished he knew Hebrew so that he could have read Maimonides in the original. He also owned a pair of Tefillin.

About twenty years ago, I became friends with a symphonic composer named Paul Schoenfield. He has been very successful and now some reviewers compare him to George Gershwin. He was raised in a non-religious home and considered himself an agnostic. A musician by profession, he would study abstract algebra on the side, as an intellectual hobby. He said that when he came upon Galois Theory (a very elegant branch of modern abstract algebra) he became convinced that if there could be something so beautiful, so elegant and intricately structured in the world, there must be a G-d Who created it. (I myself was awed by it when I first studied it.)

This brought him to believe in G-d. He decided to look deeper into Judaism, contacted the local Chabad *shul* (synagogue) and began studying Torah. Eventually, he became completely observant—a *Ba'al Teshuva.* (Recently, one reviewer described him as "a mind always at work thinking about music, the Torah, or math.")

A major difference between Einstein and Schoenfield is the times. These are Messianic times and thus there is a greater sensitivity and openness to G-d and a desire to connect with Him personally. The Chabad *shul* itself is in fact a part of this. The Rebbe MHM has covered the globe with this network of Chabad institutions with the goal of filling the world with the knowledge of G-d and making it available to everyone.

There is a strong trend among scientists now not only to recognize G-d as the Creator, but to view man as the focal point of this creation i.e. G-d created the universe for a purpose and that purpose involves man. (This is the "Anthropic Principle" that we spoke about above.) More than that science cannot know because science only understands the Creator by looking at the creation. Knowledge of the purpose of creation must come from the Creator Himself. This is the Torah, which G-d gave to Moshe Rabbeinu to instruct man on the fulfillment of the purpose of his creation, which is to make the world a place where the glory of G-d is completely revealed.

Continuous Creation

The story of creation continues: It is a fundamental teaching of Chabad Chassidus that G-d continually creates the world anew. This principle was first taught by the Baal Shem Tov, then explained in detail by Rabbi Schneur Zalman of Liadi, the *Alter Rebbe*, founder of Chabad Chassidus, in his famous book, the Tanya.

This teaching is based on a verse in *Tehillim* (Psalms 119) which says, "Forever, O G-d, Your words stand in the heavens." G-d created the universe with the 10 creative statements at the beginning of the Torah ("Let there be light" etc.). The Tanya explains that the existence of every entity in the world derives from the Hebrew letters and words of these 10 statements in some way. These letters are the life force with which G-d maintains the existence of every object even after it was created, and if He withdrew those letters the object would return to nothingness.[13]

In Chassidus, even the most abstract concepts have the most practical application. Continuous creation tells us that since G-d, from Whom only goodness comes forth, is always re-creating the world, everything that we experience is truly good. As the *Alter Rebbe* writes:

> The true belief in the Creator is…that the creation is…from G-d's wisdom which cannot be comprehended by

13. *Tanya*, "Sha'ar HaYichud V'HaEmunah," esp. chaps. 1–2.

> any created being. This creation occurs at every moment—everything that is created comes into being in a manner of *yesh me'ayin* (something from nothing) from G-d's wisdom which gives life to everything. When one thinks deeply into this and can imagine his own coming into being, literally at every moment, how could it occur to him that anything bad is happening to him....G-d's wisdom is the source of life, goodness and pleasure....It's only because this is not comprehended well that it may seem that one is experiencing evil or suffering. But the truth is that no evil descends from above. Everything is good. But it is not comprehended because of the great magnitude of the goodness.
>
> This is the fundamental faith for which man was created—to believe that there is no place unoccupied by G-d, and 'in the light of the King's face there is life.' Thus there is strength and delight in His place since there is only good all the time. Therefore, the very first thing is that a person should be happy and rejoice all the time and literally live off his faith in G-d Who gives life and is good to him at every moment....The believer will not be concerned with any suffering that he may experience, and in all worldly matters the positive and the negative are the same for him with true equality.

The Alter Rebbe concludes the exposition with the ultimate statement of optimism:

> Thus it follows that everything is perfect good; it just may be that this aspect is not comprehended. But when one truly believes this, it *becomes* good in a visible and obvious way.[14]

14. *Tanya*, "Igeres HaKodesh," chap. 11.

The Age of the Earth

"Our experiments should certainly test the fundamental principles of physics. There is a great tendency to think that these principles, once discovered, are eternal, yet history tells a different story. Almost every principle once proclaimed has been superseded. No matter how useful they are or how good an approximation they give to phenomena, sooner or later most principles fail, as experiment probes the natural world more accurately."

—Lee Smolin[1]

Science of Ignorance

In spite of all the logical proofs of creation, the point of view common among scientists for a long time was that the universe was always here. I call this a "point of view" rather than a theory since it must be made clear that ideas of the origin of the universe cannot in any sense be considered scientific theories. An idea can be called a scientific theory when:

1) the theory is based on observations,

2) a significant amount of the relevant data has been collected and analyzed and

3) the idea can be tested repeatedly to see if it holds up.

While all respectable scientific theories follow this pattern, ideas regarding the origin of the universe are fundamentally deficient in all of these areas. Firstly, the origin of the universe was not observed by anyone. All we have to go by is what we see now, a very long time

1. Lee Smolin, *The Trouble with Physics* (Boston: Houghton Mifflin, 2006), 218.

after the origin of the universe. Furthermore, we see very little of the universe even now. According to modern cosmology, much of the universe is not even observable—it *cannot* be observed. The "observable universe" is the space around us bounded by the *event horizon*—the distance to which light can have traveled since the universe originated. But the entire universe is much bigger than the observable universe. One astronomer made the following analogy: Take a dinner plate and go out into an open field. Put the plate on the ground and stand on it. Now compare what you can see up to the rim of the plate with the view up to the horizon. This is comparable to the fraction of the universe that is visible to us!

Furthermore, the popular view among scientists now is that 96% of the universe is composed of "dark matter" and "dark energy" that has *never* been observed. So a theory of the origin of the universe would be a theory based on almost no data. Furthermore, no such theory could be tested since we can't do any experiment to try to make a universe.

When dark energy first entered discussions on cosmology, two cosmologists wrote:

"Cosmology has a reputation as a difficult science, but in many ways explaining the whole universe is easier than understanding a single-celled animal."[2] The *real* reason this is so is because with the single-celled animal, the whole thing is right there before us. We have all the information. We have all the data—and there's a lot of it! In the case of the whole universe, we have almost no information and it's this very lack of information, the absence of observations that makes it relatively easy to put together a theory that cannot be contradicted by the facts—because we don't know the facts. It is Science of Ignorance.

Light from the Stars

Now, we mentioned the event horizon which raises the issue of the *time* it takes for the light of a star to reach us and the implied age of the universe. The Rebbe MHM was once asked this question and responded in a letter dated 1 Sivan, 5716 (1956):

---

2. Martin A. Bucher and David N. Spergel, "Inflation in a Low Density Universe," *Scientific American*, Jan. 1999, 63.

Regarding what you wrote that you read in the books of the astronomers about stars whose light rays must travel for much more than 6,000 years until it reaches the earth, and you ask how this fits with what our holy Torah states—that we are in the year 5716 after creation.

Now, even if you want to say that the calculation of the distance of the star is correct (for there are differences of opinion among the astronomers on this), this raises no question regarding the date of creation because just like the stars were created, the light rays were also created. Just like G-d can create a star which will first start to radiate light from the time of its creation and onwards, it is equally possible that He would create the star together with the light rays radiating from it.

This is especially evident when considering what it says in the first *parsha* of the Torah: 'It was morning the first day,' while 'Let there be luminaries' was not said until the fourth day. Thus we see that there was a presence of light [morning] before the luminaries were placed in the heavens....

A deep knowledge and an analysis of the relevant principles [upon which the conjectures of the scientists regarding the age of the earth is based] will prove to the normal mind that these are merely *conjectures* which are far from certainty. The scientists themselves are of this opinion as is mentioned in their books. Nevertheless, in the introductory texts studied in the schools, they hide the doubts that they have in these principles.[3]

From this letter we see that while G-d created the entire universe 5770 years ago, He may have created certain things to *appear to an observer* to be older.

In another letter, dated 18 Teves, 5722 (1962), the Rebbe MHM dealt with the issue of the age of the universe more extensively, explaining the logical errors in the theories of evolution. As an introduction to this letter, we explain an issue that arises in all scientific research, namely the error of extrapolation.

---

3. *Likutei Sichos* 15:478.

Extrapolation

Consider the following simple example. Suppose I want to investigate the relationship between two quantities, represented by the variables x and y. I design an experiment and take measurements. Suppose I find that for x = 1, the value of y = 4. For x = 2, y = 8; and for x = 3, y = 12. I notice that for each value of x, the value of y is 4 times the value of x and I conclude that the formula to describe the relationship is y = 4x. Right? No, this is wrong. But the error is so subtle that it eludes us. And so has it eluded many great scientists throughout history.

To discover the error, let's consider another experiment. I go out on Eastern Parkway in Brooklyn and begin to survey the street. (This is not very safe and I don't recommend trying this experiment.) I notice that between Kingston Ave. and Albany Ave. the street is flat. I continue to survey and notice that between Albany Ave. and Troy Ave. the street is also flat. Continuing further, I notice that between Troy Ave. and Schenectady Ave. the street is flat. I therefore conclude that the Earth is flat. This time the error is clear. I made a big jump in drawing my conclusion. I *extrapolated* outside the range of my observations in drawing my conclusion and this is a serious error. But there was a time when great scientific minds made even this error and concluded that the Earth was flat.

Now let's go back and correct my first conclusion. What I should have concluded was that *between the values of x = 1 and x = 3* the correct formula is y = 4x. In the case of the Earth too, it is correct to say that between Kingston Ave. and Schenectady Ave. on Eastern Parkway, the Earth can be considered flat (as an approximation). Any scientific conclusion is valid only within the range of the original observations. Drawing a conclusion outside the range of observations is an error of extrapolation.

Throughout this history of science, this error has been made many times. Two of the most famous errors of extrapolation in the recent history of science were:

- Assuming that Newton's laws of motion were valid at all velocities. This error was corrected by Einstein's Special Theory of Relativity which showed how the laws change at very high velocities.

- Assuming that the known laws of physics were valid over all distances. This error was corrected by Quantum Mechanics which shows how the laws are very different over very small distances (on the scale of an atom).

The Rebbe MHM's Letter[4]

After not having heard from you for a long time, I was pleased to receive regards from you through the young men of Chabad who visited your community recently in connection with the public lecture. I was gratified to hear that you participated in the discussion, but it was quite a surprise to me to learn that you are still troubled by the problem of the age of the world as suggested by various scientific theories which cannot be reconciled with the Torah view that the world is 5722 years old.

I underlined the word theories, for it is necessary to bear in mind, first of all, that science formulates and deals with theories and hypotheses while the Torah deals with absolute truths. These are two different disciplines, where "reconciliation" is entirely out of place.

It was especially surprising to me that, according to the report, the said problem is bothering you to the extent that it has trespassed upon your daily life as a Jew, interfering with the actual fulfillment of the daily Mitzvoth. I sincerely hope that the impression conveyed to me is an erroneous one. For, as you know, the basic Jewish principle of *na'aseh* (first) and *v'nishma* (afterwards)[5] makes it mandatory upon the Jew to fulfill G-d's commandments regardless of the degree of understanding, and obedience to the Divine Law can never be conditioned upon human approval. In other words, lack of understanding, and even the existence of "legitimate" doubts, can never justify disobedience to the Divine Commandments; how much less, when the doubts are "illegitimate," in the sense that they have no real or logical basis, such as the "problem" in question.

Apparently, our discussion which took place a long time ago, and

4. The Rebbe MHM wrote this letter in English and it has been copied in many publications over the years. All of the other letters of the Rebbe MHM in this book were written in Hebrew and were translated by this writer.

5. *Na'aseh v'nishma* (Shemos 24:7) means "we will do it and then we will understand it."

which, as I was pleased to learn, has not been forgotten by you, has nevertheless not cleared up this matter in your mind. I will attempt to do so now, in writing, which imposes both brevity and other limitations. I trust, however, that the following remarks will serve our purpose.

Basically the "problem" has its roots in a misconception of the scientific method or, simply, of what science is. We must distinguish between empirical or experimental science dealing with, and confined to, describing and classifying observable phenomena, and speculative "science," dealing with unknown phenomena, sometimes phenomena that cannot be duplicated in the laboratory. "Scientific speculation" is actually a terminological incongruity; for "science," strictly speaking, means "knowledge," while no speculation can be called knowledge in the strict sense of the word. At best, science can only speak in terms of theories inferred from certain known facts and applied in the realm of the unknown. Here science has two general methods of inference:

(a) The method of interpolation (as distinguished from extrapolation), whereby, knowing the reaction under two extremes, we attempt to infer what the reaction might be at any point between the two.

(b) The method of extrapolation, whereby inferences are made beyond a known range, on the basis of certain variables within the known range.

For example, suppose we know the variables of a certain element within a temperature range of 0? to 100?, and on the basis of this we estimate what the reaction might be at 101?, 200?, or 2000?.

Of the two methods, the second (extrapolation) is clearly the more uncertain. Moreover, the uncertainty increases with the distance away from the known range and with the decrease of this range. Thus, if the known range is between 0? and 100?, our inference at 101? has a greater probability than at 1001?.

Let us note at once, that all speculation regarding the origin and age of the world comes within the second and weaker method, that of extrapolation. The weakness becomes more apparent if we bear in mind that a generalization inferred from a known consequent to an unknown antecedent is more speculative than an inference from an antecedent to consequent.

That an inference from consequent to antecedent is more speculative than an inference from antecedent to consequent can be demonstrated very simply:

Four divided by two equals two. Here the antecedent is represented by the dividend and divisor, and the consequent—by the quotient. Knowing the antecedent in this case, gives us one possible result—the quotient (2).

However, if we know only the end result, namely, the number 2, and we ask ourselves, how can we arrive at the number 2, the answer permits several possibilities, arrived at by means of different methods:

(a) 1 plus 1 equals 2;

(b) 4 - 2 equals 2;

(c) 1 x 2 equals 2;

(d) 4÷2 equals 2.

Note that if other numbers are to come into play, the number of possibilities giving us the same result is infinite (since 5-3 also equals 2; 6 divided by 3 equals 2 etc. ad infinitum).

Add to this another difficulty, which is prevalent in all methods of induction. Conclusions based on certain known data, when they are ampliative in nature, i.e. when they are extended to unknown areas, can have any validity at all only on the assumption of everything else being equal, that is to say on an identity of prevailing conditions, and their action and counter-action upon each other. If we cannot be sure that the variations or changes would bear at least a close relationship to the existing variables in degree; if we cannot be sure that the changes would bear any resemblance in kind; if, furthermore, we cannot be sure that there were not other factors involved—such conclusions of inferences are absolutely valueless! Furthermore, as I mentioned during our conversation, in a chemical reaction, whether fissional or fusional, the introduction of a new catalyzer into the process, however minute the quantity of his new catalyzer may be, may change the whole tempo and form of the chemical process, or start an entirely new process.

We are not yet through with the difficulties inherent in all so-called "scientific" theories concerning the origin of the world. Let us remember that the whole structure of science is based on observances of reactions and processes in the behavior of atoms in their present state, as they now exist in nature. Scientists deal with conglomerations of billions of atoms as these are already bound together, and as these relate to other existing conglomerations of atoms. Scientists know very little of the atoms in their pristine state; of how one single atom may react on another single atom in a state of separateness; much less of how parts of a single atom may react on other parts of the same or other atoms. One thing science considers certain—to the extent that any science can be certain, namely that the reaction of single atoms upon each other is totally different from the reactions of one conglomeration of atoms to another.

We may now summarize the weaknesses, nay, hopelessness, of all so-called scientific theories regarding the origin and age of our universe:

(a) These theories have been advanced on the basis of observable data during a relatively short period of time, of only a number of decades, and at any rate not more than a couple of centuries.

(b) On the basis of such a relatively small range of known (though by no means perfectly) data, scientists venture to build theories by the weak method of extrapolation, and from the consequent to the antecedent, extending to many thousands (according to them, to millions and billions) of years!

(c) In advancing such theories, they blithely disregard factors universally admitted by all scientists, namely, that in the initial period of the "birth" of the universe, conditions of temperature, atmospheric pressure, radioactivity, and a host of other catalystic factors, were totally different from those existing in the present state of the universe.

(d) The consensus of scientific opinion is that there must have been many radioactive elements in the initial stage which now no longer exist, or exist only in minimal quantities; some of them—elements the catalystic potency of which is known even in minimal doses.

(e) The formation of the world, if we are to accept these theories, began with a process of colligation (of binding together) of single atoms or the components of the atom and their conglomeration and consolidation, involving totally unknown processes and variables.

In short, of all the weak "scientific" theories, those which deal with the origin of the cosmos and with its dating are (admittedly by the scientists themselves) the weakest of the weak.

It is small wonder (and this, incidentally, is one of the obvious refutations of these theories) that the various "scientific" theories concerning the age of the universe not only contradict each other, but some of them are quite incompatible and mutually exclusive, since the maximum date of one theory is less than the minimum date of another.

If anyone accepts such a theory uncritically, it can only lead him into fallacious and inconsequential reasoning. Consider, for example, the so-called evolutionary theory of the origin of the world, which is based on the assumption that the universe evolved out of existing atomic and subatomic particles which, by an evolutionary process, combined to form the physical universe and our planet, on which organic life somehow developed also by an evolutionary process, until "homo-sapiens" emerged. It is hard to understand why one should readily accept the creation of atomic and subatomic particles in a state which is admittedly unknowable and inconceivable, yet should be reluctant to accept the creation of planets, or organisms, or a human being, as we know these to exist.

The argument from the discovery of the fossils is by no means conclusive evidence of the great antiquity of the earth, for the following reasons:

(a) In view of the unknown conditions which existed in "prehistoric" times, conditions of atmospheric pressures, temperatures, radioactivity, unknown catalyzers, etc., etc. as already mentioned, conditions that is, which could have caused reactions and changes of an entirely different nature and tempo from those known under the present-day orderly processes of nature, one cannot exclude the possibility that dinosaurs existed 5722 years ago, and became fossilized under terrific natural cataclysms in the course of a few years rather than in millions of years; since we have no conceivable measurements or criteria of calculations under those unknown conditions.

(b) Even assuming that the period of time which the Torah allows for the age of the world is definitely too short for fossilization (although I do not see how one can be so categorical), we can still readily accept the possibility that G-d created ready fossils, bones or skeletons (for reasons best known to Him), just as he could create ready living organisms, a complete man, and such ready products as oil, coal or diamonds, without any evolutionary process.

As for the question, if it be true as above (b), why did G-d have to create fossils in the first place? The answer is simple: We cannot know the reason why G-d chose this manner of creation in preference to another, and whatever theory of creation is accepted, the question will remain unanswered. The question, "Why create a fossil?" is no more valid than the question, "Why create an atom?" Certainly, such a question cannot serve as a sound argument, much less as a logical basis, for the evolutionary theory.

What scientific basis is there for limiting the creative process to an evolutionary process only, starting with atomic and subatomic particles—a theory full of unexplained gaps and complications, while excluding the possibility of creation as given by the Biblical account? For, if the latter possibility be admitted, everything falls neatly into pattern, and all speculation regarding the origin and age of the world becomes unnecessary and irrelevant.

It is surely no argument to question this possibility by saying, "Why should the Creator create a finished universe, when it would have been sufficient for Him to create an adequate number of atoms or subatomic particles with the power of colligation and evolution to develop into the present cosmic order?" The absurdity of this argument becomes even more obvious when it is made the basis of a flimsy theory, as if it were based on solid and irrefutable arguments overriding all other possibilities.

The question may be asked: If the theories attempting to explain the origin and age of the world are so weak, how could they have been advanced in the first place? The answer is simple. It is a matter of human nature to seek an explanation for everything in the environment, and any theory, however far-fetched, is better than none, at least until a more feasible explanation can be devised.

You may now ask, "In the absence of a sounder theory, why then isn't the Biblical account of creation accepted by these scientists?" The answer, again, is to be found in human nature. It is a natural human ambition to be inventive and original. To accept the Biblical account deprives one of the opportunity to show one's analytic and inductive ingenuity. Hence, disregarding the Biblical account, the scientist must devise reasons to "justify" his doing so, and he takes refuge in classifying it with ancient and primitive "mythology" and the like, since he cannot really argue against it on scientific grounds.

If you are still troubled by the theory of evolution, I can tell you without fear of contradiction that it has not a shred of evidence to support it. On the contrary, during the years of research and investigation since the theory was first advanced, it has been possible to observe certain species of animal and plant life of a short life-span over thousands of generations, yet it has never been possible to establish a transmutation from one species into another, much less to turn a plant into an animal. Hence such a theory can have no place in the arsenal of empirical science.

The theory of evolution, to which reference has been made, actually has no bearing on the Torah account of Creation. For even if the theory of evolution were substantiated today, and the mutation of species were proven in laboratory tests, this would still not contradict the possibility of the world having been created as stated in the Torah, rather than through the evolutionary process. The main purpose of citing the evolutionary theory was to illustrate how a highly speculative and scientifically unsound theory can capture the imagination of the uncritical, so much so that it is even offered as a "scientific" explanation of the mystery of Creation, despite the fact that the theory of evolution itself has not been substantiated scientifically and is devoid of any real scientific basis.

Needless to say, it is not my intent to cast aspersions on science or to discredit the scientific method. Science cannot operate except by accepting certain working theories or hypotheses, even if they cannot be verified, though some theories die hard even when they are scientifically refuted or discredited (the evolutionary theory is a case in point). No technical progress would be possible unless certain physical "laws" are accepted, even though there is no guarantee that the "law" will repeat itself. However, I do wish to emphasize, as already mentioned, that science has to do only with theories but not with certainties. All scientific conclusions, or generalizations, can only be probable in a greater or lesser degree according to the precautions taken in the use of the available evidence, and the degree of probability necessarily decreases with the distance from the empirical facts, or with the increase of the unknown variables, etc., as already indicated.

If you will bear this in mind, you will readily realize that there can be no real conflict between any scientific theory and the Torah.

My above remarks have turned out somewhat lengthier than intended, but they are still all too brief in relation to the misconception and confusion prevailing in many minds. Moreover, my remarks had to be confined to general observations, as this is hardly the medium to go into greater detail. If you have any further questions, do not hesitate to write to me.

To conclude on a note touched upon in our conversation: The Mitzvah of putting on Tefillin every week-day, on the hand—facing the heart, and on the head—the seat of the intellect, indicates, among other things, the true Jewish approach: performance first (hand), with sincerity and wholeheartedness, followed by intellectual comprehension (head); i.e. *na'aseh* first, then *v'nishma*. May this spirit permeate your intellect and arouse your emotive powers and find expression in every aspect of the daily life, for "the essential thing is the deed."

The Sun and the Earth

Until now we have been talking about the universe as a whole. Now we focus on an issue closer to home—the relative motion of the sun and the earth. Now, this is an old question which most people think has been resolved: Does the sun go around the earth or does the earth go around the sun? Rambam, based on the Talmud, states that the sun goes around the earth and this is, in fact, what everyone believed for thousands of years until the astronomer Copernicus proposed another system in which the earth revolves around the sun.

There ensued a great controversy in which men of science accepted the Copernican system while men of religion continued to hold to the geocentric (earth centered) view, known as the Ptolemaic system.

Most people now think that the issue has been resolved in favor of the system of Copernicus, the scientist, and that science has answered an ancient question about which people of religion were confused.

But this is all wrong! First of all, let us remember that Rambam's statement that the sun revolves around the earth is not merely a statement about the natural world, but is actually part of a Halacha—part of Jewish Law regarding what we know about G-d's creation of the universe. As such it is part of Torah and cannot be superseded by a scientific theory. So are we stuck in a permanent deadlock, taking a position against science? Well, even if we were, we would stick with the Torah which is absolute truth. Besides, scientific theories come and go and a new scientific theory may supersede the Copernican theory.

In fact, this is exactly what has happened in the Theory of Relativity of Albert Einstein. According to this theory, the motion between two objects is relative—either of them can be taken as the frame of reference. Thus, speaking strictly from the point of view of modern science, it is equally valid to take the position that the sun revolves around the earth as it is to say that the earth revolves around the sun—and the "challenge" to the Torah from the theory of Copernicus becomes meaningless.

A student once wrote to the Rebbe MHM saying that he was told that according to science the earth goes around the sun, contrary to what Torah says. The Rebbe MHM responded that the person who said this is ignorant of modern science. He then explained how Einstein's Relativity makes the question meaningless, as explained above.

Hans Reichenbach, a disciple of Einstein, made a major point of this in his book *The Philosophy of Space and Time*. He wrote:

> The relativity theory of dynamics is not a purely academic matter, for it upsets the Copernican worldview. It is meaningless to speak of a difference in truth claims of the theories of Copernicus and Ptolemy; the two conceptions are

> equivalent descriptions. What had been considered the greatest discovery of western science compared to antiquity is now denied its claim to truth....the conquest of the Ptolemaic cosmology by Copernicus gave rise to the new mechanics, which in turn gave us the means to recognize also the one-sidedness of the Copernican worldview....
>
> If we wish to establish the equivalence of both world conceptions, we must find a theory of gravitation sufficiently general to explain the Copernican and also the Ptolemaic planetary motion as a gravitational phenomenon. Herein lies the great achievement of Einstein...Einstein has indeed found such a comprehensive theory of gravitation.[6]

Down to Earth

Now let's get down to earth. Obviously, we are able to get a lot more information about the earth than about the universe as a whole since we are here and we can get our hands on much of it—and analyze it in a laboratory. But if we try to calculate the age of the earth, we fall into the same trap that we do when we try to calculate the age of the universe—the error of extrapolation. Yet by observing the earth we have been able to develop an extensive science of geology, and here is where an amazing development has occurred. One of the most complete and accurate computer programs developed to model the earth helps us verify scientifically the age of the earth according to the chronology of the Torah—5770 years.

This computer program, called TERRA, was developed by John Baumgardner of the Theoretical division of the Los Alamos National laboratory. Baumgardner has been described as the world's pre-eminent expert in the design of computer models for geophysical convection, the process by which the Earth creates volcanoes, earthquakes, and the movement of the continental plates.

Running on Los Alamos' supercomputers, TERRA takes the spatial volume of the Earth's mantle—2,000 miles of silicate rock that surrounds the Earth's solid iron core—and divides it into 10 million three-dimensional cells. The mantle rock is in a solid state, but over

6. Hans Reichenbach, *The Philosophy of Space and Time* (New York: Dover Publications, 1957), 217.

the years it moves and flows, behaving as if it were a fluid. The mantle rock churns or "convects" for the same reasons boiling water rolls in a pot: Cooler, heavier material on top sinks while hotter material nearest the heat source is lighter, and so rises. Terra divides the mantle into hexagonal cells, assigning each one a value for heat, direction, velocity, and other values. TERRA then "runs" each piece through time and watches where each will go—giving a 3-D map of a huge mass convecting through time.

Baumgardner, who believes in the Bible, believes that the Earth, and the entire universe, is only several thousand years old. Raised as an agnostic, he became convinced that there had been a major catastrophe in the Earth's past that accounts for many of the geological features we observe at the Earth's surface today. This was the flood in the time of Noah which, according to Baumgardner's description, caused an enormous blob of hot mantle material to come rushing up at incredible velocity through the underwater mid-ocean ridges. The material ballooned, displacing a tidal wave of sea water over the continents. After 150 days the bubble retreated with equal speed into the Earth, and the continents began re-emerging above the water, sending the runoff back to the oceans at around 100 miles an hour. (A very fast river with a huge erosion capacity runs at only about 10 miles an hour.) Baumgardner says that this runoff would have been sufficient to create the Grand Canyon and other massive geologic features and to deposit the various sedimentary layers in about one week. The geological theory on which this is based is called "runaway subduction."[7]

To be precise, TERRA does not prove that the flood took place. What it proves is that the geological phenomena based on which geologists usually say that the earth is 4.6 billion years old, can be accounted for by Noah's flood and the earth being only several thousand years old. In other words, TERRA shows that if Noah's flood took place—and we know from the Torah that it did—then there is every reason to believe that the earth is only several thousand years old.

Now, this result should have caused an upheaval in the scientific community since, if there was such a flood then there is no reason to

---

7. See Chandler Burr, "The Geophysics of G-d," *U.S. News & World Report,* June 16, 1997, 55–58.

believe that the earth is more than several thousand years old in which case evolution is impossible! The evolutionists need the billions of years to be able to say that life developed through evolution. By no stretch of even their imagination could it have happened in just several thousand years.

It should also have caused some excitement in the Torah community. From time to time people come along who believe in the Torah but try to reinterpret the 6 days of creation allegorically as a reference to 6 eras of unlimited duration in order to make the Torah compatible with evolution. (This is known as "apologetics.") But they know that there was such a flood in the time of Noah since they believe in the Torah and this implies an age of only several thousand years. This makes the reinterpretation unnecessary and useless. And again, it makes evolution impossible.

So where did the other guys go wrong? What was their error? Extrapolation, of course. Baumgardner explains that they assume that mantle viscosity (a measure of the rock's liquid-like flow) *has been consistent throughout time* and so use the value that applies today. As we explained above, a conclusion based on extrapolation is not a valid conclusion. In the case of geology and the age of the earth there is an additional issue that invalidates a conclusion based on the assumption that processes have been proceeding at a steady constant rate, and that is the possibility of the occurrence of "catastrophes"—events that may have caused drastic sudden changes in the earth, such as the flood. Baumgardner says scientists wrongly assume that geology happens consistently, that there could have been no catastrophe. "If you look at the geological record there are fingerprints of catastrophe everywhere one looks," he says.

(It is significant to note that in 1950, a scientist named Immanuel Velikovsky proposed a theory of the history of the earth based on the possibility of various catastrophes having occurred.[8])

Baumgardner also points out inconsistencies in the radiometric dating methods. Different methods give vastly different ages. To date rock, geologists commonly use three types of radioactive "parent" isotopes—samarium, rubidium, and potassium—which decay into stable "daughter" elements: neodymium, strontium, and argon. Based on the known rates of decay of these elements, the ratio of parent-

8. Immanuel Velikovsky, *Worlds in Collision* (New York: Doubleday, 1950).

to-daughter elements in a rock is considered an indication of its age. [*Warning: There is an extrapolation issue here.*] But different isotopes yield different dates for the same rocks! Consider, for example, the Cardenes basalt, a volcanic rock found in the Grand Canyon's inner gorge. He says the rock has been dated at 1.7 billion years using samarium, at 1.1 billion years with rubidium, and at 0.7 billion years with potassium. "All I'm saying as a scientist, is that there's reason to question radiometric measurements," Baumgardner says.

Even those geologists that disagree with Baumgardner have great respect for him and his computer program, TERRA. "Baumgardner is seen as one of the world leaders in numerical models of mantle convection," says Brad Hager, a Massachusetts Institute of Technology geophysicist. His disagreement with Baumgardner focuses on "thermal diffusivity," the rate at which rocks conduct heat. He concedes that if "the diffusivity of the earth was different before we learned to measure it, then you could speed up these calculations. But it would mean that rocks conducted heat differently in the past." But we have already explained that this is exactly the issue of extrapolation: There is no reason to assume that the rate of any physical process that we currently observe was always the same.

So Dr. Baumgardner has raised some serious challenges to the validity of the methods of dating rocks used by the scientists. But he is not the only one. The world renowned geologist, Eugene M. Shoemaker ("father of astrogeology," professor at California Institute of Technology, namesake of the Shoemaker-Levy comet etc.) and his partner, Jeffrey C. Wynn, describe[9] their adventurous journey deep into the Empty Quarter of the Arabian Desert to investigate the site of a crater known as Wabar. This was an area that even the tough desert guides were afraid to venture into. In a previous expedition into the area by a Western explorer in 1932, it took a month to reach the site and several camels died en route. But Shoemaker and Wynn were determined to proceed. True, they were going in Hummers instead of camels, but they were completely on their own. The long distance, high temperatures and irregular dunes precluded the use of rescue helicopters or fixed-wing aircraft. No one had ever crossed the Empty Quarter in the summer. Their first expedition was in May, 1994. By the time they arrived some of their vehicles had lost their

9. Jeffrey C. Wynn and Eugene M. Shoemaker, "The Day the Sands Caught Fire," Scientific American, November, 1998.

air conditioning and it was 142 degrees in the shade (under a tarp). In a subsequent expedition sandstorms destroyed their camp twice.

The purpose of their expedition was to try survey the site and learn what they could about the nature and the history of the legendary crater. They determined that it was caused by a meteor impact and calculated that the crater was equivalent to one that would have been caused by an atomic bomb of the size that destroyed Hiroshima.

And when did this happen? Here's where things get interesting. Using "fissiontrack analysis" of glass samples taken from the site, the British Museum had dated the event to be 6400 years ago. "Field evidence, however, hints at a more recent event," Shoemaker and Wynn write. After considering the rate at which the craters filled with sand, thermoluminescence dating of some samples and reports of a certain meteor shower in the 1800's, they conclude that the event took place between 135-450 years ago.

We see then that relying on the scientific methods of dating rock samples, in addition to the problem of extrapolation, is an example of what we called earlier "Science of Ignorance." When more information becomes available, on site analysis in this case, the results of the dating are contradicted.

Inside the Earth

While scientists claim to know what's going on far out in space and in the far distant past, what lies at the bottom of our oceans and deep inside the earth remains, to a great extent, a mystery. And herein lies the lesson that we can learn from our study of the earth…

Once the Tzemach Tzedek related the following teaching of the Baal Shem Tov: G-d says to the Jewish people, "For you will be for Me a desirable land." Just as the greatest scholars and scientists will never fully know the great natural treasures that G-d has put in the Earth, so too, no one can comprehend the great treasures that are hidden within the Jews—they are the desirable land of G-d.

The Baal Shem Tov concluded, "I want to bring about that the Jews will give forth the full measure of produce that G-d's desirable land is able to give."[10]

10. *HaYom Yom* of 17 Iyar.

Challenging Scientific Authority

"Don't believe most of what you hear. When a scientist claims to have done something important, ask to see the evidence... Give it as much scrutiny as a house you would buy or a school you would send your children to."

—Lee Smolin

"The experts who are leading you may be wrong."

—Richard Feynman

We often complain about the obstinacy and closed mindedness of scientists when it comes to discussing creation or even intelligent design. If we were to state our objection to the attitude of scientists in adhering to their belief in evolution and in refusing to consider the concept of intelligent design in the creation of the universe, we might identify the following five characteristics of the scientific community:

1) Tremendous self confidence, leading to a sense of entitlement and of belonging to an elite community of experts.
2) A monolithic community, with a strong sense of consensus, whether driven by the evidence or not.
3) A sense of identification with the group, akin to identification with a religious faith or political platform.
4) A disregard for and disinterest in the ideas, opinions and work of experts who are not part of the group.
5) A tendency to interpret evidence optimistically, to believe exaggerated or incorrect statements of results, and to disregard the

possibility that the theory might be wrong. This is coupled with a tendency to believe results are true because they are "widely believed."

But these are not my words. They are the words of Professor Lee Smolin, one of the most prominent theoretical physicists of our time. And he is not objecting to the scientific community's belief in evolution (which Smolin himself appears to believe in). In his recent book *The Trouble with Physics,* after an excellent review of the history of modern physics and an explanation of the current problems of modern theoretical physics, in a chapter titled "How do you Fight Sociology?" he attacks the scientific community for its bias in favor of one theory (specifically, string theory) over alternative theories. The above five points are part of his summary of the attack.[1]

This is but one example of how the problems we have with science or scientists regarding matters of Torah and Judaism are actually problems within science itself. They are either problems that have to do with the limitations of science or with the practice of science by the scientific community. They are problems that scientists themselves complain about, usually not openly, however. Lee Smolin's book is a rare example of a window opened into the inner workings of the scientific community. A rare example indeed.

Science of Arrogance

In discussing the situation, Smolin describes what he calls the "sociology of science," which, he writes, "refers to the influence that older, established scientists have over the careers of younger scientists. We scientists feel uncomfortable talking about it because it forces us to confront the possibility that the organization of science may not be entirely objective and rational."[2]

Many people are afraid to challenge the authority of scientists. They are in awe of scientists and feel that the scientists must know what they are talking about. Of course, many scientists claim this authority for themselves. But in fact this is one of the greatest impediments to the progress of science itself. The history of science is full

1. Lee Smolin, *The Trouble with Physics* (Boston: Houghton Mifflin, 2006), 284.
2. Ibid., 267.

of accounts of revolutionary ideas first being introduced by young scholars who challenged the authority of the scientific establishment and their ideas. One of my own physics professors, in a lecture on quantum mechanics, related that it was said that Niels Bohr, one of the originators of quantum theory, was successful in doing so because as a graduate student, he spent a lot of time riding his motorcycle instead of going to lectures so he was not biased by the opinions of the older scientific generation.

The most prominent example in the history of modern science of a scientist with revolutionary new ideas who was rejected and ridiculed by the scientific establishment was the physicist Ludwig Boltzmann. In the late 1800s Boltzmann developed statistical mechanics. This was a major innovation in physics which held, since the time of Newton, that all physical laws were strictly deterministic. In his statistical interpretation of the second law of thermodynamics he introduced the theory of probability into a fundamental law of physics and thus broke with the classical point of view. Thus he is considered by some to have laid the groundwork for quantum mechanics which was finally explained by Max Born in 1926 to be a statistical theory. In 1897 Boltzmann suggested to Max Planck to use statistical methods to explain black body radiation. The resulting theory was the first step in the development of quantum mechanics. Furthermore, Boltzmann used the concept of discrete energy levels which later became the very concept of the "quantum."

But his radical break from the classical point of view brought him violent opposition from the established scientific community. What brought upon him the fiercest opposition, however, was his assumption of the existence of atoms! That's right—as late as the beginning of the 20th century, the concept that matter is made up of atoms, and the very *existence* of atoms, was controversial and generally rejected by the physicists. (Most chemists, however, accepted it.) But the existence of atoms was one of the basic assumptions of Boltzmann's statistical mechanics. He used probability to describe how the properties of atoms determine the properties of matter. He fought vigorously to promote his theories and clashed frequently with his most bitter opponent, the famous scientist Ernst Mach.

Boltzmann was inclined to have extreme mood swings between elation and depression and, in 1906, convinced that he had lost the battle for his theories and that his life's work would go to waste, he committed suicide.

Imagine that! Atomic theory—the existence of atoms—which is now universally accepted, has actually been proved by experiment, and is so fundamental to modern science that anyone opposing it would be considered ignorant or abnormal, was so opposed by the scientific community at that time that it drove the man who promoted it to suicide.

Smolin devotes an entire chapter to discussing independent thinkers whom he calls "seers, who can see through unjustified but universally held assumptions and ask new questions."[3]

Thus, it is usually the greatest of the scientists who repeatedly emphasize how much we don't know and how much we must continue to doubt our scientific conclusions. Albert Einstein and Richard Feynman were famous for this.[4] Feynman once said, "To make progress in understanding, we must remain modest and allow that we do not know. Nothing is certain or proved beyond all doubt." Again, addressing the scientists, he said, "You have to permit the possibility that you do not have it exactly right....This freedom to doubt is an important matter in the sciences...I feel a responsibility to proclaim the value of this freedom...I want to demand this freedom for future generations."

Scientists are free to investigate the entire universe—and in fact *must* investigate it—but at the same time they must understand that there are limitations on their methods and on their conclusions. Not everything is equally accessible to scientific investigation. Investigations into distant space or into the distant past are not as conclusive as an experiment done in a laboratory. And where there is more ignorance than information or where one extrapolates he cannot draw a valid conclusion. He may only make a conjecture and say, "It might be like this."

But, as Smolin writes, "They seem to be confident enough of the truth of the conjecture to present it as fact."[5] And again, "When it

---

3. Ibid., chap. 18. See the section "An Independent Character" in Rabbi Yisroel Aryeh Leib's biography (chap. II, sec. 2) for some quotes from Smolin on this.

4. See the end of "Wisdom from Below" (chap. 1, sec. 3) for quotes from Feynman.

5. Smolin, *Trouble with Physics,* 277.

comes to defending their belief in these unproved conjectures…[they] often note that something is 'generally believed'…or that 'no sensible person doubts that it's true.' They seem to feel that appeal to consensus within their community is equivalent to rational argument."[6]

There is a certain humility that a scientist must have. When dark matter was first introduced into discussions on cosmology, the cover of Scientific American read, "New observations have smashed the old view of our universe."[7] But you were so sure of your old view of the universe. And now you are so sure of your new view....

First, Get the Weather Straight

I often wonder about the fact that even with the most sophisticated scientific theories and the most advanced supercomputers (I worked with them for a while) and with all the relevant data available, scientists cannot predict the weather accurately. One scientist involved in the global warming controversy said, "To think that we can predict what will happen 10 years from now when we can't predict the weather 10 days from now, is utter nonsense."

In the early morning hours of August 8, 2007 (only hours after I had done some work on this section) a sudden, unpredicted tornado hit Brooklyn, New York, several miles from my home with 135 mile per hour winds, emphasizing the very point we are making in a most frightful way. It was totally unexpected and unforeseen. There was no tornado warning and no tornado watch. We have also seen the reverse. Several years ago they predicted a major snow storm in New York—"the storm of the century"—and the city was ready to shut down. As it turned out, there wasn't even enough snow to cover the ground; just a few flakes.

Or, how about the weather report on a Sunday evening in November, 2008, which stated, "Even though we predicted clear skies and no rain until the end of the week, the rain is here—now!" Indeed it was. The next day the meteorologists fought back and said that the clouds would break up. They did—for a few hours—but came right back and stayed throughout the week.

---

6. Ibid., 283.
7. *Scientific American*, January, 1999.

Yet they say with great certainty that *this* is how the universe began and *that* is what happened in the first few seconds, minutes and millennia etc. Don't they realize that this can only be called science fiction? So the next time a scientist tries to tell you how the universe began, you reply simply, "First get the weather straight, then you can tell me how you think the universe began."

Indeed, we see that the inability of scientists to predict the weather implies the absurdity of their trying to describe the beginning of the universe. But the implication is even more far reaching. It actually forces them to accept the existence of the Creator and acknowledge His continuous involvement in running the world. This is explained by the Alter Rebbe, Rabbi Schneur Zalman of Liadi, the first Rebbe of Chabad, in a discourse based on a couple of verses from *Tehillim* (the Psalms). He writes the following:

> 'You have set all the boundaries of the earth; summer and winter—You made them. Remember this, O adversary who has taunted G-d, and the lowly nation which has disgraced Your name.'[8] These verses express a triumphant response to those nations that taunt G-dliness. They are the enemies of G-d and are referred to as the lowly nation which has disgraced His name by their saying that G-d has abandoned the earth and the world is following a natural order only... everything is running by itself following the laws of nature.
>
> So to counter this taunting by the nations, this verse is stated as a proof and a very powerful demonstration. By saying 'summer and winter—You made them' it's saying that from the character of summer and winter there is a strong proof for the creation of the world from nothingness, and its being sustained in a supernatural way. This is because all the scientists have become weary in their attempt to understand the cause of the intense heat in the summer and the extreme cold in the winter since, according to the science of astronomy, the shift in the angle of the sun relative to the earth has no effect on this. The small size of the sphere of the earth relative to the sphere of the heavens is, for example, like a grain of mustard in a large hall. As the sun changes its angle

8. Tehillim 74:17–18.

> in the sphere of the heaven, what effect could its heat have on the atmosphere of the earth to make it so hot in the summer, or to reduce the heat in the winter when it goes lower in the heaven etc? Rather, this shows that this effect is a novelty, a creation from nothing, beyond the ways of nature.[9]

Now consider something much simpler than the weather. You open a can of nuts. If the can had been shaken a lot, as it probably was in the shipping, handling etc., you will find that the larger nuts have risen *to the top*! This counterintuitive phenomenon, known as the "Brazil Nut Effect," has been the subject of much scientific research for decades but still cannot be completely explained. Yet they claim to know how the universe began. Are they nuts?

In spite of all this, the scientists *must* continue to investigate the universe. We say this based on a statement of the Rebbe MHM. A well known Jewish professor of microbiology, Dr. Velvel Green, was working for NASA looking for life in space, especially on Mars, when he became a *Ba'al Teshuva* (an observant Jew). He asked the Rebbe MHM if he could continue to look for life on Mars or perhaps this was against Judaism. The Rebbe MHM told him that he should look for life on Mars and if he doesn't find it there to look somewhere else in space and if he still doesn't find it to continue looking elsewhere. Why? Because there is no reason to put a limitation on G-d by assuming that He could create life in only one place.

Furthermore, the Talmud says that if one has the ability to investigate the paths of the stars and constellations and he doesn't do it, then the admonishment of the prophet Isaiah, "He refuses to look at the work of G-d," applies to him.

One should not be awed by "the scientists." One should not accept their conclusions based solely on their authority but rather examine them critically, for the scientists are subject to the same laws of reasoning and logic as everybody else. As Richard Feynman said, "If they say to you science has shown such and such, you might ask, 'How does science show it—how did the scientists find out—how, what, where?'...And you have as much right as anyone else, upon hearing about the experiments…to judge whether a reusable conclusion has

9. *Siddur with Ma'amorim*, 605.

been arrived at....The experts who are leading you may be wrong."[10]

"There is no scientist, not even Newton or Einstein, who was not wrong on a substantial number of issues they had strong views about,"[11] Smolin writes.

Missing the Boat

Global warming is the most controversial scientific topic of public interest today. Everyone agrees that "climate change" is taking place but the question is whether it is a result of human activity or if it is just part of a natural cycle. Indeed, many astronomers are concerned about a recent decrease in sunspots which is typically a prelude to a global cooling cycle.[12] Once again we say, "First get the weather straight...."

So we ask a basic, fundamental question: Whatever climate change is taking place, is happening right before our eyes. How can scientists be so sure about how the universe started "so long ago and so far away" while the events of the origin of the universe were not observed, yet they cannot determine the cause of climate change where all the phenomena are readily available for us to observe?

For the record, we are not taking any position on global warming and climate change. But we deplore the fact that here too, as in the case of evolution and other issues, the scientists who challenge the prevalent view are suppressed and their views are not given fair consideration. As Smolin writes, "we professors tend almost reflexively to reward those who agree with us and penalize those who disagree....Intellectual fads are far too important, and people who ignore them have dicey academic careers."[13]

In the case of global warming and climate change, there has actu-

10. Richard Feynman, "What is Science?" in The Pleasure of *Finding Things Out* (Cambridge, MA: Perseus Books, 1999), 187. (See Note 5 on p. xxvii.)

11. Smolin, *Trouble with Physics,* 275. He also mentions John von Neumann, one of the greatest mathematicians of the 20th century, who, in 1932, published a proof that there could be no hidden variables in quantum theory (a proof that was widely quoted for three decades). But the proof was wrong, as later demonstrated by the quantum theorist David Bohm.

12. www.climatescienceinternational.org/index.php?option=com_content&view=article&id=31

13. Smolin, *Trouble with Physics,* 265.

ally been a government order to this effect! As reported by the organization Public Employees for Environmental Responsibility (PEER):

> Federal climate, weather and marine scientists will be subject to new restrictions as to what they can say to the media or in public....Under rules posted last week, these federal scientists must obtain agency pre-approval to speak or write, whether on or off-duty, concerning any scientific topic deemed "of official interest."
>
> On March 29, 2007, the Commerce Department posted a new administrative order governing "Public Communications." This new order covers the National Oceanic and Atmospheric Administration (NOAA), which includes the National Weather Service and the National Marine Fisheries Service.

PEER concludes its report by noting, "Every scientist who answers an unexpected question at a conference puts his or her career at risk by giving an honest answer."[14]

Now, in discussing climate change itself, most of the focus is on carbon dioxide, the most common greenhouse gas. But another gas, methane, has a much stronger greenhouse effect. Scientists say that one kilogram of methane warms the earth 23 times more than a kilogram of carbon dioxide (if at all).

What we will now present is the very instructive story of the discovery of the sources of the methane that is introduced into the atmosphere. It is instructive for us in that it illustrates certain deficiencies in the activity of scientific research:

1) Accepted scientific theory, taught in textbooks as fact may simply be wrong.
2) Scientists who believe in a particular theory are reluctant to consider evidence that contradicts that theory.
3) Errors of extrapolation are common.[15]
4) Even the greatest scientific authorities—the Nobel Prize winners—can be wrong.

---

14. http://www.peer.org/news/news_id.php?row_id=841
15. For an explanation of extrapolation, see sec. 4 of this chapter.

These problems exist even in honest and careful scientific research which is, after all, a human activity and as such it is not perfect. Our point is, however, as we discussed in "Wisdom from Below"[16], that no challenge can be brought from scientific research, an inherently flawed human activity, against the Torah, the perfect word of G-d.

The story of the research on methane was reported by the researchers, Frank Keppler and Thomas Rockmann, in Scientific American[17]. They begin their report by asking, "What do you do as a scientist when you discover something that clearly contradicts the textbooks?" The accepted view was that only anaerobic bacteria (bacteria that thrive without oxygen) produce methane. They discovered, however, that all green plants also produce methane. Now they had to try to convince the scientific establishment of this fact. "Difficult as this discovery had been for us to accept," they write, "trying to convince our scientific peers and the public was almost impossible."

[This unwillingness of the established scientific community to accept new ideas is a major part of the problem we are discussing. Michael Merzenich was a pioneering researcher in neuroplasticity, the ability of the brain to reorganize itself. He did experiments with monkeys in which the parts of their brains that normally controlled certain actions were disabled. The monkeys' brains were able to "rewire" themselves and assign control of these actions to other parts of their brains. Based on this he concluded that adult human brains could also rewire themselves. But he had great difficulty getting his colleagues in the research community to accept this. In reporting on this, the psychiatrist Dr. Jeffery Schwartz writes, "It appeared that the monkeys' brains were a lot more adaptable than the research community's."]

How much methane were the green plants producing; how much would it contribute to climate change? The amount of methane produced by a leaf or a plant was actually very small, but the cumulative effect over the whole earth would be very great. "We knew, of course, that extrapolating global estimates from a limited sample of laboratory measurements was open to error," they wrote, but they still came up with a very large number: "Between 60 million and 240 million

16. Chap. 1, sec. 3.

17. Frank Keppler and Thomas Rockmann, "Can Plants Cause Global Warming?" *Scientific American,* February 2007, 52–57.

metric tons of methane come from plants every year—this constitutes 10 to 40 percent of annual global emissions."

After Keppler and Rockmann first published their findings in 2006, Paul J. Crutzen, a Nobel prize winner, and his colleagues reexamined measurements made in 1998 of air samples from the Venezuelan savanna and concluded that up to 60 million metric tons of methane could be released from vegetation in these regions, supporting Keppler and Rockmann's results. Crutzen said that "looking back to 1998, we could have made the discovery, but accepting the general wisdom that methane can only be produced under anaerobic conditions, we missed the boat."

Breaking News: The Climategate Scandal

As we prepare to go to press, the news of one of the biggest scandals in recent scientific history broke. On November 19, 2009, it was announced that 160 megabytes of e-mails from the Climate Research Unit (CRU) at the University of East Anglia in England were obtained and made public. The e-mails involved communication among many scientific researchers and policy advocates worldwide who believe that there is global warming and that it is the result of human activity. They were discussing the hiding and destruction of data that did not support their global-warming claims. The focus has been on two of them, Professor Phil Jones, the head of the Climate Research Unit, and Professor Michael E. Mann of Pennsylvania State University, an important scientist in the climate debate.

For example, one email from Prof. Jones to Prof. Mann refers to a trick employed to hide data showing a decline in temperatures—*global cooling*:

"I've just completed Mike's Nature *trick* of adding in the real temps to each series for the last 20 years (i.e., from 1981 onwards) and from 1961 for Keith's *to hide the decline.*"[18] (*Nature* is one of the most highly respected scientific journals.)

18. This email, the third email that I quote, and several others can be found at: http://www.telegraph.co.uk/earth/environment/globalwarming/6636563/University-of-East-Anglia-emails-the-most-contentious-quotes.html
The italics in the email are mine.

Many of the emails discuss hiding data that show a cooling trend and refer to data that has been destroyed. For example, in another email Jones writes to Mann and two other professors:

"I'm getting hassled by a couple of people to release the CRU station temperature data. Don't any of you three tell anybody that the UK has a Freedom of Information Act!"

In a third email Jones writes to Mann that he will try to suppress scientific papers written by opponents of global warming:

"I can't see either of these papers being in the next IPCC report. Kevin and I will keep them out somehow—even if we have to redefine what the peer-review literature is! (The IPCC is the UN's Intergovernmental Panel on Climate Change.)

The most disturbing aspect of this scandal, known as the "Climategate scandal," is that it is not just a matter of scientific theory since proposed regulations, costing trillions of dollars, are being based on many of these claimed research results. Such a coordinated campaign to hide scientific information is unprecedented.[19]

As the story continues to unfold, Prof. Jones was forced to resign as head of the Climate Research Unit pending an investigation and has contemplated suicide. Prof. Mann is being investigated by Penn State University. U.S. Senator James Inhofe has called for a Senate investigation into the attempts of climate scientists to "manipulate data, vilify scientists with opposing viewpoints, and circumvent information disclosure laws." Some raw data already reexamined shows no global warming, and actually shows a cooling trend.

Smolin's Slogans

In the last pages of his book, Smolin concludes his analysis of the scientific community with the following admonition to his fellow scientists and to the general public.

To his fellow scientists: "If you are someone whose first reaction when challenged on your scientific beliefs is 'What does X think?' or

19. See http://www.foxnews.com/opinion/2009/11/24/john-lott-climate-change-emails-copenhagen/. For a technical report on the emails, see http://newsbusters.org/blogs/noel-sheppard/2009/11/25/climategate-smoking-gun-american-thinker-does-medias-job

'How can you say that? Everybody good knows that...' then you are in danger of no longer being a scientist. You are paid good money to do your job, and that means you have a responsibility to make a careful and independent evaluation of everything you and your colleagues believe. If you cannot give a precise defense of your beliefs and commitments, consistent with the evidence, if you let other people do your thinking for you (even if they are senior and powerful), then you are not living up to your ethical obligations as a member of a scientific community.

And to the general public: "Don't believe most of what you hear. When a scientist claims to have done something important, ask to see the evidence. Evaluate it as strictly as you would an investment. Give it as much scrutiny as a house you would buy or a school you would send your children to."[20]

Or, as Feynman said, "The experts who are leading you may be wrong."

20. Smolin, *Trouble with Physics,* 353–354.

A Scientific View of Creation

Rabbi Dr. Naftali Berg, z.l.

From his lecture at the Moshiach and Science conference of 1993

I want to talk now about some theories in physics within cosmology and show how, from these theories we can come to a very basic understanding of concepts of creation and the Creator.

First, however, there is a very basic point that must be made by way of introduction. We are going to be discussing various scientific theories. No scientific theory is absolute or certain. It is a basic tenet of modern science that all scientific theories by definition are tentative and it can only be claimed that they have a certain probability of being true.[1] Our goal here is to identify aspects of these theories that illustrate certain concepts in Chassidus, and, in particular, to see how they show the unity in the world that the Rebbe MHM describes in the sicha.[2]

An "Expanding Universe"

Among physicists of about eighty to a hundred years ago, the idea that the universe was created was not accepted. The prevalent theory, which was even held by Einstein, was called the "steady state theory"—that the world was always in existence and will always be in

1. See chap. I, sec. 3 for a complete discussion of this.
2. See also Dr. Berg's lecture on quantum mechanics and string theory in chap. V, sec. 5.

existence. [As we explained earlier,[3] after developing his theory of general relativity, Einstein also came to the realization that the universe had a beginning.]

Cracks in this theory surfaced about 70 years ago as a result of the discovery of something called a "red shift" by Edwin Hubble, when looking at far off stars and galaxies. From this he basically deduced that the universe was expanding.

What is a red shift? Since most of you live in New York, you are familiar with things like police cars and ambulances. When an ambulance or police car has its siren on and is moving towards you, the frequency—the pitch—becomes higher. When it moves away from you, the pitch becomes lower. In physics this is called the Doppler effect. A similar theory applies to light. If you have an object with a light on it, when it moves away from you, it shifts towards the red end of the spectrum. When it moves towards you, it shifts towards the blue. So when Hubble saw far off galaxies where the light was shifting towards the red, this implied that these galaxies were all moving away from us. In fact, everywhere he looked in space, all of the galaxies were moving away from us at a constant velocity, which implied an expanding universe.

But the theory of an expanding universe is difficult to reconcile with a steady state universe, so a new theory of the universe was needed.

Forty years ago, two scientists from Bell Labs, Arlo Penzias, who is a Jew, and Robert Wilson made an observation of the sky. They took very, very delicate instruments and they measured the background radiation (the temperature, the microwave radiation) of the sky all around them and found that the background radiation was a temperature of 3 degrees Kelvin. 0 degrees Kelvin means absolute zero, which, believe me, is very, very cold. 3 degrees Kelvin is just a little bit above that—still pretty cold. But they measured and found this background radiation to be uniform throughout all of space to within a tenth of a percent.

This observation turned out to be in accord with the "big bang theory" which assumes that the universe began from nothingness with a

3. See sec. 3 of this chapter, subsection "Einstein and Creation."

sudden event of creation sometime in the distant past. The theory assumes that after its initial creation the universe was a very small point which all of a sudden burst forth into the universe that we know today. The 3 degree background radiation was supposedly a remnant of that event. For their discovery, Penzias and Wilson got the Noble Prize.

However there was a problem with this discovery. If we look around us, the universe indeed looks pretty uniform. It is basically empty space. But there are very non-uniform places there. Galaxies and stars are non-uniform. If the uniformity in the background radiation observed by Penzias and Wilson indicated uniformity in the universe many years ago, how could the stars form? There had to be some non-uniformity in this background radiation.

In 1989, NASA sent up a satellite called "COBE," Cosmic Background Explorer, which had extremely sensitive measuring instruments to measure the background radiation much, much more carefully than could be measured forty years ago. The initial results—reported in a very big news conference—stated that they did find perturbations (small changes) of about 30 millionths of a degree, extremely minute deviations throughout this background radiation. But according to today's theories, this was enough to explain the big bang.

Currently today, big bang—in other words, the creation of the universe—is accepted by practically all cosmologists

The theory has at its core a concept of creation for it claims that the universe had a beginning. It was not always here. So we have proven our basic principle, our starting point, which is creation. And if there is a creation then there must be a Creator.

Critique[4]

However, the big bang theory is wrong in three fundamental ways:

> *Firstly, the big bang considers the universe to be billions of years old while we know from the Torah that it's only 5753 years old.*[5]

4. This postscript to Dr. Berg's lecture was written by the editor, Prof. Shimon Silman.
5. At the time that this lecture was delivered.

Secondly, any theory of the origin of the universe is an extrapolation of the laws of physics, which have been observed, at most, over only a few hundred years and over relatively short distances, to a length of time and a distance in space far beyond the domain of the original observations. This is faulty scientific reasoning and has led to many serious errors in the history of science.[6] *Thus the theory has not been proved and cannot be proved, even though many people speak of it as a fact.*

Some scientists, however, are careful about this. One prominent theoretical physicist, Prof. Michio Kaku, who has written several books on physics and cosmology, consistently uses the word "proof" with quotation marks when writing about so-called proofs of the big bang theory.[7]

Finally, according to most versions of the big bang theory, the "bang" occurred from some pre-existing point, or singularity—"a tiny singularity of some sort much smaller than an atom," as Kaku describes it.[8] *But the assumption of the existence of even the smallest stuff (matter, energy or whatever) prior to the creation of the universe—stuff that was always in existence—violates the fundamental belief of Judaism that G-d is the only existence that was always here and He created the universe from absolute nothingness.*[9] *Some scientists speak of the bang having occurred from nothing. But by "nothing" they don't necessarily mean the same thing that we do. For example, they may just mean empty space. But space itself is a creation—part of the universe that G-d created from* absolute *nothingness.*

6. This issue was discussed extensively in sec. 4 of this chapter.
7. Michio Kaku, *Parallel Worlds* (New York: Doubleday, 2005).
8. Ibid., p. 61.
9. Rambam, *Mishneh Torah*, "Hilchos Teshuva" 3:7 and the notes of the *Kesef Mishneh* ad loc. See also *Igros Kodesh* of the Rebbe MHM, 2:200.

Natural Science on the Road to Redemption—
Cold Water: A Dream Come True

by Dr. Aryeh Gotfryd

Dr. Gotfryd spoke at several Moshiach and science conferences over the years. The following sections are based on his lectures, the thesis of which is that scientific theories are now converging to the truths of the Torah.

In this section Dr. Gotfryd presents his background and explains how he came to Chabad. In it he describes two awesome dreams that he had of the Rebbe MHM and how these shaped his thought and his subsequent work.

In the next two sections he focuses on many different areas of science and scientific thinking that have come to recognize the truths of the Torah, which, he explains, is a phenomenon of the Era of Moshiach.

Dr. Gotfryd earned Canada's first doctorate in Applied Ecology, specializing in urban wildlife and habitats, and taught an award-winning, accredited course on Faith and Science at the University of Toronto. He works as an environmental services consultant and contractor in Toronto, Canada.

Waking Up Jewish

My grand awakening, the pivotal event on which my approach to Judaism hinged, was a six-hour argument with a Chabad rabbi who had answers for all my questions and questions for all my answers, leaving my secular worldview a lot less compelling than it was for my first 26 years.

Having finally woken up to the facts of life from a Jewish perspective, I suddenly realized that my lifestyle until then was very much out of tune. Those facts of life, so simple, had completely eluded me before but now that I knew them, I wanted nothing more than to learn more about them, share them, and live in harmony with them.

The simple facts that inspired me so were: that I have a *neshama*—a soul—a part of G-d; that the Torah is G-d's will and wisdom in words; that the Mitzvos provide a dignified, satisfying lifestyle for every Jew; and that G-d knows who we are and cares about what we are doing.

So, starting in January 1981, I climbed the ladder of observance, deepening my commitment to Judaism but still not attuned to the Rebbe or his unique and central role in the grand scheme of things. Meanwhile, my career as an environmental scientist was just starting to open up, with accelerated progress in my doctoral thesis in applied ecology, and a steady stream of environmental assessment contracts.

Despite these successes, I was dissatisfied. Wading through ecological journals for insight into the wonderful world of nature seemed relatively futile. I thought, "Why bother struggling to understand nature from the bottom up, when I could be instead in yeshiva learning divine wisdom about the world and its purpose from the blueprint of creation itself?" But alas, here I was in the Department of Zoology at the University of Toronto, consigned to guessing and fudging along with the rest of the intellectual prima donnas in the ivory tower world of academia. My rabbinic advisors said that I needed to finish my doctorate, so instead of following my heart to feed on the kernels of Torah wisdom, I was exiled to chew on the husks of secular science instead.

Deep Calling To Deep

Finally, just before marrying in Sivan 5742 (June 1982), I made my way to Crown Heights, New York for my first visit to the Lubavitcher Rebbe. At first, I wasn't impressed. He seemed a small, dignified man, nothing so unusual. Wise. Saintly. Scholarly. But not so astounding that an ocean of people should be continuously following him around, parting like the Red Sea into two walls of humanity whenever he would pass by. For the first time, I started to wonder whether he was all he was reputed to be.

Nonetheless when it came time to write the customary note one brings to a *yechidus* (audience), I opened my heart and wrote the Rebbe to help me find the service of G-d within applied ecology, and not just that my academic pursuits be a means to earn a degree, a position, a livelihood and so on. With my note in hand, I went in line with over a hundred other grooms and brides-to-be, who were scheduled for group audiences in the Rebbe's office. Once inside "the upper Garden of Eden," as his office is called, I took my turn to place my note on his desk. That's when it happened.

The Rebbe, holding the note, looked up to its author with a fixed penetrating gaze that gave me the distinct and intense impression that I was being "known." In what couldn't have been more than three seconds, it felt like my entire past, present and future were laid out before this man, and that in the intensity of the moment he came to know more about *Aryeh Avraham ben Chana* (me) than I would ever know about myself. So disorienting was this wordless soul to soul contact, that I did not hear so much as a single word of his five minute shower of plentiful blessings and sage advice.

The First Dream

A few weeks later, I had a dream, the most vivid of my life. It was especially notable because I rarely remember dreams and if I do, it's a blurry, gray jumble of images that evaporate as soon as I awaken. In this dream, my bride Leah and I were sitting together with the Rebbe in our living room. He presented to us a publication clearly entitled "SCIENCE" Volume 64, No. 4 (No. 1). The layout of the title was identical to the prestigious journal by the same name, which is the weekly publication of the American Association for the Advancement of Science. "Here is something to help you think," said the Rebbe to Leah.

He opened it up for us. On the first page was a blue sky with white, fluffy clouds. The page was beautifully typeset in English with a crisp, black font and wide margins on the page. I started to read what seemed a little like Psalm 104, all about wildlife, vegetation, the hydrological cycle, humanity and meaning to life in the presence of G-d—in a sense, it was a kind of poetic treatise on the science and spirit of applied ecology. As we read on, the clouds on the page started to move across the sky and a fresh breeze brought a subtle and delicious fragrance. This itself was remarkable because cigarette smoking had all but eliminated my sense of smell years ago. Orchestral music welled up featuring inspired melodies, elegantly harmonized, and from the intensity of pleasure in this dream, I fairly fainted. The Rebbe turned the page and the next page was even better, and the next better yet. Then he said, "Let me know when you get to page 7" and I woke up. With great excitement I related this fantastic dream to my wife who said, "That's interesting."

The Buried Treasure

The next reasonable step was to seek out Science Vol. 64, No. 4 (No. 1) and especially page 7. As a graduate student in the University of Toronto's Department of Zoology, it seemed likely to find such a popular journal in the Zoology library and find it I did, but only with volume numbers dating back to the 200s, decades after the envisioned Volume 64. I tried the Physics library and the Chemistry Library and the General Science Library but none of them had back issues that old. Well, forget it, I told myself. After all, it was only a dream, so I put it out of my mind. Almost.

More than a year later, the possibility that there may be something more to the dream was still nudging me periodically so I finally decided to put the matter to rest definitively, one way or the other. I went where scientists rarely go: to Fort Book—the immense concrete structure housing University of Toronto's central reference library for humanities as well as complete catalogue data on all the holdings belonging to the university's dozens of libraries. Sure enough, ancient issues of Science were listed to be in the storage stacks of the Science and Medicine library.

What a room that was. Special access was arranged to the cellar where outdated books were shelved on long parallel bookcases, mounted on tracks and rollers. There were no spaces at all between the 20-odd bookcases, so at first glance there seemed to be no way to access any of the books. The trick was that each bookcase had a kind of steering wheel mounted on the end. When a wheel was spun, several dozen times around like a ship captain's wheel, a space would slowly emerge between two bookcases so you could walk in to retrieve a reference, provided of course that no one spins the wheel to squash you while you're in there.

There it was. Volume 64, dated the summer of 1926, with its various issues. Now I would finally see if there was anything to this odd dream. The article in issue No. 4, page 7, was entitled "Science for Humanity's Sake" by William Blum. In his address upon receiving the American Chemists' Medal, Blum explains that because the study and application of science is spiritually uplifting, science and religion interface in man. Nonetheless each discipline has its own domain, with science describing how G-d works but only religion is able to address why the world works that way.

In the other dream reference, Issue No. 1 of Vol. 64 of Science, on page 7, one finds of all things a poem about the joy of discovery, quoted by B.T. Baldwin, president of the prestigious Sigma Xi Research Society. The poem reads:

> Come wander with me—
> In regions yet untrod,
> And read what is still unread
> In the manuscript of G-d.

I recalled my request of the Rebbe asking to help me find G-d through science. I remembered my beautiful dream, with its bibliographic references. And now I held in hand two buried treasures precisely related to the dream and to the encounter with the Rebbe. Immediately I thought, "This must be the finger of G-d." But my next thought was, "Maybe its just dumb luck." So I put it to a test. The alternative hypothesis: Maybe the dream was meaningless and in the 1920s it merely happened to be popular for scientists to refer to G-d in their scholarly speaking and writing.

So page by page, I checked all of Volume 64 and many other volumes in the 1920s and 30s, then 40s, 50s and 60s. No spiritual references whatsoever. Perhaps other scientific weeklies, like Nature, revealed such a trend. On the contrary, they were all spiritually mute in the extreme. Alternative hypothesis disproved.

The correlation between dream and reality was just too improbable to be meaningless, so I decided to act on the Rebbe's "instruction," in his saying "Let me know when you get to page 7." The gist of my letter to him was: Sorry to bother you, but I had this dream, and it took me a while, but I found these facts, so I'm letting you know. Was this a communication from you? Shall I act on it or forget about it? If you did send it, enclosed are copies of the articles referred to in the dream. Now what?

No Answer is Also An Answer

At the time, it seemed so strange. On this one question, the Rebbe did not answer. Yet on every other matter, whether a *simchah* (celebration) or a major turning point, he was always quick to respond to me. In fact over the following four years, I wrote the Rebbe MHM no less than eight separate times on various matters, always appending a request form him to respond about the dream and the articles, but alas, nothing. Not even the slightest acknowledgment that I had asked. Meanwhile, on my own initiative I became involved in researching and writing about the relationship between religion and science, especially with reference to Judaism.

The Second Dream and Its Interpretation

Then in the fall of 1987, on the 8th of Tishrei, 5748, I had my second dream encounter with the Lubavitcher Rebbe. He was in bed, horribly ill and emaciated. He called to me to get him a glass of water. "Hot or cold?" I asked. "Cold," the Rebbe MHM answered in that quiet definite voice that could split mountains. I ran to get some cold water, brought it to him and just as I was handing it to him, I woke up in a trembling sweat, exactly one minute before my alarm was due to ring. Could the Rebbe really be so unwell? If so, what will be? Deeply disturbed, I made my way to shul and asked my *mashpia* if he knew how to do the *HaTavat Chalom* ceremony (to transform a bad dream to good). He said I should speak to the Rav of the Lubavitch Community in Toronto, Rabbi Dovid Schochet, who can interpret dreams.

"Dream interpretation? We believe in that stuff?"

"Have you ever heard of a book called the Chumash?"

"I know. You are referring to Yosef and Pharaoh's dreams. But that's not relevant to this situation."

"Oh really?" my mashpia continued, "The Torah is just a history book?"

"Okay, fine. It's relevant in all times and places, but... the Rav?"

"I'm telling you, he knows what to do."

After *davening* (prayers) the Rabbi agreed to convene a *minyan* (quorum) for the service to rectify dreams, following which I related my dream. The Rabbi replied: "In your dream, the Rebbe was unwell but now he is fine, thank G-d. So you don't have to worry about that. Usually we are used to thinking about how we need the Rebbe so we are not so attuned to how he needs us. You have merited to see not only *that* you are needed but also *how* you are needed. Water symbolizes Torah—it flows from a high place to a low place. Water however may be hot or cold. Heat signifies emotions and cold signifies intellect. The Rebbe is telling you that he needs your intellectual service in Torah and Mitzvos."

The Rabbi's words touched me deeply. I felt as if a heavy weight was lifted off my shoulders. Relieved and with a new-found spring in my step, I thanked Rabbi Schochet and exited the sanctuary into the hall.

At that very moment, I happened to encounter the Administrator of the Chabad House, Rabbi Moshe Spalter, who said, "Aryeh, do you remember putting in a request for the book *Emuna U'Mada* at the Lubavitch Library? Well it took over a year for it to turn up, but here it is. It just turned up this morning." I did a double take. The cover was blue—a blue sky with white, fluffy clouds. The first dream. "Moshe, I know that *Emuna* means faith, but what does *Mada* mean?

"It means science," Moshe casually replied, handing the Hebrew paperback to me. I carefully leafed through the first few pages. It was a collection of letters from the Rebbe on a wide range of topics. The very first chapter was entitled "Proof for the Existence of the Creator"—a rational, step-by-step essay demonstrating the necessity of G-d's existence according to the criteria of science and common sense. I slowly rushed my way to page 7. Among the hundreds of words on page 7 of *Emuna U'Mada*, there are precisely two in bold type—*tiyur* and *biyur*—description and explanation. There the Rebbe writes that science can only *describe* how the world works but in no way can it explain *why* it works that way.

The point he was making was exactly the point raised by Blum on page 7 of Science, Vol. 64, No. 4! I stood dumbfounded. In my hands, I held the blue sky and white clouds of the first dream, delivered on the very morning of the second dream. And how strangely coincidental it was that the very same point was made on the very same page number by the two very different authors in two very different publications from very different decades. Especially considering that all of this was obviously the fulfillment of my request at the *yechidus* to find G-dliness and meaning in my work in the natural sciences!

No doubt about it. These events had been stamped with the Divine fingerprint, i.e., a combination of low-probability events synchronized in a meaningful and unified way. Still, as sure as I was, I wanted a little more confirmation and besides, I still hadn't let the Rebbe know that I had truly and fully reached "page 7." So off I went to Crown Heights to let the Rebbe know. But how? I was still pondering this question as I approached him in the dollars line on a Sunday in November, 1987. I knew that they would be pushing from the back and pulling from the front so there would not be much time to tell the whole story. Prudently, I distilled it down to two words which I blurted out as I passed before him: "*Biyur V'Tiyur*!" which must have sounded pretty strange to the Chassidim standing around. But it didn't sound strange to the Rebbe because immediately he responded with a single, loud, emphatic and resounding Hebrew word: "Amen!"

Within a few days, I was invited by Professor Yirmiyahu (Herman) Branover to participate in the first International Torah and Science Conference co-sponsored by his B'Or HaTorah magazine and Rabbi Shalom B. Lipskar's Aleph Institute. Observant Jewish scientists from all over the world were to gather during Chanukah, 5748 in Miami Beach to present their discoveries of how Torah provides incomparable insight into the issues raised by science, the arts and modern life. "Aryeh, do you have a topic in mind?" Professor Branover asked me. "As a matter of fact I do: Description and Explanation in the Sciences."

Following the conference, I was delegated the task of preparing the proceedings for publication. The resulting book, "Fusion: Absolute Standards in a World of Relativity," has as its first chapter my translation of the Rebbe's letter from *Emunah U'Mada* quoted above. And its fourth chapter is my essay called "Beyond Description: The Boundary Between Religion and Science." Since then, this "How and Why" issue has become a key theme in my books, articles and talks the world over. After all, it seems that's what the Rebbe wants of me.

Conclusion

What can we learn from all this? See G-dliness! The One Above controls the workings of this world according to Particular Divine Providence. Moreover, this Particular Divine Providence operates not only when we see how it all comes together, but also when we don't. We know this to be true because the "punch line" of a Divine Fingerprint event must be set up in advance by the One drafting the plan and coordinating various unrelated factors long in advance of that climactic glorious moment when it all comes together.

But what role does the Rebbe have in all this? Our sages say, "A Tzaddik decrees and the Holy One Blessed Be He carries it out." They also say, " The Holy One Blessed Be He may decree [something negative] and the Tzaddik cancels it." This is the power of the Rebbe, and Baruch Hashem, I have been privileged to receive his guidance, and to realize it.

May we all realize our potential in connecting with the Rebbe and doing our part to bring Moshiach NOW!

A Revolution In Academia

by Dr. Aryeh Gotfryd

One of the changes that come with Moshiach and Geulah is the revolution that is now underway in the world of academia. It is an intellectual revolution based on scientific discovery. The essential notion revolutionizing the academic world is that the basic principles of religion and the basic findings of natural science not only do not conflict, but actually harmonize exquisitely. Scientific theories are now converging to the truths of the Torah.

Here are some examples:

Modern Physics acknowledges that the universe depends on a higher conscious being[1] unlimited by space and time, that continuously creates, sustains and coordinates all the material and energy in the universe.[2] This is similar to the Chassidic notion of continuous creation by G-d.

Relativity theory[3] shows that the idea of an earth-centered planetary system is just as valid scientifically as a sun-centered solar system.[4] The Torah, of course, teaches that the universe is earth-centered.

Quantum theory has given man the central role of being the pre-

1. Menas Kafatos and R. Nadeau, *The Conscious Universe: Part and Whole in Modern Physical Theory* (New York: Springer-Verlag, 1990).

2. Paul Davies, "G-d and the New Physics," *New Scientist* 98, no. 1363 (1983): 872–874.

3. Albert Einstein, *The Meaning of Relativity*, 3rd ed. (Princeton, NJ: Princeton University Press, 1950).

4. Hans Reichenbach, *The Philosophy of Space and Time* (Dover, 1958).

planned goal of the cosmos since its beginning.[5, 6] "For my sake was the world created." Molecular Chemistry has revealed an atomic alphabet of some 100 elements, generating billions of compounds, similar to the Divine Alef-Bet generating everything.[7]

Biology has demoted Darwinian evolution from the rank of fact to the level of unrealistic speculation[8,9] leaving *Bereishis Bara Elokim* ("In the beginning G-d created...") a more viable alternative.

Ecology and environmental activism are based on the conviction that one small deed (or even a thought)[10] can make a disproportionately huge impact in the world at large, similar to our concept that even one good thought can bring the Geulah.

Chaos theory shows that a few simple rules generate a hidden order that unifies apparently random events in a wide variety of contexts.[11] This is similar to random events being guided by particular divine providence according to Torah principles.

The implications of Omega Point theory, developed by renowned physicist Frank Tipler, are summed up in the title of his 1994 book, "The Physics of Immortality: Modern Cosmology, G-d and the Resurrection of the Dead."[12]

Simultaneous with this upwelling of secular wisdom, there has been a parallel outpouring of Divine Knowledge in the form of Chassidus generally, accelerating with Chabad Chassidus more specifically, and fairly exploding with the Chassidus of the Rebbe especially.

---

5. J. Von Neumann, *Mathematical Foundations of Quantum Mechanics* [translation] (Princeton, NJ: Princeton University Press, 1955).

6. John A. Wheeler, "Recent Thinking About the Nature of the Physical World: It From Bit," *Ann. N.Y. Acad. Sci.* 655 (1992): 349–363.

7. In *Sefer Yetsirah* we find, "Seven stones build 5,040 houses," meaning that with seven letters one can form 7! = 7x6x5x4x3x2x1 = 5,040 seven-letter words. This is parallel to a finite number of atomic elements generating countless types of molecules by combining in different ways.

8. Proc. Chicago Symp, "Macroevolution," *Science* 210 (1980): 883–887.

9. For an overview and various angles on the issues, see Aryeh Carmell and C. Domb, eds., *Torah Views on Science and Its Problems* (Spring Valley, NY: Feldheim, 1976), especially Part 2, "Creation and Evolution." The famous letter of the Rebbe on creation and evolution is reprinted here.

10. Jerry Eichoefer, personal communication regarding his research published in *Princeton Technical Reports* entitled "Mind Over Matter: Margins of Reality," 1996.

11. James Gleick, *Chaos: Making a New Science* (London: Viking, 1987).

12. See Prof. Branover's discussion of this in sec. 2 of this chapter.

This synchronized and integrated blossoming of Divine and secular wisdom has been anticipated for thousands of years, in the mystical tradition of authentic Judaism. Thus the Zohar states that in the year 5600 (1840 CE), the gates of wisdom on high and the fountains of wisdom below will be opened and the world will be brought to a state of restoration, to be ushered in to the seventh millennium, that is, the coming of Moshiach and a world of good.

The Rebbe points out in a sicha (vol. 15, p. 42) that "wisdom on high" refers to Torah, especially *Torah Ohr* and *Likutei Torah* which were published around then, and that "wisdom below" refers to science. Scientific discoveries from the years around 1840 CE include: the wave theory of light, thermodynamics, the absolute temperature scale, electromagnetic induction, the conservation of energy, Maxwell's electromagnetic equations, organic chemistry, the theory of valency, and molecular structure.

But with all the spiritualization of the sciences, why aren't all the professors and textbooks calling out: "G-d is One!" or at least humbly acknowledging the validity of faith? The problem is inertia. The outdated materialistic world view of the previous age is firmly lodged in the heart of the ivory tower and it doesn't want to move!

The average scientist is spiritually numbed by the gross misconception that science provides reasonable explanations for natural events and processes, thereby removing the need for any supernatural being. The Rebbe addresses this misconception directly in his letter on the proof for the existence of the Creator:[13]

> Of course I am not unaware that all this is governed according to the "laws of nature"—but I think it is important to emphasize that such expressions provide no *explanation*, but rather give a convenient summary or *description* of the existing situation, that is, that natural phenomena are conducted according to specific patterns.
>
> But to say that a "law of nature" is an independent being in and of itself, and that this being rules over all of creation, and that there are thousands of such beings, according to the number of natural laws, is absolutely absurd. Moreover there

13. Published in *Emunah U'Mada*, 7.

> is not one scientist in the field who would say so. Rather such laws are merely convenient, summary expressions for describing a situation, so that one should not be forced to duplicate at every turn a lengthy description of the "simple" facts. Nonetheless it is clear and obvious that such an expression provides no *explanation* whatsoever.

Why do we need to know this? So that believers in science may also come to know that there is "a Boss." And if they keep thinking about it they may even come up with the same conclusions that *Avraham Avinu* (Abraham) did. They will conclude that there is only one Boss. That this Being is beyond time and space. That It is invisible, omnipotent, omniscient, without form or body. And It wants to be known by us, i.e., by such beings that are not beyond time and space, not invisible, not omni-anything and definitely not without form and body.

And when it is clear to even the logic of the animal soul that this supernatural Being creates and runs the world right now, and every detail of it, and wants us to know about Him, there will be generated an intellectually inspired love and fear. This in turn leads to more ready acceptance to do His will, which is after all the main point as the Rebbe concludes in the above letter, "because action is the main thing."

And in the words of the mathematical meteorologist who pioneered the science of chaos, "the flap of a butterfly's wings in Brazil can set off a tornado in Texas." Thus even the "secular" scientist can appreciate the reasonableness of the Rambam when he states that each mitzvah must be viewed as the one that tips the scales and saves the whole world with the coming of Moshiach NOW!

The Conscious Universe

by Dr. Aryeh Gotfryd

Among the many signs of the Redemption that the Rebbe MHM directed us to notice is the modern explosion of knowledge in the realm of science. Some of these scientific discoveries go far beyond the typical descriptive role that science normally plays in understanding nature. Instead we find unmistakably religious implications in a variety of scientific disciplines.

The tools of modern physics have yielded theories similar to the concepts of the continuous creation of something from nothing, the existence of a cosmic plan, the centrality of humankind in this plan from "before" the beginning of time, and the intimate relationship between man and a higher consciousness that sustains the universe while being itself quite beyond considerations of time and space.

To be sure, many of the proud proponents of these rediscoveries claim to be irreligious, but this may be largely because they don't understand much about religious notions generally and Chassidus in particular.

Following is a summary for the non-scientist describing how the hard-nosed physicists became so "religious." What the following discussion is intended to accomplish is to help us open our eyes to see how science is coming rapidly and naturally to "The earth shall be filled with the knowledge of G-d as the waters cover the sea."

The Littlest Bits

Quantum physics is the science of littlest bits—the size of an atom or smaller. Examples are the smallest bit of electricity, an electron, and the smallest bit of light, which is a photon.

"Wavicles"

All subatomic particles may be seen either as waves or particles, but not both at once. For example, when ejected from a "gun," electrons travel fairly straight, forming a cluster of points on a detector screen. This shows their particle nature only. However, when projected through a narrow slit, they make a diffraction pattern, which shows their wave nature.[1]

Either or Neither or Both?

Now which state is the little bit in before it is observed? Wave? Particle? Both? Or none? And does it matter? According to mainstream physics, the pre-observation state has the potential of being either but is actually neither. Moreover it is the act of observation itself that brings the thing into being in one of its possible states.

How so? Picture an electron being fired at two slits. There are two ways to observe it. One way is to set up a detector that can check which one of the two slits it went through. This method invariably reveals the electron to be a particle that has passed through only one of the slits. The other way is to record the electron's passage as a diffraction pattern on a screen, which means it went through both slits and is a wave.

It seems impossible that the electron should go through only one of the two slits, and through both of them, at one and the same time. Yet this seemed to be the case.

Delayed Choice Experiments

Strange as it seems, Professor John Wheeler has found a way to perform the two-slit experiment, making the decisive measurement after the electron has passed through the two-slit apparatus. In this situation, one chooses afterwards whether the electron passed through only one of the slits or both of them. Thus the observer is free to determine the nature of physical reality not only now but retroactively as well!

1. See Dr. Berg's discussion of this in chap. V, sec. 5a.

Observing and Reality

Before the measurement, the event was not in some determined but still unknown state. It was somehow in an omni-potential including both of its mutually exclusive states. Only by its being measured does the object enter one finite state. In the language of physics, this is usually called "the collapse of the wave function" into reality.

Many leading physicists claim that only a human consciousness can collapse the wave function. As Professor Eugene Wigner puts it, "The being with a consciousness must have a different role than the measuring device." This implies that human awareness makes sub-atomic physical reality happen.

Quantum Universe

So experiments on quanta, the littlest bits, show an observer-dependent reality under special conditions. But what about the universe at large? Does the big world also need observers like us in order to exist? Those little bits (be they electrons, photons, quarks or whatever) are also observer-dependent when not in the laboratory. And being the building blocks of all matter and energy, they are literally everywhere, and are moreover subject to the same laws of physics wherever they are.

So the same applies to the universe as a whole. All of physical reality is observer-dependent and the laws of nature exist only because we do.

Retroactive Reality

The delayed-choice experiments not only show how the observer gives reality to every place, but also to every time. The observation of the quantum gives it reality not only now but retroactively to its origin. Applying this to the universe as a whole: When we observe now, we allow the universe to emerge into reality retroactively to its beginning.

Free Will

Professors Avi Rabinowitz and Yirmiyahu Branover propose that free will is the special quality of observership that allows it to collapse the wave function. Since free will is unconstrained by nature and natural laws, it can induce them.

The Conscious Universe

In 1982 a yet more radical step took place in the spiritualizing of physics. Alain Aspect conducted special delayed-choice experiments showing that reality is non-local. This means that "the new physics seems to demand some guiding influence located as it were above nature sustaining all of existence."

How does this work? Paired particles are emitted from a single source. It is known that their spins must match. After they are emitted, a rapid decision is made on how to observe one of the pair. Then the second particle is measured. It turns out that the decision on the first particle, instantly determines the result on the second of the pair, even though the two particles were so widely removed that their remoteness precluded the possibility of electromagnetic communication, i.e., no information moving slower than the speed of light can pass between them.

This finding has given rise to an unprecedented wave of "new physics" mysticism because it implies that all physical and temporal separations are illusions and that underlying the universe is a unified consciousness that relates to man and especially to his free-willed consciousness.

In summary, we see that physics has softened its materialistic stand. It recognizes a supernatural consciousness behind everything (G-d) and promotes man to the central role of being as important to the existence of the universe as the universe is to the existence of man.

But why is all this happening now? Because the days of Moshiach are already upon us and the "secular" sciences need to deepen their insight into the greatness of G-d and the importance of man in creating a dwelling for G-d here below.

REFERENCES: CHAPTER IV

The Convergence of Science to Torah: The Trend of Science

Sec. 2: Interview with Professor Branover

Reichenbach, Hans. *The Philosophy of Space and Time*. New York: Dover Publications, 1957.

Sec. 3-4: Intelligent Design / Age of the Earth

Rabbi Bachya ibn Pekuda. *Chovas HaLevavos*, "Gate of Unity."

Rabbi Schneur Zalman of Liadi. *Tanya*, "Sha'ar HaYichud V'HaEmunah," New York: Kehot Publishing Society.

Behe, Michael. *Darwin's Black Box: The Biochemical Challenge to Evolution. 2nd ed.* Free Press, 2006.

Collins, Francis. *The Language of G-d: A Scientist Presents Evidence for Belief.* Free Press, 2006.

Kaku, Michio. *Parallel Worlds*. New York: Doubleday, 2005.

Meyer, Stephen C. *Signature in the Cell: DNA and the Evidence for Intelligent Design*. HarperOne, 2009.

Reichenbach, Hans. *The Philosophy of Space and Time*. New York: Dover Publications, 1957.

Smolin, Lee. *The Trouble with Physics*. Boston: Houghton Mifflin, 2006.

Velikovsky, Immanuel. *Worlds in Collision*. New York: Doubleday, 1950.

Sec. 5: Challenging Scientific Authority

Smolin, Lee. *The Trouble with Physics*. Boston: Houghton Mifflin, 2006.

Sussman, Brian. *Climategate: A Veteran Meteorologist Exposes the Global Warming Scam*. WND Books, 2010.

Sec. 7: Natural Science on the Road to Redemption

Davies, Paul. "G-d and the New Physics." *New Scientist*. vol. 98:1363, 1983.

Einstein, Albert. *The Meaning of Relativity, 3rd ed.* Princeton University Press, 1950.

Kafatos, Menas and R. Nadeau. *The Conscious Universe: Part and Whole in Modern Physical Theory*. New York: Springer-Verlag, 1990.

Reichenbach, Hans. *The Philosophy of Space and Time*. New York: Dover, 1958.

Von Neumann, J. *Mathematical Foundations of Quantum Mechanics*. Princeton University Press, 1955.

Wheeler, John A. "Recent Thinking About the Nature of the Physical World: It From Bit." *Ann. N.Y. Acadamey of Science*, 655:349-363, 1992.

CHAPTER V

The Quest for Unification:

The Trend of Science

"Ultimately, we will come to the realization that the fundamental point of the existence of the world is based on the unification of the two concepts of quantity and quality—matter that carries the force, and the force itself." —*the Rebbe MHM*

Theoretical physicists are obsessed with unification. But why?! Why assume that there are any unifying theories in the first place? We have individual theories that work well and make correct predictions. What drives scientists to look for unifying theories—especially the elusive Theory of Everything, one all-inclusive theory that unifies all of science? Who told them that there is such a theory?

We submit that this quest for unification is in essence a search for G-d. It is the natural tendency of the created being to search for its origin, the Creator.

As the Rebbe MHM explains in the sicha of Parshas Noach, the revelation of the Wisdom from Above—Chassidus, which teaches the knowledge of G-d—in the mid-1800s, came together with a parallel revelation of the Wisdom from Below—new scientific theories. The unification discovered in the sciences works hand in hand with the deep concepts of the unity of G-d taught in Chassidus to prepare the world for the ultimate unification—the revelation of the absolute unity of G-d in the Era of Moshiach. Thus the search for unifying theories is, in essence, a search for G-d.

In this chapter we present various unifying theories in mathematics and physics based on the lectures at the Moshiach and science conferences over the years covering topics ranging from quantum mechanics of the tiny atom to the mathematics of infinity, from the loops of string theory to the mathematics of logic.

We begin with a summary of the sicha of Parshas Noach that we have been referring to throughout this book.

Sicha of the Rebbe MHM on Parshas Noach

Summarized by Prof. Shimon Silman and Dr. Naftali Berg

The Torah says, "In the 600th year of the life of Noah...all the fountains of the depths split open and the windows of heaven opened up."[1] Referring to this, the Zohar makes the following statement: "In the 600th year of the 6th millennium [the year 5600 on the Jewish calendar which corresponds to 1840 on the secular calendar] the gates of wisdom above will open, and the fountains of wisdom below, and thus the world will be set to enter the seventh millennium," the Messianic Era.

The opening of the "gates of wisdom above" refers to the tremendous revelation of the deeper levels of the Torah—the teachings of *Chassidus*, especially the publication of the works of the *Alter Rebbe*, Rabbi Schneur Zalman of Liadi, founder of Chabad Chassidus, which give us a deeper understanding and greater knowledge of G-d. This prepares the world for the coming of Moshiach when "The world will be filled with the knowledge of G-d."

The "fountains of wisdom below" refers to the new discoveries in science and the development of new scientific theories. This requires some explanation—how could scientific developments actually play a role in preparing the world for Moshiach? [Science has always been considered the enemy of religion. How could it come about that science not only becomes a friend of Judaism but actually helps prepare the world for the greatest Jewish religious event of all time—the coming of Moshiach?!]

1. Bereishis 7:11.

The Rebbe MHM discusses three ways in which the sciences prepare the world for Moshiach. In describing the unique character of the knowledge of G-d in the Era of Moshiach, the Prophet Isaiah says, "All flesh will see that the mouth of G-d has spoken,"[2] i.e. that G-dliness itself will be visible. The first way in which modern science and technology prepare us for the Era of Moshiach is by enabling us to visualize certain concepts relating to G-d which would otherwise be understood only abstractly. As an example, the Mishneh in *Avos* tells us to be aware that there is "an eye that sees and an ear that hears." We believe that G-d is always watching everything we do. Now that we are familiar with communication technology which transmits audio and video signals across the world so that people can be seen and heard instantly on the other side of the world, it is easy to comprehend—to actually visualize and not just believe in—the omnipresent watchfulness of G-d.

Secondly, our making use of the technology derived from the sciences prepares the world for Moshiach. For example, communication technology can be used to spread knowledge of G-d (Torah) throughout the world. As we said, the Era of Moshiach is characterized as a time when "The earth will be filled with the knowledge of G-d." The electromagnetic waves which carry knowledge of G-d over radio or TV bring this information not only to listeners and viewers all over the world, but even to places where there are no listeners, since the electromagnetic waves spread throughout the whole world. Thus these broadcasts actually *fill the world* with the knowledge of G-d.

On the deepest level, the sciences themselves—the theories developed and the knowledge derived—prepare us for the Era of Moshiach in a third way: the ultimate knowledge of G-d that will be revealed in the Era of Moshiach is the absolute unity of G-d. But the world as we perceive it appears to present an opposite picture of reality—many different forces of nature and many different substances.

Chassidus, which reveals the basic, absolute unity of G-d in the world, teaches that the multitude of creations in the world is no contradiction to the absolute unity of G-d. On the contrary, such a multitude of creations can *only* come from the absolute unity of G-d.[3]

2. Yeshaya 40:5.
3. See Tzemach Tzedek, *Derech Mitvosecha*, 49a.

Now, the true existence of the world *is* the unity of Hashem. Therefore this unity must manifest itself in the physical existence of the world also. There must be a concept of unity within the physical world itself.

The true unification between secular knowledge and the Torah occurs when one sees ideas from the deeper levels of Torah in the sciences themselves, and this concept of unity within the world (which in its innermost aspect is the manifestation of the unity of G-d) has become more and more revealed because of the development of the sciences.

It used to be that people believed that each of the natural forces is a separate force in and of itself and that the matter which comprises everything that exists in the world is constituted of many different elements. But the more that the sciences advance and are developed, we recognize that the multitude and differentiation of these different elements are only an external matter resulting from various combinations of the different particles (protons, neutrons and electrons). Thus we have been able to greatly reduce the number of basic particles. Ultimately, we will come to the realization that the fundamental point of the existence of the world is based on the unification of the two concepts of quantity and quality—matter that carries the force, and the force itself. *Absolute* unity is an aspect of G-dliness; the analysis of the creation in its own terms must be based on the *unification* of two components—quantity and quality.

This deeper understanding of the unity within the physical world which is discovered through scientific research helps prepare us for the revelation of the ultimate unity—the *Achdus Peshuta*[4]—the absolute unity of G-d which is revealed in the time of Moshiach.

---

4. *Achdus Peshuta* literally means "simple unity," which means a unity that does not consist of parts. There is only one G-d and He has no subdivisions.

A System of Logic for Messianic Phenomena

The Messianic world order is an extension of the natural order. It is a world order in which natural phenomena continue to occur but there is also a prevalence of miraculous phenomena. In the Messianic world, the natural order and the miraculous order combine into one world order—a situation which is often paradoxical and requires a new system of logic to describe.

In the next three sections we present such a system, known as Three-Valued Logic. In the first of these three sections we discuss the significance of the number three in the Torah. In the second, we discuss the Messianic world order and its paradoxical aspects. Then in the third, we develop the Three-Valued Logic itself.

The Triple Thread[1]

"True strength (according to the Torah) is connected with the number 3...as it says, 'And the triple thread cannot quickly be severed'—three indicates a power that lasts forever."[2] —*The Rebbe MHM*

The Structure of the Universe

A careful analysis of the underlying structure of the universe shows that there is a fundamental significance to the number 3; on many levels we find that the world has a triple structure. In this section we will examine this triple thread that runs through the entire order of creation, from the spiritual worlds to the physical world, from the structure of man to the Torah that was given to direct the life of man.

In describing the creation of the universe, G-d said, "I created it, I formed it, I even brought it to actuality."[3] Based on this, Kabbala and Chassidus identify three spiritual worlds on which the entire order of creation is based: *Briya*, *Yetzira* and *Asiya*—the worlds of Creation, Formation and Actuality.[4]

Creation takes place from nothingness. It is the initial bringing into being of an entity, and at the point of creation the entity itself is just a point, unstructured and undeveloped. It is a point of conception—like a thought appearing in one's mind. Formation gives structure to the entity and develops it, considering all the possibilities of bringing it into actuality—like a discussion based on a thought. This is analo-

1. The primary references for this section are the sichos of *Mevorchim Sivan* in *Sefer HaSichos* 5750 and *Parshas Tisa* in *Sefer HaSichos* 5752.

2. *Sefer HaSichos* 5751, 635.

3. Yeshaya 43:7.

4. See chap. I, sec. 2, "Wisdom From Above," for a discussion of the spiritual worlds.

gous to speech. Finally, a decision is made as to what the entity will be and how it will appear in all its detail. Action is taken and the entity is brought to full actuality.

Thus the worlds of Creation, Formation and Actuality are, so to speak, G-d's thought, speech and action.

Within this triple structure there is a triple substructure, as each of these 3 worlds subdivides into a beginning, a middle and an end. The beginning is the level of creation within each world, the middle is the level of formation and the end is the level of actuality. Thus, in the world of Creation there is 1) creation in Creation, 2) formation in Creation and 3) actuality in Creation, etc. This can be best understood in terms of thought, speech and action: Within thought itself there is the initial flash of thought, the formation of the thought as it takes shape in the mind and then the finalized, fully structured thought as it sits in the mind ready to be expressed in speech. Thus there are aspects of thought, speech and action within thought itself. Similarly, speech and action subdivide into three levels.

The Kabbalistic analysis of creation is also based on three components, *olam*, *shana*, and *nefesh*—space, time and life.[5]

Our own physical world (Actuality) is composed of three basic elements—fire, water and air. (These are not to be confused with the concept of elements used in the physical sciences. Suffice it to say that each of the scientific elements is itself composed of the three basic elements of fire, water and air.[6]) There is also a fourth basic element—earth. But the fundamental Kabbalistic work Sefer Yetzira mentions only fire, water and air since earth is not an independent element; rather, it is composed of the other three.

In the scientific analysis of the world, it was found that each of the scientific elements has a triple structure since every atom is composed of three component particles—protons, neutrons and electrons. Since everything in the world is made of these atoms, we see that every physical object in the world has this triple structure.

(Later, in section 4 of this chapter, we discuss the search for a theory of everything. While string theory may be the best candidate for

5. See Rabbi DovBer of Lubavitch, *Ateres Rosh,* chap. 1.

6. See Rambam, *Mishneh Torah,* "Hilchos Yesodei HaTorah" 4:3 regarding gold, an element.

such a theory, many scientists have reservations about it. An alternative unifying theory called Causal Dynamical Triangulation (CDT) constructs space-time geometries from simple triangular structures. So again we see a triple structure.)

The Structure of Man

Man, the focal point of creation, is triple in nature from the creation of his soul to the structure of his mind and body to his manner of conduct in the world. The soul starts out in the world of Briya, descends to the world of Yetzira, and then proceeds to the world of action, Asiya. What does it do as it passes through these worlds? It connects with the Torah on the levels of thought, speech and action. Its goal is to come down into a physical body in this physical world and, by fulfilling the Torah and its mitzvos here, to bring man and the world to perfection. G-d's presence can then be completely revealed in the physical world, as the perfected world with perfected man is a proper residence for G-d. This is the ultimate goal of creation.

Man's physical body is composed firstly of the head, then the main part of the body and, finally, the legs which are external to the body—a triple structure. Similarly, he has three internal organs which direct his life: the brain, the heart and the liver. And each external limb has three components: bone, flesh and blood vessels.

A complete human act begins with a thought. This is followed by a discussion of the idea with others to clarify the issues and determine the best course of action. This consultation with others—perhaps friends or professionals—assures a higher degree of success in carrying out the plan. Then the plan must be brought to action.

On the other hand, while the main thing is action, the plan must first go through the stages of thought and speech for if one jumps straight to the action, his act will not have a solid foundation and will not be as strong or successful.

In the development of a man from early childhood, we find that thought, speech and action appear as separate stages of development. Initially a baby has only the ability to act but it can't talk or think effectively. At approximately the age of two he begins to speak. This faculty continues to develop until he grows up (Bar Mitzva), when he begins to think effectively.

Indeed, the Hebrew word for man, Adam, is composed of 3 letters: Alef represents thought ("*a'alefcha chochma*"), *Dalet* stands for *dibur*—speech, and *Mem* stands for *ma'aseh*—action.

In Chassidus, we find an interesting discussion of the kabbalistic concept of *tohu* (chaos) and *tikkun* (structure) which applies the triple structure of man to one's daily life. The Alter Rebbe explains why Yakov split his camp into two parts when confronted by his brother Esav.[7] Esav represents the nations of the world, whose source is in *tohu* and whose *sfiros* (Divine emanations) consist of two "lines," a line of kindness and a line of severity. On the human level this refers to one who relates to his fellow with one of two attitudes. Either he looks favorably on his fellow's action or he looks critically. He can switch from one attitude to the other but he will always follow only one of these two lines. Yakov, on the other hand, represents the Jewish people who have their source in *tikkun* whose *sfiros* consists of three "lines." On the human level this refers to one who relates to his fellow with three possible attitudes. He can look upon his fellow favorably, critically or with a blend of the two attitudes—the fellow may deserve criticism but I will favor him anyway. This is the attribute of mercy. Yakov's own personality was composed of three lines, but when he prepared to confront Esav he took the posture of two lines by splitting his camp into two camps.[8]

The Structure of the Torah

Since the Torah was given to instruct man on how to live a life of holiness and bring the revelation of G-dliness into the world, it would not be surprising to find that the Torah itself has a triple structure—to fit the nature of man that it is addressed to.

Indeed, the Torah has an overall triple structure and many aspects of the giving of the Torah on Mt. Sinai were connected with the number 3. This is expressed in the following statement of the Talmud regarding the giving of the Torah:

"Blessed is G-d who gave the triple Torah to the triple nation by the third person on the third day in the third month."[9] It is a triple Torah

7. Bereishis 32:8.
8. Rabbi Schneur Zalman of Liadi, *Torah Or*, "VaYishlach," 48
9. Shabbos, 88a.

because it is divided into three parts—the Torah (the "Five Books"), the Prophets and the *Kesuvim* ("Writings"). The Jewish people are a triple people since they are divided into three (tribal) groups—*Cohein, Levi* and *Yisroel*. The "third person" is Moshe Rabbeinu, who was the third child of Amram. The Torah was given on the third day of "preparation" in the third month of the Jewish year—*Sivan*.

Rav Nissim Gaon, one of the early commentators of the Talmud, extends this list by mentioning several more ways in which the giving of the Torah to the Jewish people is connected with the number 3. Among them is the fact that the Jews are descended from the three forefathers, Avraham, Yitzchok and Yakov. Moshe Rabbeinu was from the third tribe, Levi. When he was born, he was hidden for three months. The name Moshe, as well as the name Levi, is spelled with three Hebrew letters. Also, one of the names of the Torah, *Emes* (Truth), is spelled with three Hebrew letters.

On a deeper level, it is more accurate to say that man has a triple structure because the Torah does, rather than the reverse. As the Zohar says, the Torah was G-d's blueprint for creating the world: "He looked in the Torah and created the world." The fact that man and the universe have a triple structure derives from the aspect of the number 3 that is in the Torah which was used to create them.

The Alter Rebbe, in the Tanya, explains that G-d's ten creative statements in Breishis were used not just to create the world, but they continue to give life to and maintain the existence of the world. The words and letters of these statements combine to form the Hebrew names of everything in the world, and the letters of these names give life to the objects which they name.[10] Thus we would expect that even Hebrew words themselves—the language of the Torah—have a triple structure in some sense, and indeed the authorities on Hebrew grammar (specifically, the *Sefardi chachomim*) say that every Hebrew word has a root word (*shoresh*) of three letters.

What about 10?

After reading what we have written about the fundamental signifi-

10. *Tanya*, "Sha'ar HaYichud V'HaEmunah," chap. 1.
Sefer HaSichos 5751, 2:507–509.

cance of 3, the reader may be wondering about the common subdivision of things into ten parts—from the ten *sefiros* in Kabbala to mathematics and science which is usually based on the number 10. The Rebbe MHM addresses this and explains that the subdivision into ten is itself based on 3. How is it that a thing comes to have ten parts? First we note that an entity has three general parts—a beginning, a middle and an end. The beginning is the "head" which establishes the purpose for the existence of that entity and contains the object itself in a microcosm. The middle is the main part of the object through which its purpose is attained. The end is the completed object having attained its purpose.

So far we have three: beginning, middle and end. But each of these three parts also has a beginning, a middle and an end—just as we discussed earlier regarding thought, speech and action. We now have nine parts. Finally, the tenth is not a part of the thing, but rather the pure essence of the object before it is divided into any parts.

We have been explaining how the universe, man and the Torah given to instruct man, all have a triple structure. Indeed this threesome—Torah, man and universe—itself comprises a Triple Thread!

Waking Up Jewish

As a man opens his eyes and becomes conscious of reality—whether it is the consciousness of a child maturing into adulthood or the consciousness of an adult waking up in the morning—he becomes aware of three things. Firstly, he becomes aware of himself, his own presence. He exists. But where did he come from? His second awareness is that he was created by the Creator. At this point, if he is waking up in the morning, he will say "*Modeh Ani*," giving thanks to the Creator—"I give thanks before *You*," recognizing both himself and the Creator. But his awareness does not stop there. He then attains the third awareness—that there is a world around him that the Creator put him in for a purpose.

This, in fact, is the meaning of the Mishneh at the beginning of the third chapter of Pirkei Avos which says, "Look at three things...." While the obvious meaning of the Mishneh is that this phrase is an introduction to the continuation of the Mishneh: "Look at [the following] three things and you will not come close to committing a transgression," the Rebbe MHM explains that there must be a deeper message here since the Mishneh really doesn't need to tell us that there are three things if it proceeds to list the three things. We could count them up ourselves. Rather, the phrase "Look at three things..." is an instruction in itself telling us that our consciousness must always be composed of three things: ourselves, G-d and the world.

The point is that one cannot take the attitude that since he was created to serve G-d, as the Talmud says, "I was created to serve my Creator," he will take an "I and Thou" attitude—"Me and my Creator." It is not sufficient for one to isolate himself in a Beis HaMedrash and devote himself to Torah study and prayer, thereby establishing a personal relationship with G-d. He must realize that the Creator put him in the midst of a vast creation and the complete relationship with G-d involves having an effect on the creation—refining it and elevating it to make it a holy place, a "residence" for G-d.[11]

Torah, Creation and Geula

We have considered the triple structure of all of reality—from the spiritual to the physical, from the cosmic to the personal.

As we mention above, in the triple structure of "beginning, middle, end," the beginning establishes the purpose of the thing, the middle is the thing itself which carries out the purpose, and the end is the completion of the thing, having carried out its purpose.

The entirety of all existence begins with the Torah which preceded creation and which is the purpose of creation itself. The creation of the world in which the Torah is studied and observed, bringing about the refinement of the world and the revelation of G-dliness in it, is the means by which this purpose is carried out. The Geula, which is the completion of this process, is achieved when the world is completely refined so that G-d may be completely revealed in it. This is the ultimate triple thread.

These 3 stages are symbolized by the first three letters of the Hebrew alphabet, *Alef*, *Beis* and *Gimmel*: The entire Torah is contained in the Ten Commandments which begin with the Alef of

---

11. *Sefer HaSichos* 5751, 2:507–509.

Anochi ("I am the L-ord your G-d..."). The creation of the universe begins with the Beis of *Bereishis* ("In the beginning..."), and Geula begins with the letter Gimmel.[12]

The Geula of Moshiach is the ultimate purpose of creation.[13] This Geula is in fact the third Geula, the first being the Geula from Egypt by Moshe Rabbeinu after which the First Beis HaMikdosh was built and the second being the Geula from Babylon by Ezra in which the Second Beis HaMikdosh was built. In the Geula of Moshiach the Third Beis HaMikdosh will be built and the fact that it will be the third is of extreme significance, as we would expect. The Zohar says that the Third Beis HaMikdosh will last forever since it will be "the structure of G-d." This means that the Atzmus—G-d's essence—will be revealed in it. This will be discussed in more detail in the next section.

12. *Sefer HaSichos* 5752, 2:426–428.
13. *Tanya*, chap. 36.

The Messianic World Order

Classically, natural phenomena are described by a 2-valued logic. In the writings of Chassidus and Kabbala we find that the phenomena of the Messianic era are characterized by a paradoxical combined natural-and-miraculous world order. 2-valued logic is inadequate to describe such phenomena. Thus we suggest that a 3-valued logic be used as a model for these Messianic phenomena. First introduced in the 1920s, 3-valued logics are now widely studied in mathematics and have many applications. They have also been suggested as models for the paradoxical phenomena of quantum mechanics.

In this section we discuss the paradoxical nature of the Messianic world order and in the next section we describe 3-valued logic in detail.

In the classical view of the world any physical event is considered to be limited to a choice of two possibilities: either the event occurs or it doesn't. To take an example from Newtonian mechanics, the legendary apple either fell on his head or it did not. If one approaches a fork in the road, he either turns to the right or to the left. If an object is located in a given region of space, then it is not located in another region disjoint from the first. All of this is very "logical," of course. To be specific, this is called Two-Valued Logic (2VL) and is characterized by the "Law of Excluded Middle" which states that given a statement A, either it is true or its negation is true. In logical notation: $A \vee \neg A$ (read "A or not A").

Two-Valued Logic is generally adequate for the world as we now know it. However, it is known from the prophets, the Talmud and Midrash, and Kabbala and Chassidus that in the Messianic Era a different world order will prevail. For example, the prophet Isaiah talks about how the wild animals will be peaceful[1] and the Talmud talks about how trees will grow the same day they are planted[2] and loaves of bread and articles of clothing will come forth from the ground in *Eretz Yisroel*.[3] But in the extreme, the Messianic Era involves phenomena of a paradoxical nature as can be seen in the following Midrash which discusses Ahashverosh's attempt to satisfy both Mordechai and Haman:

> G-d said to him: "I don't satisfy the will of everyone, but you are trying to satisfy both of them .... If two ships leave the shore and one wants a north wind and one wants a south wind can one wind lead both of them simultaneously? It can only lead one or the other."
>
> ...In this world when the north wind blows, the south wind doesn't blow; and when the south wind blows, the north wind doesn't blow. But in the future, at the ingathering of the exiles, G-d says, "I will bring a multidirectional wind to the world that both of these ships can use."[4]

Thus we see that the classical 2VL must be extended to a system of logic in which a paradoxical statement such as "going north and going south" (simultaneously) is acceptable, i.e., a system in which the law of excluded middle does not hold. We emphasize that we want to extend rather than replace 2VL since we want the new system to describe the non-paradoxical phenomena of the Messianic Era the same way as always.

We are not concerned here with the "physics" of going north and going south simultaneously—that is part of the miraculous dimension of the Messianic Era. Our concern is to have a mechanism with which to incorporate such phenomena into our intellectual system, i.e., to be able to "talk about" such phenomena.

The Messianic World Order

The fact that there will be a new world order in the Messianic Era

1. Yeshaya chap. 11.
2. *Toras Kohanim* to VaYikra 26:4.
3. Kesubos 111b.
4. *Esther Rabba* 2:14, quoted in *Sefer HaSichos* 5752, 218.

can be seen from the sources quoted above as well as from many other sources. These sources have been analyzed extensively and explained in the sichos and ma'amorim of the Rebbe MHM, so these discourses will be our primary reference.

The Rebbe MHM's discussion of the world order in the Messianic Era is based on the interplay between the natural order and the miraculous order, and focuses on the Messianic prophecy of Micha, "As in the days of your exodus from Egypt, I will show you wonders,"[5] which refers to the wonders and miracles of the Messianic Era.[6]

G-d created the world with two world orders, a natural order deriving from a finite revelation from G-d, and a miraculous order deriving from an infinite revelation from G-d. Of course, relative to G-d Himself (the Atzmus), Who is above all limitations and is not restricted to either of the definitions of finite or infinite, these are both the same and there is just one world order. But the physical world was created in such a way that the natural order would prevail and the infinite revelation, the source of the miraculous order, would be hidden. Nevertheless, there is interplay between the two orders, as we now explain.

When a miracle occurs, it interferes with the prevalent natural order. This can happen in one of two ways:

(1) The miracle breaks the natural order. These are the obvious open miracles such as the splitting of the Red Sea or the oil of the menorah burning for eight days.

(2) The miracle is hidden within nature. This is characterized by the fact that we don't see any unnatural event. Natural events occur but the manner in which they occur is unnatural. According to the laws of nature there is no way that nature could conduct itself in such a manner. The classic example of this is the miracle of Purim where one could see that even though the events were natural, the causes of the events were above nature. This type of miracle has a natural aspect to it.

The miracles of the second type are actually more powerful. Put simply, it is a greater act to perform a miracle under the disguise of nature

---

5. Micha 7:15.

6. The main references for this discussion are *Likutei Sichos* 17:152–155; *Sefer HaMa'amorim Melukat* 4:223–234 and 5:303–312.

than simply to break nature, i.e., to interrupt the natural order and let the miraculous order take over. Thus, the Talmud attributes this type of miracle to G-d Himself, quoting the verse, "He does great wonders by Himself."[7]

We find the reverse also—that nature itself has a miraculous aspect to it. When one considers the fact that the order of nature itself doesn't change, that it remains constant throughout the millennia, with the sun and the stars maintaining their orbits around the earth, etc., one realizes that the system is being maintained by an infinite energy source since a finite energy would show weakness after awhile. Yet, a finite body cannot contain within it infinite energy,[8] so it must be that the finite natural order has in it an infinite infusion of energy from G-d. As the Rebbe MHM wrote in a discourse:

> By contemplating the fact that the order of nature is constant with out change ... and from the fact that throughout time the stability of the creations has not weakened ... through this we know the greatness of the Creator—that His power (from which energy comes into the world) is infinite and unlimited. In spite of the fact that the nature of the creations derives from a limited life force from G-d ... this regards only nature itself, but the constancy which is in the order of nature and which does not change or weaken, derives from the infinite unlimited power of G-d .... This shows that even the finite life of the world is connected with the infinite [revelation of G-d]. [9]

This observation has not escaped modern science. On the contrary, in modern science this observation has been made at the very foundations of matter—in the structure of the atom. The story begins when physicists first discovered the structure of the atom in the early part of the 20th century. They were confronted by a profound mystery. The mystery, first described by the famous Jewish physicist Neils Bohr, was stated by the physicist Paul Ehrenfest in 1931 as follows: "We take a piece of metal or a stone.

7. Tehillim 136:4.

8. This principle, stated by Rav Saadia Gaon (*Emunos V'Deos*, part 1, chap. 1) and by Rambam (*Guide to the Perplexed*, part 2, introduction 12) is also a basic principle in quantum mechanics.

9. *Sefer HaMa'amorim Melukat* 4:228.

When we think about it, we are astonished that this quantity of matter should occupy so large a volume. Admittedly, the molecules are packed tightly together, and likewise the atoms within each molecule. But why are the atoms themselves so big ...?"[10]

The question became even more dramatic than the size of the atoms. Given the strong attraction between the positively charged nucleus of the atom, due to the protons, and the negatively charged electrons around the nucleus, Ehrenfest asked, "What prevents the atom from collapsing in this way?"

The question was, how does the atom remain stable? What is it that maintains the structure of the atom? The implication of this simple question was far-reaching: If every atom would collapse, there would be no matter in the universe! In 1925 the physicist Wolfgang Pauli, inspired by the work of Bohr, formulated a principle that stated that the electrons had to remain in their orbits in a certain manner.[11] But this principle, which came to be known as the Pauli Exclusion Principle, merely stated the fact. It provided no explanation of the phenomenon based on the laws of physics. The problem came to be recognized as a fundamental problem of modern physics.

Theoretical physicists, who wanted a mathematically complete answer, attempted to derive this result from the fundamental equations of quantum mechanics, but got nowhere. Then, in the 1960s, mathematical physicist Freeman Dyson began a more intense research effort. More recently, this research has been led by mathematical physicist Elliot Lieb of Princeton University. After working on the problem for 28 years, and considering cases where matter does, in fact, collapse (as in a very strong gravitational field), he declared:

"When you consider all of the physical forces at work in matter, even in something as simple as a glass of water, it's sort of miraculous that everything doesn't just collapse and then, releasing huge amounts of energy, blow up."[12]

It was G-d's intention when creating the world that it would ulti-

10. Quoted by Dyson and Lenard in *Journal of Mathematical Physics* (August, 1967).

11. See, e.g., Robert Eisberg and Robert Resnick, *Quantum Physics of Atoms*, 2nd ed. (New York: Wiley, 1985), 309.

12. Richard Lipkin, "Stability of Matter," *Science News* 148 (14 October 1995):

mately function under a world order where the natural and the miraculous, the finite and the infinite, both function simultaneously. Yet the physical world in its original state was too grossly material to be able to handle such a paradoxical state. At any point in time one or the other order would prevail (a two-valued logic). It was left to the Jewish people to refine the physical world throughout history, by the observance of Torah and mitzvos (which causes the Atzmus to be revealed in this world), to a state of spirituality where the world could function under such a unified world order, as G-d had originally intended. This state of refinement will be attained in the Messianic Era when G-d (the Atzmus) will be fully revealed, as Isaiah says, "The glory of G-d will be revealed and all flesh will see it."[13] The infinite glory of G-d—the Atzmus that maintains the existence of the world—will be visible throughout the finite universe.[14]

Impossible of the Impossibilities

Rabbi Shlomo ben Aderet—the Rashba—was a leading Rabbinic authority of the 13th century. In his many responsa he deals with a large number of halachic and philosophical issues. In one of them, he uses the term "impossible of the impossibilities" to describe the ability of G-d to do things that are not only physically impossible according to the laws of nature (such as turning water into blood), but also logically impossibl—contradictory—such as doing an act and, at the same time, not doing it.[15] Such a contradiction was present in the Holy of Holies of the Beis HaMikdosh where the ark was found not to take up any space[16] even though it certainly took up space since it had specific physical dimensions.

As we said, such an order of things will be part of the world order in the Messianic Era. We mention here briefly three areas in which we find specific mention of this: 1) the Beis HaMikdosh, 2) the Torah, and 3) Melech HaMoshiach.

(1) *The Beis HaMikdosh*

The Zohar says that the Third Beis HaMikdosh will last forever since

13. Yeshaya 40:5.
14. *Sefer HaSichos* 5751, 587.
15. *Teshuvos HaRashba*, no. 418.
16. Yoma 21a.

it will be "the structure of G-d." The Rebbe MHM explains that the First Beis HaMikdosh didn't last forever because it involved a great revelation of G-dliness, which was not completely connected with the physical world and therefore couldn't last. In the time of the Second Beis HaMikdosh the Jews were repenting from the sins committed during the First Beis HaMikdosh. This involved a refinement and elevation of the lower, physical world so the revelation of G-dliness was more strongly connected to the lower world and the Second Beis HaMikdosh lasted longer. However, the world itself, being defined in terms of finite time and space, is limited, so the Second Beis HaMikdosh didn't last forever either. But in the Third Beis HaMikdosh, the finite limitation of the world will combine with the infinite revelation from G-d since, as the Zohar says, the Third Beis HaMikdosh will be the structure of G-d. This means that the Atzmus will be revealed in it and "the Atzmus defies all description—"negation of the finite and negation of the infinite." Thus the limitation of the lower world and the infinity of above become one entity.[17]

(2) *The Torah*

It is well known from the Talmud that in the controversies between the schools Hillel and Shamai the halacha is decided according to Hillel. However, this is not the end of the story. There is a mishneh in Pirkei Avos which states that every controversy that is for the sake of heaven will endure. As an example, the mishneh mentions the controversy between Hillel and Shamai. The question is raised that since the law was decided according to Hillel, how does the controversy endure? Only the opinion of Hillel endures, but not Shamai's. In Likutei Torah, the Alter Rebbe quotes the Ari Zal, Rabbi Yitzchok Luria, who answers this question by explaining that in the future, in the Era of Moshiach, the halacha will be according to Shamai.[18]

But the Rebbe MHM notes that this does not yet answer the question because when the law is decided in the Era of Moshiach according to Shamai, then the opinion of Hillel will not endure. Thus we are back to the original question. But, the Rebbe MHM explains, there will come a time, later on in the Era of Moshiach, when the halacha

17. *Shaarei Geula* 2:115–119.
18. *Likutei Torah*, Bamidbar 54b.

will be like both Shamai and Hillel simultaneously. Thus their controversy will endure because both of their opinions will endure.

How is it possible that the law will be like both opinions when in fact they contradict each other and are thus mutually exclusive? The Rebbe MHM explains that from the perspective of G-d, the opposite opinions coexist simultaneously: "He contains both opinions—the positive and the negative and both of them together .... As far as G-d is concerned, they can come to be revealed in the reality of the world just as they are relative to G-d's essence (Atzmus)—positive and negative together. This is because He is the impossible of the impossibilities, above the limitations of the impossible and the contradictory."[19]

(3) *Melech HaMoshiach*

In chapter I, we discussed various ways in which Melech HaMoshiach embodies opposite characteristics within himself, enabling him to unify opposing forces in the world: "The innovation of Moshiach is that through him the two concepts of 'above' and 'below' will be united."[20] Thus "A spirit of G-d will rest upon him [Moshiach], a spirit of wisdom and understanding, a spirit of counsel and power, a spirit of knowledge and fear of G-d.... And the wolf shall dwell with the lamb and the leopard shall lie with the young goat... And the nursing child will amuse himself over the hole of a poisonous snake; and into the hole of a venomous snake shall a weaned child stretch his hand."[21] This unification of opposites is a fundamental characteristic of Melech HaMoshiach which derives from his life source which is Atzmus, the essence of G-d.[22]

Thus, the life of the Rebbe Melech HaMoshiach himself transcends the finite and the infinite. There is a verse in Tehillim (Psalms) relating the request for life that Moshiach asks from G-d, and G-d's response granting him life forever: "Life he asked from You. You granted it to him—length of days forever." Here, too, we would expect that this "length of days forever" is associated with the essence of G-d, the Atzmus. This point is discussed by Rabbi DovBer of Lubavitch (the Mittler Rebbe, the second Rebbe of Chabad) who contrasts it with the life of King David of whom a similar—but not identical—statement is

19. "Kuntres Halachos shel Torah she-B'al Peh" in *Sefer HaSichos* 5752, 33–34.
20. *Sefer HaMa'amorim Melukat* 2:297.
21. Yeshaya 11:2–8.
22. See chap. I, sec. 4.
23. Rosh HaShana 25a.

made: "David, King of Israel, is alive and endures."[23]

The Mittler Rebbe discusses the difference between the "length of days forever" of Moshiach and the "alive and endures" of King David:

> Regarding Moshiach it says, "Life ...You granted it to him" which means a gift with no interruption. Regarding David—his life came only from the ten *sefiros* of *Atzilus*, but regarding Moshiach it says, "'You granted it to him, which means from Atzmus, the infinite." (Addressing G-d in second person—"You"—refers to G-d Himself, i.e., His essence, Atzmus.)[24]

It is significant that the Rebbe MHM spoke of his life from Atzmus on the eve of Rosh HaShana 5752, the Rosh HaShana before his stroke. He said, "Regarding Moshiach—in addition to the revelation of the name of Moshiach...we also have the Atzmus of Moshiach Tzidkeinu which is one thing with the infinite Atzmus."[25]

A life that comes from Atzmus is above all definitions and limitations of natural life and death as we know them. As Rabbi Shalom DovBer of Lubavitch (the fifth Rebbe of Chabad) says, "Relative to Atzmus, life and death are equal."[26] In fact, as mentioned above, Atzmus transcends all limitations. Thus, the miraculous and the natural are equal relative to Atzmus and can coexist. It follows that the life of Moshiach is completely above the laws of nature. As the Maharal of Prague writes, "Moshiach will be ...outside of nature and the way of the world."[27]

A Prayer for Moshiach

My father, HaRav Moshe DovBer Silman z.l. was a Rabbi in Minnesota for twenty five years. He passed away in 1972 from ALS which, in the later stages, made speaking very difficult. One day, a number of weeks before he passed away, he called me over urgently and asked me to write down a *drasha* which he was going to dictate.[28]

---

24. *Ma'amorei Admur Ha'Emtzai,* Devorim 1:214–15.

25. *Sefer HaSichos* 5752, 6–7. See note 49 for a reference to the Tzemach Tzedek, who discusses the life of Moshiach originating in Atzmus in a manner similar to that of the Mittler Rebbe in the previous footnote.

26. *Sefer HaMa'amorim 5670*, 100.

27. Quoted in *Shaarei Geula* 2:69 n. 29.

28. He also told me to publish it, which I did in the local Jewish newspaper in Minnesota (in a somewhat different version) for his 10th yartzeit, in Beis Moshiach Magazine for his 26th yartzeit and now in this book.

He raised the following question: At the end of the *tefilas tal* (prayer for dew on *Pesach*) and *tefilas geshem* (prayer for rain on *Shmini Atzeres*) we say, "For a blessing and not for a curse; for life and not for death; for satisfaction and not for deficiency." Why do we say it this way? Why don't we just say, "For a blessing; for life; for satisfaction?" Won't that itself imply that it should not be for a curse etc.?

The explanation, he said, is that nothing in this world is perfect. So when we have life there may be an element of death in it; in a blessing there may also be a negative element and in satisfaction or plenty there may be an aspect of deficiency. So we are asking for a complete blessing with no negative element, for life without death and for satisfaction with no deficiency.

That was the essence of his *drasha* as I understood it. As I think about it now, I realize that there is support for this explanation in the *tefilas tal* itself. For a few lines earlier we say, "*B'mashmaneinu al yehi razon* (in our abundance there should be no deficiency)."

But this state of affairs cannot be attained in the time of Golus since, as we said above, there is no perfect good in this world. Life ends in death. A blessing may have an aspect of a curse.[29] And in every satisfaction there is something missing, either physically or spiritually.

Indeed, the Rebbe MHM expresses this very concept in the sichos. He says that even after we eat a full Shabbos dinner with meat, fish and wine, we are still not satisfied because we didn't eat the *shor habar* and the *leviyasan* nor drink the *yayin hamishumar* (the meat, fish, and wine of a special feast the G-d makes in the Era of Moshiach).

It is only in the Messianic world order that we attain the full blessing of the true and complete Geula, everlasting life with no interruption, and the full spiritual and physical satisfaction of eating the *shor habar* and the *leviyasan* and drinking the *yayin hamishumar*.

Thus, according to my father's explanation, the prayer of "For a blessing and not for a curse; for life and not for death; for satisfaction and not for deficiency" is a prayer for Moshiach.

---

29. On a deeper level, the Alter Rebbe explains that the curses mentioned in the Torah are actually blessings from a very high level which may not be able to be actualized in this world in a positive way and must be transformed into blessings. See Likutei Torah, VaYikra, 48b, Sefer HaSichos 5751, 2:768, HaYom Yom of 17 Elul. The Mittler Rebbe said that when his father, the Alter Rebbe, read the curses in the Torah in Parshas Ki Savo, he heard only blessings.

Three-Valued Logic

"In G-dly intellect there are arguments of positive and negative, but the positive and negative of the G-dly intellect are not opposites. In human intellect, one of these paths is false, but in G-dly intellect, both the positive and negative are true. They are two branches of light, two rays of light from one source, each illuminating a part of the concept."[1]

—Rabbi Yosef Yitzchok Schneerson, the Previous Lubavitcher Rebbe

Shortly after the fountains of Wisdom from Below broke open in 1840, a new branch of mathematics was revealed which ultimately provided the structure to help us understand the phenomena of the Messianic Era. In the year 1847—the year that the Alter Rebbe's *Likutei Torah* was published—a British mathematician named George Boole published a paper explaining that logic has an algebraic structure. Thus began the development of a new branch of mathematics known as mathematical logic. Around this time, a mathematician named Augustus De Morgan, who was working on mathematical logic, announced that he felt that a third "truth value" should be added to the structure of logic. In other words, in addition to the two truth values "true" and "false" (i.e., the characterization of statements as being either true or false) with which classical logic works, there should be a third truth value characterizing a third state. However, he did not know exactly how to develop this idea and it remained a vague concept.[2]

1. *Likutei Diburim*, sicha of Simchas Torah, 5630 (1929), sec. 29.

2. See Richard Epstein, *The Semantic Foundations of Logic—Propositional Logics*, 2nd ed. (New York: Oxford University Press, 1995), chap. 8. This is our primary reference on 3VL.

Beginning in the late 1800's a few attempts were made to develop the concept, especially by the logician Charles Pierce who developed what he called a "triadic logic." He expressed his motivation in a letter in which he wrote, "I have long felt that it is a serious defect in existing logic that it takes no heed of the limit between two realms...In every field of thought whatsoever there is an intermediate ground between positive assertion and positive negation."[3]

Then, in the early 1920s, Jan Lukasiewicz and Emil L. Post developed this concept in a very precise mathematical way. The classical truth values "true" and "false" were represented by the numbers 1 and 0, respectively, and the third truth value, which could represent "true" and "false" simultaneously, was represented by ½. This new system retained the ability to represent all the information that classical logic did while having the new capability of representing as well the middle ground—coexistence of opposites, contradictions and paradoxes. It became known as Three-Valued Logic (3VL).

Earlier in this book, when we discussed the axiomatic method,[4] we pointed out that new systems in mathematics and science are usually discovered by challenging the axioms of the old systems. For example, non-Euclidean geometries were discovered by rejecting Euclid's parallel postulate. Here too, as we will see, 3VL, a non-Aristotelian logic, was discovered by rejecting the law of excluded middle (which asserts that every statement is either true or false).

It is interesting to note that in the 1920s, the new quantum theory was being developed and in the following years some mathematicians and physicists, such as the great mathematical physicist John von Neumann, Hans Reichenbach and Garrett Birkhoff, developed three-valued logics to deal with the paradoxical nature of quantum mechanics.[5]

Here we shall discuss the original 3VL of Lukasiewicz, which is the best known. Our emphasis will be on the comparison between 3VL and 2VL, i.e., to show exactly how 3VL extends 2VL.[6]

---

3. See Nicholas Rescher, *Many-Valued Logic* (New York: McGraw-Hill, 1969), 4–7.

4. Chap. I, sec. 3.

5. Leonard Bolc and Piotr Borowik, *Many-Valued Logics*, vol. 1, *Theoretical Foundations* (Berlin: Springer Verlag, 1992), 24, 75.

6. Standard references that discuss all of these 3VLs in detail are Rescher, *Many-Valued Logic* and, for a more algebraic treatment, Bolc and Borowik, *Many-Valued Logics*,

How is a form of logic presented? Most branches of modern mathematics are presented in terms of mathematical logic and set theory, but how should mathematical logic itself be presented? It is usually presented in terms of a *formal language*.[7] A formal language *L* is a countable collection of symbols satisfying the following conditions:

(1) There is a subset *F* of *L* called the formulas of *L*. An effective procedure is given to determine if an expression (a finite sequence of symbols of *L*) is a formula.
(2) There is a set of formulas called the Axioms of *L*.
(3) There is a finite set of rules of inference and an effective procedure to determine if a given formula follows from a set of formulas under the rules of inference.

Formal Language for 2VL
(1) The symbols are the variables A_1, A_2, ..., B_1, B_2, ... , parentheses (,), and logical connectives:

¬	"not"
∧	"and"
∨	"or"
→	"implies"

(2) Formulas: All variables are formulas. If A and B are formulas then ¬A, (A∧B), (A∨B) and (A→B) are formulas. The set F is the smallest set containing the variables, which is closed under formula formation.

(3) The axioms are:

1) $A \to (B \to A)$
2) $(A \to (B \to C)) \to ((A \to B) \to (A \to C))$
3) $(\neg B \to \neg A) \to ((\neg B \to A) \to B)$

Axiom 3 prevents contradictions from occurring since it says that if ¬B implies both A and ¬A, a contradiction, then ¬B is rejected and we conclude its opposite, B.

(4) Rule of Inference: modus ponens, also known as the detachment rule: From A and A→B, we conclude B.

---

7. Elliot Mendelson, *Introduction to Mathematical Logic*, 4th ed. (London: Chapman & Hall, 1997), 34–35.

Formal Language for 3VL

(1-2) The symbols and the formulas are the same as for 2VL.

(3) The axioms are:

1) $A \rightarrow (B \rightarrow A)$

2) $(A \rightarrow B) \rightarrow ((B \rightarrow C) \rightarrow (A \rightarrow C))$

3) $(\neg A \rightarrow \neg B) \rightarrow (B \rightarrow A)$

4) $(((A \rightarrow \neg A) \rightarrow A) \rightarrow A)$

Axiom 4 accepts contradictions.

(4) Rule of Inference: modus ponens, as for 2VL.

The axioms for Lukasiewicz's 3VL were not developed by Lukasiewicz himself, but by Mordechaj Wajsberg in 1931.

Now we must discuss the semantics and the syntax of 2VL and 3VL. The semantics are associated with determining what the theorems are, and the syntax is associated with determining what the tautologies are. The concept of *theorem* is the same for both 2VL and 3VL: We define a *proof* to be a sequence of formulas $A_1, \ldots, A_n$ where each A_i is either an axiom or follows from preceding formulas by the rule of inference. A *theorem* is the last formula in a proof.

The notion of truth is introduced using a *valuation function*.[8] The valuation functions, also known as truth assignments, will be different in 2VL and in 3VL since the number of truth values is different. In 2VL, a valuation function is a function v mapping the set of variables S to the set $\{0,1\}$,

$$v\colon S \rightarrow \{0, 1\}$$

0 and 1 are called the *truth values*, 0 representing "false" and 1 representing "true."

Any valuation function on S can be extended uniquely to a valuation function V on F, the set of formulas:

$$V\colon F \rightarrow \{0, 1\}$$

8. For a basic discussion of valuation functions, see Herbert Enderton, *A Mathematical Introduction to Logic*, 2nd ed. (New York: Academic Press, 2001), 20–23. For a more extensive discussion, see Richard Kaye, *The Mathematics of Logic* (Cambridge: Cambridge University Press, 2007), chap. 7.

The well known "truth tables" provide an effective procedure for extending a valuation function on the variables to a valuation function on the formulas:

Truth Tables for 2VL

A	¬A
1	0
0	1

A	B	A∧B
1	1	1
1	0	0
0	1	0
0	0	0

A	B	A ∨B
1	1	1
1	0	1
0	1	1
0	0	0

A	B	A→B
1	1	1
1	0	0
0	1	1
0	0	1

The situation for 3VL is similar. Here, however, a valuation function is a function

$$v: S \rightarrow \{0, ½, 1\}$$

from the set of variables to a set of three truth values. 0 and 1 represent "false" and "true" as before and the value ½ can be used to represent something which is not known (or knowable) to be either true or false, or even something which is considered to be both true and false simultaneously, i.e., a contradiction!

The surprising thing is that such a language is a smooth extension of 2VL as the truth tables for 3VL show:

Truth Tables for 3VL

¬A	
A	¬A
1	0
½	½
0	1

A∧B				
A	B:	1	½	0
1		1	½	0
½		½	½	0
0		0	0	0

A∨B				
A	B:	1	½	0
1		1	1	1
½		1	½	½
0		1	½	0

A→B				
A	B:	1	½	0
1		1	½	0
½		1	1	½
0		1	1	1

The final point in this development is to talk about the tautologies. A *tautology* is a formula A such that *V*(*A*) = 1 for every function V on the variables of A, i.e. a formula which is always true. The most famous tautology of 2VL is A∨¬A, the law of excluded middle. But A∨ ¬A is not a tautology in 3VL as the following truth table shows:

A	¬A	A∨¬A
1	0	1
½	½	½
0	1	1

When A has the value ½, A∨ ¬A also has the value ½ in which ca the formula is not "true."

What type of formula would be a tautology in 3VL? Not surprisin ly, it would be a formula involving a contradiction. So we will demo strate a formula involving a contradiction that will always be true 3VL.[9] Consider first the formula:

$$(A \rightarrow \neg A) \wedge (\neg A \rightarrow A).$$

This means that "A implies the negation of A, and the negation A implies A," or, in the terminology of logic, "A is true if and only the negation of A is true," a very contradictory statement. Yet the f lowing truth table shows that when A has the value ½, this stateme has the truth value 1, meaning "true."

A	¬A	A → ¬A	¬A → A	(A → ¬A) ∧ (¬A → A)
1	0	0	1	0
½	½	1	1	1
0	1	1	0	0

Now, let's symbolize this contradictory statement by the symbol (for contradiction). We now show that there is a tautology in t three valued system, which is built on C. The statement A ∨ (¬A) ∨ is such a tautology as shown in the next truth table:

A	¬A	C	A∨¬A∨C
1	0	0	1
½	½	1	1
0	1	0	1

9. Epstein, *Semantic Foundations of Logic*, 320–21.

We already saw that the statement A∨¬A, a tautology of 2VL, is not a tautology in 3VL. Now we see that by combining it with the contradictory statement C, we have a tautology, a statement that is always true! This tautology says that in 3VL it is always true that either A or the negation of A or the contradictory statement C is true. It is sometimes called the "Law of Excluded Fourth."

Equivalence classes of tautologies generate algebras. In 2VL they generate Boolean Algebras and in 3VL they generate Kleene Algebras[10] but we will not pursue this here. We mention only that the Post Algebras, associated with the Many-Valued Logic proposed by E. Post, were defined by our friend, the well known Lubavitcher chosid and mathematician, Professor Pesach (Paul) Rosenbloom in 1942.[11]

At this point we must ask ourselves, "What does Three-Valued Logic really mean?" This is probably best answered by a story that is told of two people who once came to a Rabbi to settle a dispute. The Rabbi listened to the first person and said, "You're right." Then the second person spoke. The Rabbi listened to him and then said to him, "You're right." At this point the student sitting with the Rabbi spoke up and said, "Excuse me, esteemed Rabbi. You just said to the first person that he's right. Then you said to the second person—who said the opposite—that he's right. How can they both be right?" To this the Rabbi responded, "You're also right!"

In the physical world we find many applications of 3VL. The Structured Query Language (SQL) used to query relational databases is probably the best known example.[12] An entire book, volume 2 of Bolc and Borowik's *Many Valued Logics*, is devoted to applications of 3VL. In addition, there is a fascinating phenomenon in the physical world itself which shows that now 3VL is essential to modeling the world. In 1996, a team of physicists working with Dr. C. Monroe at the National Institute of Standards and Technology in Boulder, Colorado, succeeded in getting an entire atom to exist simultaneously in two widely separated places. As reported in the New York Times "The real significance of the Institute's feat," Dr. Monroe said in an

10. H. T. Nguyen and E. A. Walker, *Fuzzy Logic* (London: Chapman & Hall, 2000) 63–67.

11. Bolc and Borowik, *Many-Valued Logics*, 46.

12. D. Simovici and R. Tenney, *Relational Database Systems* (New York: Academic Press, 1995), chap. 2.

interview, "is that the two states of the same atom were not only pulled apart but were separated by a relatively enormous distance—a distance large enough to represent a transition from the domain of quantum mechanics to the everyday world ...."[13]

In conclusion, we propose that 3VL is the proper logic for the Era of Moshiach. It maintains the features of 2VL so it can be used for non-paradoxical phenomena. But it also extends 2VL so it can be used to model paradoxical phenomena. It is essential in modeling certain physical phenomena (recall that von Neumann and Reichenbach used it for quantum mechanics) and it is certainly the logic of choice for Messianic phenomena.

13. "Physicists Put Atom in Two Places at Once," *New York Times*, 28 May 1996, Science Desk, 1.

The Concept of Infinity in Mathematics and Chassidus—*Mathematical Infinity*

Dr. Tsvi Yehuda Saks

Dr. Tsvi Saks was one of the most esteemed members of the Rabbi Yisroel Aryeh Leib Institute. Over the years, Dr. Saks had been a professor of Mathematics at the University of Costa Rica and at Daemen College, Amherst, NY. His field of research was topology, specializing in abstract infinite space. More recently, he was a researcher into the applications of Artificial Intelligence, and worked at IBM. He passed away on 14 MarCheshvan, 5766 (2005)

This section was compiled based on several lectures that Dr. Saks had given at the RYAL Institute's annual Moshiach and Science conferences from 5753 and on. In it, Dr. Saks discusses the concept of infinity as it is currently understood in mathematics and as it is explained in Chassidus.

The theme of this conference is "Bringing Moshiach into the World of Science."[1] It sounds to me like it's a pattern of "from above to below"—bringing G-dliness and Moshiach down into the world of science, which is what I want to do. I also want to try to do the reverse, or the dual, of this—bringing science into the world of Moshiach and G-dliness. I want to do this with some very concrete examples.

1. This was the theme of the 1993 conference.

My first career, before I began working in Artificial Intelligence, was doing research in pure mathematics—Mathematical Infinity. That is the science that I'm going to talk about today.

I'm going to actually discuss some reasonably technical mathematics. Hopefully, it will be accessible to everyone in the audience. If it isn't, just relax because I will not do it forever. I'll come back and relate it to the Chassidus.

Science and the Era of Moshiach

Before I begin, I should mention a sicha of the Rebbe MHM which has served as the inspiration for my research into the area of mathematical infinity and Chassidus.[2]

The Rebbe MHM writes that the Zohar had predicted that, starting in 1840, there will be great advances in the secular sciences and Chassidus in order to ready the world for the Messianic Era. He states that the deepest level of positive interaction between secular knowledge (science and mathematics) and Chassidus occurs when secular knowledge is used to explain and illuminate deep concepts in Chassidus.

The Rebbe MHM asks: Why is scientific advance related to the coming of Moshiach, and why does the Zohar link advances in Chassidus and science to the coming of Moshiach? This can be understood from three perspectives:

(1) Advances in science enable us to visualize Torah concepts and understand them more deeply. For example, the telephone and radio, which enable us to hear events all over the world, provide us with palpable models that enable us to visualize the concept of "an Eye that sees and an Ear that hears." This makes the concept of Divine scrutiny of our deeds more real to us. Utilizing this knowledge for the service of G-d provides us with a certain glimpse of the level of perception that will be attained in the Era of the Redemption: "And all flesh will see that the mouth of G-d speaks," that is, all flesh will then enjoy perceptions of Divine service with physical, sensory vision.

---

2. This is the sicha of Parshas Noach, discussed in the Introduction and in sec. 1 of this chapter.

(2) When technology (the radio, television and internet, for example) is used to disseminate Torah knowledge worldwide, it pre-echoes the universal diffusion of knowledge in future when "...the earth will be filled with the knowledge of G-d, as the waters cover the ocean bed." Moreover, it foreshadows the promise of the above-quoted verse, that "all flesh will see ...," for the image and sound is seen and heard simultaneously around the world. The electronic waves literally fill the earth and its atmosphere with the knowledge of G-d.

The Rebbe MHM says, however, that these examples merely *utilize* scientific applications and technology. However, the prophecy of the Zohar speaks of the *knowledge* of science interacting with Torah and Chassidus.

(3) The true synthesis of Chassidus and science occurs when the knowledge of science is used to explain, support and illuminate Chassidic concepts.[3] In this way, science is contributing to our ability to perceive G-dliness within the created universe, which we will be able to do completely when Moshiach comes. The Rebbe MHM brings the concept of unity as an example, that the advance of scientific knowledge and understanding is increasingly revealing the inherent unity in the universe, as expressed in the forces of nature.

With G-d's help, and the Rebbe MHM's inspiration, we will follow this program for the concept of infinity, that is, to use the theory of mathematical infinity to explain and illuminate the concept of infinity as discussed in Torah and Chassidus.

The concept of infinity is used in Torah and Chassidus in several fundamentally different ways. First, G-d is referred to as *Ain Sof* (literally, Without End, or Infinite). On the other hand, the Torah has revealed the fact that the creation is quantitatively infinite. For example, G-d created infinitely many troops of hosts that serve Him[4] and infinitely many spiritual worlds.[5] Now surely the concept of infinity as it measures and describes the creation is vastly inferior to the con-

---

3. More precisely, it says in the sicha that "the true unity between the secular knowledge and Torah occurs when one sees ideas from the deeper levels of the Torah in the sciences themselves."

4. Chagiga 13b.

5. *Sefer HaMa'amorim 5710*, 132.

cept of infinity as it applies to G-d Himself. In fact, all of G-d's infinitely many troops are considered as absolutely nothing before Him. These insights lead to this one question: How can infinity be limited, and how can there be different levels of infinity?

Introduction to Mathematical Infinity

As a brief introduction to the subject of mathematical infinity, let me pose a very simple question. Suppose you are making a *melave malka* (dinner following Shabbos). Everybody has to have bread for the melave malka, so you make sandwiches. You really don't know how many people are coming and you don't really know how many sandwiches you are making. But you make a rule that everybody takes exactly one sandwich. At the end of the melave malka there are no sandwiches left. Everybody took exactly one sandwich. Now what inference can we make? We know that everyone took exactly one sandwich and every sandwich was taken by exactly one person. Do we know how many people came to the party? We certainly do not. There is something that we do know, however. That is, we know that the number of people that came to the melave malka and the number of sandwiches are the same. Because we have what is called a "one-to-one correspondence."

This notion of one-to-one correspondence was used by Georg Cantor in the 1870s to formulate the modern theory of mathematical infinity. The basic definition is that two sets of entities have the same number of elements if there is a one-to-one correspondence between the entities of the two sets, like in the example of the people and the sandwiches. For finite sets, this definition conforms to what we would expect, but for infinite sets, there are many surprises, some of which I would like to discuss.

The most important and simplest infinite set is the set of natural numbers, i.e., the positive integers, 1, 2, 3, ... The '...' means ad infinitum.

The way we generate the natural numbers is to start out with 1. Then there is a principle or a rule that allows us to generate new numbers from the old numbers. The rule is that you can always add 1 to a number. So you start out with 1, then add 1 to it and you get 2. Now you add 1 to the 2 you get 3, and so on. For any number n, you add 1 to it and you get $n + 1$. What I want to do, though, is to utilize the concept of one-to-one correspondence to show you a very strange example involving mathematical infinity.

First, consider the set of squares 1, 4, 9,... It is obviously smaller than the set of positive integers, because the set of squares is part of (a subset of) the set of positive integers, but with many numbers—all of the non-squares like 2, 3, 5 etc.—left out. (See Diagram 1.)

Positive Integers

1	2	3	4	5	6	7	8	9	...	N	...
↓			↓					↓		↓	
1			4					9		N^2	

Squares

Diagram 1

Now, let's set up a correspondence so that each number is associated with its square. The number 1 will correspond to 1, 2 will correspond to 4, 3 to 9, ... In general, n corresponds to n^2. What we now have is a one-to-one correspondence between the sequence "1, 2, 3 ..." and the sequence "1, 4, 9, ..." (See Diagram 2.)

1	2	3	4	5	...	N	...
↓	↓	↓	↓	↓		↓	
1	4	9	16	25	...	N^2	...

Diagram 2

It is clear from the diagram that the set of squares is the same size as the set of positive integers since they can be matched up in a one-to-one correspondence.

Each positive integer has exactly one square, and each square has exactly one positive integer which is its square root, very similar to the people and the sandwiches in our earlier example. Thus the set of positive integers and the set of squares have the same number of elements, which should surprise you, since there seem to be far fewer squares. In fact, this is characteristic of infinite sets, namely, any infinite set has the same number of elements (via a one-to-one correspondence) as some of its subsets which are distinctly smaller than itself.

The one-to-one correspondence between the positive integers and the squares was first observed by Galileo in the early 1600s. Galileo was quite confused by his discovery, and, therefore, he rejected the notion of infinite numbers. Cantor, on the other hand, explored the notion of one-to-one correspondence with fresh insight, and in the 1870s, developed the theory of infinite numbers. Note that over 250 years passed from the initial discovery to development of a formal, coherent theory, which is an enormous length of time, especially considering the vast scientific progress during this period. It is noteworthy that the Theory of Infinite Numbers was developed after the date of 1840 which was specified by the Zohar.

Before Cantor, mathematical infinity was regarded as "infinity as a potential." For example, the set of positive integers 1, 2, 3 ... is infinite because there is no last number. But it was infinity as a potential because there was no perspective or claim that the set of all the positive integers actually exists as a complete object. Cantor's basic contribution was that mathematically infinite objects, such as the set of positive integers, can be considered to be well-defined objects that actually exist and can be manipulated in many of the same ways as finite objects.[6]

Cantor made a major breakthrough when he showed that not all infinite sets have the same number of elements, by proving that the set of positive integers has fewer elements than the set of all numbers between 0 and 1. He used the symbol $\aleph_0$ to represent the positive integers, and the letter c (for continuum) to represent the set of numbers between 0 and 1, and he was able to prove that $\aleph_0$ is less than c.

Remember our example of the people and the sandwiches. If every person had taken exactly one sandwich, and there were sandwiches left over, then we would have concluded that there were more sandwiches than people. Similarly, Cantor explored all possible one-to-one correspondences between the set of positive integers and the set of numbers between 0 and 1. He proved that for any such correspondence, there is some number between 0 and 1 which is not paired with any positive integer. Thus, the assumption that there is a one-to-one correspondence between the positive integers and the set of numbers between 0 and 1 is false, and we must say that there are more numbers between 0 and 1 than there are positive integers. Here is the proof:

6. See Rudy Rucker, *Infinity and the Mind* (Princeton, NJ: Princeton University Press, 2004) for an extensive discussion of many aspects of mathematical infinity.

Suppose the following is an arbitrary one-to-one correspondence between the set of positive integers and the set of number between 0 and 1, expressed as decimal expansions.

1	→	X_1	=	$.X_{11}$	X_{12}	...	X_{1n}	...
2	→	X_2	=	$.X_{21}$	X_{22}	...	X_{2n}	...
⋮								
N	→	X_n	=	$.X_{n1}$	X_{n2}	...	X_{nn}	...
⋮								

Diagram 3

Cantor now defines a number between 0 and 1 which is not paired with any positive integer. Let's call this number $j = .j_1 j_2 \cdots j_n \cdots$. If x_{11} is not 4, we will define j_1 to be 4; if x_{11} is 4, then we will define j_1 to be 2. Note that we already know that j is not equal to x_1, since they differ in the first decimal place.

For every positive integer *n*, we similarly define j_n. If x_{1n} is not 4, we define j_n to be 4; if x_{1n} is 4, then we define j_n to be 2. Clearly as before, j is not equal to x_n, since they differ in the *n*-th decimal place. Inductively we have a defined a number $j = .j_1 j_2 \cdots j_n \ldots$which is between 0 and 1 and is unequal to x_n for every *n*. We have thus proved that for any one-to-one correspondence between the set of positive integers and the set of numbers between 0 and 1, there is some number between 0 and 1 which is *not* paired with any positive integer. Therefore, the set of numbers between 0 and 1 is *bigger* than the set positive integers, or in the language of mathematical infinity, $c > \aleph_0$.

In addition to these two different mathematical infinities, there are vastly more levels of mathematical infinity which are subject to the laws of mathematically infinite numbers. However, it can also be proven that there are levels of mathematical infinity which are too great to be subject to the same laws. Thus mathematical infinity is limited.

It is implicit in the fact that mathematical infinity is limited, that since G-d is Absolutely Infinite, He is outside of the realm of mathematical infinity. This can in fact be proven. To begin with, let's refer to everything that exists as the *Totality*. We will prove that the

Totality is above mathematical infinity in the sense that it is not a set, and cannot be assigned a (mathematically infinite) number.

Let's assume that the Totality was a set. One of the basic axiomatic assumptions of mathematical infinity is that for any set X, there exists a set called the *power set* of X, denoted by P(X), which is the set of all subsets of the original set. It can be proven that the number of elements in the power set P(X) is distinctly greater than the number of elements in the original set X. More precisely, if X has N elements, then P(X) has 2^N elements, and 2^N is greater than N. Thus for any set, there is always a bigger set. However, since the Totality already includes everything that exists, it is impossible that there is a bigger set. Thus our assumption is false and the Totality is not a set, and is not subject to the confines of mathematical infinity. But then, of course, neither can G-d—Who is certainly much greater than the Totality of everything that exists—be subject to the confines of mathematical infinity..

Thus mathematical infinity gives us a framework to deal with limited infinities which model the quantitatively infinite aspects of the creation, and acknowledges that there are unlimited infinities which are beyond its realm.

A Model for Creation

Dr. Tsvi Yehuda Saks

In the previous section Dr. Saks explained the basic mathematics of infinity, introducing $\aleph_0$, the mathematical symbol for an infinite quantity. In this section he develops the mathematics of infinity further and shows how it can be used as a model for the creation of the universe.

The Creation, as discussed in Chassidus, is not that G-d just said, "Poof" and the universe came into being. Rather, He created an infinite panorama of spiritual worlds. There is a concept of an infinite descent of worlds. In each world there is a concealment of G-dliness giving rise to the next lowest world in which there is more concealment and in which the world and spiritual beings that inhabit that world have more sense of self, more self-definition and more individuality. The culmination of this process of infinite descent of spiritual worlds is the existence of the physical world.[1]

This description of the creation of the world by G-d is very abstract and complex, with many references to infinity. I feel that the mathematics of infinity can help us to understand and visualize some of the processes through which G-d created the world. This should not be surprising. One of the basic tenets of this conference is that not only can there be no conceivable contradiction between true science and Torah, but that science can be helpful in understanding Torah.

1. See chap. I, sec. 2, "Wisdom From Above," for a detailed discussion of this.

True science is an investigation into the world, which is the creation of G-d. Torah is the blueprint that G-d used for creation, meaning that He looked into the Torah (which preceded the world) and created the world according to the Torah. With that view, we see that science can help us reveal the creation and thus help us understand Torah. Since G-d used infinite processes when He created the world, as will be discussed at length, it is logical that the science of infinite processes, namely the mathematics of infinity, would be useful to help us understand those portions of the Torah which deal with the creation.

A mathematician develops an intimate relationship with his mathematical work and with the mathematical objects with which he works. Thus, although the concepts discussed are quite abstract, to me they are very real. The personal context for this paper is that I am a *ba'al teshuvah*, a Jew who was not raised in a Torah-observant home but made a personal decision as an adult to observe Torah and mitzvos. At that time I was already a mature professional mathematician. My field of expertise is topology, and the best way that I have of describing my work is to say that I did research into abstract infinite space. In fact, I used to say that my doctoral thesis and the research that followed it had no known relationship with the physical world. Many people used to ask me to speculate on what possible application there might be for this work, and I had no meaningful answer. Thus it came as a great surprise to me when I began to study Torah, and in particular Chassidic philosophy, and specifically the description of the creation of the world by G-d, that I felt that this was indeed the actual application of the mathematical work which I had done.

Mathematics is perhaps the purest and most rigorous science. The assumptions that one is making must be clearly specified. In addition, the acceptable methods for constructing new objects and proving new propositions are clearly defined. Thus mathematics is as objective as any science can be. Nonetheless, the truth of any mathematical system depends on the truth of its assumptions, and hence can be said to be true only relative to its assumptions. When I began to study Torah, the great difference between the relative truth of mathematics and the absolute truth of Torah, which puts us in contact with G-d Who is the Ultimate Truth, became very clear to me. My view of the relationship between mathematics and Torah is that mathematics allows us to construct abstract models which can be useful in helping us to understand and visualize some of the processes through which G-d Almighty created the world.

Where Do I Come From?

The term used to characterize creation in Jewish Philosophy is *yesh me'ayin*—creation of something from nothing (*ex nihilo*). What does something from nothing mean? And how is this creative process described? When G-d Al-mighty created the world, He did it by putting a spiritual life force in every created being which is the source of the existence of that being and continues to keep it in existence after it was originally created. In other words, He continuously recreates it. Thus everything that exists as a creation has G-dliness in it which is the source of its existence. So the question is asked in Chassidus, why is it called "something from nothing"? The physical universe, which has only a dependent existence, should be called "nothing." G-dliness, which is the source, is a true existence and should be called "something." So creation should be called "nothing from something." Why is it called "something from nothing"? Chassidus explains that it's called "something from nothing" because the created being thinks it's a something because it is unaware of the fact that it has a source. It is unaware of the G-dliness within it which created it and keeps it in existence. So the concept of *yesh me'ayin*, of something from nothing, is that the something cannot trace its existence back to its source.

This is the critical condition for creation *yesh me'ayin* that I am focusing on here: that the created being cannot trace itself back to its source, that it is unaware of its dependence on the source for its existence, and perhaps even unaware that it has a source which is outside of itself. Note that this stands in strong contrast with the concept of creation of "something from something," like a work of art or a building. A building is made from an architect's plans with physical materials. No matter how complex the building may be, we can trace the construction back to the raw materials that were used and the plans that were followed.

My claim here is that the mathematical model that we described earlier is a model for this because the number $\aleph_0$ is similar to *yesh me'ayin*. It cannot trace itself back to its source. Its source is the entire infinite process which is coming into it. To explain this, let's introduce the concepts of *predecessor* and *successor*. The sequence of positive integers 1, 2, 3,... is generated by starting with the number 1, and repeatedly adding 1 to get the succeeding numbers. Thus each number n has an immediate successor which is $n + 1$, the number following it. For example, the successor of 2 is 3; the successor of 3 is 4, etc. Also, each number n, except for 1, has an immediate predecessor which is $n-1$, the number preceding it. In this process, each number comes from a definite source, its immediate predecessor: 3 comes from 2, 4 comes from 3 and n comes from $n-1$. But $\aleph_0$ has no immediate predecessor. (Any number n, no matter how big, will be the predecessor of $n + 1$, not of $\aleph_0$.) It does not come from any specific source. It requires the entire infinite sequence to generate $\aleph_0$. Thus the whole infinite sequence is its source.

From the perspective of $\aleph_0$, if it looks back and tries to see where it came from, it is going to have a very hard time realizing that there is an infinite sequence which comes into it and generates it. It is completely unaware of this infinite process and thus cannot trace itself back to its source. Since it doesn't have a source that it can observe, it says, "I am independent."

Similarly, we see in the physical world that people who don't believe in G-d and are not aware of the concept of the creation see themselves as being independent because they don't have any way to relate to the infinite process which is coming into them at every moment and continuously recreating them. Even though we are talking about conscious human life, where intellectual abstraction is most pronounced, and daily life revolves around abstract realities, a person may still not recognize the complex spiritual structure through which his life is created and guided.

So we have a basic model for creation which models the infinite descent of spiritual worlds culminating in a physical world, the *yesh*, which is not aware of its origin, which is therefore called *ayin*—nothing. $\aleph_0$ represents this physical world.

Infinite Gaps

This is a very simple model for the creation, and in fact it's too simple for a number of reasons. For example, we know that there are four major spiritual worlds preceding the physical world. These are *Atzilus*, the world of emanation; *B'riya*, the world of creation; *Yetzira*, the world of formation and *Asiya*, the world of action. There are infinite gaps between each of these worlds. *Atzilus* is infinitely higher then *B'riya*; *B'riya* is infinitely higher than *Yetzira*; and *Yetzira* is infinitely higher than *Asiya*. But we don't have that here in our model because in the progression of the numbers of the sequence, each time we attain a new object—a successor—there is a uniform gap of 1. There are no infinite gaps along the way.

Thus we need to expand our model into a more complex one which will include in it the notion of an infinite gap. The simplest way to expand our model is to look at $\aleph_0$ not only as the limit point of the sequence of positive integers, but also as the beginning of a new sequence. In order to do that, I need to talk about the concept of *ordinal numbers*. Ordinal numbers are numbers which are based on an ordering, as opposed to *cardinal numbers*, which indicate only quantity. For finite numbers, the two categories are basically the same. But for infinite numbers there is a remarkable distinction. The ordinal numbers are much more subtle than the cardinal numbers in the following sense: If we add 1 quantitatively to $\aleph_0$, the first infinite number, then it's still $\aleph_0$:

$$\aleph_0 + 1 = \aleph_0.$$

In fact, the Tzemach Tzedek makes exactly this point in *Derech Mitzvosecha*. He says that an infinite quantity will not increase by adding a finite quantity to it.[2] The above equation is a mathematical representation of that concept.

Ordinal numbers, however, are different. With ordinals we can add 1 to $\aleph_0$ and get something new because we are considering the way that the number in question relates to what came before. $\aleph_0$ is the end point of the whole infinite sequence of numbers 1, 2, 3,...and, as we explained above, it has no immediate predecessor. But $\aleph_0 + 1$ has an immediate predecessor, namely $\aleph_0$. So from the point of view of ordinality these two are very different despite the fact that quantitatively they are identical.

(To express this qualitative difference between $\aleph_0$ and $\aleph_0 + 1$ in more technical terms: $\aleph_0$ is identified with the sequence 1, 2, 3,... which leads up to it. This is a sequence without a limit point. But $\aleph_0$

2. Derech Mitzvosecha, "Ha'amonas Elokus," chap. 11, based on Tanya chap. 48.

+ 1 is identified with the sequence 1, 2, 3, …, $\aleph_0$ which leads up to it, and this sequence has a limit point, namely $\aleph_0$.)

Let's continue this process and add 1 repeatedly to $\aleph_0$. We will get the following new sequence:

$$\aleph_0, \aleph_0 + 1, \aleph_0 + 2, \ldots, \aleph_0 + n, \ldots, \aleph_0 + \aleph_0 = 2\aleph_0$$

We can repeat this process and add 1 infinitely many times to $2\aleph_0$ until we get

$$2\aleph_0 + \aleph_0 = 3\aleph_0$$

Similarly, we can get $4\aleph_0$, $5\aleph_0$, etc. by constructing a new sequence each time. This gives rise to a sequence of multiples of $\aleph_0$ whose end-point is $\aleph_0 \cdot \aleph_0$:

$$\aleph_0, 2\aleph_0, 3\aleph_0, \ldots, \aleph_0 \cdot \aleph_0$$

While each of these numbers is different from the others as ordinal numbers (since each has a different predecessor), they all represent the same infinite quantity so they are equal as cardinal numbers. But eventually we will get to something which is quantitatively bigger then all of the others. This is called $\aleph_1$.

Now, let's put all this together into one long sequence:

$$1, 2, 3, \ldots, \aleph_0, \aleph_0 + 1, \aleph_0 + 2, \ldots, 2\aleph_0, 2\aleph_0 + 1, \ldots, 3\aleph_0, 3\aleph_0 + 1, \ldots, \aleph_0 \cdot \aleph_0, \ldots, \aleph_1$$

This is an infinite sequence that also has infinite gaps in it before $\aleph_0$(which has no predecessor), before $2\aleph_0$ (which also has no predecessor) and similarly before all the multiples of $\aleph_0$. Recall that this is precisely what was missing in our first model. We needed the infinite gaps to represent the infinite transition (*tzimtzum* or contraction) between *Atzilus* and *B'riya*, between *B'riya* and *Yetzira*, and between *Yetzira* and *Asiya*. The final endpoint $\aleph_1$ represents the physical universe. So this is a much better model for the creation as it is explained in Chassidus.

We still don't have the exact mathematics to model *yesh me'ayin* completely, but at last year's conference a statement was made that Rabbi Yisroel Aryeh Leib could discuss the whole *seder hishtalshelus*— the whole order of creation—from *Briya* down to the physical world in everyday language so that anybody could understand it. So, G-d willing, may he be together with us immediately with *Techiyas HaMeisim* (revival of the dead) so he can do that. I honestly believe that if he would do that, then we could gather a group of mathematicians and really give very precise mathematics to completely describe this process as a model.

Actual *Infinity*

Until now we have been working from the bottom up, i.e., starting with mathematics and then using it to help us understand something about G-d. Now I want to go the other way, starting off with the Chassidus and using it to answer an important question in mathematics and philosophy.

Earlier in this lecture, we quoted from the Tzemach Tzedek's discussion of infinity in chapter 11 of the section "Ha'amonas Elokus" (Belief in G-d) in his book *Derech Mitzvosecha*. Over the years, I've been asked a question by many Lubavitchers based on a statement that the Tzemach Tzedek makes there. He says that it is impossible that many finite, limited individuals could join together to form an actual infinity. Infinity cannot be composed of finite individuals. So the question is, how can there be mathematical infinity? Is this not precisely what we have done to define mathematical infinity? We said that you can take these finite individuals—the positive integers—one at a time, and join them together to form the infinite set {1, 2, 3, ...}. But the Tzemach Tzedek says that this cannot be. This seems to declare that mathematical infinity does not exist.

Actually, however, the nature of mathematical infinity has been a matter of debate among the mathematicians themselves with some, such as Gauss (one of the greatest mathematicians of all time), saying that there cannot be actual infinity, and that mathematical infinity is just a potential infinity; i.e., the sequence 1, 2, 3,... continues indefinitely but is never actually infinite. Other mathematicians, including Cantor, on the other hand, held that the infinite sets were infinite in actuality. Does it follow from what the Tzemach Tzedek says that Cantor was wrong?

One of those who had asked me this question was Rabbi Simon Jacobson, and he was the one who helped me work out the answer. He showed me certain letters of the Rebbe MHM and learned them with me, and the answer is based on those letters. It turns out that a similar question had been asked of the Rebbe MHM based on sources in Torah. There are numerous references in Chassidus which refer to G-d having created infinitely many worlds. It is also stated in the Talmud[3] that G-d created infinitely many "regiments" of "troops" (i.e. angels) to serve Him, where each regiment is finite. These sources would seem to contradict the statement of the Tzemach Tzedek, because each world or troop is finite, yet there are infinitely many of them. How could this infinity possibly be attained?

To resolve this seeming contradiction, the Rebbe MHM writes that according to human logic, it is impossible that an actual infinity composed of finite individuals should exist. The Tzemach Tzedek's statement is made based on the fundamental principle that G-d prefers to create the world so as to conform to human logic. (In fact, Chassidus explains that the reason that G-d created the universe through a succession of worlds, as we explained above, rather than by just saying, "Poof!" is so that the creation could be understood by the human mind.) The Tzemach Tzedek rejects the notion of an actual infinity composed of finite entities because it contradicts human logic. However, G-d is above all limitations and contradictions, and when He so chooses, He can and does create the world with qualities that contradict human logic. In the case of the infinitely many worlds and regiments, which Chassidus and the Talmud have revealed to us, G-d used His unlimited supra-rational power to create infinitely many of them. So for these particular cases, the limitations of logic can no longer veto the existence of an actual infinity of limited entities. The point is that the Torah would have to tell us explicitly that G-d is creating in this exceptional manner. Otherwise we would assume that He is creating in the usual way, i.e., in a manner understandable by human logic.

I believe that this explanation of the Rebbe Melech HaMoshiach can be used to explain how great geniuses could disagree about whether or not the world is infinite. Cantor, who had a strong belief in G-d, understood that G-d could create an infinite world. Other geniuses, such as Aristotle and Gauss, who did not have a strong belief in G-d, came to the conclusion that the world must be finite. Since they were limited to their human understanding, the conclusion that the world is finite was the only conclusion that they could reach.

I think this is the best example that I have of Torah and Moshiach illuminating mathematics and philosophy.

3. Chagiga 13b.

An Infinite Experience

Dr. Tsvi Yehuda Saks

In the previous two sections, Dr. Saks discussed the concept of mathematical infinity, which is represented by the symbol $\aleph_0$; the concept of infinity in Chassidus; and the application of mathematical infinity to model creation. In this section, he gives a personal account of the events leading up to his discovery of the applicability of mathematical infinity to creation, and some personal issues he has had as a Chassidic mathematician.

I would now like to talk about the ideas presented earlier on a personal level and explain how I came to all this because it was a big issue in my life. My Ph.D. thesis and the subsequent research that I did had literally no relationship to the physical world. No one knew of anything that the mathematics I did would model, and I always wondered, like anyone who did such work would wonder, what application could there possibly be for this mathematics.

To put this in context, let me explain how I came to be observant and a Lubavitcher. Before I became observant, I was a product of the 60's—a so-called "spiritual seeker." Several people I knew from the old days, who had become observant before I did, told me that any kernel of truth that I may have thought I had found in any of the spiritual disciplines that I studied can all be found in the Torah. I heard this enough times that I believed it.

I wrote to the Rebbe MHM that I was going to learn in yeshiva. I left my position as a professor in a university and went to learn for three years, in Hadar HaTorah. I would really like to express appreciation to the yeshivah because I lived there and had my three meals a day and excellent instruction from dedicated teachers and didn't have to pay anything.

While sitting in yeshivah, I started to learn in Tanya and Chassidus about the descent of the worlds from the spiritual to the physical, and I was completely overcome by the realization that this was the actual application of the mathematical work that I had worked on for so many years that had no other application.

I expected to find the truth that I was seeking in the Torah, and I did. It was much, much more than I ever had experienced before. But that was all that I expected. To find that the actual application of the mathematics I did was in Torah—that was something that I was totally unprepared for. I could never have imagined that, and it was a truly awesome and inspiring experience.

So I became involved in applying mathematics to Chassidus. I spoke at some B'Or Ha'Torah conferences in Miami, which gave me a chance to develop the ideas that were germinating inside of me. These are not easy concepts. It took a long time to get to the point where I could take the ideas I had, which were very, very abstract and very difficult, and bring them down to a point where I could actually talk about them, more or less in English, to a non-mathematical audience.

A Mathematical Ma'amar

Then something extraordinary happened in 5749 (1989). The Rebbe MHM published a ma'amar for 22 Shevat, Rebbitzen Chaya Mushka's yartzeit.[1] In that ma'amar he really opened my eyes in a new way to the importance of the work that I had been doing in this area. He discusses the concept of *yisron ha'or min ha'choshech*—the superiority of light over darkness.[2] He says not just that light is better than darkness, but that there is a certain quality of goodness that comes into being when the light is the result of a process of the transformation of darkness into light. The light that comes from the darkness is a superior light. The Rebbe MHM mentions this in connection with Yisro, the father-in-law of Moshe Rabbeinu, who came to the awareness that the Jewish G-d is greater than all of the gods and

1. *Sefer HaMa'amorim Melukat* 3:57–63.
2. Koheles 2:13.

powers. Yisro was in fact a great scholar in "external wisdom"—philosophy and science. In making his declaration that the Jewish G-d is greater then all the other gods, he brought about a refinement of the external wisdom that he had mastered, and a transformation of darkness into light. This was the final critical event that had to take place before G-d could give the Torah to the Jewish people.

But the Rebbe MHM goes even further, referring to King Solomon's statement, "I saw that there is an advantage to wisdom over nonsense like the advantage of light over darkness." The Zohar asks why we need King Solomon, the wisest of all men, to tell us this. Doesn't everyone know that wisdom is better than nonsense? The Rebbe MHM explains that King Solomon is referring to the superior light that is attained by transforming darkness, and saying that in the case of wisdom vs. nonsense, something even greater happens. "Nonsense" in this context refers to the sciences, which are nonsense relative to the deep wisdom of the Torah. But when this nonsense—the sciences—is transformed into light, not only does a superior light come into being, but the sciences themselves can actually become part of Torah!

This can happen in two ways—on two levels. The lower level is where one uses the scientific knowledge he possesses to increase his knowledge of Torah. In this case the scientific knowledge becomes an intermediary to aid in his understanding of the Torah. This extends the influence of Torah into the world and brings the Geula closer. Then the Rebbe MHM says that there is an even higher level where the science actually becomes part of Torah, such as when Rambam uses Greek mathematics and astronomy to explain issues in the Laws of *Kiddush HaChodesh* (declaring the new months and years). Rambam uses the actual calculations of the Greek mathematicians to decide Halachos in Torah. He thereby refined these Greek sciences, elevated them and transformed them into Torah. The mathematics is no longer an intermediary through which our understanding of Torah is enhanced, but actually becomes a part of Torah. They are now part of Rambam's book on Jewish Law, Mishneh Torah.

The possibility to transform external wisdom into Torah was the result of Yisro's own declaration, by which all of his knowledge of philosophy and science was transformed. In fact, his name "Yisro," meaning "superior," refers to the superior revelation in Torah brought about by this transformation.

When this ma'amar was published, I had already been thinking that doing this work of using mathematical models for Chassidus would bring about an elevation of the mathematics itself. That was clear to me. But that such an application would bring out a new revelation in the Torah—that was something that I could never have conceived of. Yet that is exactly what the Rebbe MHM says in this ma'amar. So that had a tremendous effect on me, pushing me harder to work on these issues and to think of them more and make time for them. It also had the effect of bolstering my belief—though the Rebbe MHM doesn't say this explicitly in the ma'amar—that *baal teshuva* scientists have a critical role in bringing Moshiach and the Geula, which is to take the knowledge that we have from our scientific work and use it to amplify the Torah in new ways. Just as Yisro's declaration influenced G-d Almighty to give the Torah, so too we have to accomplish this transformation of the sciences so that G-d Almighty will be influenced to bring the Geula now.

This conference of the Rabbi Yisroel Aryeh Leib Research Institute on Moshiach and the Sciences and similar ones, I think, are very important for accomplishing this end. As I said earlier, G-d willing Rabbi Yisroel Aryeh Leib will be here with us and the Geula will be here. I really do think that when this happens—it should be now—we will be able to give mathematically precise constructions that will accurately model the creation.

Lessons in Avodah

The spiritual implications of the mathematical infinity that I have been discussing here are many. The concept expressed by the verses in Tehillim, "How *numerous* are your creations, Oh G-d"[3] and "How *great* are your creations, Oh G-d"[4] is a fundamental principle that we have to meditate upon. It is explained in many places in Chassidus and in Rambam that contemplating the greatness of G-d leads a person to proper love of G-d. So if people will have attained a slightly greater appreciation of the greatness of G-d as a result of this lecture, we will have accomplished an important goal.

3. Tehillim 104:24.
4. Tehillim 92:6.

I want to comment specifically on the existence of *actual* infinity that we discussed earlier. This fact, and the fact that for any level of mathematical infinity there is always a greater one, makes more concrete the concept that no matter how far we have progressed in our own personal service of G-d, we can always go farther. We can always go infinitely farther. Even in the basic sequence of infinite numbers that we presented, we can see that there are infinitely many different levels of how infinitely far one can go. We can always reach a new level which is incomparably higher than where we were before.

Several years ago an event occurred which I'd like to share with you. It does not directly have anything to do with mathematical infinity. Rather, it is something that I experienced through learning a sicha of the Rebbe MHM, and it changed my life in an important way in a critical area.

The sicha that I'm referring to was published by the Rebbe MHM for *Parshas Shemos*, 5751,[5] the year that he was revealed as Melech HaMoshiach. In the sicha he talks about the concept of *bitochon*—trust in G-d. Basically the concept of trust in G-d is that if we have a problem, a threat or a critical situation, we should trust in G-d that He will help us. He will get us out of that situation in a good way. The Rebbe MHM describes it as an *avoda*—an active dynamic service of G-d—of placing oneself in G-d's hands and truly believing and living with the positive outcome and the fact that G-d is taking care of it and He will help me. This is not necessarily easy to do. The Rebbe MHM calls this "hard labor within one's soul."

It was right around that time that I was in danger of getting laid off from my job. I was working in a research unit in a company, and our funding would be disappearing in the following few months. The chance of getting new funding was dismal. I was over-specialized, and getting another job was not easy, especially since the economy was in terrible shape. So this was pretty serious. In fact, people did get laid off at the time, and some were out of work for a long period before they got another job.

The sicha, however, absolutely changed my whole attitude towards life at that time because instead of being depressed, which I know

---

5. Printed in *Likutei Sichos* 36:1–6.

would have been the normal reaction, I was tremendously upbeat. I worked with the sicha and lived with it, and in fact I did get another job—from my old boss whom I had left a few years before.

This is how it happened: I wrote to the Rebbe MHM on my birthday, 2 Teves, which is in the early winter, and I received the answer (a letter signed by the secretary) a few months later, after Pesach. I had been talking to my old boss during this period about a job. A week after receiving the answer from the Rebbe MHM, I spoke to my old boss on the telephone. He said, "We just got a major contract. Come right in." So I came right in, and by that time we were just discussing the details of the offer he was making me. He said, "It was a week ago Friday that we got the official confirmation of this contract." That was *exactly* the day that the letter from the Rebbe MHM arrived, more than three months after I had written.

I want to conclude with the notion that in our current situation, *bitochon*, trust in G-d, is a very important concept. The Rebbe MHM says in the sicha that this is the response to make to a crisis which, in some sense, we are in. This is something that people should be encouraged to work on—to relearn the sicha and to live with it. As the Rebbe MHM concludes the sicha, in the merit of the *bitochon*, the Geula will happen immediately. And the Rebbe Melech HaMoshiach will be completely revealed.

A Theory of Everything

"Once, it was thought that each of the forces of nature was an independent force, and that material objects were composed of many different fundamental components. But the more the development of the sciences advanced, the more they realized that this multiplicity and individuality of these components is only an external appearance...and they reduced the number of the fundamental components. It will be realized that the basic existence of the world is built from the unification of two concepts: quantity and quality—matter carrying a force, and the force."[1]

—The Rebbe MHM

The unification of the forces of nature began in the mid 1800s, when the fountains of wisdom below broke open, with the unification of electricity and magnetism, giving rise to the new theory of Electromagnetism, as we explained at length in the section "Wisdom from Below."[2]

The discovery of the periodic table of the elements at that time, and the later discoveries of atomic structure and quantum mechanics in the early 1900s, facilitated an understanding of the chemical properties of the elements, and the structure of substances composed of the elements, based on the distribution of electrons in the atoms of these elements. The reduction of matter to three basic components—protons, neutrons and electrons—was accomplished.

In the meantime, two new forces were discovered: 1) the Strong Force which holds the protons and neutrons together inside the nucleus of the atom, and 2) the Weak Force which is involved in

1. Likutei Sichos 15:47 (the "Sicha of Parshas Noach").
2. Chap. I, sec. 3.

radioactive decay. So in the early 20th century four forces were known to scientists: 1) the Strong Force, 2) the Weak Force, 3) Electromagnetism and 4) Gravity. It was time to unify them all.

But unification did not come easily. Einstein spent the last 30 years of his life trying to develop a Unified Field Theory to unify quantum mechanics and general relativity (his theory of gravity) and show that these two forces are really manifestations of one grand underlying principle. But he was not successful. He was isolated and alone while the majority of physicists pursued more fruitful topics. He was acutely aware of his failure and wrote to a friend in the early 1940s, "I have become a lonely old chap who is mainly known because he doesn't wear socks and who is exhibited as a curiosity on special occasions." With his death in 1955, the quest for unification faded into the background.

The main problem is that general relativity and quantum mechanics, as currently understood, are mutually exclusive. The general relativity theory of gravity works very well on the large (astronomical) scale but breaks down on the small scale of atoms and molecules, while quantum mechanics is very successful in dealing with the atomic scale but is incompatible with the description of space and time that underlies general relativity. Essentially, general relativity is a new theory of space and time but it retains the classical (Newtonian) view of the relationship between the observer and the observed, while quantum mechanics is a new theory of matter and energy with a new perspective on the role of the observer, but it retains the classical view of space and time. What is needed is a unifying theory of "quantum gravity" which incorporates both of them. Much progress has been made in this direction but it is not yet complete. A prominent contemporary theoretical physicist wrote in 2001, "As we enter the 21st century, there is no more challenging problem in science than the completion of this theory."

Meanwhile, in the 1960s, the unification of the electromagnetic force and the weak force was accomplished, giving rise to "electroweak theory." This theory, combined with the theory of the strong force, comprised what is known as the Standard Model of physics, but the unification with the force of gravity—general relativity—was still missing.

In the 1980's, some major successes in developing theories of unification gave a new impetus to the effort to find one all-encompassing theory. The search was on for Grand Unification Theories (GUTs) and Theories of Everything (TOEs), as they were called. The most exciting of these new theories was known as "string theory" which, in the 1990s, was extended to "M-theory."

Before discussing string theory itself we should take note of a sociological factor in the community of scientists (yes, they are sociological beings just like everybody else) that indicates how pervasive the excitement over string theory became: Back in the 1980s a young physicist would have trouble getting a position at a research university if his main interest was string theory. In recent times, many young physicists have complained that if they are *not* interested in string theory they have trouble getting such a position.

Points and Strings

The first point to make about string theory is that, in this theory, the fundamental particles of nature are no longer seen as "points" but rather as "strings." As described by Prof. Brian Greene, one of the major theorists and proponents of string theory:

> According to string theory, if we could examine these particles with even greater precision—a precision many orders of magnitude beyond our present technological capacity—we would find that each is not point-like but instead consists of a tiny, one-dimensional *loop*. Like an infinitely thin rubber band, each particle contains a vibrating, oscillating, dancing filament that physicists have named a *string*....
>
> Although it is by no means obvious, this simple replacement of point-particle material constituents with strings resolves the incompatibility between quantum mechanics and general relativity (which, as currently formulated, *cannot both be right*)....This is a tremendous achievement, but it is only part of the reason string theory has generated such excitement....
>
> Einstein [in his search for a Unified Field Theory] was simply ahead of his time. More than half a century later, his

dream of a unified theory has become the [goal] of modern physics. And a sizeable part of the physics and mathematics community is becoming increasingly convinced that string theory may provide the answer. From one principle—that everything at its most microscopic level consists of combinations of vibrating strands—string theory provides a single explanatory framework capable of encompassing all forces and all matter.

String theory proclaims, for instance, that the observed particle properties—that is, the different masses and other properties of both the fundamental particles and the force particles associated with the four forces of nature (the strong and weak nuclear forces, electromagnetism, and gravity)—are a reflection of the various ways in which a string can vibrate. Just as the strings on a violin or on a piano have resonant frequencies at which they prefer to vibrate—patterns that our ears sense as various musical notes and their higher harmonics—the same holds true for the loops of string theory. But rather than producing musical notes, each of the preferred mass and force charges are determined by the string's oscillatory pattern. The electron is a string vibrating one way; the up-quark is a string vibrating another way, and so on....

[Thus,] particle properties in string theory are the manifestation of one and the same physical feature: the resonant patterns of vibration—the music, so to speak—of fundamental loops of string. The same idea applies to the forces of nature as well. Force particles are also associated with particular patterns of string vibration and hence everything, all matter and all forces, is unified under the same rubric of microscopic string oscillations—the "notes" that strings can play.

For the first time in the history of physics we therefore have a framework with the capacity to explain every fundamental feature upon which the universe is constructed. For this reason string theory is sometimes described as possibly being the "Theory of Everything" (TOE) or the "ultimate" or "final" theory. These grandiose descriptive terms are meant

> to signify the deepest possible theory of physics—a theory that underlies all others, one that does not require or even allow for a deeper explanatory base.[3]

In the next section, we present Dr. Naftali Berg's discussion of quantum mechanics and string theory, compiled from his lectures at the Moshiach and Science conferences of 1993-1994.

String theory is not without its critics, however. It has been criticized primarily because: 1) Strings are too small to be detected by any experiment that we could perform with any current experimental apparatus, and 2) While it accomplished its goal of unification beautifully and elegantly, it has not yet predicted anything that we do not already know. But these are not fundamental objections in our opinion.

It has even been criticized as "recreational mathematical theology." But mathematical theology is exactly what we are looking for—a mathematical theory unifying all the forces and matter in the universe as a preparation for the revelation of the absolute unity of G-d in the Era of Moshiach. As for the "recreational" part, well, aren't we supposed to enjoy our work? It is a fundamental teaching of Judaism that all our service of G-d—which includes everything that we do—should be done with happiness!

So, is string theory the theory that we are looking for—the one that the Rebbe MHM said will show that "the basic existence of the world is built from the unification of two concepts... matter carrying a force, and the force," i.e., a string and its vibration? We don't know yet. And we are careful not to jump and say, "This is it!" (But we certainly hope that it will be.)

By the way, regarding M-theory, the extension of string theory, mysteriously, nobody knows what the M stands for. It has no official meaning. But if it is successful and truly accomplishes the unification that we are looking for, then we will say that the M stands for Moshiach.

---

3. Brian Greene, "A Theory of Everything?" www.pbs.org/wgbh/nova/elegant/everything.html

Unifying Theories in Physics
Quantum Mechanics and String Theory

By Dr. Naftali Berg

This section is a compilation of Dr. Berg's lectures at the Moshiach and science conferences of 1993 and 1994 which took place on the yartzeit of Rabbi Yisroel Aryeh Leib. In 1993 he discussed quantum mechanics and in 1994 he discussed string theory. Although he was very sick in 1994, he wanted very much to deliver his lecture and made a heroic effort to come. Indeed, his lecture generated tremendous excitement among the audience. He passed away two months later. (In the next section we present a longer and more technical version of the 1994 lecture.)

Once, Dr. Berg presented an early edition of the Tanya to the Rebbe MHM. Upon accepting it, the Rebbe MHM said to him, "When are you going to write a book like this?" Dr. Berg was awestruck and had no reply. We hope that the lectures of Dr. Berg presented in this book will fulfill the Rebbe MHM's request.

I would like to start by reviewing the famous sicha of Parshas Noach in which the Rebbe MHM discusses how the developments in modern science are a prelude to the Era of Moshiach. Hopefully what we do here will help the Rebbe MHM shlita have a complete recovery.

In the sicha, the Rebbe MHM talks about the statement in the Zohar that in the year 5600 (1840) there will begin great revelations in the sciences and in Chassidus. He shows how the two are connected and work hand-in-hand. He identifies three levels of interaction. On the first level he explains how the sciences can be used as an example for Torah concepts. For example, the existence of video technology makes the concept of "the eye of G-d that sees" more real for us. On the second level he explains how one can use the scientific technology like radio or satellite communication for transmitting the knowledge of G-d—Chassidus—to the entire world. This actually purifies the air and space that the radio waves travel through.

The third level is to actually see the concepts of Chassidus expressed in the sciences. I now want to give a free translation of what the Rebbe MHM says about this:

> The true unity between the secular knowledge with Torah occurs when one sees ideas from the deeper levels of the Torah in the secular studies themselves.
>
> One of the main areas where one sees this readily is in the deeper levels of the Torah—Chassidus—which reveals the basic, absolute unity of G-d in the world. It teaches that the multitude of creations in the world is no contradiction to the absolute unity of G-d. On the contrary, such a multitude of creations can *only* come from the absolute unity of G-d (as is explained in Chassidus).
>
> Now, the true existence of the world *is* the unity of Hashem. Therefore this unity must manifest itself in the physical existence of the world also. Thus there must be a concept of unity within the physical world itself.

Let me just make one comment here regarding the theories that I am about to examine. The Theory of Everything that I will discuss expresses the desire of scientists to make everything as simple as possible. The point that Chassidus makes and that we will make an effort to show is that the unity in the universe is a direct result of the fact that it was created by the Creator Himself, Who is a perfect unity. So we try to find unity in the world because this is an indication of the unity of G-d Who created the world.

Let us continue to explore the sicha:

> "This idea, the oneness which is in world, which in its inner most aspect is a manifestation of the unity of G-d, is more

revealed [or, I would say, it is easy for us to appreciate and perceive] because of the development of the secular studies."

Now the Rebbe MHM gives an example, and this is the example we will amplify today:

> It used to be that people thought that every one of the natural forces is a separate force by itself and that the matter of everything that exists in the world is constituted of many different elements. But the more that the sciences advance and are developed, we recognize that the multitude and differentiation of these different elements are only an external matter resulting from various combinations of the different particles, etc. Thus we have been able to greatly reduce the number of basic particles. Ultimately, we come to the realization that the fundamental point of the existence of the world is based on the unification of the two concepts of quantity and quality—matter that carries the force, and the force itself. *Absolute* unity is an aspect of G-dliness; the analysis of the creation in its own terms must be based on the unification of two components—quantity and quality.[1]

Rabbi Yisroel Aryeh Leib wrote a very fundamental paper discussing eigenvalues and eigenvectors[2] which is the mathematics of quantum mechanics. Thus, I think it is appropriate that we should talk about quantum mechanics. Be warned that quantum mechanics is a very, very difficult subject because it discusses things at a fundamental level where we don't have any experience. Richard Feynman, a famous Jewish physicist who was born in Brooklyn, once remarked that he read in the paper that there are twelve scientists who understand relativity. He said he is not sure whether or not that is true; there may be more than twelve or less than twelve. One thing he is positive of—no one understands quantum mechanics. So be warned that this is a very, very difficult subject.

What was the state of scientific thought prior to quantum mechanics, approximately 100 or 150 years ago? After Newton put forth what we now call the classical laws of mechanics, physicists were so confident of themselves that they felt that given the initial conditions

1. *Likutei Sichos* 15:47.

2. "The Location of Eigenvalues and Eigenvectors of Complex Matrices," reprinted in Appendix 3.

they could predict the course of everything in the entire world. This philosophical outlook is called "determinism" because it maintains that everything is completely determined by the laws of physics. It is often called Laplacian determinism after the physicist Pierre Simon Laplace who was a major proponent of this idea.

Quantum mechanics, on the other hand, takes a very different point of view. Instead of things being certain, things have probabilities. It deals with what we call a wave function which is part of Schroedinger's equation. The eigenvalues are the energy levels of the wave function. The wave function is actually a probability function of where a particle may be. A particle has a certain probability of being in a given place. But it is a rather fuzzy description. Another very fundamental part of quantum mechanical theory is what is called the Heisenberg uncertainty principle which states that there is an inherent uncertainty in what can be observed in the world. You cannot determine or measure exactly two "conjugate variables." For example, you cannot measure exactly the momentum and the position of a particle. If you can measure one very accurately, you can't know the other accurately at all. This uncertainty is not something that comes from a lack of our knowledge—a lack of our ability to measure—it's a fundamental uncertainty in the world. So determinism now came to be replaced with uncertainty.

However, this principle was not accepted universally. Einstein never accepted this principle of uncertainty, and he made the famous statement that he does not accept it because "G-d does not play dice with the universe" so there cannot be a basic uncertainty in the world. Niels Bohr, another Jewish physicist, who argued with Einstein on this, responded that it is not our business to tell G-d how to run the universe. Now, who was right in this disagreement? We will answer this soon when I relate to you a *yechidus* (private meeting) that I had with the Rebbe MHM on this topic.

Quantum mechanics, as we said, deals with probabilities. A particle by itself has a *wave function* and a *state vector*. When it interacts—it comes in contact with something or a measurement is taken—its splits into various different states and becomes a summation of the various different states. It could be either here or there, and where it is actually is rather uncertain. However when a human being, a conscious observer, observes the event—the measurement—he produces a concrete, real result. So according to this, we come to understand that the human being in quantum mechanics plays a very important role. He determines the reality of where the particle is or what the state of the particle is. When the experiment was first set up it was uncertain, and now when the human being, the conscious observer, becomes involved, the state becomes clearly defined. Not only does he have free will to do things, but by his action, he actually determines the reality of the universe. This is a very strong statement about free will—that a human being actually determines reality. He makes reality.

Let me now show you a very simple experiment called "the two slit experiment" which I think will make some of these concepts easier to understand. This is a very famous experiment in quantum mechanics

We start off on one side with a source. At some distance from the source we put a plate with two slits. (That's why it's called "the two slit experiment.") We label one of them "slit 1" and the other "slit 2." And at some distance past the double slit we will put a screen or a backstop on which we can observe what passed through the slits.

Let's consider three different situations. In the first, the source would be a machine gun firing bullets, which are distinct, individual pieces. If the machine gun is wobbling, the bullets going through the hole are flying in all directions. So some of the bullets will go through slit 1 and some will go through slit 2. But no bullet will go simultaneously through both slits. That is evident to everyone. Therefore if we draw a graph of the "probability distribution function" of the number of bullets, P, that are arriving at the screen, it will be the combination of what went through slit 1 and what went through slit 2.

We can verify this by repeating the experiment with slit 2 closed and see how many bullets go through slit 1 alone. Call this P1. Then we repeat the experiment with slit 1 closed and see how many bul-

Diagram 1

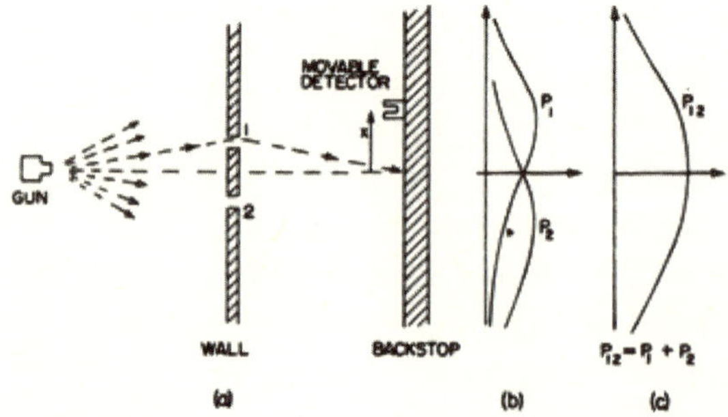

lets go through slit 2 alone. This will be called P2. Indeed, we find that P = P1 + P2. (See Diagram 1)

Now, let's say we replace the machine gun with a wave. We might have a water tank here. Touching the surface of the water repeatedly would create ripples. So now the ripples are passing through the slits. And a ripple, being a wave, would indeed go through both holes at the same time. So instead of seeing the previous distribution function, what we now see is called an "interference pattern," which shows how the waves interfere with each other. In the middle the two waves come together and reinforce each other, making a large wave, a "peak." This is called constructive interference. At other points there are "troughs" where the waves cancel each other out, as in Diagram 2.

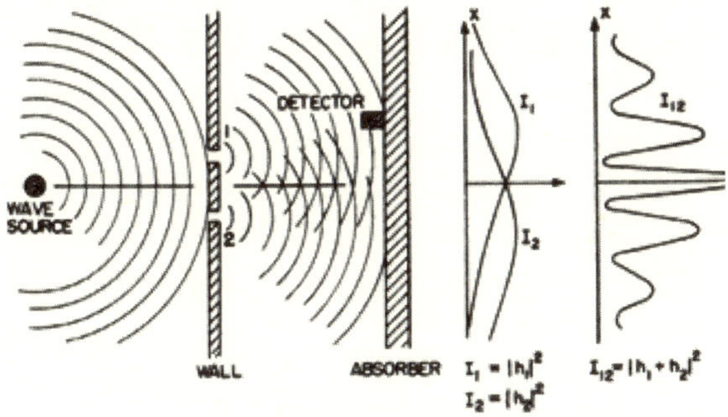

Diagram 2

What happens when, instead of using something which we know is a wave, like water, we use quantum mechanical particles like electrons? We take a bulb which contains a filament and we heat it up so that it shoots off electrons. Now an electron is essentially a particle. We place detectors which click each time an electron is detected.

You would say that since this is a particle we should get what we got when we shot off the particle-bullets. But that is not what we get. What we get instead is the interference pattern. (See diagram 3.) So somehow, on this quantum mechanical level, the electron is, in a way, a particle which also behaves like a wave. This is explained by the statement that I made earlier. The wave function of the particle splits into two states. The resultant is now a summation of these two states. It can be shown with very simple mathematics that this is indeed the interference function, the same as the interference pattern of the water waves.

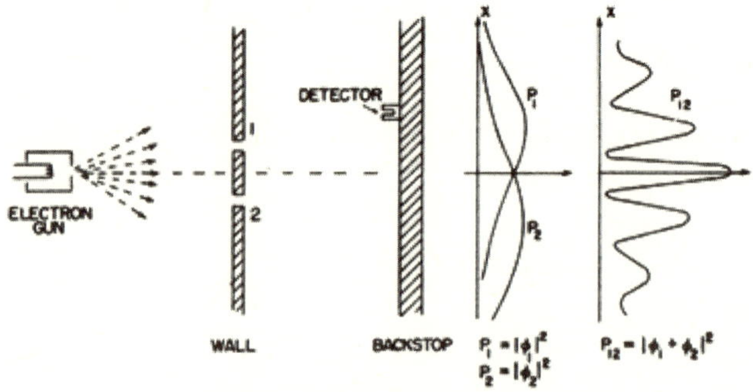

Diagram 3

However, what happens if we introduce into the system a human observer who tries to determine where the particles came from? There are various clever ways to do this. If an electron passes through slit 1, we would see a flash of light there and we would say, "1." If it would pass through slit 2 we would see a flash of light there and we would call that, "2," etc. As soon as we start counting the particles here and there—as soon as the human being becomes involved—we've changed things and the interference pattern disappears and is replaced by the distribution function of particles! This shows that as soon as a human being enters into the experiment and makes an observation with his conscious mind, he perturbs nature and determines that it's no longer a wave which can exist here and there simultaneously, rather these are now particles which either pass through one slit or the other. The situation is actually much more complicated than what I just explained, but essentially that is the experiment.

There are some scientists who have taken an extremely strong approach to interpreting this. Eugene Wigner, a Jewish physicist from Hungary, made a very Kabbalistic or metaphysical type of statement. He said that the human conscious observer does not obey the laws of quantum mechanics. When he enters into the experiment, by some power he puts the particle into one state or another. He creates reality. This is a very strong manifestation of free will.

Another very great physicist, John Wheeler, wrote a paper called, "Genesis and Observership"[3] and stated the following near the close

3. *Foundational Problems in the Special Sciences: Part two of the proceedings of the fifth International Congress of Logic, Methodology and Philosophy of Science, London, Ontario, Canada* (Boston: D. Reidel, 1977), 3–34.

of the paper: "It is often stated that the development of life is accidental and unimportant in the scheme of things. Quantum mechanics has led us to take seriously and explore the directly opposite view—that the observer is as essential to the creation of the universe as the universe is to the creation of the observer."

I said earlier that I would discuss what the Rebbe MHM said about the uncertainty principle and about the controversy between Einstein and Bohr. I had a unique *yechidus* with the Rebbe MHM a number of years ago when it was still possible to have a *yechidus*. I asked him, "Was Einstein correct in saying G-d doesn't play dice with the universe or not?" The Rebbe MHM said that of course Einstein wasn't correct because G-d is above the laws of physics, He created them. He can do whatever He wants. But there is no uncertainty as far as G-d is concerned.

Nevertheless, the Rebbe MHM said that the uncertainty principle of quantum mechanics, which is related to probability, is incorrect for the following reason:

We have probabilities. Probabilities mean that certain events occur with a greater or lesser probability. For example, you could win the lottery and become a millionaire. This is not a miracle; it just has a very low probability of occurring. In quantum mechanics, every event has some probability of occurring, so everything is possible. An obvious miracle would also be possible [as a natural event] except with a very, very low probability of taking place. The splitting of the Red Sea would also be possible but the probability is very small. So according to the uncertainty principle, a miracle is not breaking nature; it's an event of low probability. The Rebbe MHM said that it is our belief that G-d reveals Himself in the world in a way that breaks nature. So to say that the miracle of the splitting of the Red Sea is consistent with the laws of physics is against our belief.

Now, it would seem that this is clear and unequivocal—certainly after the Rebbe MHM said it. But two months ago before *Pesach* (Passover), when I was in *Eretz Yisroel*, we had a Shabbaton and I discussed ideas similar to what we are talking about now. Some yeshiva boys from *baal teshuvah* yeshivas came over to me complaining, "How can you say this?" Their teachers had told them that the splitting of the Red Sea was consistent with the laws of nature; it was just an event that had a very small probability of occurring. Thus, we certainly see that we need the guidance of the Rebbe—of our Moshiach—to tell us what is right and what is wrong. Things are not so obvious. We must have the guidance of Moshiach for things which seem so clear to actually become clear.

A Theory of Everything

Scientists have been trying to discover a theory of everything that would encompass both the forces of nature and matter. There are four basic forces of nature. There is the electromagnetic force—electricity, which we are using here and which was explained about 120 years ago. There is the force of gravity which has been known for a very long time, but Einstein quantified it about 80 years ago in his general theory of relativity. There are two additional forces in nature which were discovered as a result of quantum mechanical theory—the strong nuclear force and the weak nuclear force. The strong nuclear force holds the protons and neutrons, the basic elements of the nucleus together; and the weak nuclear force is relevant for radioactive decay. Both of these forces, unfortunately, have been used for very destructive purposes. The audience will learn later, in the second part of this presentation, about a peaceful use of these forces in a thorium reactor.[4] And, G-d willing, the reader will learn from my presentation how we can actually tap into a much more primitive, more basic force in nature itself, which hopefully will be tapped into when Moshiach comes.

Scientists have been trying to find a unifying theory for these forces, as the Rebbe MHM has said, and also a unifying theory for all of the matter of the universe. Again, as the Rebbe MHM has said, in the last century people thought that matter consisted of many, many different things which they organized into a table of elements. There were 90 some known elements, and now we have made some ourselves, so there are over 100 elements. But in recent decades we have found elements or constituents of matter that are even more fundamental. We found the neutron, the proton, the electron, and

4. This is the Radkowsky thorium reactor. See chap. III, sec. 7 for a description of the reactor and our interview with Prof. Radkowsky. At the Moshiach and Science conference in 1994, Mr. Joel Grae, president of the Radkowsky Thorium Power Corporation (now known as Lightbridge Corporation) and one of the company engineers spoke about the reactor.

we have gone far beyond even these. We have come to what we think are the basic constituents of matter. But still, we have several constituents of matter and several forces. The search for a theory of everything, a unifying theory, is the attempt to tie this all together in one theory which shows a basic unity in the entire world.

The Rebbe MHM *shlita* said, based on the Zohar, that this whole development, which is a Messianic development, started in 1840, and while the origins of this theory of everything can actually be traced back to that time, we will see, very interestingly, that the actual real work on this started in the 1950s, which corresponds exactly to the time that the Rebbe MHM assumed the leadership of Chabad. He is the one, of course, who brought the complete realization of Moshiach into actuality. And it is in the time of Moshiach that the culmination of this unity is attained. Finally, in the last several years this has really gained a lot of momentum, again corresponding to the time when the Rebbe MHM has really put a lot of emphasis on the topic of Moshiach.

Dr. Berg proceeded with a very technical review of the history of modern physics since the year 1840, discussing non-Euclidean geometries, Riemannian geometry, unifying theories in physics, Maxwell and electromagnetic theory, Einstein and relativity, quantum electrodynamics (QED), quantum chromodynamics (QCD), symmetry, gauge symmetry and parity.

He then began to talk about string theory, emphasizing the connection between string theory and Chassidus and Kabbala.

String Theory

We are getting to the punch line now. In the late 1960s, and especially later in the 1980s, a new theory was developed. This is the theory of strings or superstrings. String theory turns out to have relevance in 26-dimensional space and in 10-dimensional space. Now those who know Chassidus and Kabbala know that the numbers 26 and 10 are very, very relevant.[5] The Rebbe MHM said, in a sicha, that

5. Twenty-six is the numerical value of the Hebrew name of G-d. He created the universe through a process of *tzimtzum* in which He emanated the ten sefiros of *Atzilus* and through them brought the universe into actuality. See chap. I, sec. 2, "Wisdom From Above," for a more extensive discussion of this.

the world was created in this fashion to show that the Creator put His imprint on the creation for us to know Him from creation. Those in the audience who know more Chassidus than I do will probably be able to amplify this much better.

We have talked about particles. Each of these things—electrons, quarks etc.—are point particles. A point particle is very limited as to the different vibrations it can have. The new theory came and put forth the concept that perhaps the basic matter is not particles, but strings. When I refer to a "string," I'm talking about some manifestation of the fabric of space-time. What it is, I don't know. I'm not sure if anyone knows. But the idea is that instead of having a point particle, we have a string. A string is very, very small—something like a Planck length which is 10^{-35} meters. That is extremely small. It's a fraction with a 1 on top and a 1 with 35 zeros after it on the bottom.

Now if one sets this string to vibrate and keeps the ends fixed, it will vibrate in different modes, just as if one takes a violin string and vibrates it, there are only certain notes, certain vibrations called tones or frequencies, at which it can vibrate. The theory was that the string is the fundamental element of nature. The different vibrations are the different particles. If one has a different vibrational mode, this corresponds to another particle. Thus all of the particles were supposedly explainable based on the vibrations of the strings. The electrons, the quarks and everything else were these different vibrations of the string.

When scientists started examining this theory, it was found that the better explanations involved a string that is closed like a circle which can also vibrate, called a "heterotic string". One can have vibrations traveling clockwise or counterclockwise. It turns out that the only dimensions in the multidimensional space in which this theory is consistent (it is solvable and there are no infinities there) is where the counterclockwise vibration is in 26 dimensions and the clockwise vibration is in 10 dimensions. So again we have the 10 and 26....

Dr. Berg concludes by suggesting that now, in the Era of Moshiach, we will be able to tap into the energy of this deep level of matter just as we have been able to utilize the nuclear energy hidden within the nucleus of the atom.

Ten Dimensions: A Physical Theory of Everything

By Dr. Naftali Berg

In this section we present a complete—and rather technical—version of Dr. Berg's lecture at the 1994 Moshiach and Science conference. In it he reviews the sicha of Parshas Noach, discusses unifying theories in physics in general, and speaks of string theory as a specific unifying theory.

There is a story behind this edition of the lecture. At the time of the 1994 conference Dr. Berg was very sick. (He passed away two months later.) It was with tremendous mesiras nefesh *(self sacrifice) that he managed to come and speak at the conference—which he did because of the importance of Moshiach and science research to him personally. Because of his condition, he was unable to edit the lecture himself so he asked a colleague of his at the Army Research Laboratory, Dr. Bruce Weber, with whom he had discussed string theory and who was very familiar with the issues, to edit the paper for him so that it could be published in this book.[1] I have done only minor editing of Dr. Weber's draft.*

1. In a cover letter, Dr. Weber described his editing of the paper as follows: "I have worked continuously for 3 days to derive the attached draft. The draft is complete if not as clear as I would like it to be. It is much closer to the physical meaning I believe Dr. Berg was seeking. In my editing I have not altered any of the religious text....This draft, as edited, addresses Dr. Berg's spoken wishes to me as to what he wanted me to achieve. Basically, he wanted me to clarify, expand and integrate the many ideas as presented....Dr. Berg and I had spoken at length [about] the subject of the presentation. Much material used by both of us is in my possession and I have referred to it in the editing."

In the famous sicha of Parshas Noach,[2] the Rebbe MHM points out that studying the sciences releases the holy sparks that are there and brings us closer to the ultimate redemption.

In the last part of the sicha, the Rebbe MHM discusses the fact which the Zohar reveals—that in the year 5600 (1840) there will be a revelation of external wisdom and the wisdom of Chassidus. We see how the two are connected and can work hand-in-hand. The Rebbe MHM shows how secular studies can be used as an example from which we can understand what the Torah means when it says, "an ear which hears and an eye which sees..." This is easily understood today with television and radio, which can transmit over the whole world. The Rebbe MHM also discussed how one can use scientific technologies, such as satellite communications, for simultaneously transmitting Chassidus to the entire world. Since the radio waves permeate and penetrate the entire world, this actually purifies the air and the space through which these radio waves travel. This matter will be fully revealed in the times of Moshiach.

Then the Rebbe MHM talks about using the secular studies themselves to actually clarify and enhance concepts of Torah. In the time of Moshiach we will see with our eyes of flesh the revelation of G-dliness. Science and technology deal with the physical world, and we relate to the physical world much more easily than to the concepts of Chassidus, because of the fact that we have eyes of flesh and a mind of flesh which are bounded by the space-time continuum in which we exist. Therefore when we use the ideas of science and technology, to which we relate better, to understand the Chassidus concepts, we elevate those ideas of science and technology.

The Rebbe MHM further explains that the true unity of the secular studies with Torah is manifest in the fact that in the secular studies themselves are seen concepts of the inner dimension of Torah. One of the main points readily seen is that Chassidus comes to

2. *Likutei Sichos* 15:42–48.

reveal the basic, primordial unity of G-d in the world. G-d is a perfect unity. The fact that there is a multitude of creations is not a contradiction to G-d's absolute unity; just the opposite is true. This multiplicity of creatures and creations comes specifically from G-d's unity, as is explained in many places in Chassidus. Since the true existence of the world is from the unity of G-d, we must say that the unity of G-d manifests itself also in the existence of the world—that the world itself testifies to G-d's unity.

The theories which I discuss here—the theory of everything and its related theories—stem from the scientist's desire to explain the universe as simply as possible. The scientist finds *beauty* in the mathematics and the formulae when he can explain everything in the universe upon their basis, especially if everything can be explained with a single equation. But from a Torah point of view, showing that there is a unifying force in the universe testifies to and is a direct result of the fact that the universe was created by G-d, who is Himself a perfect unity. This is a very subtle distinction, but is very important; I believe this constitutes the basic distinction between secular studies and Torah. The scientist seeks unity—a theory of everything, via a single equation or idea, because in such simplicity there is great beauty. In my humble opinion, the pursuit of secular studies is egoistic; it enhances the ego of those who are involved in it, whereas the purpose of Torah is to *nullify* one's ego before G-d's unity. When Moshiach comes, the status of science will hopefully also be at that level.

The Rebbe MHM continues to explain that since the true existence of the world is the perfect unity of G-d, this manifests itself in the basic unity of the world—i.e., in the fact that the world itself is a manifestation of unity. This idea—the oneness which is in the world, which in its innermost aspect is a manifestation of the unity of G-d—is easier for us to perceive and appreciate due to the development of the sciences, as we will discuss later.

It used to be that people thought that every one of the natural forces is an independent force by itself and that the basic matter of the world is constituted of many different elements. But as time has gone on and science has developed further and further, we have come to the recognition that this multitude and differentiation of these various elements is merely an external manifestation, and in actuality this is just a combination of different parts. And we have been able to reduce this number of basic particles and forces tremendously—until we've reached the realization that we can recognize that the basic point of the existence of the world constitutes the *unification* of the two concepts of quantity and quality—matter that carries the force, and the force itself. *Absolute* unity is an aspect of G-dliness; the analysis of the creation in its own terms must be based on the unification of two components—quantity and quality.[3]

Four Forces

Now let me present a very basic overview of what I want to talk discuss. Then we will go into the details. Scientists have long been searching for a theory of everything or a basic equation or phenomenology to explain the forces of nature, the matter these forces control and the physical conditions which gave reason to the separation of this one unifying force or principal into four forces. Only in the past few years have the mathematical and the physical ideas become sufficiently powerful to make this goal possible. There has been tremendous progress this century; the goal is in sight but the obstacles yet to overcome are large.

There are four basic forces of nature: the electromagnetic, gravity, the strong nuclear, and the weak nuclear forces. The electromagnetic force was explained by Maxwell almost 140 years ago. The force of gravity was first described by Newton in his treatise *Principia* in 1686 using the newly developed (by Newton and Leibniz) mathematics of Calculus, and then finally quantified by Einstein about 230 years later in his "general theory of relativity." The last two forces of nature, the strong and weak nuclear forces, were discovered this century as a result of experimental studies into the nature of the atom and theoretical developments in quantum mechanical theory. The strong nuclear force is what holds together protons and neutrons, the basic components of the atomic nucleus. The weak nuclear force is involved in radioactive decay. Both of these forces have unfortunately been used for very destructive purposes. A comparison of the relative strengths of the four forces reveals that the strong nuclear force is the strongest, the electromagnetic force is 100 times weaker, the weak nuclear is 1000 times weaker, and the gravitational is 10^{42} (1 followed by 42 zeros) times weaker. Human familiarity with these

3. Ibid., 47.

forces is in reverse strength order, with gravitation the most familiar. This is so because it takes energy to overcome the strength of a force. As increasing amounts of energy are necessary to overcome the stronger forces, the event necessary to provide the energy becomes more and more rare. I will show how we can tap a much more primitive and basic force in nature itself, a force which will hopefully be tapped when Moshiach comes.

Scientists have been trying to find a unifying theory for these four forces, as well as a unifying theory for all the matter in the universe. If such a theory can be found, all of the forces would be considered special cases of a single force. As the Rebbe MHM said, in the last century people thought that matter consists of many, many different things and they made an atomic table of over 90 elements. Now we have made some elements ourselves, so there are over 100. Science has discovered the proton, neutron, and electron and has gone further and further until it has found what we think are the basic constituents of matter. But still we have many different types of matter—constituents of matter—and many forces. The search goes on for a unifying theory to explain how we tie all these particles together—one theory which shows a basic unity of the entire world. The Rebbe MHM said this development started from 1840. The developments I am about to talk about started around that time.

The real work on the unification of the four forces in nature started in the 1950s, which corresponds exactly to the time of the Rebbe MHM's leadership, who of course actualized the realization of the issue of the Moshiach—who will be the culmination of this unity. In the last few years this has gained much momentum.

Force Equals Geometry

Faraday, an English inventor, introduced the concept of electric and magnetic *field* which refers to a force originating from a source, such as electrons, and propagating (at the speed of light, as Maxwell later showed) to some distance and affects other particles there. This is the basis of radio and television, etc. So the idea of a field began with Faraday. Conceptually, this was a very important step, as we will see.

The next very important step, from a historical point of view, was achieved by the German mathematician Riemann. Until the 1850's,

Euclid's geometry, developed two thousand years ago, reigned supreme. But, as anyone who has studied it knows, Euclidean geometry applies only in a flat 2-dimensional space.[4] Riemann's teacher, Gauss, asked him to formulate an equation which will transform geometry into higher dimensions not limited to a *flat* space but applicable to a curved space, as well. The outcome of Reimann's work provided the formalism for the development for a non-Euclidean geometry.

Remember that if you have a flat 2-dimensional surface with two parallel lines, those lines will remain parallel infinitely in both directions. But Riemann's geometry was that of curved spaces such as elliptical or hyperbolic spaces. Physically the term space here is meant to define the locations and directions along which objects are constrained (by the local geometry) to move. For example the surface of the earth is a spherically curved space. If one has lines there, such as the lines of longitude, they may be parallel at the equator, but they will not remain parallel. In fact they intersect at the two poles. This is called "positive curvature." You can also have "negative curvature" such as in a horse-saddle. Parallel lines there would never intersect; they would diverge.

This was the basis of Riemann's proposed non-Euclidean geometry. What is of interest to us here is that Riemann formulated the mathematics for expressing the curvature of space, in his Riemann "metric-tensor." His formalism allowed mathematicians and physicists to show the curvature of the local field at every point in space. The metric tensor is not a single number but an array of numbers. For example, for a four-dimensional space this would be "G_{11} x G_{41}"—a 4x4 matrix with 16 elements. It turns out that only 10 of these 16 elements are unique. So for a four-dimensional space you really need ten unique numbers to describe the curvature.

Riemann's work under Gauss was mathematical, but influenced by the physicist Weber; he was also interested in explaining physical forces and fields. To connect forces, fields, and geometry let us consider the following: The easiest example is a concept called "Flatland," a concept popularized by E. A. Abbott in 1884. Flatland is a two-dimensional space, much like a piece of paper. The creatures in Flatland are also two-dimensional (such as circles and squares) and are confined to their two-dimensional space. Let's however con-

4. It can be extended to three or more dimensions but this is not yet *curved* space.

sider a very important, deep principle. If the creatures move along this space and the space is flat, they are not subjected to any other force while moving straight. Now what if we crumple the paper? When the two-dimensional creatures try to move in a straight line through the "hills" and "valleys" we've created, they will be tilted. They do not realize that their space has become crumpled, since they are aware of only two dimensions. So they don't see the reason why they're moving at an angle. In fact, they don't even perceive it.

Riemann proposed that the force which is acting on these creatures and which the creatures feel is due to the geometry of the space which they are in. This was a very revolutionary concept. In fact, this turned out to be the basis of Einstein's General Theory of Relativity. Faraday had said that charges, such as electrons, are the sources of the forces but Riemann said that geometry is the source. This is a tremendously new idea. The problem with Riemann's being a mathematician was that he did not account for the "real-world" source of this "crumpling" or change in geometry. He merely said that if we could explain this, we could explain electricity, magnetism, and gravity, the only forces that were known at that time.

In the 1860s Maxwell proposed the first theory to unify electricity and magnetism. Until then, electricity and magnetism were considered to be two separate forces, but Maxwell showed that one is a manifestation of the other.

A Four Dimensional Force

A tremendous new development came with Einstein. The theories of relativity that Einstein proposed were so radical that they really go against our common sense and are counter-intuitive. He presented his Special Theory of Relativity in 1905, and his General Theory of Relativity in 1915.

A special point which Einstein brought out in his Special Theory of Relativity was that the speed of light is a constant in any frame of reference independent of the frame's velocity. This means to say that if I stand still (i.e., with a constant velocity of zero), and light goes by me at the speed of 186,000 miles per second and impinges on the windshield of a car traveling at 100,000 miles per second in the same direction (a very fast car), one would normally say that within the car the light is traveling only 86,000 miles per second (186,000–100,000 = 86,000). But this is wrong. No matter what the velocity of the frame of reference may be, as long as it is moving at a constant velocity, the speed of light is also a constant. So the speed of light in the car is also 186,000 miles per second. This has been experimentally verified many, many times using accelerators and sub-atomic particles.

Now the question is, how is this possible? It is completely counter-intuitive. Einstein proposed that time must be slowing down in the frame of reference which is moving at some velocity. So if the car is moving at 100,000 miles per second, time has slowed down in that car, so the light in the car is still moving at 186,000 miles per second.

One of the implications of this is called the "twin paradox." If you have two twins and one travels in a spaceship at close to the speed of light, it is proposed that his biological clock will slow down. So if he leaves and then returns twenty years later, his twin on earth will have aged twenty years, while the brother from space will hardly have aged at all. When Professor Branover met with the Rebbe MHM once some years ago and this topic came up, the Rebbe MHM said that even if that's what will happen, when it comes time for the twin who looks younger to pass away from this world, his years will be reckoned by earth years.

One of the things that Einstein proved in his Special Theory of Relativity is that time is actually another dimension. But the Special Theory of Relativity does not include accelerated motion and in particular the acceleration due to gravity.

To resolve this, Einstein developed another theory to include accelerated motion. This became his General Theory of Relativity, which was published in 1915. This theory is based on what Einstein called the principle of equivalence. What he said is that the physics which takes place in an accelerating, not constant, frame of reference, is equivalent to the force of gravity, This means that if people are riding in an elevator and, G-d forbid, the cable snaps, the elevator falls (as an accelerating frame of reference) and the people in the elevator feel a force just like gravity. Thus, when astronauts go up in a rocket, the pressure they feel is denoted in "g" forces, i.e., how many times the force of gravity the acceleration is.

Einstein then investigated how a light beam would behave in an accelerated frame of reference. We know that if the elevator is falling, the light beam must seem to curve. But by the theory of equivalence, light must also curve under the force of gravity. How could light possibly curve, unless the space-time through which it moves is also curved? On this basis, Einstein proposed that space-time is curved and this curvature causes light to bend.

Let's explain this with a very simple example. Consider a rubber membrane stretched across a drum. If you drop a steel ball onto this rubber membrane, the membrane will become deformed with the steel ball in the middle pushing down the membrane. The area around the ball will be curved like a funnel. Now onto this funnel shaped space, roll a marble across the membrane. Instead of going straight across, the marble will curve along the geometry of the membrane. So Einstein proposed that the same thing must happen to light, only light is the marble and space-time is the membrane. In the vicinity of a huge mass, such as a star, space-time is deformed just like the ball on the membrane. When light coming from a distant star passes close to a nearby star, the light must bend, making the distant star's position appear to have shifted. This bending was observed during an eclipse in 1919—a dramatic confirmation of Einstein's theory. In 1915 Einstein proposed this theory in four dimensions—three of space of one of time.

Physics in Higher Dimensions

Six years later, in 1921, Theodor Kaluza, a German mathematician, solved an Einsteinian metric-tensor in five dimensions. He showed that not only does it obey the theory of relativity for the force of gravity, but Maxwell's equations concerning electromagnetism are also accounted for in the fifth dimension. Very simply, a 4x4 matrix is sufficient for Einstein's gravitational theory. If you add another one, creating a five-dimensional metric-tensor, you have solutions for both gravity and the electromagnetic force: *unification*! The problem is that if we do not observe this additional dimension experimentally, and we cannot visualize it in our heads, how can it be a correct description of reality?

To answer this, Oskar Klein, in 1926, suggested that this additional fifth dimension was "compactified." It somehow curved up to such a small size that we can't see it, but it's really there. In the 1920s this didn't make much sense so few physicists took notice. Besides, quantum mechanics had taken center stage and people were thinking in terms of particles and subatomic particles. So Riemann's notion that "force equals geometry" fell into disuse. We will come back to this notion of compactification later.

Quantized Forces

Quantum mechanics was a theory developed to explain the electromagnetic interaction of light and charged particles such as the electron and the proton, and the observed spectroscopic structure of atoms. The electron and proton are charged particles and, according to Faraday, the *source* of the electromagnetic force. In contrast, the photon, a single particle of light, is the *carrier* of the force.

Briefly, for every particle in nature there is a corresponding anti-particle. So for the electron with a negative charge, there is an equal-but-opposite particle called a "positron." When a positron meets an electron, mutual annihilation occurs and two equally energetic and oppositely directed photons are released in their place. The quantum mechanics of the atom, with anti-particles substituted for the normal-matter particles, is indistinguishable from that of normal matter.

Starting in the 1920s, scientists began working on the problem of describing the electromagnetic interaction of electrons, both positive and negative, with light. Using and developing quantum mechanics, they struggled for about 30 years, until the problem was solved in the early 50s by Richard Feynman—a very famous Jewish physicist. He called this theory Quantum Electrodynamics (QED), and for this work he received the Nobel Prize in 1965. As one of his innovations Feynman proposed a diagram—the "Feynman diagram"—as an accounting method to list all possible interactions of these particles. By combining many such Feynman diagrams any electromagnetic interaction could be accurately calculated. By the 1950s QED explained everything we know about electromagnetism of the quantum. QED is the most accurate theory in all of physics, being able by theory alone to accurately predict the measured electron magnetic moment to over 10 decimal places. The closest other physical attribute of a particle that can be predicted by theory alone is accurate to less than four decimal places.

Physicists have been trying to explain how the observed particles were actually built from more fundamental particles. To do this they needed to determine which are the most fundamental of the particles and how many of them exist. So, beginning in the 1930s and continuing even today, physicists have built particle accelerators to uncover the basic building blocks. But when they used these accelerators to break apart protons and neutrons, they found they would release many other particles. In fact, instead of reaching a limiting number of building blocks, they were just finding more and more—reaching 500-600! They were moving further away from finding a theory of everything since it was becoming more and more difficult to incorporate the hoards of additional particles as they were discovered. Nevertheless, a theory has been developed in an attempt to explain all that is observed, and next I will try to explain two theories which are generally accepted as the best explanations to date.

As we have said, there is a strong nuclear force and a weak nuclear force. The strong force binds individual neutrons and protons together to form the atomic nucleus. When the scientists started bombarding nuclei and breaking them up, they found other particles more basic than these. To make a long story short, the basic building blocks are called quarks. Different types of quarks are characterized by a combination of three "colors" and six "flavors." Quarks have never been detected one at a time; only packets of three quarks have been observed. If one puts three quarks together in a certain fashion, one will produce a proton, while if one puts them together differently, one will produce a neutron, and similarly for each of the other basic particles observed. What holds these quarks together is called a "gluon" (glue-on). A gluon has a very interesting property: normally, when one moves away from the source of a force—like electricity or gravity—the force diminishes. Gluon-gluon interactions, on the other hand, are like a rubber band. When one stretches it, the force becomes stronger and stronger. So it is nearly impossible to pull quarks apart. The harder one pulls, the stronger the gluons hold the quark together. This theory of the strong nuclear force is called Quantum Chromodynamics (as contrasted with Quantum Electrodynamics).

The previous discussion then serves as an introduction to the state of physics in the 1970s. However, resurgence in interest in the connection between forces and geometry began in the 1950s, only to culminate today in two "theories of everything." One is called the "Standard Model," which is a completely quantum mechanical theory of the two nuclear forces and electromagnetism but excludes gravity. The other is called "Superstring Theory" and it includes all of the four forces. The point to remember from all of this is that if geometry can be a cause of forces, multidimensional spaces may be necessary to express their connection.

Symmetry and Forces

Around the 1960s physicists started making progress in organizing the observed hundreds of particles and the forces between them. But in order to understand the organizing principles and the evolution toward a "Theory of Everything" we will have to understand the mathematical concept called "symmetry" and the role of symmetry on the nature of these forces. We will learn that critical to understanding the difference between these forces is the notion that the breaking of a symmetry causes the separation of a force from the whole.

Symmetries can be geometrical and thus easy to imagine, but some are very abstract and harder to grasp. Let's start with a geometrical symmetry—the case of a circle. One has a circle which is allowed to rotate around its center point. Given any amount of rotation, the circle will look exactly the same. So the circle is considered to have complete "rotational symmetry," i.e., it looks the same even after one rotates it. If instead of a circle, we tried this with a square, the only way it could be rotated and still look the same is if we rotate it about its center in integer multiples of 90 degrees. Then it would look the same as it did when we started. So its symmetry is in steps of 90 degrees. If we have an equilateral triangle we also have rotational symmetry, but of 60 degrees. From this we can imagine all sorts of 2 and 3-dimensional symmetries. A sphere, for example, has complete rotational symmetry about its center in 3 dimensions. A physical vacuum, like the sphere, has complete rotational symmetry about a point—all directions appear similar since "nothing" is there.

These are concepts which we can relate to, but now, if by analogy time is a fourth dimension, it should be possible to make "rotations" that convert space into time and vice versa. Such 4-dimensional "rotations" are precisely the distortions of space and time demanded by Special Relativity. Thus space and time must be considered as different parts of the same physical entity—"space-time." Such ideas that combine apparently separate physical notions are considered to "unify" our understanding of nature, and as such are termed "beautiful." A unification combines different pieces of physical information so that some new physical information is uncovered and all old physical information is retained.

Another simple physical symmetry is that of charge symmetry. If all of the charge in a system of particles such as an atom is changed to the opposite type, exchange positive for negative, and vice versa, the physics should not change. Thus the light emitted by an excited hydrogen atom with a positively charged proton in the nucleus and circled by a negative electron should be the same as that emitted by a negatively charged anti-proton in the nucleus circled by a positron (anti-electron).

A more complicated geometrical symmetry is that due to reflection about a plane. For example, if you look at an object in the mirror and it looks exactly the same as when you look at it directly, we would say the object has reflection symmetry. The capital letters A, I, O, and U possess reflection symmetry. Visually human beings do not have an exact reflective symmetry. Unless you look exactly the same right and left, when you look in the mirror, you see yourself differently than other people see you, because of the difference between right and left. Functionally, human beings are either left or right "handed." In physics a reflective symmetry is called "parity," and the handedness is termed as left or right handed "chirality."

All the forces in nature except one exhibit parity, i.e., upon reflection they don't change. The one exception is the weak nuclear force. It exhibits "chirality," i.e., it has a preference to a particular handedness. Now chirality is a property of the number of dimensions in a space. Chirality can only exist if the space contains an odd number of dimensions. Thus the 4 dimensions of space-time cannot account for the observed handedness of the weak nuclear force. When a neutron decays into a proton, an electron and a neutrino, there is a preferred electron spin-handedness with respect to the direction in which the electron is emitted. So there is a difference between right and left-handedness for that interaction. This is only true for the weak force.

Thus the handedness of the weak force due to its odd dimensional requirement is termed to "break" the symmetry of parity that the other 3 forces possess. Such symmetry breaking is a physical marker indicating the cause of the manifestation of a separate force.

Both nuclear forces are not gauge invariant. "Gauge" symmetry is a quality that depends only on the magnitude of a change in that quality. But this quality is much more difficult to explain. Money, gravitational energy and voltage possess gauge symmetry. Each quality depends on some point of reference. The value of the dollar is tied to the value of the yen or mark or gold. It's a relative quality. Similarly, the amount of energy you must expend against the force of gravity to move an object from the floor to the table is also relative. It depends only on the weight of the object and the height you must move it. It does not depend on where you started; starting from the table and moving it an equal height to a shelf requires exactly the same amount of energy. Another example often given in books is as follows: You have a voltage difference between two objects, say 2 volts on one and 10 volts on another. On another pair the voltage difference is say from 4 to 12 volts. In both cases the difference is an increase of 8 volts. Therefore, the two voltages have gauge symmetry—an equal relative change.

There are many more types of symmetry that are relevant to the study of the forces of nature but by now we have a general picture. Combining these different types of symmetry determines the laws of physics. In quantum electrodynamics, the electromagnetic field is symmetrical to rotation into itself. This symmetry is called U1 where the "1" signifies the electron and "U" represents the rotational group. For quantum chromodynamics, there are three colored quarks—red, green, and blue—so we have a different kind of symmetry, called "SU3." For the weak force, there are two particles involved—the electron and the neutrino. This symmetry permits the exchange of these two particles for one another by the rotational group "SU2." So we have 1, 2, and 3-particle symmetries. Combining these symmetry groups results in a super-symmetry noted as SU3 x SU2 x SU1. Together these quantum theories of the two nuclear forces and electromagnetism is called the "Standard Model." However, keep in mind that none of these theories made through the early 1970s take gravity into account. In addition, attempts to quantize gravity have led to a number of unsolved problems including a universe with an infinite energy density (which is clearly not observed).

Einstein believed that geometry would determine the particles as well as the dynamics of all forces. One side of the field equations of general relativity depended only on the geometry of space while the other contained mass quantities which he had to add in an arbitrary way. For this he would say that the geometrical side was beautiful but that the other side was ugly. To correct this he spent much of the remainder of his life trying to convert the matter qualities to geometrical qualities. He never succeeded in the sense that he could never understand matter as geometry. But that is exactly what String Theory does.

Multi-Dimensional Superstring Theory

The "Theory of Strings" was developed sporadically in the time between the 60s and the early 80s. It is actually now called "Superstring Theory." But the theory of superstrings turns out to be consistent only in spaces of 26 and 10 dimensions. Now those who have studied Chassidus or Kabbala know that the numbers 26 and 10 are very, very relevant to creation.[5] The Rebbe MHM has said that the world was created in a manner to show that the Creator put His imprint on the creation, in order for us to know Him from the creation.

We have talked about particles. Electrons, quarks, and all the other things here are "point particles." A point particle is very limited in the various vibrations it can exhibit. So the theory of strings suggests that perhaps the basic matter is not particles, but strings. By a "string" we are referring to a certain manifestation of space-time. Exactly what the manifestation is, we don't know, and we are not sure if anybody knows. But the concept is that instead of being a point, a particle is a string.

The string is very, very small. It has what is called a Planck length, or 10^{-35} meters. That's extremely small. It's 1 divided by: a 1 followed by 35 zeros. Now if we set this string to vibrate and keep the ends

5. Twenty-six is the numerical value of the Hebrew name of G-d. He created the universe through a process of tzimtzum in which He emanated the ten *sefiros of Atzilus* and through them brought the universe into actuality. See chap. I, sec. 2, "Wisdom From Above," for a more extensive discussion of this.

fixed, it will vibrate in different modes. To illustrate this, if we vibrate a violin string, it will only vibrate in certain modes—certain vibrations—which are called tones, or frequencies. Different particles are just different vibrational or rotational modes of these multi-dimensional strings.

Thus String Theory maintains that the string is the fundamental element of nature, that different vibrations correspond to the different "particles," and that the proper number of dimensions in nature is greater than 10. Recently, a closer examination of this theory suggested that these strings should be closed, like a circle, which can also vibrate. Such strings are called "heterotonic strings", and can have traveling-wave vibrations traveling clockwise or counterclockwise around the circle. Furthermore, it turns out that there are only two levels of multi-dimensional space for which this theory is consistent (i.e., it is solvable—with no infinities). For counterclockwise vibration it is 26 dimensions and for clockwise vibrations it is 10 dimensions. So we see again the "10" and the "26."

In addition, the theory predicts that 16 of the 26 dimensions of 26-dimensional space are invisible, leaving 10. It turns out that the lost 16 dimensions contain all the symmetry necessary to explain all four of these forces. It doesn't tell us how, but it has enough symmetry to explain them, and potentially explains everything in our world. The *vibrations* of the strings correspond to the particles and the *symmetry* of the strings corresponds to the four physical forces which we discussed. This is a very simple yet powerful description of nature and though it is a long way from being fully worked out, it has many admirers.

In the physical world, we don't see 26 or 10 dimensions—we see only four. So the scientists theorized, following the ideas of Oskar Klein, that the ten dimensions became compactified, or curled up into very tiny unique spaces. These dimensions do not take up room in normal 3-dimensional space but are orthogonal to it. How can this be? The idea is as follows. Imagine a line in 3-dimensional space, and then imagine that if you could look very closely you would see that the line is really a compactified cylinder. Walking along such a curled up 4th dimension, you would return to the starting point because geometrically the local space is circularly shaped (and globally it is cylindrical). Furthermore at every [point][6] along this long thin cylinder a different 4-dimensional universe would be attached. As you would move from one [point] along this cylinder to the adjacent [point] a new 4-dimensional universe would appear. In all, an infinite number of 4-dimensional universes would be available for you to explore in your inter-dimensional travels.

Let's examine a fundamental problem with this theory, and then its implications. The problem is that the theory appears to be scientifically untestable. The reason for this is that the small spaces in which these superstrings reside can only be probed at tremendous energies, like the energies which existed at the time of creation. This is troubling because through the history of physics, theory and experiment always went hand-in-hand. Thus physicists are seeking predictions of this theory which can be tested. The problem is that no one can solve the equations of Superstring Theory even approximately, thus making prediction impossible. The physics of superstrings is at an impasse. We have no way to harness these energies now (even a thorium reactor[7] would not come close to these energies) and no way to solve the equations. Why are physicists continuing to support this theory? The physicists say it's *mathematically beautiful* and can—if it can ever be solved in a single equation—explain how (not why) the universe functions the way it does. It is too appealing to ignore.

The Anthropic Principle

Two final points regarding the ramifications of Superstring Theory are seen in the Anthropic Principle.[8] First, if we look at the constants of nature, we see that they are exactly right for human life to exist in this world. As the Nobel laureate physicist Steven Weinberg has stated it, "The world is the way it is, at least in part, because otherwise there would be no one to ask why it is the way it is." This is a definition of the so called "weak anthropic principle" and most scientists

6. In Dr. Weber's manuscript, the word "line" appears here, but it appears that the intention is a point.

7. This is the Radkowsky thorium reactor. See chap. III, sec. 7, for a description of the reactor and our interview with Prof. Radkowsky. At the Moshiach and Science conference in 1994, Mr. Joel Grae, president of the Radkowsky Thorium Power Corporation (now known as Lightbridge Corporation) and one of the company engineers spoke about the reactor.

8. See chap. IV, sec. 3 for a discussion of the Anthropic Principle.

would not take issue with it. The "strong anthropic principle," much more contentious amongst physicists, says that there had to be an intelligent entity—a Creator, Who created the world with constants commensurate to the existence of life. Thus such a "strong anthropic principle" suggests that the tremendous coincidence of these constants' being right for life is in fact no coincidence at all, and that there must have been a Creator who made them all right in the first place.

Second, if you can learn to control these forces, you can actually obtain energy which was in existence at the time of the creation of the universe. What could be done with this? One of the things which have been proposed is what is called "worm holes." Normally, to get from one place in space to another, one cannot go faster than the speed of light. A worm hole is rupture in the fabric of space-time. If one could go through it, one could actually travel from one place to another in space without being bounded by the speed of light. A worm hole can, in theory, be created. Mass and energy are equivalent. If one can bring together enough mass, one can perturb space-time. If one can bring enough energy to bear, one can actually see the fabric of space-time. This allows one to see from one space-time in the universe to another without the limitations of the speed of light. One can see instantaneously. The Midrash, in describing the first day of creation, says that G-d created the light—let's interpret it as energy—with which Adam could see from one end of the world to the other.

The Future

I believe that, with the help of G-d, in the times of Moshiach, G-d will give us the power to control these energies that prevailed at the time of creation. As the Gemara and Midrash tell us, G-d is hiding this light until the time of Moshiach, when He will reveal it again. That means that we will then have the ability to control the space-time fabric, because then we will have the ability to solve the equations which control it, so we will be able to see from one end of the universe to the other instantaneously, and perhaps other very wondrous things.

I now want to conclude by talking about how this explains some concepts in Chassidus. We started off with the sicha of the Rebbe

MHM from the Parsha of Noach where he talks about the multiplicity of creatures which were created from the perfect unity of G-d. The Rebbe MHM said that by studying and understanding these attempts by physicists to unify all the forces of nature and all the matter of the universe, we can actually come closer to an understanding of this unity in an absolute sense—one can have a multiplicity of creatures from a basic unity, as discussed in Chassidus.

It seems very obvious that if one has a single basic element which is the string, and all the forces and all the matter which we see are merely its different vibrations, then we have exactly this concept—that from the basic unity of this string, one can see all of the forces which are in all the elements of matter which are in the universe. As Chassidus explains, the actual multiplicity comes as a result of the ten *sfiros* of *Atzilus* and so on all the way down to the ten *sefiros* of *Asiya*—that these are a means through which G-d, the perfect unity, can manifest Himself in a multitude of creations. The manifestation comes through the ten *sefiros*.

We can see a very specific connection. The numerical equivalent of G-d's name is 26, which is the basic force of creation. As the Alter Rebbe explains in *Tanya* and elsewhere, this comes down and is clothed in the ten *sefiros* ("The *yud* is wisdom..."). The equivalent of *sefiros* from the physicist's point of view is the ten dimensions. The ten dimensions allow this basic unity—for the string to be able to manifest itself in these various vibrations and exhibit all the forces of nature and all the particles which we observe.

This was a very brief attempt to show how understanding a concept in physics can help us understand Chassidus. The Chassidic and Kabbalistic notions of 26, the name of G-d, and the ten *sefiros* etc., and how they manifest themselves down here in a world of our dimensions, should certainly be examined much further. Possibly through understanding these theories we can reach a further understanding, which we shall definitely have when Moshiach reveals himself.

Auditory Hair Cell Regeneration

By Professor Levi A. Reiter

At the conference on Moshiach and Science of the RYAL Institute in 1995, Professor Reiter delivered the keynote address. He discussed the recent discovery of the regeneration of hair cells in the ear and explained how it can be considered a model to help understand Techiyas HaMeisim. He lectured on this a few more times over the years, most recently at Touro College in 2009, to keep us informed of the latest research. This section is based on those lectures.

Techiyas HaMeisim (revival of the dead) is completely above the order of nature. Chassidus explains that Techiyas HaMeisim comes from Atzmus, the essence of G-d Himself, relative to which life and death are equal.[1] Thus Techiyas HaMeisim is even above the order of miracles. The "second era" within the Era of Moshiach, when there will be innovations in the world order including the fusion of the natural order[2] with the miraculous order, is called the Era of Techiyas HaMeisim.

Nevertheless, as with all aspects of the Era of Moshiach, we are given preparation for it in advance, as the Rebbe MHM explained in the sicha of Parshas Noach that the discoveries in the sciences in recent times prepare us for the revelations of the Era of Moshiach.

Professor Reiter is Professor of Audiology and Hearing Sciences, and Audiology Program Head at Hofstra University in Hempstead, New York. He is also on the faculty council of the Long Island Doctor of Audiology Consortium, and maintains a private practice in audiological testing and hearing aids in Crown Heights. He was recently in the news for explaining a mysterious audiological phenomenon which earned him the informal title of "the guru of hearing."[3]

1. *Sefer HaMa'amorim* 5670, 100.
2. See "the Tel Aviv Lecture," chap. VI, secs. 1b and 1c.
3. See the CBS News video online at http://www.youtube.com/watch?v=ZrL4iU5dFX8

Chassidus explains that everything in the world of physicality has a corresponding point in the heavens which makes the physical event possible.

Therefore at a time such as now when the Lubavitcher Rebbe has stated as a biblical prophecy that "The time of your redemption has arrived," there should be evidence in the sciences, as in all aspects of physical life, of such a radical change which is about to take place in the world order.

In the time of Redemption one of the significant miracles which goes against all of our current scientific principles—yet is one of the 13 fundamental principles of the Jewish faith—is the resurrection of the dead. All the Jewish people who have ever lived and died will arise from their graves and be alive again in physical bodies.

The Rebbe, uttered this prophecy 18 years ago, and at about the same time period an amazing process of scientific discovery had just begun. The initial discoveries were made quite accidentally in the field of hearing science—discoveries which reflected faithfully the Rebbe's predictions that Moshiach, redemption and indeed resurrection are truly in the air.

The Process of Hearing

Before going any further, a little background about sound and the hearing mechanism itself is in order.

We are all surrounded by a medium known as air. Air may seem to be almost spiritual in nature, but it is quite physical. It consists of billions of tiny chunks, known as "air particles". In fact, in just one cubic inch of air, there are 400 billion billion air particles. Air is that dense.

But, do we have to take it on faith that there really are invisible air particles here? No. When we drive to the country at about 70 mph, if we put our hands out the car window, we can actually feel the air particles coursing through our fingers because our driving speed is condensing them further, and they become tangible.

Very generally, sound is nothing more than the movement of these same air particles against our ear drums and auditory system.

What follows is a highly oversimplified example of how the process of hearing works. It will, however, serve to convey the point of this lecture.

The ear consists of three sections: the outer ear, the middle ear and the inner ear.

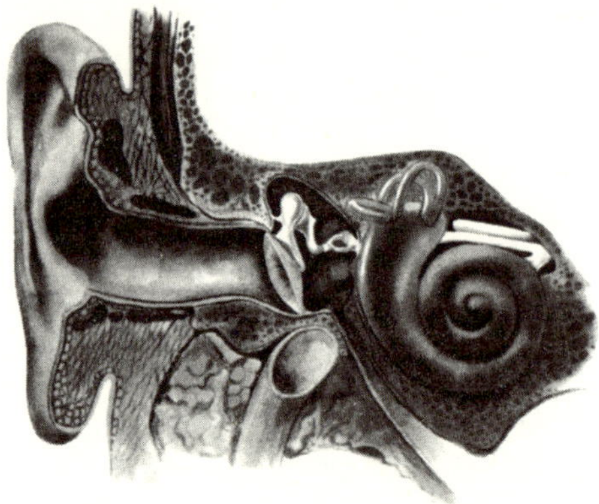

Figure 1

The outer ear consists of the *pinna*—a large funnel-like device, and the ear canal—which is a one inch long tube capped at the end by the ear drum.

Sound is heard because a vibrating object, like your vocal cords while speaking, or a car horn, etc. is disturbing the surrounding air particles, causing them to move or vibrate at the same rate. Now, if the vibrating air particles happen to be caught by the pinna and directed through the ear canal, they will vibrate the ear drum itself.

In other words, the ear drum would now be vibrating at the same rate or frequency as the air particles—which are vibrating at the same rate as the vocal chords, etc.

Is that enough to account for hearing? Not yet. If the process would end here, we would all be deaf. That is, the outer ear is important only in helping to direct air particles to the drum, but the outer ear does no hearing itself.

What about the middle ear? This is merely a hollow chamber containing the three smallest bones of the human body. The bones—known as *ossicles*—actually form a chain which bridges the gap between the ear drum at one end and the inner ear, or *cochlea*, at the other. When air particles vibrate the ear drum back and forth, the three ossicles of the middle ear vibrate as well, and the last bone, known as the *stapes*—pushes in and out of a little hole leading into the inner ear or cochlea. The inner ear is a fluid-filled system. Thus the piston-like action of the stapes pushes the inner ear fluids back and forth.

Now, bathed in this fluid are thousands of highly specialized cells known as "hair cells." These are wonderful, microscopic cells with remarkable little hair-like structures projecting upwards into the fluid.

Above is a diagram of a hair cell. Note that the body of this cell is similar to all of our bodily cells—with one exception. The hair cell is crowned with a unique *cuticular plate* on its top, into which are stuck many projections known as hairs, or *stereocilia*. When the inner ear fluids vibrate from sound, they flow back and forth over the hair cells, causing the projections to bend. Whenever they do bend, they release a drop of a neurochemical transmitter substance onto a corresponding nerve fiber which is connected to each hair cell and ultimately travels to a specific spot in the brain. When the brain receives the impulse, we hear sound.

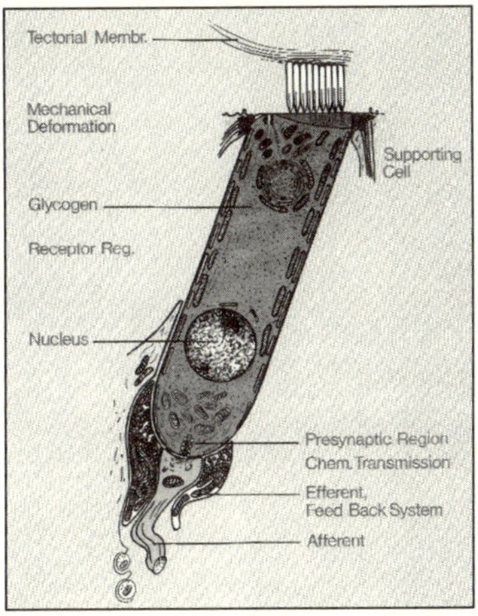

Figure 2

This works in two ways:

1) Specific sets of hair cells are tuned to bend or to vibrate in response to certain specific sounds. For example, high pitched sounds will bend hair cells located at the beginning or "base" of the cochlea, while low pitched sounds will bend

cells at the end or "apex" of the inner ear. Thus, the parti ular spot in the brain which receives the nerve impulse ide tifies which hair cells were stimulated, and therefore, which sound frequencies or pitches are being heard.

2) In addition to the *place* of hair cell vibration, the *rate* at which the hair cells vibrate may also provide the brain with information. This is because the rate of hair cell vibration follows the rate of the air particles' vibrations, which follow the rate of vibration of the vocal tract, etc., and so another code may be sent to the brain for interpretation. If the brain receives 300 nerve-impulses/second, it knows that the vocaization had a frequency of 300 cycles per second.

In short, the single structure most critical to the presence of hearing is the hair cell. It is the link between sounds in the environment and the brain.

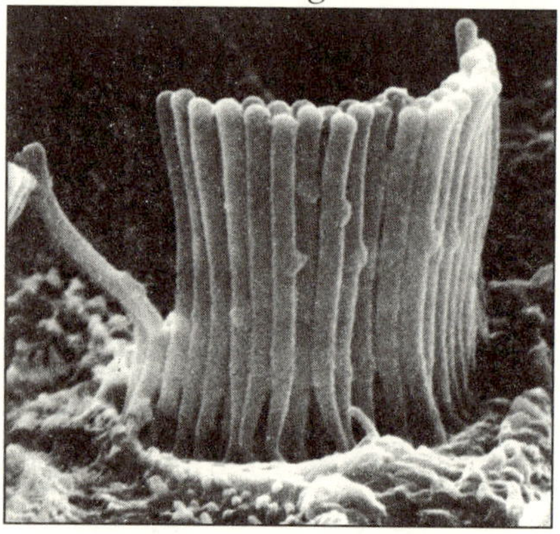

Figure 3

Unfortunately, hair cells are also the most easily damaged structures of the ear, and when that damage occurs, it normally results in permanent sensory-neural hearing loss. For example, noise from power tools and loud music may be deadly to our hair cells. The same is true for certain antibiotics, cancer drugs, and water pills. Aging, disease, and hereditary factors may also cause deafness by destroying the hair cells of the inner ear.

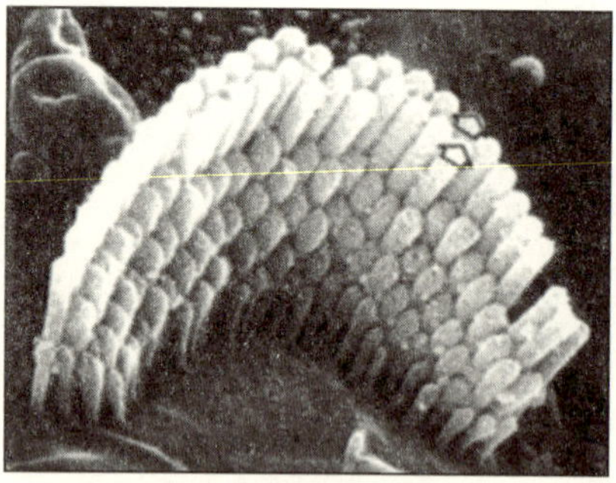

Figure 4

Up until now hair cell damage and destruction was considered to be irreversible. After all, if one loses the nervous tissue of an eye, G-d forbid, one loses sight forever. The same holds true for the sensory hair cells of the ear. They do not grow back.

Let's take a closer look at a hair cell.

Figure 3 shows an electron microscopic view of a hair cell magnified 12,000 times. Note the beautiful geometric symmetry of these projections.

Figure 4 shows a "bird's eye view" of another hair cell. In this view, we see that the cilia are arranged in a definite U-shaped pattern, in 3-4 parallel rows with the outer row being the tallest.

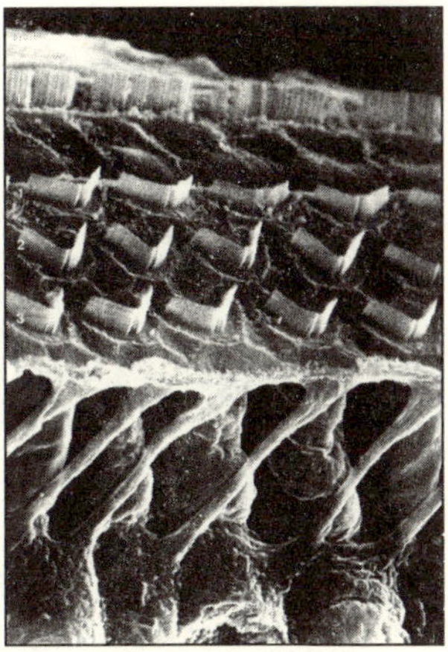

Figure 5

Figure 5 presents a more distant look. Instead of one hair cell, we see the tops of many—peeking through a flat supportive tissue termed the *reticular lamina*. Note that beneath each "W"pattern of hair projections, there is a vertically oriented hair cell body. In addition to the hair cells, Figure 5 shows other cells that serve only to provide structural support for the delicate hair cells. *Deiter cells*, for example, can be seen to be holding or cupping the base of each hair cell, securing its rigid positioning. The Deiter cells, and other supportive tissue help to form the sensory body of the cochlea known as the *organ of corti*.

Hair Cell Regeneration and Techiyas HaMeisim

Before the time the Rebbe began speaking intensively about redemption and Techiyas HaMeisim, there was not even the faintest glimmer of hope that the nervous tissue of the inner ear (or for that matter any sensory nervous tissue) would regenerate and return to the world of the living. But it is this very regeneration which has now been observed as we are now going to describe.

Several years ago, Dr. Raul Cruz (1987) of the University of Washington, was doing an experiment on *ototoxic* medications—medicines that can harm hearing while trying to help heal a person. We know that certain antibiotics, especially the group ending in the suffix "mycin"—streptomycin, kanamycin, gentamycin, neomycin, tobramycin, etc.—are highly ototoxic. But don't worry, people have little to fear from erythromycin.

Dr. Cruz (1987) had a good question. He wanted to know how much time it would take for the ototoxic drug gentamycin to ravage the entire cochlea, ultimately causing deafness by eliminating the cochlea's hair cell population. He used chickens in his research, because their cochleas are somewhat similar structurally to our own. In his experiment he gave a large group of chicks a 10 day dosage of gentamycin injections—10 days being a conventional regimen for antibiotic therapy.

Following the 10 day gentamycin treatment which was given to all chicks, the animals were divided into several groups. Each of these groups was then "sacrificed" and examined microscopically a different number of days after the gentamycin treatment had ended.

Group # 1 was "sacrificed" right after the 10th day of gentamycin-treatment. The cochleas of this group were dissected and the hair cells counted. The results showed that in the very first segment of the cochlea (i.e., in the basal turn) there was nearly total elimination of the hair cells. This would translate to near total deafness for the high pitched sounds, which is indeed what we find in many people who have had to take this kind of ototoxic medication.

Group # 2 was "sacrificed" a week after the first group, that is, one week after the 10-day regimen had ended. When the cochleas were studied microscopically, hair cell damage was found further along the length of the cochlea—as would be expected. However, Dr. Cruz was shocked to find that the basal region of the cochlea—which had been depleted of hair cells initially—was now partially repopulated with new budding hair cells. By the end of the experiment 32 days later, the last group's cochleas were entirely repopulated with new hair cells.

At the same time period, a similar unexpected finding emerged when Dr. Douglas Cotanche (1986) of Harvard University studied noise-induced deafness in chicks. It has long been known that loud music, power tools, gunfire, and so on, can permanently damage hearing by destroying the cochlea's delicate hair cells. Dr. Cotanche was interested in understanding exactly where along the length of the cochlea the damage would be located when it is overstimulated by powerful tones of various frequencies. Several days after observing the devastating hair cell destruction that resulted, he was amazed to find that newly regenerated hair cells were beginning to appear.

In more recent years, much research has been focusing on the mammalian cochlea in the attempt to relate the experimental findings more directly to hearing in humans.

Professor Yehoash Raphael (2005) of the University of Michigan identified a gene—which he appropriately termed *Atoh 1*—which when injected into the cochleas of deafened chinchillas caused the regrowth of fully functional hair cells. Amazingly, once the new hair cells developed, nerve fibers from the auditory nerve managed to find their way to the new cells, enabling them to communicate with the brain. In short Raphael's deaf chinchillas began to hear once again.[4]

The fact that in this day and age the field of biological science is suddenly shocked to find evidence of regeneration of dead nervous tissue in the ears of vertebrates and warm blooded mammals, should warm the hearts of each and every one of us, for this is truly the Rebbe's prophecy of *Higia Z'man Geulaschem* ("the time of your redemption has arrived") finding expression in the world of science, as predicted by Kabbala and Chassidus. And as the Rebbe tries to impress upon us, if only we will open our eyes we will see Moshiach and live with the redemption as it unfolds.

As icing on the cake, in Halacha it states that if one injures another, he must compensate the injured party according to his current worth in the market place. For example, if he lost an eye, how much less can he now earn? This determines how much he must be paid. However, if the injury destroyed his hearing, then he must be paid the worth of his life. Because hearing is such a vital sense, its total loss is equated with death itself.

In this light, hair cell regeneration is truly similar to Techiyas HaMeisim—may it be truly now.

4. See P. V. Viastarakos, et al., "Sensory cell regeneration and stem cells: What we have already achieved in the management of deafness," *Otology & Neurotology* 29, no. 6 (2008): 758–68 for the latest review of the literature.

REFERENCES: CHAPTER V

The Quest for Unification

Sec. 2: A System of Logic for Messianic Phenomena

Rabbi DovBer of Lubavitch, the Mittler Rebbe.

Ma'amorei Admur HaEmtzai, Devarim vol.1. New York: Kehot Publication Society.

Rabbi Menachm Mendel Schneerson, MHM.

Likutei Sichos, vol. 17. New York: Kehot Publication Society.

Sefer HaSichos, 5750 - 5752. New York: Kehot Publication Society.

Sefer HaMa'amorim Melukat, vol. 4-5. New York: Kehot Publication Society.

Sha'arei Ge'ulah, vol. 2. Jerusalem: Heichal Menachem, 1992.

Rabbi Yosef Yitzchok Schneerson.

Likutei Diburim. New York: Kehot Publication Society.

Bolc, Leonard and Piotr Borowik, *Many-Valued Logics, vol. 1, Theoretical Foundations*. Berlin: Springer Verlag, 1

Eisberg, Robert and Robert Resnick, *Quantum Physics of Atoms, 2nd edition*. New York: Wiley, 1985.

Epstein, Richard. *The Semantic Foundations of Logic - Propositional Logics*, 2nd edition. New York: Oxford University Press, 1995.

Enderton, Herbert. *A Mathematical Introduction to Logic*, 2nd edition. New York: Academic Press, 2001.

Kaye, Richard. *The Mathematics of Logic*. Cambridge University Press, 2007.

Mendelson, Elliot. *Introduction to Mathematical Logic, 4th edition.* London: Chapman & Hall, 1997.

Nguyen, H.T. and E.A. Walker, *Fuzzy Logic*. London: Chapman & Hall, 2000.

Rescher, Nicholas. *Many-Valued Logics*. New York: McGraw-Hill, 1969.

Simovici, D. and R.Tenney, *Relational Database Systems*. New York: Academic Press, 1995.

Sec. 3: The Concept of Infinity in Mathematics and Chassidus

Rabbi Menachem Mendel of Lubavitch, the Tzemach Tzedek.

Derech Mitzvosecha (Hebrew). New York: Kehot Publication Society.

Rabbi Menachm Mendel Schneerson, MHM.

Likutei Sichos, vol. 36. New York: Kehot Publication Society.

Sefer HaMa'amorim Melukat, vol. 3. New York: Kehot Publication Society.

Rucker, Rudy. *Infinity and the Mind.* Princeton University Press, 2004. This book has an extensive discussion of many aspects of mathematical infinity.

Sec. 4-5: Theories of Everything / Unifying Theories in Physics

Rabbi Menachm Mendel Schneerson, MHM.

Likutei Sichos, vol. 15. New York: Kehot Publication Society.

Foundational Problems in the Special Sciences, ***Part two of the proceed ings of the Fifth International Congress of Logic, Methodology and Philosophy of Science, London, Ontario, Canada.*** **Boston: D. Reidel Publishing Co., 1977.**

Greene, Brian.

"A Theory of Everything?" at www.pbs.org/wgbh/nova/elegant/everything.html

The Elegant Universe: Superstrings, Hidden Dimensions, and the Quest for the Ultimate Theory. W. W. Norton & Company, 2003.

Smolin, Lee.

The Trouble with Physics. Boston: Houghton Mifflin, 2006.

Sec. 6: Auditory Hair Cell Regeneration

Cotanche, DA. Regeneration of hair cell stereociliary bundles in the chick cochlea following severe acoustic trauma. *Hear Res.* 1987; 30:181–195.

Cruz, R.M., Lambert, P.R., & Rubel, E.W. Light microscopic evidence of hair cell regeneration after gentamycin toxicity in chick cochlea. *Arch Otolaryngol Head Neck Surg.* 1987; 113(10):1058-1062.

Raphael, Y. Auditory hair cell replacement and hearing improvement by Atoh1 gene therapy in deaf mammals. *Nature Medicine,* 2005, vol. 11, n°3, pp. 271-276.

Viastarakos, PV, Nikolopoulos, T.P., Tavoulari, E., Papacharalambous, G., et al. Sensory cell regeneration and stem cells: what we have already achieved in the management of deafness. *Otology & Neurotology, 2008;* 29(6): 758-68.

CHAPTER VI

Special Lectures

The Tel Aviv Lecture

On Rosh Chodesh Shevat, 5756 (1996), an international conference on Moshiach and Science took place at Tel Aviv University. It was a joint project of the Igud L'maan HaGeulah Ha'Amitis V'Hashleima (Organization for the True and Complete Geula) directed by Rabbi Zimroni Tzik, and the Rabbi Yisroel Aryeh Leib Research Institute on Moshiach and the Sciences. Four major lectures were delivered by renowned Rabbis and scientists: 1) Rabbi Yitzchok Ginsburgh, 2) Rabbi Shmuel Chefer, 3) Prof. Yirmiyahu Branover and 4) Prof. Shimon Silman.

We present here Prof. Silman's lecture in four parts. We have preserved the original text of the lecture as much as possible, updating it occasionally in the footnotes. Many of the topics covered in the lecture are also discussed elsewhere in this book at greater length. References to sections in this book where some points are developed further are also given in the footnotes. In a sense, this lecture can be considered a summary of the book.

Part 1: Deeper Research

The first part focuses on the visualization of G-dliness in the Era of Moshiach, while addressing many basic questions in the areas of faith, science and the Era of Moshiach.

The declaration that we opened the conference with, *"Yechi Adoneinu Moreinu V'Rabeinu Melech HaMoshiach L'Olam Voed,"* states the fact that the Rebbe Melech HaMoshiach is alive, one hundred percent, a soul in a body and continues to lead the Jewish People through the Geula that we are already in. One of the manifestations of this that we see on a daily basis is that we are able to write a letter to the Rebbe MHM, open up one of his volumes of *Igros Kodesh* (published letters), put the letter inside and receive an answer—a prophetic answer—directly to the point of the question we asked and sometimes even going beyond the question we asked.[1] It happened to me some weeks ago that I put a note in the *Igros Kodesh* and then opened it to the place where I had placed the note and saw the following words: *"Certainly you participated in the international conference on research...."*

At that time I didn't know of any international conference on research in my profession (mathematics), or in any other area. But I knew that I had better keep my eyes and ears open. The next morning at 8 o'clock the telephone rang. It was a call from a Lubavitcher *shaliach* in Israel whom I had never met before, Rabbi Zimroni Tzik. He told me that he was organizing a conference in Israel for research in the area of Moshiach, Geulah and Science and he wanted it to be a joint project of his organization and the RYAL Institute. I knew right away that this was the conference that the Rebbe MHM was referring to!

This is one of many day-to-day events that are happening all the time to all of us—not necessarily only to Jews who are *tzadikim* and scholars, but to everyone. Just as when the Rebbe MHM gave out

1. For more on Igros Kodesh see the end of Chapter I, sec. 6.

dollars the opportunity to receive a dollar and a *brocha* was open and available to everyone, advice and miracles of the Rebbe MHM are now available to everyone.

The RYAL Institute

I am honored to represent the Rabbi Yisroel Aryeh Leib Research Institute on Moshiach and the Sciences which is named for the youngest brother of the Rebbe MHM. Rabbi Yisroel Aryeh Leib was a great chosid who was famous for his brilliance and diligence in the study of Chassidus. He was a mathematician by profession and he held a position in the Department of Theoretical Physics at the University of Liverpool, England. He passed away on 13 Iyar, 5712 (1952) at the age of 45.[2]

By 1992, the year of Rabbi Yisroel Aryeh Leib's 40th yartzeit, the Rebbe Melech HaMoshiach had already announced, *"Higia Zman Geulaschem!"*—that the time of our Redemption had arrived and that we were already in the Era of Moshiach. All of our activities were to be directed to advancing further into the Era of Moshiach. Thus, on Rabbi Yisroel Aryeh Leib's 40th yartzeit, 13 Iyar of that year, I organized a conference in Minnesota (where I was then living) on the theme "The Role of Science in the Era of Moshiach." The keynote speaker was the famous Lubavitcher mathematician Professor Pesach (Paul) Rosenbloom, to whom the Rebbe MHM had given a manuscript of a mathematical work of Rabbi Yisroel Aryeh Leib to edit and publish 17 years earlier.

Later that year I moved to Crown Heights, and before Rabbi Yisroel Aryeh Leib's next *yartzeit* I asked the Rebbe MHM if I should organize another conference on Moshiach and science for the 41st yartzeit. He answered "Yes" and we had an exciting conference on "Bringing Moshiach into the World of Science." The keynote speaker was Rabbi Dr. Naftali Berg, *z.l.*, Director of Research of the Advanced Technology Office of the Army Research Laboratory of the Pentagon, who spoke about the Rebbe MHM's statement that the prophecy of Isaiah —"They shall beat their swords into plowshares" —had already begun to be fulfilled. Dr. Berg brought examples from

2. For the biography of Rabbi Yisroel Aryeh Leib, see Chapter II, sec. 1

various research activities at the Pentagon which illustrate this fact.

During that conference I received the *bracha v'haskama* (blessing and consent) of the Rebbe MHM to establish the Rabbi Yisroel Aryeh Leib Research Institute on Moshiach and the Sciences to carry out research and hold an annual conference on Moshiach and Science.

Moshiach, Torah and Science

Our topic tonight concerns one of the most fundamental prophecies of the Era of Moshiach, the prophecy of Isaiah that "The glory of G-d will be revealed and all flesh will see [it]."[3] This entails the unification of the infinite with the finite—the infinite, miraculous revelation of G-d with the finite natural world. With G-d's help we will discuss this and its implications, and present a mathematical structure to model this paradoxical phenomenon.

Our starting point will be another paradox—the dilemma of many Jewish scientists who feel torn between the teachings of the Torah and the teachings of the scientists.

The truth is, however, that there cannot possibly be such a conflict. There is a famous statement in the Talmud that "Everything that G-d created in His world, He created only for His glory."[4] This includes not only material creations, but also intellectual creations such as science. This means that science was also created for G-d's glory. So how can there be a conflict between science and the Torah? There can't possibly be a conflict between something which G-d created for His glory—science—and the glory of G-d itself—the Torah!

The Talmud in Sanhedrin goes even further. It says that "The world was created only for Moshiach"[5]—who will bring about the *full revelation* of the glory of G-d (as stated in the prophecy of Isaiah that we just quoted). So everything in the world—including science—was created for the purpose of this ultimate revelation of G-d's glory.

What role does science play in preparing the world for the Era of

3. Isaiah 40:5.
4. Pirkei Avos 6:11.
5. Sanhedrin 98b.

Moshiach? This question is answered in a famous *sicha*[6] of the Rebbe MHM which explains the statement of the Zohar that in the 600th year of the 6th millenium (1840), "The gates of wisdom above will open and the fountains of wisdom below, and thus the world will be set to enter the 7th millennium" (similar to the opening of the windows of heaven and the fountains of the earth in the 600th year of the life of Noah, bringing the flood). The opening of the gates of wisdom above refers to the extraordinary revelations in Chassidus at that time especially the publication of the works of the *Alter Rebbe*, Rabbi Schneur Zalman of Liadi, founder of Chabad Chassidus. The fountains of wisdom below refers to the new discoveries in science and the development of new scientific theories.

Now, it's clear how the revelations in Chassidus are a preparation for the Era of Moshiach when "the world will be filled with the knowledge of G-d" (this is the subject matter of Chassidus), but what role do the developments in the sciences play in preparing the world for Moshiach?

The Rebbe MHM discusses several ways in which the sciences prepare the world for Moshiach. The primary function of the new developments in the sciences in preparing the world for the Era of Moshiach is to demonstrate the unity that exists within nature. In earlier times it was thought that there were many different, independent, forces in nature and many different natural elements, but the more science advances, the more it is realized that underlying the apparent diversity of forces and elements there is a basic unity, and that the various forces and substances are merely combinations and rearrangements of a few basic forces and particles. The awareness of this unity in nature is a preparation for the revelation of the absolute unity of G-d in the Era of Moshiach.

Why then does there sometimes appear to be a conflict between science and the Torah? To answer this, let's focus on a letter that the Rebbe MHM once wrote to a conference of religious Jewish doctors in the United States. In it he writes:

"True science, whose purpose is only to seek the truth—its conclusions cannot contradict the Torah which is the Torah of truth. Rather, on the

6. *Likutei Sichos*, vol. 15:42-48.

contrary, the more that [the scientists] will deepen their research, the more they will increase their recognition of the truth of both the principles and the details of our religion."[7]

Thus if a Jewish scientist feels that there is a conflict between the Torah and science, it's because his scientific research is not deep enough.

Creation from Nothing

As an example, the various theories of evolution have caused a great deal of confusion in the minds of some people, as these theories conflict with the Torah which says that the universe was created *Yesh Me'ayin*—omething from nothing—in six days. However, these theories are not even scientific, but are rather purely speculative in nature and will eventually be rejected on both common sense and scientific bases.

Someone once wrote a letter to the Rebbe MHM asking for a proof that there is a Creator of the world. The Rebbe MHM replied:

"The proof of this is very simple. Certainly, if one saw a book containing a few printed pages, there would be no doubt in his mind that it was made by a conscious intellect, who arranged the letters, printed them, etc., rather than it having happened by accident that a bottle of ink turned over on the paper, and as the ink spilled it happened that the shapes of letters were formed, composing lines and pages in such a manner that they explain an intellectual concept, etc. Anyone who says that this happened by accident by the ink spilling from the bottle, is not in the category of a man of intellect and not even in the category of a small child. How much more so regarding the entire universe which has a wondrous order—many thousands and millions of times more than the order in a book of a few pages...."[8]

This is a compelling proof which can only get stronger by doing more and deeper scientific research because "the more that they will deepen their research" the more structure becomes revealed in the natural world on deeper and deeper levels (from molecules to atoms to neutrons, protons and electrons to quarks, etc.) Here we see very

---

7. *Igros Kodesh* vol. 11:202
8. *Igros Kodesh vol.* 15:378

clearly how science was created for the glory of G-d since any advancement in science necessarily reveals more of the complex structure of the universe, thus proving the existence of the Creator. A Jewish scientist who approaches his profession with this mindset can never go wrong. His research will never take him away from the truths of the Torah because the deeper he delves into the depths of scientific research, the more he will become aware of the awesome wisdom of the Creator who designed, set up and maintains this magnificent structure.

To illustrate this point, let me mention a personal experience. Ten years ago, I was working as an applications consultant for a supercomputer company (ETA Systems, a subsidiary of Control Data Corporation). I worked in a consulting office at the University of Minnesota advising scientists and engineers on the use of the Cyber 205 supercomputer. There were two other consultants in the office, one black and one Chinese. The black consultant would often ask me questions about Judaism. One day he said to me, "I know that it says in those books over there," referring to the *Chitas* (*Chumash*, *Tehillim* and *Tanya* in one volume) and *Siddur* that I had on the shelf, "that G-d created the universe...." He went on to say that a physicist friend of his told him that the universe was not created; it just evolved. He wanted to know how one can prove that the Torah is right. (It was a sincere, non-confrontational question. He really wanted to know. On another occasion, I told him about the 7 Noahide Laws and his response was very positive: "You mean that's all I have to do...?")

"Look," I said to him, "the universe has such a complex structure. How could it have just evolved? Someone must have designed it and set it up." He agreed that that made sense but wanted to know how we could actually *prove* it. "Take, for example, the Cyber 205," I responded. "Do you agree that it is a very complicated computer?" The entire purpose of our office was to explain the complexities of the Cyber 205 to the users. "Of course!" he exclaimed. I then continued, "What if someone told you that the Cyber 205 had evolved? No one designed it. No one built it. It just got there by itself—from an accident in some engineering laboratory. What would you say?" To this he did not respond. He just burst out laughing at the absurdity of such a notion. "So how much more so," I concluded, "the entire universe—of which the Cyber 205 is just a small part!"

In fact, as scientific research continues, the theories of evolution become more and more discredited as they are shown to be contradictory, logically inconsistent and absurd. As an example, data recently obtained from the Hubble Space Telescope indicated a new value for the so-called Hubble constant. The implications are that, according to current cosmological theory, there are stars in the universe that are "billions of years" older than the universe itself—of which these stars are a part![9]

Thus, the Rebbe Melech HaMoshiach completely rejects apologetics—the attempt to smooth over the differences between Torah and science by reinterpreting statements in the Torah to make them consistent with current scientific theories. This is absurd because what is going to happen is the exact opposite. Deeper research in the sciences will lead to the recognition of the truth of the Torah as the Rebbe MHM writes in the above letter. Now that we are in the Era of Moshiach, this is happening quickly.

Implications For Jewish Education

All this has far-reaching implications for Jewish education, which is one of the most important applications of what we are discussing. Jewish children who study science in school are usually not being taught science from the perspective of Torah, but rather the way secular scientists want it taught. They are being taught the scientific theories of the origin of the universe, evolution, etc. as statements of fact rather than as speculative theories that can't possibly be verified by experiment. No scientist observed the origin of the universe, and it can't be duplicated in a laboratory.

The Rebbe MHM is often asked why it appears to scientists that the universe is billions of years old while in fact it is only 5756 years old. In many private audiences and letters he responded to this question. A typical example is the following excerpt from a letter dated 1 Sivan, 5716 (1956):

> "Regarding what you wrote that you read in the books of the astronomers about stars whose light rays must travel for much more than 6,000 years until they reach the earth, and

9. "Hubble Space Telescope Eyes a Younger Universe" in Science News 146 no. 18, 278; 148, no. 11, 166

you ask how this fits with what our holy Torah states—that we are in the year 5716 after creation:

"Now, even if you want to say that the calculation of the distance of the star is correct (for there are differences of opinion among the astronomers on this), this raises no question regarding the date of creation. [The point is] that just as the stars were created, similarly the light rays were created. Just as G-d can create a star which will first start to radiate light from the time of its creation and onwards, it is equally possible that He would create the star together with the light rays [already] emanating from it.

"This is especially evident when considering what it says in the first parsha of the Torah: 'And it was morning, the first day,' while 'Let there be luminaries' was not said until the fourth day. Thus we see that there was a presence of light before the luminaries were placed in the heavens...."

"A deep knowledge and an analysis of the principles [upon which the conjectures of the scientists regarding the age of the earth is based] will prove to the normal mind that these are merely *conjectures* which are far from certainty. This is the *opinion of the scientists* themselves, as is mentioned in their books. Nevertheless, in the introductory texts studied in the schools, they hide the doubts that they have in these principles."[10]

In this letter the Rebbe MHM teaches us a basic principle which can be applied to all questions regarding the age of the universe: G-d created the entire universe 5756 years ago as stated in the Torah. If certain things in the universe appear to an observer to be older than that, he must consider: 1) the possibility that his calculation is in error, and 2) he must understand that G-d may have created them to appear that way to begin with.[11]

Once the children are taught the doctrine of evolution, it becomes integrated into their minds and its very hard for them to abandon it. Even when they do abandon it, the initial impression remains.

---

10. Likutei Sichos, 15:478

11. See chapter IV, sec. 4 for a full discussion of these issues.

Chinuch HaKasher (kosher Jewish Education) is one of the Rebbe MHM's major projects. Jewish children should be taught creation exactly as it is stated in the Torah—that G-d created the entire universe in six days. They should not be taught any theory of science that is against the Torah, and the teachers should not be afraid to say that while the scientist may have a certain opinion at this time, the Torah tells us the truth and that eventually science will also recognize that the Torah is right.

Part 2: The Union of the Finite and the Infinite

A major aspect of the future revelations of the Era of Moshiach is that the miraculous order and the natural order will become one unified world order. In this part we discuss new results in mathematical physics concerning the nature of matter which parallel this Messianic revelation and prepare us for it.

The Miracle of Atomic Stability

When the fountains of wisdom from below broke open and scientists started to investigate the internal structure of matter which led to the development of atomic theory and quantum theory, they found themselves faced with a wondrous property of matter. The mystery, first described by the famous Jewish physicist Neils Bohr, was stated by the physicist Paul Ehrenfest in 1931 as follows: "We take a piece of metal or a stone. When we think about it, we are astonished that this quantity of matter should occupy so large a volume. Admittedly, the molecules are packed tightly together, and likewise the atoms within each molecule. But why are the atoms themselves so big?..."[1]

The question became even more dramatic than the size of the atoms. Given the strong attraction between the positively-charged nucleus of the atom, due to the protons, and the negatively charged electrons around the nucleus, Ehrenfest asked, "What prevents the atom from collapsing in this way?" Thus the question was: How does the atom remain stable? What is it that maintains the structure of the atom? The implication of this simple question was far-reaching: If every atom would collapse in this way, there would be no matter in the universe! In 1925, the physicist Wolfgang Pauli, inspired by the work of Bohr, formulated a principle which stated that the electrons had to remain in their orbits in a certain manner.[2] But this principle,

1. Quoted by Dyson and Lenard in *Journal of Mathematical Physics* (August, 1967).
2. See, e.g., Robert Eisberg and Robert Resnick, *Quantum Physics of Atoms, Molecules, Solids, Nuclei and Particles* 2ed (New York: Wiley, 1985), 309

which came to be known as the Pauli Exclusion Principle, merely stated the fact. It provided no explanation of the phenomenon based on the laws of physics. The problem came to be recognized as a fundamental problem of modern physics.

Theoretical physicists, who wanted a mathematically complete answer, attempted to derive this result from the fundamental equations of quantum mechanics but got nowhere. Then, in the 1960's, mathematical physicist Freeman Dyson started to do deeper research which has continued until the present time. Most recently, this research has been led by mathematical physicist Elliot Lieb of Princeton University. After working on the problem for nearly 30 years, and considering cases where matter does in fact collapse (such as in a very strong gravitational field), he declared, *"When you consider all of the physical forces at work in matter, even in something as simple as a glass of water, it's sort of miraculous that everything doesn't just collapse and then, releasing huge amounts of energy, blow up."*[3]

Two Types of Miracles

Exactly what kind of miracle this is, he may not know, but the Rebbe MHM explains this in the discourse "*Kuntres Achron Shel Pesach* 5750,"[4] which discusses the Messianic Prophecy "As in the days when you came out of Egypt, I will show you wonders."[5] In this discourse, the Rebbe MHM defines two categories of miracles: 1) revealed miracles that are obviously higher than nature, and 2) hidden miracles that work within nature and are concealed by nature. While the miracles of the first type have the advantage that the miraculous aspect is evident, those of the second type have a more profound advantage. With these miracles the infinite revelation from G-d—the source of miracles—combines with the finite revelation from G-d—the source of natural phenomena. In these miracles the infinite and the finite, the miraculous and the natural, coexist in the same event. In the miracles of the first type, on the other hand, the natural order must be broken in order for the miracle to occur.

To illustrate this paradoxical union of the infinite and the finite, the

3. "Stability of Matter," *Science News* 148 (Oct. 14, 1995):252-253
4. *Sefer HaMa'amorim Melukat* vol. 4:225-234
5. Micha 7, 15

Rebbe MHM explains that this wondrous state is actually found in nature itself:

> "By contemplating the fact that the order of nature is constant without change...and from the fact that throughout time the stability of the creations has not weakened...through this we know the greatness of the Creator—that His power is infinite and unlimited. In spite of the fact that the nature of the creations derives from a limited life force from G-d...this applies only to the order of nature itself, but the constancy within the order of nature—the fact that it does not change or weaken—must derive from the infinite unlimited power of G-d....This shows that even the finite life of the world is connected with the infinite revelation of G-d."[6]

The natural order itself is a finite expression of G-d's power, but in maintaining the stability of the order of nature—to keep the system running the same way, following the same laws throughout time without breaking down—requires an unlimited infinite expression of G-d's power. *The stability of nature is miraculous.*

The scientific discovery of this aspect of matter now, like all of the scientific discoveries since "the fountains of wisdom from below" broke open is a preparation for the revelations of the Era of Moshiach when "the union of the infinite revelation of G-d with the finite nature of the world will occur"[7] since "nature itself will be a receptacle for the infinite revelation of G-d which will be embedded in it."[8] And the miracles which are concealed by nature will become visible as miracles at the same time that they coexist with nature. This is what the prophet means by saying, "I will *show you* wonders."

This change in the natural order will come about as a result of the elevation and refinement of nature brought about by the dedicated service of G-d and observance of G-d's Mitzvos even when it may go against one's own personal nature.[9] Changing our personal nature

---

5. Micha 7, 15
6. *Melukat* 4:228
7. *Ibid.*, p. 233
8. *Ibid.*, p. 231
9. *Ibid.*, p. 233

brings about a change in the nature of the world. We can see from this the extraordinary importance of the practical observance of the Mitzvos.

So we see that the fusion of the finite and the infinite is to be found: 1) in the constancy and stability of nature, 2) in the miracles that are embedded within nature, and 3) later on in the Era of Moshiach, it will be visible throughout the entire world order.

At this point one may ask a very natural question: This union of the finite and the infinite is impossible! It's not only physically impossible, but even logically impossible—it's self contradictory! How can there be such a paradoxical combination? How can it be that in the Era of Moshiach this will be the order of the world?

The Rebbe MHM addresses this and explains it in the discourse:

"The finite and the infinite are two opposites. Therefore, in order for the infinite revelation of G-d to exist in the finite world, the infinite revelation must come from *Atzmus*, the essence of G-d," which transcends all limitations—*both the limitations of the finite and the limitations of the infinite.* Relative to *Atzmus*, finite and infinite are equal and can thus coexist. We will develop this concept further in Part 3 of this lecture. First, however, we must make some general comments regarding the Geula.

Seeing the Geula

We discussed earlier (in Part 1) the internal conflict between faith and intellect[10] experienced by some Jewish professional scientists. A similar internal conflict is now going on in the minds of some people regarding the Geula.

As is well known, in the year 5751, the Rebbe Melech HaMoshiach announced, *"Higia Zman Geulaschem!"*—"The time of your Redemption is here!" Now, more than five years later, some people are wondering exactly where we stand in the revelation of the Geula. On the one hand the Rebbe MHM clearly stated that the Geula is here, so we must believe it. But must we rely solely on belief? In characterizing our present time, the Rebbe MHM did not say that an extraordinary belief in the Geula is required in order to

10. See Chapter I, sec 1, for a discussion of faith and intellect in general.

accept the fact that the Geula is here. In fact, he said quite the opposite:

On 19 Kislev, 5752 (1992) the Rebbe MHM made the following very clear and powerful statement:

"In these days—the Days of Moshiach—which we are now in, all we need to do is open our eyes and then we see that the true and complete Geula is already here, literally."[11]

Thus we don't have to rely on faith to maintain that the Geula is here. We can see it with our eyes. Just as in the case of the Jewish scientist his internal conflict was resolved by doing deeper scientific research, here too the internal conflict is resolved by doing deeper research into the world—"opening the eyes." This involves looking at the world in a more analytical and penetrating manner. But in order to do this one must be professionally trained, just as the scientist must be professionally trained in order to carry out scientific research. Thus the first instruction that the Rebbe MHM gave after announcing that the Geula was here was that we study all the topics of Moshiach and Geula in order to be able to "live with the times"—to understand the world in this new Era of Moshiach.[12] In other words, we were instructed to professionally train ourselves in the area of Moshiach and Geula.

To be sure, faith still has a fundamental role in "seeing the Geula" since there are aspects of the Geula that go beyond what we are able to see, and in these areas we must maintain and strengthen our faith in what the Rebbe MHM said. Furthermore, when we do see an aspect of the Geula, we do not yet see it in its full infinite glory. We only see some finite expression of it. Thus, again, we must maintain the faith that there is more—infinitely more—of the Geula here than what we see.

In several cases, the Rebbe MHM himself opened up our eyes to see the Geula in the world. Throughout the year 5751, he explained many times that the Persian Gulf War was the war predicted in detail by the Midrash *Yalkut Shimoni*[13] as the war that occurs in the year

11. *Sefer HaSichos* 5752, 1:173
12. *Sefer HaSichos* 5751, 2:691-2
13. Yalkut Shimoni on Isaiah, sec. 499

that Moshiach is revealed. While the war at first frightened the world and threatened Israel, the Rebbe MHM said that the Jewish people should not be afraid—that there would be a miraculous victory and that the Land of Israel would remain safe throughout. He emphasized that this was the sign that the time of the Geula had arrived. As the Midrash says, when this war occurs Melech HaMoshiach will arise and announce, *"Higia Zman Geulaschem!"*—"The time of your Redemption is here!" This was exactly the announcement that the Rebbe MHM made at that time—this was the revelation of Melech HaMoshiach.

It is important to realize just how far-reaching this is. The Rebbe MHM explained that according to the natural course of events, the Persian Gulf War would have led to a world war.[14] Presumably, such a world war would have been a global nuclear war which would have destroyed the whole world, G-d forbid. Thus we are now enjoying the benefits of the miracle of this war *not* having occurred. Clearly, all other conflicts remaining in the world are mere "child's play" compared to a global nuclear war.[15] It may be a bit hard to appreciate this since we are talking about a war that didn't happen. But all we have to do is think back to a world war that did happen—World War II—and imagine: What if the revelation of Moshiach had occurred in 1939 and had miraculously brought World War II to a sudden halt and prevented the Holocaust from happening?...

The freedom of the Jews to leave Russia and go to Israel, as many thousands of them did, was the beginning of *Kibutz Galuyos*—the Ingathering of the Exiles. The characteristic of this exodus which clearly defined it as part of the fulfillment of the prophecies of the Ingathering of the Exiles was the fact that they were free to leave in a pleasant and peaceful manner (as opposed to having to flee or escape) and with the assistance of the Russian government![16]

The declaration by the heads of state of the major world powers in February 1992 that they were going to reduce arms production and increase food production for the world was the beginning of the ful-

14. *Michtav Klali* (open letter) of 25 Adar, 5751

15. In fact, the number of wars has decreased since the "Swords Into Plowshares" declaration of 1992. See "The Human Security Report," Chapter III, sec. 10

16. *Sefer HaSichos* 5752, 2:441 and several earlier sichos of 5752.

fillment of the Messianic prophecy of Isaiah, "And they shall beat their swords into plowshares."[17] At our Moshiach and science conference in 1993 on the theme "Bringing Moshiach Into the World of Science," Rabbi Dr. Naftali Berg *z.l.*, who was the Director of Research of the Advanced Technology Office of the Army Research Laboratory of the Pentagon, spoke at length about how the beating of the swords into plowshares—the transformation of military technology into peaceful technology—is an ongoing process at the Pentagon. He focused particularly on the Clinton administration's new "dual use technology" policy which requires that any new technology researched for military use must be simultaneously researched for peaceful use.[18]

Another example of beating swords into plowshares is a new design for a non-proliferative nuclear reactor, designed by a nuclear engineer at Tel Aviv University, Prof. Alvin Radkowsky. He was the scientist who designed the nuclear submarines, aircraft carriers, etc. for the U.S. Navy. Before moving to Israel some twenty years ago, he met with the Rebbe MHM for 45 minutes, during which time they discussed many matters of Torah and science and he received the blessing of the Rebbe MHM. In Israel he began the design of this new nuclear reactor which uses thorium as its primary fuel. It can supply cheap nuclear energy to any country with no possibility of the materials being used to make weapons. Radkowsky himself sees this as part of the fulfillment of the prophecy of "Swords Into Plowshares." I interviewed him in connection with our Moshiach and science conference in 1994 (on the theme "Seeing Moshiach Through the Eyes of Science"), and the president of the Radkowsky Thorium Power Corporation spoke about the project at our conference.[19]

In addition to the specific information about the presence of Geula in the world that the Rebbe MHM conveyed in each of these cases, he was actually opening our eyes—training us how to look at the world with eyes of Geula.

We, as professional scientists, have the imperative to look deeply

17. *Sefer HaSichos* 5752, p. 362-376

18. This particular program was later discontinued.

19. See Chapter II, sec. 7, "A Nuclear Reactor for the Era of Moshiach" for a transcript of the interview and a description of the project. The company is now called Lightbridge.

into the course of the development of science in recent years and see how the sciences themselves are converging with the "truth of both the principles and the details" of the Torah and preparing us for the revelations of the Era of Moshiach.

Part 3: The Coexistence of Opposites

We discuss how the revelation of Atzmus, which is responsible for the union of the finite natural order with the infinite miraculous order, actually lies at the foundation of all aspects of the Geula. Special emphasis is put on the everlasting life of the Rebbe MHM.

The Revelation of Atzmus

"The ultimate perfection of the [Era of Moshiach] is that the infinite Atzmus will be revealed in this material world."[1]

Now that our eyes are more open to seeing the Geula in the world, we are ready to discuss those aspects of the Geula characterized by the coexistence of opposites, and which consequently require the revelation of Atzmus, G-d's essence, in order to occur. In fact, the very transition from *Gola* (exile) to *Geula*, which are complete opposites, requires the revelation of Atzmus relative to which *Gola* and *Geula* are equal![2]

This makes it possible for *Gola* to actually be transformed into *Geula* (which is represented by the fact that, in Hebrew, when the letter *alef* is added to the word "*Gola*", it spells "*Geula*.")

The key concept in all this is that Atzmus transcends all limitations—it is not limited by any definition, even the definitions of logic and reason. Thus opposites such as "yes" and "no", infinite and finite, miraculous and natural, can all coexist since the definitions which make them exclusive of each other are no longer a limitation. Furthermore, having no limitations, Atzmus never changes:

"The Atzmus defies all description—negation of the finite and negation of the infinite."[3]

---

1. *Sefer Ma'amorim Melukat* 2:330
2. *Sha'arei Geula* 1:262
3. *Sha'arei Geula* 2:119.

From the perspective of Atzmus, the infinite and the finite, and similarly the miraculous and the natural, are really one thing—the expression of G-d's perfection. In fact, G-d's original intention in the creation was that there would be one world order in which the miraculous and the natural would function simultaneously.[4]

It is only in the Era of Moshiach, however, that this is actually achieved since it is then that the world attains the refinement necessary to run under such a paradoxical order—which requires the revelation of Atzmus.

As would be expected, once the world does attain this refinement and Atzmus is revealed, we will experience the coexistence of opposites in all areas of the Geula. We will discuss three areas of the Geula where the coexistence of opposites introduces a level of reality never before possible. These are the Torah, the *Beis HaMikdosh* (Holy Temple) and the life of Melech HaMoshiach.

Beis Shamai and Beis Hillel Simultaneously

We start with the Torah. It is well known (from the Talmud) that in the controversies between Beis Shamai and Beis Hillel, the Halacha is almost always ruled according to Beis Hillel's opinion. However, this is not the end of the story. There is a Mishneh in *Pirkei Avos* which states that every controversy which is "for the sake of heaven" will endure. As an example, the Mishneh mentions the controversy between Beis Hillel and Beis Shamai. The question is raised as to how it is that *the controversy* endures—since the Halacha was decided according to Beis Hillel's opinion, evidently only *their* opinion endures. In *Likutei Torah*, the Alter Rebbe quotes the Ari Zal who answers this question by explaining that in the future, in the Era of Moshiach, the Halacha will be ruled according to Beis Shamai's opinion.[5]

(In a famous *sicha*[6] the Rebbe MHM explains how this will happen: Beis Shamai was in fact sharper and saw more deeply into the Halacha than Beis Hillel. The Halacha was decided according to

4. *Likutei Sichos* 17:152
5. *Likutei Torah*, Bamidbar 54b
6. "Kuntres B'Inyan Torah Chadoshoh Me'Iti Tetzei" in *Sefer HaSichos* 5751, 566-582.

Beis Hillel, however, because the majority of the scholars, who were not on the level of Beis Shamai, saw the Halacha as did Beis Hillel. In the Era of Moshiach all Jews will be great scholars, and the *Sanhedrin* of Moshiach will be the greatest scholars of all time. They will see the Halacha with the depth of Beis Shamai and thus will decide the Halacha according to Beis Shamai's view.)

The Rebbe MHM notes that this does not yet answer the question because when the Halacha is decided in the Era of Moshiach as Beis Shamai rules, the opinion of *Beis Hillel* will not endure. Thus we are back to the original question. To answer this, the Rebbe MHM explains that there will come a time, later in the Era of Moshiach, when the Halacha will be according to both Beis Shamai and Beis Hillel simultaneously. Thus their controversy will endure because both of their opinions will endure. How is it possible that the Halacha will be ruled according to both opinions when in fact they contradict each other and are thus mutually exclusive? The Rebbe MHM explains that from the perspective of G-d, these opposite opinions coexist simultaneously: "He contains both opinions, the positive and the negative, and both of them together....As far as G-d is concerned, they can come to be revealed in the reality of the world just as they exist relative to Atzmus—positive and negative together. This is because He is *Nimna HaNimnaos* ("impossible of the impossibilities")—above the limitations of the impossible and the contradictory."[7]

Eternity of the Third Beis HaMikdosh

Regarding the third Beis HaMikdosh, the Zohar says that it will last forever since it will be "the structure of G-d." The Rebbe MHM explains that the first Beis HaMikdosh did not last forever because it involved a great revelation of G-dliness from above which was not completely connected with the lower, physical world and therefore couldn't last. In the time of the second Beis HaMikdosh the Jews were *ba'alei teshuva* (returning from the sins committed during the first Beis HaMikdosh) which involves a refinement and elevation of the physical world. Thus the revelation of G-dliness was more strongly connected to the lower world and the second Beis HaMikdosh

7. "Kuntres Halachos Shel Torah SheBe'al Peh," in *Sefer HaSichos* 5752, 33-34.

lasted longer. However, the world itself, being defined in terms of finite time and space, is limited; thus the second Beis HaMikdosh didn't last forever, either. But in the third Beis HaMikdosh, the finite limitation of the world will unite with the infinite revelation from G-d since, as the Zohar says, the third Beis HaMikdosh will be "the structure of G-d." This means that the Atzmus will be revealed and "the Atzmus defies all description—'negation of the finite and negation of the infinite.' Thus the limitation of the lower world and the infinity of above become one thing."[8]

The Everlasting Life of Melech HaMoshiach

"Relative to Atzmus, life and death are equal."

Similarly, the life of Melech HaMoshiach transcends the finite and the infinite. There is a verse from *Tehillim* (Psalms) which tells of the request for life that Moshiach asks from G-d, and G-d's response granting him life forever: "Life he asked from You. You granted it to him—length of days forever."[9] This "length of days forever" is called *chaim nitzchi'im*—everlasting life. Here too, we would expect that this "length of days forever" is associated with the essence of G-d, Atzmus.

This point is discussed by Rabbi DovBer of Lubavitch, the *Mittler Rebbe,* who contrasts it with the life of *Dovid HaMelech* (King David) of whom a similar—but not identical—statement is made: *"Dovid Melech Yisroel Chai V'kayam"*—Dovid, King of Israel, is alive.

(The Rebbe MHM notes that this means that *Dovid HaMelech* is alive literally in this world, for if it meant that he is alive in *Gan Eden* there would be no point in making the statement since every Jew is alive in *Gan Eden* after this world. "From the moment he was anointed as king of Israel in Hevron until the end of all generations he continues to live—and continues to live as the King of Israel!"[10])

The *Mittler Rebbe* discusses the difference between the *"chaim nitzchi'im"* of Melech Moshiach and the *"chai v'kayam"* of Dovid HaMelech:

---

8. *Sha'arei Geula* 2:115-119.
9. Tehillim 21:5
10. Sicha of 7 MarCheshvan, 5746

"Regarding Moshiach it says, 'Life...You granted it to him' which means a gift which has no interruption. Regarding Dovid—his life came only from the ten *sefiros*...of *Atzilus*, but regarding Moshiach it says, '*You* granted it to him,' which means from Atzmus, the infinite."[11] (When G-d is addressed in second person—"You"—it refers to G-d Himself, i.e. His essence, Atzmus.)

It is significant that the Rebbe Melech HaMoshiach spoke of his life from Atzmus on Erev Rosh HaShana, 5752, the Rosh HaShana before the stroke. He said, "Regarding Moshiach—in addition to the revelation of the name of Moshiach...we also have the *Atzmus of Moshiach Tzidkeinu* which is one thing with the infinite Atzmus of G-d."[12]

A life which comes from Atzmus is above all definitions and limitations of natural life and death as we know it. As Rabbi Shalom DovBer, the 5th Rebbe of Lubavitch, writes, "Relative to Atzmus, life and death are equal."[13] In fact, as mentioned above, Atzmus transcends all limitations. Thus the miraculous and the natural are equal relative to Atzmus and can coexist. It follows that the life of Moshiach is completely above the laws of nature. As the *Maharal of Prague* writes, *"Moshiach will be...outside of nature and the way of the world."*[14]

Yakov Avinu Did Not Die

While it is impossible to characterize a life which is above definition, some comprehension of it can be achieved by considering the characteristics of the lives of other *tzadikim* whose lives are not said to have come from Atzmus, but which are not completely subject to the laws of nature either, such as Dovid HaMelech, mentioned above. The most well-known of these is *Yakov Avinu* (Jacob), of whom the Talmud says, "Yakov Avinu did not die."[15] This statement is discussed and analyzed extensively by the commentaries of the Torah and the Talmud. To begin with, we note that Rashi, whose commen-

11. *Ma'amorei Admur HaEmtza'i*, Devorim 1:214-215
12. *Sefer HaSichos* 5752, 1:6-7. In footnote no. 49, the Rebbe MHM quotes the Tzemach Tzedek from *Derech Mitzvosecha* who also talks about the life of Moshiach coming from Atzmus.
13. *Sefer HaMa'amorim Eter* (5670), 100.
14. Quoted in *Sha'arei Geula* 2:69, note 29.
15. Talmud, Ta'anis 5b

tary on the Torah is *pshat*—the simple, logical meaning, which is directed to the understanding of the five year old child who is beginning his study of the Torah, states quite simply that "Death is not mentioned regarding him. Our Rabbis have said that Yakov Avinu did not die."[16] This is a literal statement intended to be understood by a child. In his commentary on the Talmud Rashi states further (explaining why there was a burial etc.), "It appeared to them that he died but he was alive."

What is the nature of the life of Yakov Avinu? One of the commentaries on the Torah, the *Alshich*, provides the technical structure.[17] As is well-known from the Midrash, Kabbala and Chassidus, there are five parts of the soul which give life to the body. The first three, called *Nefesh, Ruach* and *Neshama,* are contained in the body itself (*Nefesh* is centered in the liver, *Ruach* in the heart and *Neshama* in the brain). The last two, *Chaya* and *Yechida*, provide a general life force which is higher than the specific life of the first three—they encompass the entire body uniformly. The Alshich explains that when death occurs, G-d forbid, the three parts of the soul that are contained in the body leave the body. The passing of tzadikim is different, however. When a tzadik passes away only the *Ruach* and *Neshama* leave the body but the *Nefesh* stays in the body. Thus tzadikim are not called dead, but rather "sleeping in the dust." In fact, it has been observed that the bodies of certain tzadikim that had to be moved from their original grave to another gravesite many years after they were buried, remained complete and unchanged from the time they were buried.

The statement "Yakov Avinu did not die" goes beyond this, the Alshich continues to explain. It means that in the case of Yakov Avinu all three parts—the *Nefesh, Ruach* and *Neshama*—together with all the faculties of the soul remained in his body. This distinguishes Yakov Avinu from other tzadikim.

One who is skeptical about the possibility of the soul remaining in the body after all signs of life have ceased should consider the following:

A certain woman was found to have an aneurysm (swollen artery)

16. Rashi on Breishis 49:33
17. Quoted briefly in *Likutei Sichos* vol. 4:1261.

in her brain. Conventional brain surgery could not be performed because if doctors operated on the artery while the blood was still circulating in the brain, the artery could burst, causing certain death. As the surgeon explained it, it would be like operating on a tense balloon. So she was referred to a certain medical team at Columbia University that has been successful at performing a radically different type of surgery.

In their procedure they deflate the artery by stopping the blood circulation and draining the blood from the brain. They achieve this by circulating the blood of the patient out of the body and through a refrigeration unit, thus lowering the patient's body temperature to the point that the heart stops and even *the brain waves cease!* At this point the patient is scientifically and legally dead. They then perform the operation. After the operation they warm the blood again as they circulate it through the body, starting the brain waves and the heart again. (CBS News reported on this on their news program "60 Minutes" on April 2, 1995. At the time of their report, this medical team had performed such operations 40 times and in 37 cases the patients had survived.)

The patients in these cases are regular human beings, not tzadikim. If a regular human being, who is subject to the laws of nature, can still be alive for a length of time even after all motion and all signs of life—*including brain waves*—have stopped, it should be easy to accept what the Talmud, Kabbalah and Chassidus tell us about certain tzadikim, who (to varying degrees) are not subject to the laws of nature, that "it appeared to them that he died but he was alive."

There is a practical application of this on a daily basis. The *Midrash Rabba* tells the story of Yakov Avinu blessing his sons. Before he blessed them, he asked them if their belief in G-d was complete, to which they responded, *"Shema Yisroel Hashem Elokeinu Hashem Echod!"* Addressing their father by his title Yisroel, they said, "Hear us, Yisroel, G-d is our G-d, G-d is One!" Their belief in G-d was complete. The *Midrash* continues that to this very day when we say *"Shema Yisroel"* every day, we are talking to Yakov Avinu—who is alive—and saying, "Listen, Yisroel our father in the Cave of *Machpelah,* we are still faithful to that which you commanded us regarding the unity of G-d."[18]

18. Bereishis Rabbah, 98:3

A statement similar to "Yakov our father did not die" is made by the Talmud regarding Moshe Rabbeinu (Moses): "Moshe did not die."[19] And, of course, this is true regarding Dovid HaMelech—*"Dovid Melech Yisroel Chai V'Kayom."* The Rebbe MHM said that the same thing applies to the Previous Lubavitcher Rebbe as well: "The *Nasi* (leader) of our generation...who did not die, just like Yakov our father."[20]

On the night of Simchas Torah 5711 (1950), the Rebbe MHM instructed the chassidim going out to the various shuls for *hakofos* to deliver the message that "Moshe did not die" (referring to the Previous Lubavitcher Rebbe, who had "passed away" eight months earlier, as well as to Moshe Rabbeinu) , and that if they have difficulty accepting it, to tell them—just like a teacher tells a child—"When you grow up you will understand."

The unique gift of *chaim nitzchi'im* granted to the Rebbe Melech HaMoshiach which derives from Atzmus transcends even this, as discussed above. The Rebbe MHM himself referred to it: "In this generation...*chaim nitzchi'im* without any interruption."[21] In connection with this, the Rebbe MHM raises the question that it is explained in Chassidus that there is actually an advantage to fulfilling the verse, "And to dust you shall return."[22] Thus the Baal Shem Tov said of himself that he could have gone up to heaven alive as *Eliyahu HaNavi* (Elijah the prophet) did, but he wanted to fulfill the verse "And to dust you shall return." How is this advantage attained when there is uninterrupted *chaim nitzchi'im*? The Rebbe MHM explains that the same advantage can be attained in another way—by the fulfillment of the concept expressed in the *Shemoneh Esrei* prayer "And my life shall be considered as dust by everyone." (Note the similarity to the Talmudic statement regarding Yakov Avinu, "It appeared to them that he died but he was alive"—his life was "like dust" to all appearances.)

19. Talmud Sota 13b.

20. *Sefer HaSichos* 5752, 1:242.

21. *Sefer HaSichos* 5751, 496. The Rebbe MHM made similar statements many times throughout 5751 and 5752. These statements are printed in the volumes of *Sefer HaSichos* 5751 and 5752.

22. Breishis 3:19.

Gimmel Tamuz

While life is usually associated with motion, Atzmus, which is higher than life, is associated with stillness. As the Rebbe MHM says, "Any motion, whatever it may be, is not the essential *(Aetzem)* existence."[23] Atzmus transcends definition; every motion has a description, a definition, and thus a limitation.

It is my understanding that based on this we can gain some insight into the event of *Gimmel Tamuz* (3 Tamuz, 5754 (1994)—the day of the concealment of the Rebbe MHM). The essence of the life of the Rebbe MHM, which derives from Atzmus and thus transcends all definition—including the definitions of motion—became revealed. This life is higher than nature. The Rebbe MHM himself wrote about this in a *Gimmel Tamuz* discourse published for *Gimmel Tamuz*, 5749, based on the verse in Tehillim, "He (G-d) places our soul into life...,"[24] which the Rebbe MHM explains to mean "The life into which he places our soul is higher than nature." The discourse concludes with a discussion of miracles that occur within nature and the union of the natural with the miraculous.[25]

Atzmus, however, is not limited to stillness; it is not limited to anything. As we mentioned above, it is characteristic of the Era of Moshiach that Atzmus will be revealed in all aspects of the world order. Thus Atzmus will be revealed within motion and speech, as well. We will see the Rebbe MHM again—immediately—in full motion even as we see Atzmus revealed in these actions.

Are we ready for this? Has the world attained the state of refinement necessary for Atzmus to be revealed? The Rebbe MHM answered this question in the sicha of Parshas VaYetzei 5752 (1991). He writes:

"It is understood that we now exist in a state in which the physical body and even the physical world have been completely refined and are receptacles ready for all the light and spirituality including, primarily, the light of Moshiach Tzidkeinu—the light of the true and

23. *Sefer HaSichos* 5752, 1:128.
24. Tehillim 66:9.
25. *Sefer HaMa'amorim Melukat* 3:201-207.

complete Geula; including also the revelation of the absolute Atzmus—the Atzmus and essence of G-d—as it is revealed in the absolute Atzmus of Moshiach Tzidkeinu....Through this it is revealed also in every single Jew.

"The only thing missing is that a Jew should open his eyes properly and see how everything is already ready for the Geula!"[26]

A Lubavitcher chosid[27] approached the Rebbe MHM on the night of Simchas Torah, 5754 (over a year and a half after the stroke) with the well-known photograph[28] of the Rebbe MHM waving both arms and asked if we would see him again as in the photograph. He responded "Yes." After the *yom tov* the chosid wrote a letter to the Rebbe MHM repeating the question: "Will the Chassidim again see the Rebbe *Shlita* doing as in the enclosed picture, <u>with the same physical capabilities (standing, waving both arms, *mitoch simcha*) and even more? We ask the Rebbe Shlita for his blessings that this occur soon."</u> The Rebbe Melech HaMoshiach again answered, "Yes."

26. *Sefer HaSichos* 5752, 1:151-152. Italics in the original sicha.

27. Mr. Meir Schreiber of Crown Heights, who told me the story and provided me with a copy of the letter. The underlining is as in the original letter.

28. The photograph is reproduced in the opening pages of this book

Part 4: Three Valued Logic and the Third Geula

In this final section we present the mathematical structure of three valued logic and explain how it is used to attain a logical understanding of the revelation of Atzmus *in the world and the coexistence of opposites that it entails. Real world applications to computers and physics are described.*

In Part 1, we discussed the confrontation between faith and intellect. A similar issue arises when discussing the world order of the Geula: We have been speaking of phenomena that are not only beyond the limits of nature (such as miracles) but also beyond the limits of logic—the coexistence and simultaneous functioning of opposites, the combination of the natural with the miraculous, etc. Must we rely on faith to accept these matters, or can we also understand them with our intellect? In other words, must we accept them on faith or can we "open our eyes" and see them?

In fact, the coexistence of opposites in the Era of Moshiach applies not only to the order of nature, but also to intellect and logic itself, as the Rebbe MHM says, "The intellect itself will be permeated and filled with the knowledge of the level of G-dliness which is higher than the limitations of intellect," so that the paradoxical coexistence of the natural and the miraculous, the infinite and the finite, etc. will not only be experienced in the Era of Moshiach but will also be clearly understood by the intellect. Intellect itself will comprehend things that are now outside the limitations of intellect. Atzmus itself will be comprehended: "Within the realm of understanding and comprehension, there will be the comprehension of the essence of the infinite Atzmus."[1]

We also discussed the sicha of the Rebbe MHM which explains how the revelations in the sciences since the year 5600 (1840) prepare the world for the revelations of the Era of Moshiach. So it is nat-

1. *Hadranim on the Rambam and Shaas*, 33

ural to ask: Has anything been discovered in mathematics or science that prepares us for an intellectual understanding of the concept of the coexistence of opposites? The answer, quite surprisingly, is "Yes!"

Three Valued Logic

The story starts shortly after the fountains of wisdom from below opened up. In the year 1847, around the time that *Likutei Torah* was published,[2] a British mathematician named George Boole published a paper explaining that logic has an algebraic structure. Thus began the development of a new branch of mathematics known as mathematical logic.[3] In the same year, a mathematician named De Morgan, who was working on mathematical logic, announced that he felt that a third "truth value" should be added to the structure of logic. This means that in addition to the usual two truth values "true" and "false" (i.e. the characterization of statements as being either true or false), to which heretofore classical logic had been limited, there should be a third truth value characterizing a third state. However, he did not know exactly how to develop this idea and it remained a vague concept until the 1920s.

In the early 1920s a Polish mathematician named Lukasiewicz developed this concept in a very precise mathematical way. The classical truth values "true" and "false" were represented by the numbers 1 and 0, respectively, and the third truth value, which could represent "true" and "false" simultaneously, was represented by ½. This new system retained the ability to represent all the information that did classical logic while having the new capability of representing as well the coexistence of opposites, contradictions and paradoxes. It became known a Three Valued Logic (3VL).

It is interesting to note that around the same time the "new quantum mechanics" was being developed and it was found that classical logic was not able to deal with some of the physical phenomena at

---

2. This was preceded by the printing of *Torah Or* about 10 years earlier.

3. See Pesach (Paul) C. Rosenbloom, *The Elements of Mathematical Logic* (New York: Dover Publications, 1950), especially ch.2, sec. 4 on many valued logics. See also L. Bolc and P. Borowick, *Many Valued Logics 1: Theoretical Foundations* (Heidelberg: Springer-Verlag, 1992). Our own discussion is based on Richard L. Epstein, *The Semantic Foundations of Logic—Propositional Logics*, 2ed (New York: Oxford University Press, 1995), sec. VIII: "Many Valued Logics."

the quantum level. Specifically, it was found that in quantum systems the distributive law of logic no longer held true. This stems from the fact that in a quantum system, observables are represented by self-adjoint operators instead of functions. The failure of the distributive law of classical logic to hold true is a result of the non-commutativity of matrix multiplication. A new Quantum Logic was thus developed for quantum mechanics.[4]

It is no accident that in this era, not only were new theories of physics being developed, but even new systems of logic! We know from the sicha of Parshas Noach[5] that all the developments in the sciences since 1840 are intimately connected with and actually derive from corresponding developments in Chassidus which prepare the world for the knowledge of G-d in the Era Moshiach. An extraordinary event must have occurred in the world of Chassidus, causing a breakthrough in preparing the world for the Geula of Moshiach, in order for there to have been the parallel event of the development of new systems of logic and new theories of physics.

This event was the Geula of the Previous Rebbe on 3 Tamuz and 12-13 Tamuz, 5687 (1927). As the Rebbe MHM writes concerning this Geula, "It was a revealed miracle higher than nature. Nevertheless, the miracle was clothed in nature...in a way that nature did not hide the miracle....Since in this Geula there was the revelation of the infinite Atzmus...the Geula was in a manner that the miracle and nature joined together." "Through this, the barrier between the lower and the higher worlds was nullified." "Through the union of the miraculous and the natural that occurred in the Geula, the power was given to all those who follow the ways of the one who was redeemed [the Previous Rebbe] that the faith which is higher than intellect (the miraculous) be drawn down also into intellect (the natural), in a manner that the intellect will not obscure the faith." "Through this, the power was then given to every Jew to nullify the barrier between his dealing in worldly matters and the observance of Torah and Mitzvos." This was the formal end to the conflict between

4. Paul Bamberg and Shlomo Sternberg, *A Course in Mathematics for Students of Physics*, 2 (Cambridge: Cambridge University Press, 1991), "Quantum States and Quantum Logic," 831-835. See also Yu. I. Manin, *A Course in Mathematical Logic* (New York: Springer-Verlag, 1977), "Quantum Logic," 82-95.

5. *Likutei Sichos*, 15:42-48

faith and intellect—between Torah and science—that we have been discussing! The Rebbe MHM concludes, "This becomes an immediate preparation for the fulfillment of the prophecy that 'The glory of G-d will be revealed and together all flesh will see that the mouth of G-d has spoken,' that the flesh will see G-dliness (not just because of the power of the revelation of G-d's glory, but also) from the flesh itself."[6]

Three valued logic provides for us a model—a mathematical framework—within which to describe and discuss phenomena of the Geula that result from the revelation of Atzmus and thus involve the coexistence of opposites. The first thing to note is that now that the barrier between the lower and the higher worlds is removed, the coexistence of opposites that we have been discussing is a standard part of science and mathematics. Most recently, as mentioned in Part 2, it was realized that the very stability of matter involves the coexistence of opposites, the infinite power of G-d in the finite atom. This was always part of the world order even though for centuries it was unknown. Now that it's known, it's also part of science, i.e. part of intellect. In the next section we outline the structure of 3VL and describe one of its most common everyday applications—computers.

A Short Course in Three Valued Logic[7]

In mathematical logic statements are represented and analyzed as follows: We consider first an initial set of statements, called atomic sentences. This is the collection of the basic statements from which more complex statements are constructed. The atomic sentences are represented by symbols A, B, C, etc. Algebraic operations are then defined on these symbols forming new, more complex sentences. The operations are defined in such a way as to correspond to common logical reasoning. We will define four operations which will correspond to the concepts of negation ("not"), conjunction ("and"), disjunction ("or"), and implication ("if...then") represented symbolically as follows:

- Negation: $\neg A$ represents "not A"
- Conjunction: $A \wedge B$ represents "A and B"

6. The quotes in this paragraph are from *Sefer HaMa'amorim Melukat* vol. 3, "Kuntres Gimmel Tamuz 5749," p.207, and from *Sefer HaMa'amorim Melukat* vol. 5, "Kuntres Chag HaGeula 12-13 Tamuz 5751," 325.

7. See Chapter V, sec. 2c for a fuller, more technical discussion of three valued logic.

- Disjunction: A∨B represents "A or B"
- Implication: A→B represents "If A then B" or "A implies B"

For example, let's consider two atomic sentences: "The light is on" represented by the symbol A, and "The bell is ringing" represented by the symbol B. The four operations and their meanings are listed below:

- Negation: ¬A means "The light is **not** on"
- Conjunction: A∧B means "The light is on **and** the bell is ringing"
- Disjunction: A∨B means "The light is on **or** the bell is ringing"
- Implication: A→B means "**If** the light is on **then** the bell is ringing"

Now, any of these statements may be either true, in which case it is assigned the value 1, or false, in which case it's assigned the value 0. The truth value (1 or 0) of a complex statement will depend on the truth values of the atomic sentences. We must now define how the truth value of a complex statement is determined from the truth values of the atomic sentences. For example, consider the statement above, "The light is on and the bell is ringing." If it is true that "The light is on" but false that "The bell is ringing," then the compound sentence "The light is on and the bell is ringing" is false. Symbolically we are saying that if A is true and B is false, then A∧B is false. Stated mathematically, when A has the truth value 1, and B has the truth value 0, the conjunction A∧B has the truth value 0.

It should be clear that this is a property of the abstract structure of these operations and is independent of the statements which the symbols represent, i.e., A being true and B being false will result in A∧B being false no matter what sentences the symbols A and B represent. The truth values for statements formed by negation, conjunction, disjunction and implication are usually summarized in tables called "truth tables". (See figure 1).

Truth Tables for 2VL

A	¬A
1	0
0	1

A	B	A∧B
1	1	1
1	0	0
0	1	0
0	0	0

A	B	A ∨B
1	1	1
1	0	1
0	1	1
0	0	0

A	B	A→B
1	1	1
1	0	0
0	1	1
0	0	1

Figure 1

We note that there are certain statements that are always true such as A∨¬A which means either A is true or the negation of A is true, and ¬(A∧¬A) which means that it is not true that both A and its negation are true. Here too, it makes no difference what statement A represents. These statements are called tautologies.

The innovation in three valued logic is the introduction of a third truth value, ½, to represent the notion of being both true and false simultaneously. For this concept to be logical, truth values must be assigned to complex sentences in a consistent manner and in such a way that the system of three valued logic is an extension of two valued logic. This means that in a case where the third value ½ is not used, the three valued system gives the same results as the two valued system. This is in fact the case as can be seen from the truth tables for three valued logic given in Figure 2.

Truth Tables for 3VL

¬A	
A	¬A
1	0
½	½
0	1

A∧B				
A	B:	1	½	0
1		1	½	0
½		½	½	0
0		0	0	0

A∨B				
A	B:	1	½	0
1		1	1	1
½		1	½	½
0		1	½	0

A→B				
A	B:	1	½	0
1		1	½	0
½		1	1	½
0		1	1	1

Figure 2

What about tautologies in 3VL? This is where things get interesting and the results are surprising. Firstly, let's consider the tautologies of the two valued system and see if they are still tautologies in the three valued system. Remember that to be a tautology it must always be true, i.e. its truth value must always be 1. A∨¬A is a tautology in 2VL. But, as Figure 3 shows, when A is assigned the value ½, the statement A∨¬A has the value ½ so it is not a tautology in 3VL. So in this system it is not valid to say that either A or its negation will always be true.

A	¬A	A∨¬A	¬(A∧¬A)	A∧¬A
1	0	1	1	0
½	½	½	½	½
0	1	1	1	0

Figure 3

Figure 3 also shows that when A has the truth value ½, the statement ¬(A∧¬A) has the value ½ so this statement also is not a tautology. In this case we notice that the contradictory statement A∧¬A also has the value ½ (instead of 0) which shows that contradictions are not rejected in this logical system!

Now it gets even better. We will show that there is a contradictory statement in 3VL that always has the truth value 1, i.e. it is a tautology. First, consider the statement:

$$(A \rightarrow \neg A) \wedge (\neg A \rightarrow A).$$

This means that "A implies the negation of A and the negation of A implies A," or, in other words, "A is true if and only if the negation of A is true," a very contradictory statement. Yet the truth table in Figure 4 shows that when A has the value ½, this statement has the truth value 1, meaning true!

A	¬A	A → ¬A	¬A → A	(A → ¬A) ∧ (¬A → A)
1	0	0	1	0
½	½	1	1	1
0	1	1	0	0

Figure 4

Now, let's symbolize this contradictory statement by the symbol C (for contradiction). We now show that there is a tautology in the three valued system which is built on C. The statement A ∨ (¬A) ∨ C is a tautology as shown in Figure 5.

A	¬A	C	A∨¬A∨C
1	0	0	1
½	½	1	1
0	1	0	1

Figure 5

We already saw that the statement A∨¬A, a tautology of 2VL, is not a tautology in 3VL. Now we see that by combining it with the contradictory statement C, we have a 3VL tautology—a statement which is always true! This tautology says that in three valued logic it is always true that either A or the negation of A or the contradictory statement C is true.

We went through the details of this to show that three valued logic is in fact logical and consistent. We now go even further and show that this is not just an abstract concept but actually has down-to-earth applications every day in computers.

Applications of Three Valued Logic in the Real World

When computers were put to the task of dealing with very large amounts of data, it was found that the way in which the data was organized, or "modeled," was an issue of great importance which would affect the efficiency and accuracy of the processing of the data. The best and the most popular model, called the "Relational Model," was introduced by the mathematician E. F. Codd in 1970. It organized data into related tables. It had a very solid foundation in theoretical mathematics so that any question regarding the model could answered mathematically.[8] Then a language had to be developed to interact with this model. A language based on mathematical logic, called the "Structured Query Language" or SQL, was developed at IBM in the 1970s.[9]

SQL is based on three valued logic. The third truth value, instead of being used to represent contradictory information, is used to represent unknown information.[10] For example, suppose we are dealing with a database of employees of a company and the computer program must evaluate the truth value of the statement: "(profession is engineer)∧(age is greater than 55)." If both parts of the statement are true then the whole statement is true; if one part is true and the other is false then the whole statement is false in accordance with the rules of ∧ ("and") in both two valued and three valued logic. But what if part of the information is unknown? For example, what if it is known that the profession is engineer but the age is unknown? Since SQL is based on three valued logic, the unknown information "(age is greater than 55)" is assigned the value ½. Then, in accordance with the rules for ∧ in three valued logic (in figure 2), the whole statement

8. For a mathematical discussion of the Relational Model see Dan A. Simovici and Richard Tenney, *Relational Database Systems* (New York: Academic Press, 1995), ch. 2.

9. Simovici and Tenney, ch 3. Many books have been written on SQL. See for example, J. Hursch, SQL: *The Structured Query Language*, 1st ed (Blue Ridge Summit: TAB Professional Books, 1988), especially ch. 11, "Logic and SQL" for an introduction to SQL based on mathematical logic.

10. Simovici and Tenney, pp. 104-106. (The use of 3VL to represent unknown information in SQL has been a matter of lively discussion in computer journals over the years.)

is assigned the value ½ and the program continues.

It is an interesting fact that the computers we use every day for our information needs use three valued logic—the "logic of Atzmus"—to process data. But when we consider the fact that we are now in the Era of Moshiach when Atzmus is out in the open, it is not so surprising that three valued logic is part of our daily lives.

A New Sense of Reality

An interesting application of three valued logic in engineering is switching theory in electronic circuits, but we will not discuss it here. We conclude our discussion of three valued logic with a fascinating example from quantum mechanics.[11]

Actually, quantum mechanics has shown that at the foundation of physical reality—at the microscopic level of atoms and subatomic particles—paradoxes and contradictions are commonplace. Now, in the Era of Moshiach, the paradoxical aspects of this hidden level of reality are coming out in the open. In fact, these paradoxical aspects of quantum mechanics are now being applied to the design of computers in a new field known as "quantum computing."

Much of the paradoxical nature of the world at the microscopic level is due to the fact that at this level, all the information about a particle cannot be known precisely. This is expressed by the well-known "uncertainty principle." The information must be given in terms of probabilities, and the determination of which of the possible events actually occurs is made by the observer of the event! The relationship between this concept and the teachings of Chassidus has been analyzed and written about extensively by Professor Yirmiyahu Branover.[12]

What is the nature of the reality, i.e. what is the state of the particles involved, before the observer views the event? This question was debated openly by two Jewish physicists, Albert Einstein and Neils Bohr, in the early years of the development of quantum mechanics. The opinion that prevailed, that of Neils Bohr of Copenhagen, came

11. For some background in quantum mechanics, see chapter V, sec. 5a.

12. See Rabinowitz and Branover, "The Role of the Observer in Halacha and Quantum Physics" in *Fusion - Absolute Standards in a World of Relativity* (Jerusalem: B'Or HaTorah Publications and Feldheim Publishers, 1990), 91-109.

to be known as the "Copenhagen Interpretation" of quantum mechanics. According to this view, two contradictory, mutually exclusive possibilities can exist simultaneously in reality until the observer makes his decision as to how he will observe the event (e.g. as a particle or a wave, spin up or spin down, etc.).[13]

Now this level of reality has been brought out into the open. A team of physicists working with Dr. Christopher Monroe at the National Institute of Standards and Technology in Boulder, Colorado succeeded in getting an entire atom to exist "in the bizarre state of being in two well-separated positions at once" i.e. to exist simultaneously in two different places. "Imagine a marble in a bowl, rolling back and forth," Dr. Monroe explained. "At one point, the one marble appears as two marbles rolling back and forth in opposite directions, passing through each other and appearing simultaneously at each edge of the bowl. This situation defies our sense of reality."[14]

The New York Times[15] quoted Dr. Monroe as saying, "The real significance of the institute's feat is that the two states of the same atom were not only pulled apart but were separated by a relatively enormous distance—a distance large enough to represent a transition from the domain of quantum mechanics to the everyday world." But for us, the significance of the event is the appearance in the everyday world of an aspect of physical reality involving a contradiction (thus requiring three valued logic to deal with it), an event in line with the current Era of Moshiach when Atzmus—and coexistence of opposites—is out in the open.

We see from the above examples that the mathematical structure that serves as a model for the revelation of Atzmus in the world, which not so long ago was a mystery even to the mathematicians, is now, in the Era of Moshiach, part of everyday math, science and technology.

We have presented a logical structure within which we can under-

13. Robert Eisberg and Robert Resnick, *Quantum Physics of Atoms, Molecules, Solids, Nuclei and Particles*, 2nd ed (New York: John Wiley & Sons, 1985), 77-80.

14. Richard Lipkind, "Two Atoms in One," *Science News*, 149, no. 21 (May 25, 1996), 325. Originally reported in *Science*, May 24, 1996.

15. *New York Times*, "Physicists Put Atom in Two Places at Once," New York Times, May 28, 1996, sec. C, p.1, See www.nytimes.com/1996/05/28/science/physicists-put-atom-in-2-places-at-once.html?pagewanted=all.

stand the presence of Atzmus in the world. But any logical structure is finite so its usefulness as a model for the presence of the infinite Atzmus in the world is limited. As the Previous Lubavitcher Rebbe wrote, "The greatest understanding has a limit. But...belief is a feeling without a limit. Melech HaMoshiach will enable us to understand the greatness of the sincere, earnest service of G-d."[16]

Where logic leaves off, faith takes over. In fact, faith must be the foundation upon which any logical structure must be built.[17] Rabbi Levi Yitzchok Schneerson, the father of the Rebbe MHM, explains this in a letter[18] to the Rebbe MHM and Rebbitzin Chaya Mushka. Rabbi Levi Yitzchok writes that the *Etz HaChaim* (Tree of Life) and *Etz HaDa'as* (Tree of Knowledge) refer to faith and intellect. Faith is life, as the Prophet Habakuk says, "The tzadik lives through his faith." Rabbi Levi Yitzchok writes, "If the faith precedes the knowledge...then the knowledge is also life....These two trees must be joined together....This is what it says [in the prophecy of Isaiah] about the Era of Moshiach—that 'Faith will be the belt of his body.' Only at the end of the prophecy does it mention knowledge—'The earth will be filled with the knowledge of G-d.'"[16]

This gives us guidance and direction in living a life of Geula. Our foundation must be the *sichos* and statements of the Rebbe MHM that declare that we are in the Era of Moshiach and that the Geula is already here. Based on this foundation, we must refine our perception and deepen our understanding to look at and analyze the world in a new way. We realize that everything that is happening in the world and in our personal lives is part of the unfolding of the Geula and we come to understand the world and ourselves differently. We see Geula in the world and we develop a Geula mentality. In this way we truly live a Geula life and our "knowledge is also life."

16. Quoted by the Rebbe MHM in *HaYom Yom*, 5 Teves.
17. See chapter I, sec 1, "Faith and Intellect," for a full discussion of these issues.
18. *Likutei Levi Yitzchok, Igros Kodesh*, 413-14. (Quoted in *Sefer HaSichos* 5750, 2:629)

Professor Branover at Touro College

Prof. Branover at Touro College.

Professor Branover has lectured frequently at the Moshiach and Science conferences which are especially meaningful to him because of our themes of Moshiach, Torah and science and because they are in memory of Rabbi Yisroel Aryeh Leib.

His lecture at the conference held at Touro College in 2003 was unique, however, because it was actually a philosophical autobiography. In this lecture he discusses his transformation from a child raised under Communism to becoming an observant Jew and a Chabad chosid, the many responses he received on Torah and science issues from the Rebbe MHM, his own personal experiences with the Rebbe MHM's wisdom and prophecies, and his dedication to doing everything that can be done to bring about the full revelation of the Rebbe Melech HaMoshiach.

He concludes with the suggestion that "Every night, when we say the Kerias Shema al HaMita, we should think about our day, and ask ourselves, 'What did I do today, practically and visibly, to accelerate and bring the immediate revelation of the Rebbe Melech Moshiach?'"

Prof. Branover's lecture went well past the scheduled time. The students simply would not leave. It ended only when the custodial staff arrived to close the building.

From Communism to Chabad

More than anyone else, Rabbi Yisroel Aryeh Leib represented in his personality the relationship between Torah and science. He was known for being completely devoted to the study of Chassidus while working professionally as a mathematical physicist.[1]

On an occasion like today, Rabbi Yisroel Aryeh Leib's yartzeit, I think it's appropriate to speak about Torah and science in general. Maybe this issue is not important to some people and they are not disturbed by it. If someone with simple faith is dedicated to Torah and *Yiddishkeit* and is not concerned with any philosophical or metaphysical questions about the so-called conflict between Torah and science, that's wonderful.

I once asked the Rebbe MHM in *yechidus* whether or not it was necessary to continue doing what I was doing—giving lectures and writing books and articles about the issue of Torah and science. Maybe it's enough, I said, that there is already so much material written and spoken about on the subject in various institutions and publications. Why should we be preoccupied with analysis of the relationship between Torah and science when we could be sitting down and learning Torah? By any standard, science is *chochma chitzonis* ("external wisdom")—it's not real.

Of course, the Rebbe MHM refers to what the Alter Rebbe says in Tanya: that many *gedolim*—Rambam and Ramban and others—did pay attention to this matter, and spent time writing about it. The Rebbe MHM himself, as well as the Tanya and other sources, understands that this involvement is permissible and necessary for those outstanding representatives of Jewish history who would know how to properly use the discussions and knowledge of Torah and science to strengthen Yiddishkeit and Jewish lifestyle. But not just anyone who, by *hashgacha pratis* (Divine providence), has become a professor in a university needs to spend time writing or speaking about

1. See Chapter II for a biography of Rabbi Yisroel Aryeh Leib

Torah and science. What is permissible and necessary for Rambam and Ramban and others may not be applicable to them, and for people of our level perhaps it's a waste of time to have those discussions and publications.

The answer of the Rebbe MHM was very simple and very understandable. The Rebbe MHM said that if someone has the gift from Above of being completely undisturbed by this issue—if he is completely dedicated to spending all his time learning Torah—then of course it would be unthinkable to disturb his concentration and his peace of mind by bringing him into the conflict. But it's a different story with those who are *ba'alei teshuva*—Jews who until a certain age had lived while knowing absolutely nothing about Torah or Yiddishkeit. I know about this quite well because that's what happened to me....

From Communism to Chabad

I was unfortunately brought up in the former Soviet Union under communism, and I received my world outlook from a communist school and a communist university. Much later, thank G–d, I was helped in secret by Lubavitcher chassidim while still in the communist Soviet Union. They gave me the opportunity to open my eyes and understand what is important and what is not.[2]

(It is questionable whether such a person can really be called a *baal teshuva*, because the true meaning of the term *baal teshuva* is someone who is very advanced in Torah learning and practice, and for whatever reason suddenly distances himself from Torah and becomes a non–believer. At some point, he reconsiders his whole situation and comes back to Torah and mitzvos. This is probably the baal teshuva that the Torah refers to as standing higher than *tzadikim*.)

So the Rebbe MHM said that for all those in this situation (i.e., those who did not grow up with Torah and *Yiddishkeit*) it is extremely important to delve into the subject, analyze it, explain it, and show that everything written and spoken about the sciences' negating and

2. For the full account of Prof. Branover's experiences growing up in Russia and his return to Judaism and Chabad, see his book *Return: The Spiritual Odyssey of a Soviet Scientist* (Jason Aronson, 1996).

refuting belief in G–d is nonsense. Those who are confused by this alleged contradiction have to be helped. But they can only be helped by using a profound knowledge of science as a basis. Only someone who exhibits knowledge in science can help people rid themselves of their illusions that there are conflicts and contradictions in this area. Of course, one also needs substantial knowledge of Torah, and one must know how to learn. But one also has to know everything related to the sciences in great depth because only then can one be convincing enough to influence those one has to influence

This is in keeping with what we learn in *Pirkei Avos* ("Ethics of the Fathers")—that we have to know what to answer a non-believer. (Some argue that there are now no true non-believers (*apikorsim*) because those who are considered non-believers are really what is called "a captive child" because he or she never had the opportunity to learn anything about Torah and *Yiddishkeit*.)

I came to Israel 30 years ago after a long and exhausting struggle with the communist regime. I had been in prison there many times. Jail in itself was not so bad, but they took care to always put me with most dangerous people—like thieves and murderers—and it was not very pleasant to spend time in such a jail. But ultimately I was released.

I'm sorry to be telling you personal things before I get to the actual subject of Torah and science. But I believe it will give you a feeling and understanding of why I am preoccupied with this issue and how I came up with many of my ideas about it.

When I came to the Rebbe MHM for the first time, I had a *yechidus* (private audience) with him for two hours. I've had the great privileged of spending a total of more than 35 hours in private audience with the Rebbe MHM over the years. I wish I had written down everything I heard in those 35 hours. By writing several books and many articles, I am trying to record it all, but it's not even close to everything I heard from the him. Some call it a *chutzpa* to bring a tape recorder into the Rebbe MHM's room, but other chassidim say that it was a very big mistake not to—I should have recorded what the Rebbe MHM said because every word that is forgotten is truly a pity.

It wasn't until I was 33 years old that I put on *tefillin* for the first time—20 years after the proper age. That happened in the Soviet Union under circumstances where studying Torah could only be done alone. If people studied in a group, it was a punishable offense, and in the times of Stalin it was punishable literally with a death sentence.

We know, unfortunately, how many thousands of Jews were killed just because they wanted to study Torah and continue a Jewish lifestyle. There was not even a trial. At that time there was a Troika of three commissars—three communists who were the judges—and if there was a verdict of a death sentence, they also shot the victim at that time. All this took half an hour, and of course there was no appeal.

Imagine how people felt under such conditions. Yet they studied Torah and did not give up. Chabad also never gave in. Chabad continued its activities all those years under the communist regime in the Soviet Union, not only studying Torah and keeping the Jewish home and Jewish lifestyle, but maintaining institutions of Jewish study: *yeshivos, Talmudei Torah, chadorim,* in the full sense of the word. This did not happen in the European part of the Soviet Union; it was mainly in Asia—Uzbekistan, Kazakhstan and Tajikistan. This story has yet to be written, but it should be, to show what Jews are able to do to remain Jewish and maintain the Jewish image.

The children who studied in those *yeshivos*—usually 10 twelve-year-olds—were real heroes. Not only were the teachers and organizers deserving of praise, but so were the children, who underwent great hardship, leaving their parents' home for eleven months a year to learn in underground *yeshivos*.

Of course they were completely disconnected from their families. They were not allowed to write a letter to their parents, because that would immediately reveal to the KGB where their institution was, and the school would, of course, be destroyed in no time. Nor could they telephone their parents for eleven months! Imagine 10 twelve-year-old boys, sometimes more, enduring all those terrible conditions for the sake of learning Torah.

The people who organized those *yeshivos*, of course, put themselves in even greater danger. They organized a system of *yeshivos* in a very, very careful and well thought-out way. The duties were distributed:

there would be one person to collect the children from several towns, put them on the train, and travel with them for four to five days to Tashkent, for instance. This man did not have the slightest idea where the *yeshiva* in Tashkent was. That was known to another man, who met this group in the railway station in Tashkent. He knew exactly where the *yeshiva* was, but had no idea where the children came from. Why such precautions? Obviously, and unfortunately, anyone could be caught in a minute, imprisoned and interrogated. Under such horrific torture, even the bravest and most heroic people would not always have enough strength to withstand the temptation to give the communists the information they wanted about where these *yeshivos* where located.

Thank G-d, for the most part, the *yeshivos* were successful for years and years. Many people who received their education in those underground *yeshivos* never went to public school. This was against the law in the Soviet Union. If parents didn't send their children to public school, their child could be taken away from them.

It was under these circumstances that I was first approached by several Chabad chassidim, who educated me and explained to me the meaning of being Jewish. But they didn't let me attend a *farbrengen* right away; they checked me out for about a year or so to see how reliable I was. After being allowed to attend *farbrengens*, I made further commitments to Judaism.

With all of the dangers involved, the spiritual mood and the atmosphere of *ahavas Yisroel* (love of a fellow Jew) at the *chassidishe farbrengens* were very strong and tangible. Having finally been invited to a *farbrengen*, I discovered a new world—Jews whose purpose in life was to help other Jews.

Imagine what it was like going out on the street to attend one of these *farbrengens*. Usually it was winter, physically cold, with a lot of snow. There was also the chilling fear that one might be followed. He would have special instructions that if he discovered he was being followed, he should turn around and go home—not go to the farbrengen—ecause that could bring with him the KGB agent.

On his way to the farbrengen, he was told to use side streets where he could stand for a minute or two to see if he was being followed. When he was sure no one was following him, he could proceed to the *farbrengen*. Would that the *farbrengens* here in the free world would be with the same level of feeling, emotion, and dedication that we had there. One would hope that it wouldn't take the fear of a KGB agent behind us to create such an intensity for Judaism, an intensity of Jewish feeling and dedication, of desire to add something and be instrumental in accelerating the arrival of Moshiach.

All of this, of course, was accomplished by our Rebbe Melech HaMoshiach, despite the fact that those chassidim never saw the Rebbe MHM and never heard the Rebbe MHM. There was hardly one picture of the Rebbe MHM in the entire town, because even though it was very desirable to see the *tzadik* (because when you see the face of a *tzadik*, your prayers and your desires are purified and elevated), it was very dangerous to keep such a picture at home. So there was one picture in the town of Riga, the capital of Latvia. It was passed from one chassid to another, and everyone had the privilege of having the picture for a few days or weeks.

Communication with the Rebbe MHM was impossible. We did study *sichos* of the Rebbe MHM from the *farbrengens* (in 770), but we never saw a printed *sicha*. We got the *sichos* by way of Israel. In Israel, special people were appointed to write the entire sicha by hand, which was then presented as a private letter. It began, "Dear so and so, how are you doing, how is your family, etc.?" And then the *sicha* followed. At the end, the writer would write, "Best regards, best wishes, write to me more frequently, etc.," but the bulk of the letter was a *sicha*. If the *sicha* was a long one, it would be divided into four or five letters and sent sequentially. So we had a single copy of the *sicha*, and we really valued it.

Nowadays, thank G-d, we come into *shul*, and sichos are everywhere—on the floors, on benches—and no one pays any attention. In those times we had one *sicha*. We could make a carbon copy or copy it by hand, but we couldn't make a Xerox copy of it because all the xeroxing had to be done under the supervision of the KGB, which made it impossible for us to do.

The only condition or event that justified sending a request to the Rebbe MHM for a blessing or for advice was for a wedding or a *shidduch*. It was sent through Kfar Chabad or Nachlas Har Chabad. It came back from the Rebbe MHM with a signature in Russian, *zedushka* (grandfather). Here again, it was just unbelievable how chassidim would try to follow the tiniest detail of instructions from the Rebbe MHM without ever having seen him or without the possibility of corresponding with him. Unbelievable.

As time went by, I gained quite a good understanding of what it meant to be a Jew and a chassid. It was now years after I had met Lubavitcher chassidim and had gotten educated to some degree. At a certain point I felt that I couldn't continue living in the Soviet Union, and I knew I would have to leave. So ultimately I applied for the right to emigrate to Israel.

Of course, leaving was always a dilemma. If you left, you would be missed in the town where you were working because other Jews needed your involvement. The Jews even had a kind of fundraising system in the Soviet Union. It was amazing how Jews responded to requests to give money for a Jewish purpose.

To organize a successful fundraising event in America, you have to prepare a dinner for nearly a year in advance, invite all kinds of famous people to attend and speak there, and present the institution in the best possible way. There, it was completely different. There were many government stores in the marketplace because at that time everything belonged to the Russian government. But many of the people who worked in the stores were Jews. So we would go to the marketplace and approach those Jews working in the stores and ask them to donate money. And all I could say was, *"Es gait far a yiddishe zach"* (it's for a Jewish cause). Remarkably, and surprisingly, the Jews did not ask exactly what we were doing, or ask for any evidence that what we were saying was true; they gave quite generously.

At the time I was working in the Institute of Physics of the Latvian Academy of Science, and when I applied for the right to emigrate to Israel, I was thrown out of the Academy. My wife, who happens to be a medical doctor, also lost her job, and so we underwent serious financial difficulties.

As can be imagined, the government was very angry because they believed I was a loyal Soviet citizen. Of course, I was never a member of the communist party, but they still trusted me as a loyal Soviet citizen. And suddenly I was demanding that they let me go to Israel! It was announced that I would have to wait at least 15-20 years before there would even be a possibility of leaving. Why 15-20 years? Because I had been involved in military research, so it would be at least fifteen years before I would forget everything that I researched, or my knowledge would become obsolete. Of course, I could not make peace with this situation.

As a result of this, they started arresting me. They put me under administrative arrest, which was very unpleasant. They would arrest me for a week or ten days, then let me go, and then arrest me again in two weeks. This happened again and again. They told me they would continue this until I stopped demanding to go to Israel. At that moment I decided that I had to do something drastic: I had to place a telephone call to the Rebbe MHM's office in New York. From the Chabad siddur that I had, I copied the telephone number of 770 that was printed there, and I went to the post office to ask to have a conversation with someone in the United States.

My friends told me I was insane! They said that if I called from the post office, there was absolutely no question that I would be arrested. But I couldn't place a call from my own house because the KGB had cut off my phone-line. My friends said, "You'll be taken away and imprisoned, so what will you have achieved?" I said, "Well, they arrest me every two weeks anyway, so what can I lose? At least I will know that I did something, that I tried to speak to the Rebbe MHM." In reality, I had a connection to New York in ten minutes, but to my grave disappointment I could not speak to the Rebbe MHM. One of the secretaries (Rabbi Binyamin Klein) answered the phone and asked me to tell him what the problem was and he would tell the Rebbe MHM immediately! I was so frustrated and in such despair that I continued pushing and begging. He said, "The Rebbe MHM doesn't speak on the phone!"

I found out later that the Rebbe MHM didn't speak on the phone even when the President of the United States wanted to greet him on his 85th birthday. But at that time I just thought that maybe Binyamin Klein didn't want to transfer me. So I got completely nasty and demanding and pushy. He told me, "I will transfer you to the chief secretary, and you can speak to him." He connected me to Rabbi Hadokov, who said "Listen, it's impossible. I can't ask the Rebbe MHM to take the phone. Tell me precisely what you want, I'll write it down, and we will get you a *brachah*."

Of course in retrospect, I feel a little uncomfortable that I acted that way, but here's what happened. At a certain moment, I suddenly heard a second voice on the telephone, besides R. Hadokov's voice. I don't know how, but I knew exactly whose voice it was: the feeling was as if an electrical current of high voltage was going through my entire body. I was trembling. And the Rebbe MHM said very few words. Not to me, because the Rebbe MHM doesn't speak on the phone. But he spoke to Rabbi Hadokov, telling him, *"Zogt em az er hot shoin aleh brochos, un bald vet er aroisforen."* "Tell him that he already has all the blessings, and soon he will be released."

Of course, I went home more than happy; I was dancing in the streets. Strange things happened in the next three weeks. I was called to the government office, and the clerk who was dealing with my file was extremely angry. He started yelling and screaming, "How long will you continue this nonsense? Why are you not sitting quietly? We told you that you would have to wait a minimum of fifteen years, and if you don't stop, we will arrest you, not for two weeks, but for several years."

I went home absolutely miserable. Soon after I came home and locked the door, there was a knock at the door. I opened it to see a messenger they had sent from that same office. They couldn't call me on the phone because they had cut off my telephone lines. The messenger told me that I had been asked to come to the office immediately. "Hurry up," he said, "because it is already half past four, and at five o'clock they close the office. Make sure you are there before then."

I went to the office, knowing full well that I would come and be arrested and that would be the end of it. My wife even wanted to give me some food or something but I didn't take it. I told her that if they arrest me, the food wouldn't help me. How much food could I take with me, anyway?

When I arrived there, the clerk was even angrier than before; he was absolutely furious. He told me, without even looking at me, that the management decided I could go. Still not looking at me, he pushed a pile of papers toward me and said, "Just fill out those documents, the questionnaires, and you can go." That was three weeks after the conversation with the Rebbe Melech HaMoshiach.

I will never know, of course, what happened there. I felt that they themselves didn't know why they were letting me go. Something happened that I can't explain. But I understood, because before that I had listened to many chassidic stories similar to this, beginning with tales of the Baal Shem Tov and other *Rebbeim*. Yet for the first time in my life I felt what a *brachah* of the Rebbe MHM means. A nation can have the largest army in the world, an army with tanks and airplanes, a KGB, a secret police. (They say there were about five million people working for the KGB, most of them in the Soviet Union but in many other countries as well, because they tried to dominate the whole world, and they did succeed to a great degree.) And all this is transcended with a *brachah* from the Rebbe Melech HaMoshiach.

The Rebbe Melech HaMoshiach's Prophecies

So that's how I left the Soviet Union. No power in the world can reject or ignore the Rebbe MHM's *brachah*. When I came to New York, I had a two-hour long *yechidus*, which was marvelous, of course....

I had many other experiences with the Rebbe MHM, among them witnessing the fulfillment of the Rebbe MHM's prophecy that the communist regime would fall. People who lived in the Soviet Union couldn't believe that this terrible thing called the communist regime would ever come to an end because it had been so strong, and while it was in power, nothing could possibly overcome it. The regime had not only been fighting people, they had been fighting G-d Himself. The only thing we could believe was that maybe, if the entire world would attack the Soviet Union with nuclear bombs, only then might it be defeated.

But what happened in reality? The Soviet communist regime collapsed in a peaceful way! No one suffered, no one was killed, nothing happened! Interestingly, in 1985, five years before the fall of communism, I got a phone call from another secretary of the Rebbe MHM, Rabbi Leibel Groner, telling me that the Rebbe MHM had asked me to come to his office.

Parenthetically, it's interesting that the Rebbe MHM would invite —through a message from Leibel Groner—a simple, non-educated Jew like me saying, "Tell him (meaning me) that it should not conflict with his other commitments. Maybe he is busy and this time is not convenient for him." When the Rebbe Melech HaMoshiach, who is managing and guiding the entire world, millions and billions of people, asks a Jew to come to his office, he thinks not only of the necessity of sending him the message, but also takes into consideration all his circumstances so that it should not harm the person's standing in his place of work, etc., and whether or not it's convenient for him. Personally, I see the Rebbe MHM's greatness much more in his attention to mundane details like this than in some gigantic matter that he is involved with. When you think that the Rebbe MHM, who is really on top of the world, the entire cosmos, pays attention to such things, it's really a wonder.[3]

Of course I didn't have anything more important to do; I rushed to the Rebbe MHM's office. When I entered, the Rebbe MHM was standing, not in his office but in the next room; it was a standing *yechidus*. It lasted not more than five or ten minutes, but the Rebbe MHM was standing.

The Rebbe MHM asked me to telephone whomever I could reach in the Soviet Union, among those Jews who were still there, and encourage them and ask them to be in a good mood because in a few years communism would disappear. I spent a whole night trying to understand this, but I didn't have the *chutzpa* to ask the Rebbe MHM how it would happen and exactly what would happen, so I just gave this message to whomever I could reach on the phone.

Some of them had difficulties with this news. All of them already had contact with Chabad and they knew, of course, who the Rebbe MHM is, but still, the situation and the circumstances of their everyday life were a severe contradiction to the news that the communists would disappear in a few years. One of them told me, "Listen, of course I have to believe what the Rebbe MHM says. I will try to believe and try to live with this idea, but it's very difficult for me! My wife was taken to be interrogated by the KGB three days ago and never came back. I don't know what happened to her." Another person told me, "If the Rebbe MHM says it, that's wonderful—but I have to say that the circumstances are getting worse! A KGB guy is

3. See the subsection "Majesty and Modesty" in chapter I, sec. 4.

standing next to my door, and they are probably recording every word that we are saying right now."

So I wrote a note to the Rebbe MHM telling him that I succeeded in getting through to this number of Jews in Moscow, Leningrad and other cities, but some of them had severe difficulties accepting this idea. The Rebbe MHM wrote back in exactly half an hour, saying, "Please, call them again if you can. and tell them that the process is underway. It has started. Even though you cannot see it or touch it, it's there. And even though all the newspapers are printing articles saying the exact opposite, it doesn't mean anything." The Rebbe MHM said. "They don't know what they're doing."

What happened? Who was right, the newspapers or the Rebbe MHM? The communist regime disappeared exactly as the Rebbe MHM predicted. Later on I had a very interesting experience with Gorbachev, who was the leader of the communist party and President of the Soviet Union at the time when everything was turned around.

When he had already been thrown out of the Kremlin and was no longer president, he started going around the world collecting money. I worked in a university, and I know everyone thinks that universities have the most clever people in the world (while it's exactly the opposite and universities are doing a lot of harm...). The heads of the university decided they needed Gorbachev to visit our university and speak to the professors and teachers. They paid an enormous price—a fee of $70,000—to have him come and visit.

Since I am in the university all the time and I speak Russian, the president of the university asked me to take care of Gorbachev and show him around for the entire day that he would be at the university, which I did. At a certain point I said to him, "You know, there is a Jew, the greatest Jew of our generation, the greatest *tzadik*, the Lubavitcher Rebbe MHM, and he told me back in 1985 (it was now 1991) that communism would disappear." And I told him, "You were probably instrumental in bringing the communist regime to an end." Now Gorbachev, who was not a clever or intelligent person whatsoever, but was in fact a very simple, poorly educated person, was shocked, speechless for several minutes. He said, "First of all, I don't understand. I don't believe that anyone in the world could know that communism would disappear and that the Soviet Union would fall apart. It was a great surprise to *everyone*." I said to him, "You can believe me or not believe me, but I got the information first hand. I didn't hear it through somebody, I heard it from the holy words of the Rebbe MHM himself."

"Secondly," he said, "it couldn't have been in 1985. In 1985 I had just succeeded in becoming president. At that moment I didn't have the slightest idea what I would do. I did not have a plan. A year, a year and a half later I had a plan, but what happened was *not* according to my plan. I, in fact, wanted to *preserve* communism. My intention was just to make it appear more favorable to the U.S. and other countries so that they would give us money, because we had tremendous economic difficulties. Then the regime just fell apart. And I personally suffered, because before I was the president, the chief of the party, and now I'm just an ordinary person."

So you see what happens? Here's the leader of one of the most powerful countries in the world who didn't have any idea what he was going to do, even in his secret thoughts. But the Rebbe MHM already knew precisely what would happen.

For those who know the Rebbe MHM well, this is not surprising. It's the prophetic vision of the Rebbe MHM, the special vision of the *tzadik* from the higher spiritual worlds. That's what has to be understood properly.

Here's another example: My laboratory in Beer Sheva is now the largest laboratory in the world in the special field of magneto–hydrodynamics, which I am continuing to develop. The Rebbe MHM arranged things in such a miraculous way that the American navy was running after me and begging me to sign a contract with them for certain research. So I had funding from the American navy for the next eight years. This was completely unbelievable.

Later, when the laboratory was already very successful and it had been recognized by the whole world, we made a certain invention. We arranged for patents and I went on a tour of Europe and America and gave lectures and seminars in universities just to check how our new idea, and the new device we were going to build, was perceived. I gave seminars and presentations and everyone was very happy.

After that. I was again given an opportunity to see the Rebbe

MHM privately. As usual, when I came in, the Rebbe MHM asked right away, "First of all, let's hear what is happening in your profession." And I answered in a few sentences that we had invented such and such a device and it does this and that and it's very desirable for industry. The Rebbe MHM smiled and said, "No, that's not what I asked you. I want you to give me a seminar precisely as you are giving it in the universities."

I was very confused because I thought, to stand in front of a great *tzadik*—the Rebbe Melech HaMoshiach—and speak about velocities, two-phased flows, efficiencies, and electromagnetic interaction, is nonsense! When you speak to the Rebbe MHM, you have to speak about spiritual things, at least about practical things that are necessary to help and to save other Jews.

But, of course, you do whatever the Rebbe MHM asks. So I made the presentation shorter than what I gave in the universities, and in 20-25 minutes I gave over the highlights of the seminar. I spoke about the device, and so on, and since the Rebbe MHM's room doesn't have a blackboard, I quoted the equations orally.

When I finished, the Rebbe MHM smiled his very special smile and told me two things: "First of all, your device is good, but if you introduce slight changes in the design, it will have 50% more efficiency." When the Rebbe MHM explained why, it became very clear. The only thing I couldn't understand was how I had spoken in front of dozens of the world's greatest experts and nobody had ever made this point; nobody had noticed; the Rebbe MHM said it in half a minute without having even a pencil or a piece of paper.

The second point was even more striking. The Rebbe MHM said, "You mentioned the efficiency of your facility, and the velocity of two–phased flows. The numbers you gave are erroneous; they contain errors." I told the Rebbe MHM, "I don't know, we used the best theories we could develop. We used a computer in Paris by phone (Israel did not have supercomputers at that time), and that's what we got."

The Rebbe MHM smiled again and said, "Okay, I do respect experts in every field. I insist that in Israel, the ministers should be experts, not just politicians. The Minister of Agriculture should understand agriculture, and the Minister of Defense has to understand defense, not just be developed and educated."

But the Rebbe MHM said there was a mistake here. I got very, very nervous about this, and when I went home I told my team to check all their theories and calculations and everything. They came back several days later and said there was no mistake. We lived in this atmosphere of confusion for about two years.

Ultimately, I received a new doctoral student. For his thesis, he had to analyze all the previous materials and develop his own addition to them. One morning he came and said, "Listen, I have some difficulties. You have to explain to me why in this equation a certain quantity is written in such and such a way." We sat down and I saw suddenly that there was a mistake. One equation was written incorrectly, and that caused all the errors. Somehow we hadn't been able to find it; we didn't pay attention. Sometimes when you're doing work, you get used to your own ideas and you don't notice mistakes. Since he was a new student and had to go through everything line by line, equation by equation, he was able to find it.

Of course, the question is, how did the Rebbe MHM know? You could say that it was the special vision of the Rebbe MHM, the difference between us simple, ordinary Jews and the Rebbe Melech HaMoshiach.

Torah, Science and Moshiach

Maybe I have been speaking too long. It reminds me of a joke: There was a certain conference, and one of the speakers spoke for a very long time, much longer than he was supposed to. Gradually, one after another, the people left. Eventually, only one participant remained in the hall. After several minutes he approached the speaker and said, "You can continue; just take the key and don't forget to lock the door when you leave."

In short, I don't want to get into the situation where I will be given the key to this room and be asked to switch off the lights....

Regarding the question of Torah and science, in addition to all that the Rebbe MHM told me in yechidus about Jewish education and events in the former Soviet Union, the Rebbe MHM also spoke to me a lot about Torah and science. But often, in middle of speaking about it, the Rebbe MHM would interrupt himself and say, *"Ihr zolt nit trachten az ich koch zich in eiyer mada; eiyer mada is vichtig far mir az tzulib dem noch a yid vet leigen tefillin*—Don't think that I'm so excited about your science; what's important to me is that through your scientific achievements, one more Jew will begin to put on *tefillin*."

The whole conflict between Torah and science started with the *maskilim* —members of the "Enlightenment Movement." The *haskala* ("enlightenment") movement started 150 years ago with Moshe Mendelsohn and other Jews in Germany, France, and other countries. Jews had been given the opportunity to "break free" and get involved in business, culture, and the social life of those countries. A number of Jews came up with the idea that sitting in a ghetto and concentrating only on Torah and Gemara is not enough; we have to absorb the world culture. We have to understand the culture of the non-Jews—how they live and what they treasure—and this will enable us to enter the general society and be successful and earn money.

All the Rebbeim fought against the *maskilim*, especially the Tzemach Tzedek, particularly because the *maskilim* wanted to introduce secular studies into the *yeshivos* and *chadorim*. Mixed into the practical side of these secular subjects was the ideology of atheism and nihilism. That's why the Rebbeim were so opposed to the *haskala* movement and secular studies for Jewish children and youngsters.

What happened? It is known that the Alter Rebbe was against Napoleon's winning the war against Russia. He explained that under the Russian Czar, things would be very difficult for the Jews materially, but spiritually, the Jews in Russia would flourish. Under Napoleon, on the other hand, the Jews would prosper materially, but spiritually, it would mean their defeat.

Over the years, what happened in Germany was that all those who became *maskilim* simply disappeared through intermarriage. Those families ceased to exist over a span of a mere few generations. It didn't happen this way in Russia and Poland, however, because in those countries non-Jews wouldn't intermarry with Jews. Instead, there was a different type of tragedy there: so many Jews who were educated in the *maskilim*'s institutions became communists. It is known that the primary leaders of the communist revolution in Russia were Jews. One can list hundreds of them. Trotsky, for example, the founder of the army of the Soviet Union, was a genius, but unfortunately what he did was terrible. His real name was Leib Bronshtien. There were many other Jews who did horrific things, including, for example, the founder of the terrible KGB, which invented special methods to torture people. Those same methods of the KGB were studied and utilized later by the Nazis, may their names be erased. And of course, the Germans perfected these methods to an even greater degree because the Germans were great specialists in committing atrocities.

This illustrates that when a Jew leaves the millennia-long style of Jewish education, G-d forbid, it can only lead to tragedy.

The Truth About Science

All of science is now converging with Torah, as opposed to years ago when it seemed that all of science was leading people away from Torah. Now it's exactly the opposite.

I think it's worthwhile for us to dedicate some time to analyze the contemporary philosophy of science. Philosophy of science is the discipline that deals with the definition of science, what science is and what it isn't.

Science can never reach the truth. It can only give generalizations about the experimental facts—the best possible generalization—which may be refuted a generation later. What has happened in our time is that since we are so close to the revelation of Moshiach, science is converging with great speed with the Torah outlook.

I will refer mainly to the philosophy of Carl Popper. The main idea of his philosophy is based on the philosophy of David Hume, who discovered that a scientific theory can never be *proven* by experiment because if you come to a certain conclusion, one negative observation contradicting the theory nullifies the many positive observations supporting it.

Hume himself was frightened by his own discovery. He believed that this conclusion was so terrible that he was destroying all of science because he discovered that you cannot prove any theory. So he didn't publish his results for many years.

Carl Popper took this one step further. He said that scientists should build their theories in such a way as to be open to refutation. Then via one observation he himself can refute it. (In order to make it better, you have to try to refute it.) If he does not succeed in refuting it, then for the time being he has a good theory. But it may be refuted later.

What does all this prove? This proves that all conclusions of science are temporary and inevitably will be changed. (But in many universities and high schools, unfortunately, they are still teaching that science is absolute and determines truth. This is a primitive, outdated notion.)

Popper's attitude was endorsed by the Rebbe MHM. The Rebbe MHM says that Popper actually came to the conclusion and demonstrated that there is no such thing as scientific truth. Scientists are not looking to discover the truth—the absolute structure and behavior of the universe. Sciences can only make the best possible generalizations of presently known, empirical facts. There is no doubt that scientific theories will ultimately be changed—sometimes from one extreme to another—and replaced by other theories.

That means that if we would adjust our philosophy and world outlook, which is based on Torah and Chassidus, to scientific statements which jump from one extreme to another, it would be a catastrophe. We know that the Rambam writes that Torah is eternal and will never be replaced by another Torah, G-d forbid. This indicates how inferior science is in comparison to Torah. What would happen if we started dealing with *halacha* by jumping from one extreme to another the way science does?

Light and Quantum Mechanics

Take, for example, a philosophy of metaphysics, of light. Greek scientists believed that light is made up of very tiny particles, emanated by a source of light, which penetrate our eyes.

Fifteen hundred years later, they found that one beam of light can nullify another beam of light, which would be impossible if light were composed of particles. According to the particle theory, two intersecting beams should give a brighter point—which sometimes does happen. But sometimes exactly the opposite happens.

If you throw a stone into a pond, you will see waves. If you throw two stones into the pond, you will see two systems of waves, and they will interfere with one another. If, at a certain point, one wave is high and the other is low, they will cancel each other out and there will be no movement at all. This is similar to what the scientists observed with the light beams, so they decided that the light beams were waves, not particles.

Then they asked: waves in what? So they invented the concept of the "ether." Then the ether was found to be non-existent, so they switched to electromagnetic waves. And ultimately they came to the conclusion—through quantum mechanics established eighty years ago—that light was simultaneously particles and waves. That's already beyond the ability of the human mind to comprehend, because we can't imagine something that has such a dual nature. A particle is a grain of matter; it may be so small that you cannot see it with the naked eye and you need an electron microscope, but it's still a grain of matter. A wave is a form of movement. How could one entity simultaneously be a wave and a particle?

Quantum mechanics will probably be refuted by other theories, but it hasn't happened yet. More remarkable is that in quantum mechanics, it is assumed that if a particle is not observed, it will always behave as a wave. In other words, we need a human being to observe it and then it will become orderly and behave as a grain of matter. Such experiments have been performed. They were first proposed by Nobel Prize winner Richard Feynman.[1]

I personally saw this experiment performed at the Weitzman Institute in Rehovot, Israel. They took an electron gun, emitting one electron at a time, and directed it at a screen. When the electron hit the screen, a flash of light was emitted.

Now, if you put a partition with two holes between the gun and the screen, if we're dealing with a particle, it can go through either the right hole or the left hole. But if it's a wave, when you put a partition with two holes between the gun and the screen, the waves will penetrate both holes and create patterns on the screen. In the experiment they did, there was a sensor by each hole, and the sensor gave a signal when a particle went through the hole. The sensors indicated that each particle always went through one hole or the other. When they removed the sensors and there was no observer, it immediately gave an interference pattern on the screen—indicating waves! The presence or absence of an observer determines the reality. How could this be?

Our Sages knew this perfectly well 2,000 years ago. There is a well-known question first raised by the Rambam and elaborated on by the Rebbe MHM in a number of *sichos*. What happens if two witnesses who are needed to establish the appearance new moon both make an erroneous statement? The answer is that it is impossible for the testimony of the witnesses to be erroneous because they have to be questioned by the *Beis Din* (court of Jewish law), and if the *Beis Din* questioned them, their testimony would become *halacha*, and *halacha* dominates the physical world. So if the *Beis Din* would rule that this is a reliable statement, then nature would adjust itself to the testimony of the two witnesses. Thus the above question of their testimony being erroneous is meaningless. Their testimony cannot be in error if it was accepted by the *Beis Din* because nature will immediately readjust. Here, too, we see how the observer determines reality.

1. See chapter V, sec. 5a for a detailed discussion of these experiments by Dr. Naftali Berg.

So physics, after 2,000 years, thank G-d, ultimately came around, through experimentation, to ideas that have been described in the *Gemara*.

The number of publications discussing G-d and science is growing exponentially from year to year. Ten years ago there were a few articles on this topic that appeared in print each year; now there are hundreds. Even a popular magazine like *Newsweek* had a cover story a few years ago titled "Scientists Find G-d," and they brought many examples of theories and hypotheses accepted by modern scientists that led directly to the existence of a Creator.

Another American magazine, *Forbes*, published an article a year ago in which journalists investigated a number of the leading and most famous scientists in the U.S. and found out that more than half of them not only believe in G-d, but pray to Him. Still, they do so mostly behind closed doors, because they are embarrassed and not brave enough to do it out in the open.

In the Days of Moshiach

When G-d created the world, the goal was that it would ultimately come to the state of Redemption with Moshiach.

The Baal Shem Tov, as he describes in a letter to his brother-in-law, once went up to the spiritual worlds, and when he came to the chamber of Moshiach, he asked Moshiach when he was coming. Moshiach answered, "When the wellsprings of your Chassidus will reach the outside." We are now, as the Rebbe MHM teaches, not *close* to the revelation of Moshiach; we are *in the days of Moshiach*.

Some people believe that Chabad invented the idea of Moshiach. That's nonsense of course, because the idea has always been one of the foundations of Yiddishkeit—it's in the Torah of Moshe Rabbeinu! But what is true is that the Baal Shem Tov and each generation of the Rebbeim of Chabad were acting to bring the world closer to the days of Moshiach. That gave the concept of Moshiach, of the ultimate Redemption, a feeling of reality, not just something to dream about. This feeling has to be in every one of us—that it's in our hands to make it happen.

The Rebbe *Rayatz* (the Previous Lubavitcher Rebbe) stated many times that we should be completely ready to greet Moshiach, and he used the following analogy: Before being reviewed by the king, soldiers in the army would have to be sure that they were presentable—that everything was in order, their uniforms clean and pressed, etc. Their final preparation was to polish the buttons of their uniform. So, the Previous Rebbe said, the only thing remaining for us to do to bring Moshiach is to "polish the buttons." Our Rebbe, Melech HaMoshiach, said it's even closer in our generation—that all that's remaining is for us to go out, open our eyes, accept and greet Melech HaMoshiach.

The Rebbe MHM gave many examples of why this is true, using as proof the colossal miracles that happened in our generation: the miraculous victory of the Israeli army in 1967, the collapse of communism and the Soviet Union, and the wondrous things that happened in the Persian Gulf War in 1991, which, the Rebbe MHM said, was the sign that the time of the Redemption had arrived.

The Rebbe MHM told us that all these statements were part of his prophetic vision. And when a prophet says that something good is going to occur, it has to materialize. (Only a *negative* decree can be nullified—by prayers and repentance, etc.)

Another proof is that not only have tens—if not hundreds—of thousands of Jews returned to Jewish practice, but science itself, once removed, cold and hostile to anything spiritual, is also speaking up today, louder and louder, saying, "Yes, there is a Creator, there *is* Someone running this world."

I'll bring a few examples: John Baumgardner of the Los Alamos National Laboratory speaks about his very precise calculations of all kinds of phenomenon in the crust of the earth. He uses software that he developed that has been evaluated in geological circles as being among the best in the world. Using it to calculate the age of the earth, he came up with a number of less than 6,000 years which, of course, corresponds exactly to what we believe.[2]

There's another scientist, an astrophysicist and geologist named Thomas Gould, who investigated the formation of oil. The formation of oil is another one of the atheists' "proofs" that the world is billions of years old, because it would take billions of years for organisms living in the depths of the sea to be converted into oil.

2. See chapter IV, sec. 4 for a full discussion of Baumgardner's work.

But Thomas Gould developed a completely new approach. There had been several oil wells in which the more that was pumped from the wells, the more oil remained in the well. No one could understand how that happened. They investigated and found one such well in Louisiana and one in Sweden. They discovered that oil is generated not in billions, millions, or hundreds of years, but in real time. You pump, and there, in the same layer, quite deep in the earth, deeper than formerly assessed, there are carbon and oxygen and hydrogen and other elements which, under certain conditions (pressure and temperature, and so on), spontaneously generate oil. Thus, oil is not of biological origin.

They repeated the process in the laboratory and got the same results they had found in nature. There was one problem, however, because those who had been promoters of the atheistic theory asked, "How could you say such a thing? Oil is a transparent liquid. If you pass a beam of light through the plane of polarization, it makes a quarter turn. That's how you know that it's an organic, biological substance."

But Gould found that there is a layer about three or four kilometers thick in the depths of the earth that is full of living bacteria. Even though they are extremely tiny living organisms, there is such a tremendous amount of them that in their totality they occupy more space than all the forests and vegetables and plants on earth. So Gould is saying that oil is generated deep in the earth, then it ascends through the layer of those bacteria where it is saturated with bacteria so, of course, on tests it will behave as an organic liquid, but it is not.

The Final Frontier

We mentioned the promise the Baal Shem Tov received from Moshiach that he will come when the Baal Shem Tov's teachings will be spread to "the outside." Geographically, this already happened a while ago.

There are wonderful Chabad Houses all over the world, established according to the instructions of the Rebbe Melech HaMoshiach. There are Chabad Houses in Nepal, India and Thailand. We know that, unfortunately, many Israeli boys and girls go to the Far East to seek spirituality after their army service. At least half of those youngsters wind up making their final stop at Chabad Houses: they spend several days or weeks there and return to seek spirituality in Yerushalayim, Tzfas or other cities in Israel.

In Siberia, the previous site of many prison camps under the Soviets, there are now about one hundred religious schools established by Chabad emissaries. This happened literally in the last few years. And one can't help but feel that geographically there is nowhere else to go to spread the teachings of Chassidus. The whole "outside" has been reached already.

But maybe we should come to see that the term "outside" refers to scientific, anti-religious ideology. We see now, thank G-d, that the teachings of the Baal Shem Tov and our Rebbe MHM have reached the most distant regions in human achievement: the scientific profession. That's enough to give us real optimism that all the conditions for Moshiach to come have already been fulfilled.

Although it is very painful that we cannot see the Rebbe MHM with our physical eyes and we cannot hear his voice directly with our ears, there is not the slightest doubt that, to the thousands and thousands of people who ask for the Rebbe MHM's help and guidance, the Rebbe MHM gives answers through the Igros Kodesh and in other ways. Chabad, thank G-d, has expanded in the last few years from 1,800 institutions to over 4,000. And those *shluchim* know that everything they accomplish is due to the guidance and assistance which they get from the Rebbe MHM every day. They appeal to the Rebbe Melech HaMoshiach through proclaiming *Yechi Adoneinu* and this gives them, and us, the strength to do what the Rebbe MHM demands of us.

The Rebbe MHM said that he already did everything he could do to bring Moshiach. There is no doubt that to the Rebbe MHM, Moshiach is already here. But we don't want to simply accept this fact or even understand it with our intellect. We need to see and hear the Rebbe MHM in a physical sense, not just through spiritual means. There is no doubt that if each one of us does, to the utmost of his capability, what the Rebbe MHM wants us to do, the Rebbe Melech HaMoshiach will reveal himself immediately.

Let me conclude with a practical suggestion. When I was a small boy, during World War II, we succeeded in escaping from the Nazis (may their names be erased) and came to Siberia. In every railroad station and every store, I saw a picture of a man stretching forth his finger and demanding, in the caption below: "What did you personally do today to make the victory over the Nazis closer?" And, thank G-d, they did conquer the Nazis, and at least part of European Jewry was saved. Otherwise who knows what could have happened?

My suggestion is that every night, when we say the *Kerias Shema al HaMita* (prayer before going to sleep), we should think about our day, and ask ourselves, "What did I do today, practically and visibly, to accelerate and bring the immediate revelation of the Rebbe Melech Moshiach?" I think that will certainly help.

It's not easy. We need to work hard and every day invent new methods and new approaches. But it's impossible, unacceptable, that it be delayed any longer.

Some of us have difficulty accepting or comprehending the whole idea of the coming of Moshiach when the presence of G-d will be sensed physically. We think we should continue going our way perceiving everything practically and logically. But we know that to those who left Soviet Union, it seemed absolutely impossible that the communist regime would ever come to an end. It was beyond anyone's ability to imagine it. Nevertheless, the Rebbe MHM promised it would happen, and it happened.

The Rebbe MHM promised us that this generation, the Seventh Generation from the Baal Shem Tov, will be the one to greet Moshiach and see the days of the complete Redemption. There is no doubt that very, very soon the Rebbe Melech HaMoshiach will lead us into *Eretz Yisroel* (Israel) and build the Third Beis HaMikdosh, after which *golus* (exile) will be a thing of the past forever.

CHAPTER VII

A Look to the Future

What will the sciences be like when the Geula is complete? We have discussed the "Final Theory"—the Theory of Everything—and have described its character: a fundamental concept of force and a fundamental concept of matter carrying the force (quality and quantity).

But there may be some surprises ahead. A number of prominent theoretical physicists are challenging quantum mechanics and even relativity theory. And the Rebbe MHM has even deeper challenges to these theories, as we will explain.

But for the Jews there is an even bigger surprise ahead. For, as the Rebbe Melech HaMoshiach has said…

"When Moshiach Tzidkeinu comes, all the sciences will return to the domain of the Jewish Torah scholars (and they won't need the scientists of the nations of the world). The knowledge of these sciences will come from the Torah itself."

The Science of Moshiach

Throughout this book we have been discussing "Scientific Thought in Messianic Times," discussing aspects of the Geula that can be seen in developments in modern science and technology.

But modern science itself is not the last word. Recall the Rebbe MHM's letter to the conference of Jewish doctors in which he wrote:

"True science whose purpose is only to seek the truth—its conclusions cannot contradict the Torah which is the Torah of truth. Rather, on the contrary, the more that [the scientists] will deepen their research, the more they will increase their recognition of the truth of both the principles and the details of our religion."[1]

Modern science is indeed converging to Torah as we have seen throughout this book, but some of the sciences still have a distance to go before they reach this end. Consider string theory, for example. We like it because it fits the Rebbe MHM's description of the final theory which must be built on two concepts—a concept of force (or energy) and a concept of matter carrying the force. This may be the vibrating string of string theory where the two concepts are the vibration and the string carrying the vibration. But string theory is far from fully developed, as explained at length in Lee Smolin's book.[2] It may not turn out to be a successful theory.

There are other theories that are well-established but faulty for other reasons. Relativity theory, for example, restricts anything from moving faster than the speed of light. The Rebbe MHM objects to this, stating that we cannot put a limitation on G-d by saying that He cannot make anything go faster.[3] (Particles, called "tachyons," that

1. *Igros Kodesh* 11:202 .
2. Lee Smolin, *The Trouble with Physics* (Boston: Houghton Mifflin, 2006).
3. *Igros Kodesh* 12:394.

can *only* go faster than the speed of light have been hypothesized but have never been observed.)

The Torah's statement that the earth is the center of the universe (geocentricity), once thought to contradict science, is consistent with relativity theory, as we explained earlier.[4] But here too some work remains because according to relativity both perspectives (that the earth is the center or that the sun is the center) are *equally* valid. But the Torah says that it is specifically the earth that is the center.[5] Thus a revised theory of gravity and celestial motion is needed to make science consistent with Torah.

Perhaps the dark matter and the dark energy would be useful to this end, but it is not certain that they even exist. Indeed, a theory proposed by physicist Mordehai Milgrom of the Weizmann Institute of Science in Rehovot, Israel, called Modified Newtonian Dynamics (MOND), that involves a slight modification to Newton's inverse square law of gravitation, would make the postulation of dark matter unnecessary.[6] In any case, the theory of gravity is not complete. (There are also problems with relativity as a *theory* and we will discuss this shortly.)

Unpredictable vagaries in the weather are another important issue which we have yet to fully understand. We quoted the Alter Rebbe's statement that the extreme heat in the summer and the extreme cold in the winter have no scientific explanation, giving clear proof of the existence of G-d Who determines the weather as He pleases.[7] But the scientists are not yet all convinced of the existence of G-d. They admit that even with all the sophisticated models and advanced supercomputers, weather cannot be accurately predicted, but they claim that this is due to the complexity of the situation. They still think that in theory it is predictable. Thus improvements must be made in the theories of meteorology and climatology so that they make completely accurate predictions—which will then be seen to fail. "On that day, G-d will be one and His name will be one."

In mathematics, the theory of infinities must be refined. In dis-

---

4. Chapter IV, section 4.
5. Rambam, *Mishneh Torah,* "Laws of the Foundations of the Torah," 3:4.
6. Mordehai Milgrom, "Does Dark Matter Really Exist?" *Scientific American* (August, 2002).
7. Chapter IV, sec. 5.

cussing infinities, the Rebbe MHM has said that since there are infinitely many points on a line, and infinitely many lines on a plane, the infinity of points in the plane is greater that the infinity of points on the line. But current mathematical theory does not distinguish between these two infinities; it considers the line and the plane to have the same infinity of points.

Beyond Quantum Mechanics

Quantum mechanics has been a very successful unifying theory, making very accurate predictions with many applications in the use of modern technology in everyday life. But according to quantum mechanics the laws of nature are not precisely determined. All one can say is that there is a certain probability that a given action will have a particular result. This one "flaw" has been accepted as a fact of life, and its paradoxical conclusions tolerated as an inherent kind of "weirdness" in nature.

To be sure, the Rebbe MHM has pointed out that since this is the way modern science *defines itself*—that all of its statements are only probabilistic—no challenge can be raised from science against Torah, whose statements are certainties.

However, the Rebbe MHM has an objection to quantum mechanics. It is not the objection of Einstein who insisted that the laws of nature must be precise—not probabilistic—because, as Einstein said, "G-d does not play dice with the universe." Regarding this, Dr. Naftali Berg once asked the Rebbe MHM if Einstein was correct in this statement, and the Rebbe MHM responded that Einstein was not correct since G-d can do whatever He wants to.

The Rebbe MHM's objection has an entirely different basis. It is based on the difference between nature and miracles. He once explained in a letter to Prof. Herbert Goldstein, a prominent nuclear physicist, that the point of view of modern science that the laws of nature are only probabilistic "is at variance with the concept of nature and our own knowledge of it (science) as espoused by the Torah since the idea of miracles implies a change in a *fixed order* and not the occurrence of the least probable event."[8]

8. Reprinted in *Mind Over Matter* (Jerusalem: Shamir, 2003), 38-44. (Italics are mine.)

Thus quantum mechanics is not the last word on physical law. Ultimately, it will be modified or replaced by a theory that gives precise statements and precise predictions.

Here, surprisingly, many physicists are in agreement.[9] Einstein's objections were overruled by the majority of physicists in his time who sided with Neils Bohr and the Copenhagen school, and the "Copenhagen interpretation" prevailed. The scientific community suppressed the opposing "hidden variables" theory according to which there is a precise hidden reality behind the equations of quantum theory. As Lee Smolin writes, "Since at least the 1950s the leading journals have only very selectively published papers on this subject, while several journals have excluded such papers by stated policy. The granting agencies and major government foundations have typically not supported this work, and university departments have tended not to hire people who are doing it."

But there is a growing movement among many of the world's leading theoretical physicists to question quantum mechanics. "Many leading physicists admit private misgivings about quantum mechanics," writes Smolin. "Among them are the Nobel laureate Gerard 't Hooft, who believes that quantum mechanics will be replaced with a deterministic theory, and the famous mathematical physicist Roger Penrose who, like 't Hooft, has been motivated in much of his work of the last two decades by his conviction that quantum mechanics is wrong. And like 't Hooft, Penrose has a vision of what should replace it."

Smolin himself is of this opinion: "I am convinced that quantum mechanics is not a final theory. I believe this because I have never encountered an interpretation of the present formulation of quantum mechanics that makes sense to me. I have studied most of them in depth and thought hard about them, and in the end I still can't make real sense of quantum theory as it stands... Quantum mechanics must then be an approximate description of a more fundamental physical theory. There must then be *hidden variables*, which are averaged over to derive the approximate, probabilistic description which is quantum theory."

Another dissenter regarding quantum mechanics is Holger Bech Nielson, one of the inventors of string theory. He believes that, "Everything we think of as intrinsically true, such as relativity and the

9. See Smolin, 315-324.

principles of quantum mechanics ...are just accidental facts that are emergent from a fundamental theory so beyond our imagining that we might as well assume that its laws are random." This seems to me reminiscent of the statement of the *Chacham Tzvi* that what we call nature is just a repeated sequence of acts of G-d, or miracles.[10]

Science and the Absolute

We mentioned above some specific difficulties regarding relativity theory. But there is also a basic methodological problem inherent in it as a scientific theory insofar as it addresses the concepts of absolute space and absolute time (which it rejects). The Rebbe MHM criticizes this in a letter to a professor, explaining in detail that concepts of the absolute are beyond the realm of science. In the letter, from 1956, he writes:

> "All those who explain relativity theory emphasize its most surprising conclusions and present them first and foremost: the negation of absolute space and the negation of absolute time—that both space and time are relative. It is as if they have proved it in such a way that it is unquestionable—by the Michelson experiment and later, at the solar eclipse, by the [light from] a star, etc. So they incorporate these conclusions into the realm of scientific knowledge, putting them on the same level as Newton's three laws (before it was proved that there were inaccuracies in them).
>
> "To my great surprise, in all the books on relativity theory that I have seen, I have not found anyone who expresses any reservations about the theory. And certainly no one who studies these books expresses any doubt about this assumption.
>
> "In my opinion, however, more than just *reservations* is required because science as it is now understood, i.e. exact science that reaches its conclusions solely by making measurements etc., has no place in its realm or even in the domain of its research for dealing with absolute concepts. The professional tools of exact science cannot comprehend the absolute. It follows that science cannot negate its existence—just as it cannot confirm it.

10. *Responsa of the Chacham Tzvi*, No. 18; quoted in *Sefer HaMa'amorim Melukat*, 5:108.

"Furthermore, normally, when a scientist completes his experiments and the statement of the conclusions drawn and the laws derived from the experiments, he attempts either to *propose* a new theory or to improve upon an existing theory in the field of his work. The goal is to add the new conclusions to the realm of scientific knowledge in this field and to explain it overall in a simpler manner and, as much as possible, in a manner of straightforward logic.

"Now, in establishing the theory of relativity which, as we just explained, is only a *proposal to attempt a possible explanation* of some natural phenomena, there is no justification to bring in concepts of the absolute. In fact, this goes against the goal of a simple explanation for the description of the phenomena is now harder to understand. Is there anyone who can understand in terms of plain logic and can visualize these concepts of relativity: there is no absolute space and time, the speed of light is equal in all directions and relative to all observers, space is deformed by placing matter into it, the space of the universe is like a balloon that continually expands or contracts, the velocity c (the speed of light) has the unique property that it is absolutely impossible to increase it, etc.?

"All scientific experiments that were ever done, including those related to relativity, are based on the following: measurement, instruments of measurement and the person making the measurement, and recording the results.

"From this it follows that all the conclusions that are reached relate only to the *measurement* of space, matter and light; the "behavior" of the instruments of measurement, and the characteristics and ability of the experimenter.

"Regarding the experiments of Fizeau and Michelson etc., the explanation proposed and the theory closest to the experiments, using the fewest assumptions external to the experiments, must make use of the explanation[11] of Lorentz."[12]

---

11. See D. Halliday, R. Resnick and J. Walker, *Fundamentals of Physics*, extended 6th edition (New York: John Wiley, 2001), "The Lorentz Transformation," 932-934, especially the note on the bottom of p.932.

12. *Igros Kodesh*,12:395-396. Taken together with the quotes from the Rebbe MHM on science at the beginning of this book, in the section "Wisdom From Below," (Ch. I, sec. 3) we can get a full picture of the Rebbe MHM's characterization of science.

Is the scientific world ready for this challenge to relativity? Quantum mechanics was controversial for a long time, and even though the matter was considered settled in favor of the Copenhagen interpretation, the opposition—while suppressed—was out in the open. But we have never heard any opposition to relativity.

Now, however, the opposition is here! As Lee Smolin writes, the seers—great scientific thinkers—who challenge relativity "are hiding in plain sight. They can be recognized by their rejection of assumptions that most of the rest of us believe in." He goes on to list several of them, including Holger Bech Nielson whom we mentioned above, and Ted Jacobson who, together with his students at the University of Maryland, are leading the search for an experimental test of special relativity; and Robert Laughlin, a Nobel Prize winner.[13]

The Science of the Torah

Everything that we have been discussing concerns the scientific view of the universe and its relationship to the Torah view of the universe. But there is a deeper issue that concerns the Torah itself. Sometimes we need knowledge of science or mathematics in order to fulfill a mitzva in the Torah. The best example of this is *Kidush HaChodesh*—the laws of determining the new months and the new years of the Jewish calendar, which require knowledge of mathematics and astronomy. From where are we to get this scientific knowledge? Apparently, we must get it from the secular scientists. That is what Rambam did when he wrote his famous section on the "Laws of Declaring the New Month" in Mishneh Torah. He used the mathematical and scientific works of the Greeks for his calculations. (He pointed out, however, that it was not the Greeks themselves whom he relied on, but rather on the mathematical proofs backing up these calculations.) In using these works on mathematics and astronomy, he elevated these sciences to a higher spiritual level and, by incorporating them into his work on Halacha, actually transformed them into Torah.

But the Torah is a complete self-contained wisdom. It is G-d's own wisdom and the source of all knowledge. How could it be that we need to go elsewhere to find some of the knowledge we need to understand the Torah?

13. Smolin, pp. 314-315.

The Rebbe MHM discusses this issue in connection with a story in the Torah about Yakov and his sons. There was a famine in the Land of Israel in the time of Yakov Avinu. Yakov and his sons still had grain while the other peoples of the area had already run out of food. Thus Yakov's sons made no special effort to look for more food. Nevertheless, Yakov challenged his sons and asked them, "Why are you giving the appearance that you have plenty of food…?"[14] He wanted them to go down to Egypt to buy more food before their supplies ran out.

The Rebbe MHM explains that the basis of the approach of Yakov's sons to the issue of getting food was that they saw that G-d had been providing them with food while everyone else had none. They understood that G-d was sustaining them in a miraculous manner just as He had done for their grandfather Yitzchok in a time of famine. They maintained that since they were the children of Avraham, Yitzchok and Yakov whose lives were conducted in a supernatural G-d was always doing miracles for them—they did not have to go out and look for food like the other nations did. They had absolute trust that G-d would continue to provide for them miraculously and they didn't have to do anything on their own.

Thought for Food

The Rebbe MHM continues to explain this on a deeper level. Wisdom and knowledge are called "food" in the abstract sense. Just as physical food is taken internally, is digested and gives life to the body, so too an intellectual concept is taken internally, is processed by the mind and gives a person life. The fact that Egypt was the source of food at that time was the result (in a spiritual sense) of its being the center of knowledge and science at that time (intellectual food).

But we Jews don't need the science of Egypt as a source of knowledge. We have our own source of knowledge—the Torah. When things are the way they should be, we can get all the knowledge we need—even scientific knowledge—from the Torah, and the Jewish scholars have this knowledge readily available. This was the case in

14. Breishis 42:1 with commentary of Rashi.

the time of the prophets when the Jewish scholars from *Shevet Yisachar* wrote books on the calculations of astronomy which they derived from the Torah. In fact, the Kuzari states that these sciences were originally in the possession of the Jews, and the early nations whose people studied astronomy, etc. received this knowledge from us.[15]

Thus the position taken by Yakov's sons was essentially correct. We don't need to go to the nations of the world for food or thought. It is only during the time of Golus that we may have to use the sciences of the nations (such as the books of the Greeks on mathematics and astronomy that Rambam referred to in writing the Laws of Sanctifying the New Months), just as Yakov's sons eventually had to go down to Egypt to buy food.

When the Geula of Moshiach is complete, however, all wisdom and science will return to the Jewish scholars who will derive this knowledge directly from the Torah itself. Thus Rambam writes[16] that in the Era of Moshiach "there will be an abundance of knowledge and wisdom" and "the Jews will be great scholars...for the earth will be filled with the knowledge of G-d as the waters cover the sea."[17]

Wiser Than King Solomon

It is evident that the man who brings all this about must be a man who knows what he is looking for and knows how to find it, i.e., a man who has mastered the sciences of the world and at the same time has a knowledge of Torah deep enough and vast enough to derive the "Torah's science" from it. Thus, when Rambam writes about the wisdom of Moshiach he says that Moshiach will be even wiser than King Solomon, who had mastered the sciences of ancient Egypt and had surpassed them.[18]

As we have seen in many other areas, we see here too that the Rebbe Melech HaMoshiach is the one who fits this description. During the 1930s the Previous Rebbe sent him to Berlin and Paris to study the sciences. There he mastered both theoretical science such as mathematics and physics (he attended the lectures of Erwin

15. *Kuzari* 2:66
16. Rambam, *Mishneh Torah*, "Laws of Kings," 12:5.
17. *Likutei Sichos* 30:190-197.
18. I Kings 5:10.

Schroedinger on quantum mechanics) and applied science and engineering.

His phenomenal knowledge of the sciences is evident from his encounters with many world renowned scientists. As an example, Professor Yirmiyahu Branover of the University of the Negev relates that he once presented the plans for a new invention to the Rebbe MHM. Branover presented the value of the "two-phase flow velocity" and the value of the efficiency of the system and the Rebbe MHM immediately responded that those two values were inconsistent. Branover took the plans back to Beer Sheva where his research team reviewed their calculations but found no error. It wasn't until many months later that their error was discovered by a new doctoral student who had joined their team and went through the calculations carefully line by line to learn what the team had done.

The Rebbe Melech HaMoshiach's knowledge of Torah is vast and deep beyond description. This has been attested to by all the Torah authorities of the world. For example, the former Chief Rabbi of Israel, Rabbi Mordechai Eliyahu, z.l., said of the Rebbe MHM: "There is no Rabbi as great as he in all areas of the Torah in this generation. In the logical understanding of the Torah no secret is hidden from him.... Similarly in the hidden and Kabbalistic levels of the Torah.... It is absolutely correct to say of him that he is the greatest in the generation. There is no one second to him. But what is most important is his humility....There is no one like him."

"The Jew becomes a "partner with G-d"...in bringing the true and complete Geula. In order for G-d to carry it out completely He needs (so to speak) the participation of every Jew alive as a soul in a body, for it is through our actions and our service that the Geula comes. G-d needs the Jew to agree to this and even more—that he should want it and announce that not only has the time for the Geula come, but that the Geula is already here, literally."

—The Rebbe Melech HaMoshiach, 28 Sivan, 5751

REFERENCES: CHAPTER VII

A Look to the Furture

Rabbi Menachem Mendel Schneerson, MHM. *Igros Kodesh*. New York: Kehot Publication Society.

———. *Likutei Sichos*, vol. 30. New York: Kehot Publication Society.

———. *Sefer HaMa'morim Melukat*, vol. 5. New York: Kehot Publication Society.

Rambam, *Mishneh Torah*.

Halliday, David, Robert Resnick and J. Walker. *Fundamentals of Physics*, extended 6th edition. New York: John Wiley, 2001.

Milgram, Mordehai. "Does Dark Matter Really Exist?" *Scientific American* (August, 2002).

Mind Over Matter. Jerusalem: Shamir, 2003.

Smolin, Lee. *The Trouble with Physics*. Boston: Houghton Mifflin, 2006

APPENDIX

The Doctors' Plot

In addition to his prophecies, the Rebbe Melech HaMoshiach has performed countless miracles, both for individuals and for the entire world. One of the most dramatic of these was in fact a fulfillment of the prophecy of Isaiah regarding Moshiach: "And with the breath of his mouth he will put the evil man to death." This happened in January, 1953 when the evil Stalin arrested a group of Jewish doctors, accusing them of killing some government officials. This was the infamous "Doctors' Plot." It triggered a wave of intense anti-Semitism throughout Russia. Jews were attacked in the streets and many prominent Jews disappeared or mysteriously "died of heart attacks."

On the night following Purim that year, the Rebbe MHM held the customary farbrengen in "770." Late in the night, in the middle of the Purim celebration, his face suddenly became very serious and he began to tell a story. He said that after the fall of the Czar, the government held elections in Russia. The Rebbe of Lubavitch at that time, the Rebbe Rashab, instructed the chassidim to vote in the elections. There was one chosid who was totally removed from worldly matters and had no idea what an election was. But he followed the Rebbe's instruction and went to vote. When he got to the polls, however, he didn't know what to do—how to vote or who to vote for. He met other chassidim there and they told him what to do. After the elections, everyone started shouting "Hoorah! Hoorah!" But this chosid thought that they were shouting the Hebrew words *"Hu ra! Hu ra!"* meaning "He is evil! He is evil!" So he began to shout, *"Hu ra! Hu ra!"*

The chassidim listening to the Rebbe MHM tell this story understood that he was telling them that they should also shout "*Hu ra! Hu ra!*" So they all stood up and began shouting "*Hu ra! Hu ra!*" Around this time, back in Russia, Stalin had a stroke, and a few days later it was announced that he was dead. Subsequently, the doctors were freed and the case against them was dropped.

It is important to note that it was not just the lives of the Jewish doctors which were saved by this miracle. Stalin had planned to deport all the Jews of Russia to remote regions of Russia and was in the process of building concentration camps for them. He planned to transport them in open cattle cars so that thousands would freeze to death on the way. Krushchev later reported that when Stalin revealed his intentions at a meeting of Soviet leaders, two of them, Mikoyan and Voroshilov, protested and said that such conduct was worthy of Hitler. Stalin flew into a furious rage in response.

Indeed, Rabbi Levi Yitzchok, the Rebbe MHM's father, once used a *gematria* (numerical calculation of Hebrew letters) to express the fact that Stalin and Hitler were basically the same.

A miracle similar to this Purim miracle of the Rebbe MHM was once performed by the Tzemach Tzedek. One year on Ta'anis Esther, the afternoon before Purim, when he was davening *Shemoneh Esrei*, he stopped after the words *melech meimis* ("a king who puts to death") in the second bracha. He waited a long time. A chosid standing nearby who was watching this commented, "A strike from a Rebbe." They later found out that at the exact time the Tzemach Tzedek was saying *melech meimis*, the cruel anti-Semitic Czar Nikolai was killed.

The Alter Rebbe once said that in every generation where there is an extraordinary holiness in the world, there is also an opposing extraordinary evil. He gave the examples of a certain evil priest in the time of the Baal Shem Tov, and the emperor Franz Joseph of Austria in the time of the Maggid of Mezeritch. He concluded by saying that Napoleon was an extraordinary evil and that they had to have *mesiras nefesh* to defeat him.

In our time, the generation of Melech HaMoshiach, I believe it is correct to say that Stalin was such an extraordinary evil. But it was not Stalin alone; Sadam Hussein said that he modeled his dictatorship after Stalin's. He was the other extraordinary evil of our time, and it is well known that if he had been given a free hand, he would have perpetrated much more of the same kind of atrocities Stalin committed, G-d forbid.

Indeed, there are striking similarities between the way the Rebbe MHM "put the evil man to death with the breath of his mouth" and the way he defeated Sadam. He edited and published the very same *ma'amar* which he had said at the 1953 Purim farbrengen—which put Stalin to death—for Purim 1991, the exact date that Sadam was defeated in the Persian Gulf War.

The Doctors' Plot

Dr. Yakov Rapoport, one of the doctors arrested during the Doctors' Plot, wrote in his memoirs, "The Doctors' Plot...defies common sense. All one can do is attempt to analyze the series of events that led up to it." In this paper we will attempt to do just that—analyze the series of events leading up to the case, the case itself and the miraculous sudden death of Stalin that resulted in the case being dropped and charges brought against those responsible for prosecuting the doctors. It was a total reversal or, to use the term of the Megillah—*V'Nahapoch hu."*

The doctors' plot was a complex fabrication spun from many threads, connecting events—some of which happened and some of which never happened—with people who had little or no relevance to the alleged events. The "evidence" was either a false confession extracted under torture or a mere accusation, and guilt was determined in advance. As the senior investigator once told a prisoner, "The question of your guilt is decided by the fact of your arrest."

The threads were spun by Stalin himself, and the intended victims were not just the doctors, but all the Jews of Russia. It was a genuine Purim story, complete with a Mordechai, a couple of Hamans, and even an "Esther."

While several books have been written on the Doctors' Plot, they are all incomplete and flawed, partly because the authors did not have access to sufficient information and partly because there are aspects of it for which there exists no information. The most complete account appears to be that of Jonathan Brent[1] and Vladimir Naumov, an American and a Russian, who were given access to almost all the KGB files on the matter, including many original documents which had not been seen by anyone for fifty years. In 2003 they published a book on the Doctors' Plot[2] based on this information, and our account here is based primarily on their book.

And it was in the dark days of Stalin, he was Stalin who reigned over all of Russia, from Baku to Yakutsk, many states and provinces....

In the early days of his reign, in the 1930's, he carried out purges in which millions of people were killed. Now, in the 1950's, he was about to execute a repeat of those horrible purges.

Andrei Zhdanov was a powerful member of Stalin's Politburo. He was considered by many to be the second most powerful man in Russia. He was the architect of Soviet postwar ideology and cultural policy and he was the one whom Stalin assigned to present Soviet ideology at various international conferences. Also, his son Yuri married Stalin's daughter.

In July 1948, Zhdanov, who suffered from heart disease, went to Valdai, a health resort for the Soviet political elite northwest of Moscow. While he was there, he suffered some sort of "heart attack." The exact nature of this attack was later to become a major issue. An EKG was taken by Dr. Sophia Karpai, a Jew. Her diagnosis was that he had an intraventricular blockage that *may have been* caused by a heart attack (myocardial infarction). Accordingly, the doctors treating Zhdanov ordered that he have absolute bed rest.

On August 7, Dr. Karpai was "sent on vacation." After this time the doctors relaxed their treatment of Zhdanov and he was allowed to get out of bed, walk around in the park, go to movies, etc. On August 28, Dr. Lydia Timashuk, manager of the electrocardiograph unit at the Kremlin hospital, was called to Valdai to take an EKG of Zhdanov, the results of which, she claimed, showed that Zhdanov had actually suffered a heart attack (myocardial infarction), and she objected to the doctors' relaxation of the requirement of strict bed rest.

---

1. In the very informative video *Declassified—Stalin*, Dr. Brent and others discuss the horros of Stalin's regime in general. See http://www.youtube.com/watch?v=S69s8YyJMIM

2. Jonathan Brent and Vladimir Naumov, *Stalin's Last Crime: The Plot Against the Jewish Doctors 1948-1953* (New York: HarperCollins Publishers, 2003).

The next day Zhdanov suffered another attack. Timashuk was called back but was told not to take an EKG until the next day. She then wrote a letter to Lt. General Nikolai Vlasik, head of the Kremlin Bodyguards, accusing the doctors of neglecting the traetment of Zhdanov who, she said, should not have been allowed to leave his bed and walk around freely since he had suffered a heart attack. Such mistreatment, she concluded, could lead to the patient's death. Two days later, on August 31, Zhdanov died.

Timashuk's letter, which was delivered to Vlasik and shown to Stalin, was the first piece in the puzzle that was to become the Doctors' Plot. Timashuk herself has been accused by some of being an anti-Semite. Whether or not this was the case, at this point in the story the issue does not appear to have been anti-Semitism since none of the doctors treating Zhadanov were Jewish. The only Jewish doctor involved was Dr. Sophia Karpai, the EKG technician. She was not directly involved in Zhdanov's treatment and she was sent away from Valdai on August 7. It was later discovered that Timashuk was an agent of the MGB (the KGB of that time) and was possibly just informing on those doctors whom she believed were negligent in the treatment of an important government official.

One of the mysteries of the Doctors' Plot is that while Timashuk's allegation later became central to the government's case against the doctors, in 1948 Stalin took no action on her letter. He just sent it "to the archives." Timashuk herself was dismissed from her position and transferred to another hospital.

Why was no action taken on her letter? What was really going on here? Brent and Naumov connect this with an incident that took place a few months earlier. A Soviet scientist named Lysenko had been promoting a theory of wheat farming which, he claimed, would yield tremendous wheat production for Russia and alleviate their recurrent shortages. Russian scientists considered the theory nonsense and it was found that much of his data was faked. But Stalin liked the theory, so it was dangerous to speak against it. Most of the Russian scientists who did oppose it were dismissed from their positions and sent to the gulag. At a certain conference Zhdanov's son Yuri spoke against it, and this angered Stalin. In April, 1948, Stalin said at a meeting at which Zhdanov was present that those responsible for this provocation should be punished in "exemplary fashion."

He held Zhdanov—the father—responsible. "It was necessary to question the father and not the children," Stalin said at the conclusion of the meeting. Zhdanov was now a marked man. His trip to Valdai was actually a banishment: Stalin told him to go there "to improve his health." His heart attack in July followed a phone call in which he was apparently informed that his position had been taken over by Malenkov.

While Stalin gave no orders for Zhadanov's death, his intention was clear. He almost never issued a direct order. Rather, he would approve or have others approve orders issued by others. In his system, loyalty and intuitive knowledge of his wishes counted for more than any other quality. Loyalty to Stalin meant knowing Stalin's enemies and what to do regarding them. This system was known as "vigilance." Thus, Brent and Naumov suggest that the doctors at Valdai knew that Stalin wanted Zhdanov dead and understood that they should let him die. As Dr. P. I. Yegorov, the head doctor treating Zhdanov, said, "A heavy job has fallen to us…the fate of Zhdanov is predetermined."

The date of August 7 was the turning point in carrying out the "heavy job." On that day a letter written by Yuri Zhdanov, in which he recanted on his criticism of Lysenko, was printed in Pravda. This was the first mention in public of the rift between Stalin and Zhdanov and was the signal to the doctors that Zhdanov had fallen out of favor with Stalin. They knew what they had to do. After August 7, they allowed Zhdanov to walk around freely, which was extremely dangerous for him. The nursing care became sloppy. His personal physician went fishing. Dr. Karpai was sent away from Valdai on August 7 so no EKG's were taken. The official medical record was manipulated so that Zhdanov's eventual death would appear sudden and unexpected.

This may be why Dr. Karpai was sent away on that day. She was Jewish and she was not part of their conspiracy. She knew too much about Zhdanov's condition and would not acquiesce to their neglect of the patient.

The Second Thread: The Arrest of Dr. Ettinger

The second major stream of events leading to the Doctors' Plot was set in motion with the arrest of Dr. Yakov Ettinger, a professor at the

Moscow Medical Institute, on November 18, 1950. Ettinger, however, was not arrested in connection with any charge of medical sabotage, but rather as a "Jewish nationalist" and a supporter of the state of Israel.

Since 1947, Stalin had been waging a war against the Jews of Russia under the title of "the anti-cosmopolitan campaign," charging that Jews considered themselves superior to Russians and were sabotaging Russian culture. When the state of Israel was established in 1948 and a lot of pro-Israel feeling was expressed by Russian Jews, it was claimed that Jews could not be trusted to be loyal to the Soviet Union, and the campaign intensified. As Israel began to develop a close relationship with the United States, Russia's new post war enemy, official anti-Semitism became even more vicious. Thousands of Jewish professionals, scientists, political leaders and security personnel, as well as private individuals, were mercilessly persecuted, dismissed from their positions and imprisoned. Many were shot.

A major event of this campaign was the execution of the members of the Jewish Antifascist Committee which had been organized by the Soviet government during World War II for propaganda purposes. After the war, it promoted the establishment of a Jewish republic in the Crimea.

Dr. Ettinger, a very prominent, highly-educated physician, was considered the greatest Russian diagnostician of his time. He had many contacts in the U.S. and was very open about his political views. He criticized Stalin for his anti-Semitic campaign and for his lack of support of Israel. In a phone conversation which was bugged by the MGB he remarked, "Whoever could liberate the country from such a monster as Stalin would be a hero."

Ettinger had been interrogated 37 times when suddenly, in January 1951, an order came from Stalin to halt the interrogations. It appears that that it was at this point that Stalin conceived the Doctors' Plot. He wanted to redefine Ettinger's crime as being part of a conspiracy of Jewish doctors to commit medical sabotage. The interrogations would have to take a different direction.

The order to stop the interrogations was delivered by V.S. Abakumov, the minister of state security to the senior investigator M.D. Ryumin. But Ryumin violated this order and, despite warnings from the medical staff that Ettinger might die from the stress of the interrogations, continued to interrogate Ettinger another 39 times until March 2, 1951 when Ettinger collapsed and died following an interrogation.

Ryumin was a Haman in his own right. He was described by his own colleagues as "a primitive anti-Semite" and "dimwitted, an egoist by nature, a deceiver, capable of showing any fact in a way that served his own plan." He was cruel, vulgar and vindictive. His MGB colleagues hated him and feared him. In his interrogation of one Jewish officer he said, "In Moscow there live more than a million and a half Jews. They have seized the medical posts, the legal profession, the union of composers and the union of writers. I'm not even speaking of the trade networks. Meanwhile, of these Jews, only a handful are useful to the state. All the rest are potential enemies of the state." He said further that he was going to propose the expulsion of the Jews from Moscow....

Ryumin was now in serious trouble for violating his orders and causing the death of such a valuable prisoner as Ettinger. To cover for himself, he wrote a letter to Stalin on July 2, 1951 denouncing Abakumov for "concealing the terrorist aims of Dr. Ettinger's medical treatment of Kremlin leaders." Of course, Ettinger had not mistreated any Soviet leaders and had not even been accused of it. (A "confession" that Ryumin extracted from Ettinger on this matter was rejected by Abakumov as nonsense.) This was a complete fabrication on the part of Ryumin that was going to be extended to the full blown "Doctors' Plot." If Ettinger, the Jewish nationalist, could now have a charge of medical sabotage pinned on him, this could be extended to a conspiracy of Jewish doctors—and Jews in general—to murder Soviet government officials. The conspiracy would link alleged medical sabotage with MGB complicity and incompetence which would necessitate a purge of the MGB.

Abakumov was arrested and accused of planning a coup d'etat. Abakumov, the anti-Semite who had overseen the execution of the Jewish Antifascist Committee, was now being accused of conspiring with the Jewish doctors and MGB officers to overthrow the government. In the purge of the MGB that followed, tens of thousands of officers were arrested or dismissed. Most of them were Jewish. Many of the Jewish officers were Abakumov loyalists who were directly involved in the falsifications in the case of the Jewish Antifascist Committee.

This brings to mind the warning that the Previous Lubavitcher Rebbe gave to Nachmanson and Lulav, the two GPU officers who arrested him in 1927. He said that under the current regime in Russia, those who come to arrest others can rest assured that the time will come when they too will be arrested.

There were two reasons why Stalin needed to link the alleged conspiracy of the Jewish doctors to an accusation of a conspiracy within the MGB. Firstly, this would make the case against the Jews that much stronger since it would mean that the medical sabotage of the Jewish doctors was part of a conspiracy to overthrow the entire Soviet government. Secondly, to achieve the maximum effect, he wanted the case against the doctors to be a public trial. This would rile up the Russian population against the Jews. In order for this to be successful, the case against the doctors had to appear credible. It was thus necessary to get rid of those in the MGB who knew too much, i.e. those who knew that the alleged conspiracy was a complete fabrication.

The Need for an Enemy

For evil dictators such as Stalin, anti-Semitism was more than just an expression of their evil hate. In order to consolidate and maintain absolute power, they needed to invent an enemy and a crisis. As Brent and Naumov write, "The utility of the purge was that it was a vehicle, not for destroying enemies, but for creating them. It did not produce stability. Rather, its function was to destabilize the country, because supreme power could be achieved and held only in crisis conditions." We have seen this combination of anti-Semitism and setting up the Jews as a national enemy more recently in Arab countries. Indeed, as we mentioned above, Sadam Hussein used Stalin as the model for his own dictatorship.

For Stalin the external enemy was the United States, against whom he could not easily take action. But he created an internal enemy—the Jews—whom he could easily persecute. The accusations against the Jewish doctors claimed that they were carrying out their medical sabotage at the direction of United States (and British) intelligence.

The Jews became the channel through which American subversion flowed directly into the Soviet government, threatening to destroy it. As Stalin declared to a meeting of the Presidium, "Every Jew is a potential spy for the United States."

The Persecution Intensifies

In July, 1951 Stalin put his plot against the doctors into action. On July 11 the Central Committee of the Communist Party issued a decree which stated that there exists a "real conspiratorial group among the doctors who were fulfilling the instructions of foreign agents in terrorist activities against the leadership of the party and the government." A purge of the MGB (accused of covering up the conspiracy) began immediately.

Stalin himself directed the implementation of the plot. As Khrushchev later described it: "There was no need in Stalin's mind for an investigation....He spoke and they were arrested....He himself called the investigators, himself instructed them, himself ordered the method of the investigation and there was only one method—beat them." But Stalin concealed his role in the investigation. Rather, it was made to look as if the investigators were *discovering* many "facts" and connections as they carried out their investigations. Orders from Stalin would be attributed to others. He kept his name out of it so that no one could testify against him. The day-to-day direction of the plot was supervised by S.A. Goglidze, deputy minister of state security and a close associate of Beria.

The first Jewish doctor to be arrested after the decree of the Central Committee was Dr. Sophia Karpai, on July 16. It is not completely clear why she was arrested, but she was the only Jewish doctor who had treated Zhdanov and it appears that they wanted to extract a confession from her regarding medical sabotage in Zhdanov's treatment. She continued to maintain, however, that Zhdanov's EKG did not necessarily indicate that he had suffered a heart attack.

Karpai was a young woman, and the interrogators may have thought that they could get an easy confession out of her. But the opposite turned out to be the case. Although she was treated with great cruelty and kept in a damp refrigerated cell, she demonstrated great courage. She was not intimidated, and they did not succeed in breaking her. Brent and Naumov call her a "latter day Esther" (as will be explained later).

On November 12, 1952 Stalin's policy of torturing and beating the imprisoned doctors was implemented. The investigators were ordered to beat the doctors with "death blows." "Beat them to a pulp and grind them into powder," were Stalin's words according to Krushchev. The next day Ryumin was removed from his position, apparently because he was not effective in getting the required confessions. Also, he knew too much.

A special torture chamber was set up in Lefortovo prison, fitted with special screens to muffle shrieks of pain. A special detail of "helpers" was recruited to carry out the torture. This new method produced many extraordinary "confessions" from all the doctors—except for Karpai.

A Cold Dark Winter

It was January 1953. The torture of the doctors was in full force. False confessions were flowing like the blood of the victimized doctors. And the city of Moscow was morose....

The winter in Moscow was unusually snowy and cold. There were no thaws and the ice on the streets of Moscow grew thicker and thicker. Reports circulated of a certain "Day X" when millions of Jews throughout Russia would be deported. Each day brought new reports of threats, plots and expulsions, and lists of Jews were being drawn up in Moscow, Leningrad and other major cities. As far away as Kiev, a Jewish doctor was advised by one of his non-Jewish patients to leave the city and move to a small town, if possible.

More Jewish plots and conspiracies were being "discovered" and Jews were being arrested *en masse*.

Meanwhile, Stalin worked tirelessly to ensure that every detail of the plot and its exposure would achieve maximum impact while appearing completely credible.

On January 13, an article in Pravda titled "Spies and Murderers Under the Mask of Professor-Doctors" revealed the Doctors' Plot to the public. It described a "band of doctor-poisoners" who were aggressively involved in murdering Soviet leaders. It listed the names of the Jewish doctors and claimed that they were working for U.S. and British intelligence. The Jews in general were identified as enemies of the state, along with the Americans and British.

More newspaper articles were printed to incite the public against the Jews. They told people to look for enemies and referred to the purges of the 1930s, an indication that they were about to be repeated. Anti-Semitic meetings at offices and factories vehemently demanded that the criminals be put to a terrible death. Many volunteered their services.

Concentration Camps

On February 17, a memorandum was sent to Malenkov recommending the construction of four prison camps in remote areas of Russia: one in Kazakhstan, one in Irkutsk and two in Komi. It was to these camps that the Jews were to be deported. The construction of the camps began immediately.

Krushchev later told a Communist Party meeting how, during the Doctors' Plot, Stalin became inflamed with hatred against the Jews. His rage grew until, shortly before his stroke, "he told a meeting of Soviet leaders that he had decided to gather all the community together and transport them to a northern region within a new pale." Krushchev told his audience that when Mikoyan and Voroshilov protested and said that such conduct was worthy of Hitler, Stalin worked himself into a fury.

Another cynical and sadistic element was introduced into the story. It was not enough for Stalin to just deport the Jews. He wanted them to *request* the deportation! A letter was drafted in which the Jews would appeal to Stalin for "protection from the justifiable anger of the Russian people" by transferring them to a far off region. This was to be an open letter to Stalin signed by prominent Jewish personalities. The letter, however, was not publicized at that time. (A copy of the letter was found and published in a Russian journal in 1997.)

Meanwhile, the interrogation of the doctors continued. By mid-February all the doctors had confessed except for Dr. Sophia Karpai. On Feb. 18 they brought Karpai to a face-to-face confrontation with Dr. V. N. Vinogradov, who had worked with her at Valdai. He had just confessed to murdering Zhdanov and the purpose of the confrontation was to pressure Karpai to confess also. They could not go to an open trial with the doctors until each told the same story. But it did not succeed. Karpai continued to refuse to admit any guilt. "No, I do not confirm it," was her response.

None of the other doctors held out so long. Her face-to-face confrontation with Vinogradov certainly made her realize that she was truly alone. But although she was brutally beaten and kept in a refrigerated cell without sleep to compel a confession, she did not confess. Her heroism in the face of all this is truly astounding. The delay in the trial caused by her courageous stand may have saved the lives of millions of Jews, for in the meantime Stalin died and the decree was annulled. As Brent and Naumov write, "The fate of the Jews of Russia might have depended on this latter-day, unknown Esther."

The Death of Stalin

"The Doctors' Plot ended much as it began, in mystery," write Brent and Naumov as they begin to discuss what is known about the death of Stalin. They continue, "There is more that we don't know about Stalin's death than that we do....Each published version of his death significantly contradicts every other."

Another author writes, "It is a paradox that while the details of his final illness were broadcast to the whole world, the atmosphere of mystery shrouding the circumstances of the death of Stalin has never been dispersed. A number of people, satisfied with the information given, accept the fact that Stalin died of cerebral hemorrhage. Many, suspecting that his end was altogether too opportune, speak of it as a miracle that saved Russia from a new reign of terror. Some are of the opinion that the 'course of nature was assisted.' Others, dismissing his illness as fictitious, believe that Stalin was murdered."[3]

As we mentioned at the beginning of this section, the death of Stalin was brought about by Melech HaMoshiach—"By the breath of his mouth he shall put the wicked to death"—at the farbrengen of Purim, 5713. We will now trace the sequence of events leading up to Stalin's death and try to correlate them with the events occurring in "770" at the same time, noting an eight-hour time difference.

3. *The Death of Stalin: An Investigation by 'Monitor'* (London: Allen Wingate, 1958), p. viii.

While the various reports of Stalin's death also contradict each other regarding the timing of certain events, Brent and Naumov have attempted to sort out and analyze this information. In addition to documents available to earlier researchers, they have had access to the report of the doctors who treated Stalin while he was dying. This report was submitted to the Central Committee in July 1953, stamped "Top Secret" and filed away. It was apparently unread for fifty years until 2003 when Brent and Naumov were given access to it.

Purim 5713 was Sunday, March 1, 1953. On the night of Purim (Saturday night, February 28), Stalin sat down to a party in his dacha with Beria, Malenkov, Krushchev and Bulganin. Sometime around 4:00 a.m. [8:00 p.m., the night of Purim, in New York], in the middle of an angry tirade about the slow progress of the Doctors' Plot, Stalin suddenly broke off in the middle of a sentence and went into his room. This was to be his last tirade.

Stalin did not come out of his room the next day, nor was there any sign of movement in his room. Though worried, the staff could not enter his room without permission. At 6:30 p.m. [10:30 a.m. – *Shacharis,* Purim morning, in New York] a light went on in Stalin's room. At 10:30 p.m. or later [2:30 p.m. – *Mincha* time – in New York], a guard found Stalin sprawled on the floor near his writing table. Around 11:00 p.m., Beria phoned the dacha and told the guards not to say anything about Stalin's sickness. Doctors did not arrive until 7:00 a.m., or later, the next morning [11:00 p.m. in New York, during the Purim farbrengen]. Why was there such a delay in calling the doctors? There are two possible reasons: 1) They may have been afraid that they would be held responsible for any mistakes made in Stalin's treatment or, more likely, 2) They wanted him to die.

In any case, it appears that the declaration of *"Hu Ra"* at the Purim farbrengen occurred around the time that the doctors began treating Stalin. They found him in a deep unconscious state from which he never recovered and diagnosed a cerebral hemorrhage. His condition steadily worsened until Thursday, March 5 when, at 9:50 p.m., the evil Stalin died. Haman was dead: "When the wicked are gone there is rejoicing."

Many have suggested that Stalin was poisoned. If so, the man who would have done it, or at least would have been responsible for it, would have been Beria. And this would have been with Krushchev's complicity. There is, in fact, reason to believe this. Molotov said that Beria told him, "I did him in! I saved all of you!" claiming to be responsible for Stalin's death. In fact, Beria had good reason to kill Stalin since Stalin was in the process of preparing a case against Beria with the goal of having him eliminated.

Brent and Naumov describe a scenario, suggested by a Dr. Lawrence S. Cohen, in which Beria would have slipped a poison, such as warfarin crystals, into the wine that Stalin was drinking at the party on Feb. 28. Warfarin is a tasteless and colorless blood thinner which was administered to patients with heart disease. A high dosage over several days could induce hemorrhaging and a stroke in a patient with acute arteriosclerosis, which Stalin had. (Another Russian historian, Edvard Radzinski, interviewed the guard that found Stalin on the night of March 1. He said that after Stalin's party was over an order was given by the head guard, Khrustalev, for all the guards to go home. Radzinski speculates that it may have been at this time that Stalin was poisoned by Khrustalev, at Beria's orders.)

Following this line of thought, we must also consider the possibility that Stalin was poisoned, at Beria's orders, by the doctors who came to treat him. Note that they began their treatment of Stalin around the time that everyone was calling out "Hu Ra!" at the Purim farbrengen in "770".

"And It Was Reversed"[4]

Immediately after Stalin died, the Doctors' Plot was reversed. Beria, the head of the newly-created MVD (Ministry of Internal Affairs), in one of his first official acts, ordered a "complete review" of the Doctors' Plot. To expedite the process, he appointed a special commission. On March 17, Ryumin (the other Haman) was arrested and accused of falsifications and perversion in the Doctors' Plot.

On March 31 a "Decree on the termination of criminal prosecution and the freeing of the prisoners in the Doctors' Plot" was issued by the MGB under Beria's direction. It stated in part, "Taking into account that all the arrested doctors in the present case were illegally imprisoned...we decree: In view of the absence of crimes...to free

4. Esther 9:1.

from custody and with full rehabilitation those imprisoned in this case."

On April 3, during Pesach, the doctors were freed. The same day, the Soviet press published a communique issued by the MVD which said, "On the basis of the finding of the investigation commission specially set up by the Ministry of Internal Affairs of the USSR to verify the case, the above-mentioned [doctors] and others implicated in this case have been fully cleared of the charges preferred against them and...have been released from custody. The persons guilty of the improper conduct of the investigation have been arrested and are criminally held responsible."

On April 6, Pravda reported that "The result of a review of the Doctors' Plot showed that the doctors had been arrested by the former Ministry of State security (MGB) incorrectly, without any legal basis....The examination showed that accusations brought forward against the accused were false." It called Ryumin "a secret enemy of the government" and a "criminal adventurist." He was later executed.

"And he [Melech HaMoshiach] will judge the poor with justice and he will admonish with fairness the humble of the earth; and he will strike [the evil ones of] the earth with the rod of his mouth and with the speech of his lips he will put the wicked to death." (Isaiah 11:4)

Yechi Adoneinu Moreinu V'Rabeinu Melech HaMoshiach L'Olam Voed!

Education Day U.S.A.—and Russia?

In the sicha of Parshas Mishpatim 5752 (1992), the Rebbe MHM refers to "the effect of Melech HaMoshiach on the nations" *resulting in the Swords Into Plowshares declaration. In detailing the background of this, he discusses the spreading of "the concepts of goodness, fairness and justice among the nations of the world by their observing the Seven Noachide Laws." In this section we document how the Seven Noachide Laws and their related concepts have received recognition at the highest levels of the U.S. government as expressed in presidential proclamations, joint resolutions of Congress, etc.*

President Ronald Reagan

On April 3, 1982, President Ronald Reagan received a delegation of Chabad Rabbis in his office at a ceremony where he signed a presidential proclamation declaring April 4, the Rebbe MHM's birthday that year, as "National Day of Reflection."[1] The declaration reads:

Amid the distractions and concerns of our daily existence, it is appropriate that Americans pause to reflect upon the ancient ethical principles and moral values which are the foundation of our character as a nation.

We seek, and steadfastly pursue, the benefits of education. But education must be more than factual enlightenment—it must enrich the character as well as the mind.

One shining example for people of all faiths of what education ought

1 All of these presidential proclamations can be viewed at "The American Presidency Project" at www.presidency.ucsb.edu/proclamations.php . The images and text of several of President Reagan's proclamations, as well as some of the correspondence between the Rebbe MHM and Reagan, can be found at www.chabad.org/therebbe/article_cdo/aid/142535/jewish/The-Rebbe-and-President-Reagan.htm

to be is that provided by the Lubavitch movement, headed by Rabbi Menachem Schneerson, a worldwide spiritual leader who will celebrate his 80th birthday on April 4, 1982. The Lubavitcher Rebbe's work stands as a reminder that knowledge is an unworthy goal unless it is accompanied by moral and spiritual wisdom and understanding. He has provided a vivid example of the eternal validity of the Seven Noahide Laws, a moral code for all of us regardless of religious faith. May he go from strength to strength.

In recognition of the Lubavitcher Rebbe's 80th birthday, the Senate and the House of Representatives of the United States in Congress assembled have issued House Joint Resolution 447 to set aside April 4, 1982, as a "National Day of Reflection."

NOW, THEREFORE, I, RONALD REAGAN, President of the United States of America, do hereby proclaim April 4, 1982, as National Day of Reflection.

Note the explicit reference to the Seven Noahide Laws.

The next year, on March 25, 1983, the U.S. Congress presented the "National Scroll of Honor" to the Rebbe MHM, again honoring him on the occasion of his birthday and signed by the members of Congress and President Reagan. It reads:

Presented by the president and the Congress of the United States of America in tribute to the vision and spiritual world leadership provided by the Lubavitcher Rebbe, Rabbi Menachem Mendel Schneerson, on the occasion of his reaching the Eightieth Year

***Whereas**, the Lubavitcher Rebbe, Rabbi Menachem M. Schneerson, has reached the eightieth year of his life which is devoted to the service of world Jewry and humanity in general; and*

***Whereas**, his venerated vision, wisdom and leadership have contributed greatly to the promotion of education and the betterment of mankind; and*

***Whereas**, the President and both houses of Congress of the United States of America have accordingly recognized his accomplishments by proclaiming "Education Day, U.S.A." and "National Day of Reflection" on his birthday; and*

***Whereas**, the Lubavitcher movement, through its scores of educational centers in this country and abroad, dedicates itself to preserve, protect and foster universal values that all free men hold dear; now, therefore, be it **Resolved**, on the occasion of his birthday, March 25, 1983 corresponding to the 11th of Nissan 5743, we the undersigned do present the Lubavitcher Rebbe, Rabbi Menachem Mendel Schneerson the National Scroll of Honor recognizing his brilliant achievements and wishing him health, long life and many more years of leadership to crown his celebrated career:*

Every year since then, the Rebbe Melech HaMoshiach's birthday has been declared by presidential proclamations by every U.S. president, as "Education Day U.S.A." It is important to note that presidential proclamations are not just for ceremonial purposes. They actually have the status of American law. As explained by the American presidency Project,[2] "A presidential proclamation is 'an instrument that states a condition, declares a law and requires obedience, recognizes an event or triggers the implementation of a law (by recognizing that the circumstances in law have been realized).' In short, presidents 'define' situations or conditions on situations that become legal or economic truth. These orders carry the same force of law as executive orders—the difference between the two is that executive orders are aimed at those inside government while proclamations are aimed at those outside government. The administrative weight of these proclamations is upheld because they are often specifically authorized by congressional statute, making them 'delegated unilateral powers.'"

Reagan's 1984 proclamation reads in part:

The Lubavitch movement, headed by Rabbi Menachem Mendel Schneerson, has provided people of all faiths a shining example of the true value of education. The Lubavitcher Rebbe's work is a living reminder that knowledge is worthy only when accompanied by moral and spiritual wisdom and understanding. In fostering and promoting a tradition of ethical values that can trace its roots to the Seven Noahide Laws, which have often been cited as universal norms of ethical conduct and a guarantee of fundamental human rights, the Lubavitch movement and its greatly respected leader have shown Americans of every

2. See their website referred to in the previous footnote.

faith that true education involves not simply what one knows, but how one lives.

In recognition of Rabbi Schneerson's contributions and in honor of his 82nd birthday on the 11th day of the Jewish month Nisan, which falls this year on April 13, the Congress, by House Joint Resolution 520, has designated April 13, 1984, as "Education Day, U.S.A.," and has authorized and requested the President to issue an appropriate proclamation.

Reagan's 1985 proclamation reads in part:

Very few Americans have done more to promote these ethical values as the basis of civilization than Rabbi Menachem Mendel Schneerson, the leader of the worldwide Lubavitch movement. The word "Lubavitch" comes from the name of a Russian city and means city of love. That is very appropriate because, of all the ethical values which inform our civilization, none is more important than love—love of wisdom, love of our fellowman, and love of our Creator.

These are the values which Rabbi Menachem Mendel Schneerson exemplifies. And they are the values, with their roots in the Seven Noahide Laws, which have guided the Lubavitch movement throughout its history. They are the essence of education at its best, and we should be certain that we pass on this precious heritage to all young Americans.

In recognition of Rabbi Schneerson's contributions and in honor of his 83rd birthday, which falls this year on April 2, the Congress, by House Joint Resolution 186, has designated April 2, 1985, as "Education Day, U.S.A." and authorized and requested the President to issue an appropriate proclamation in observance of this event.

NOW, THEREFORE, I, RONALD REAGAN, President of the United States, of America, do hereby proclaim Tuesday, April 2, 1985, as Education Day, U.S.A., and I call upon the people of the United States, and in particular our teachers and other educational leaders, to observe that day with appropriate ceremonies and activities.

Reagan's 1986 proclamation reads in part:

Education is like a diamond with many facets: it includes the basic mastery of numbers and letters that give us access to the treasury of human knowledge, accumulated and refined through the ages; it includes technical and vocational training as well as instruction in science, higher mathematics, and humane letters. But no true education can leave out the moral and spiritual dimensions of human life and human striving. Only education that addresses this dimension can lead to that blend of compassion, humility, and understanding that is summed up in one word: wisdom.

"Happy the man," Scripture tells us, "who finds wisdom .... Her ways are ways of pleasantness, and all her paths are peace. She is a tree of life to those who come to possess her."

The Congress has sought to call attention to these durable values by adopting resolutions that pay tribute to the example of Rabbi Menachem Mendel Schneerson, a man who has dedicated his life to the search for wisdom and to guiding others along its pathways. He exemplifies the rich tradition of the Seven Noahide Laws, which have been the lodestar of the Lubavitch movement from its inception.

In recognition of Rabbi Schneerson's noble achievements and in celebration of his 84th birthday, the Congress, by House Joint Resolution 582, has designated April 20 as "Education Day, U.S.A." and authorized and requested the President to issue an appropriate proclamation in observance of this event.

Now, Therefore, I, Ronald Reagan, President of the United States of America, do hereby proclaim Sunday, April 20, 1986, as Education Day, U.S.A., and I call upon the people of the United States, and in particular our teachers and other educational leaders, to observe that day with appropriate ceremonies and activities.

Again, note Reagan's repeated reference to the Seven Noahide Laws in each of his proclamations.

We now present excerpts of several of the presidential proclamations of Presidents George Bush, Bill Clinton, George W. Bush and Barack Obama:

President George Bush, 1989

Ethical values are the foundation for civilized society. A society that fails to recognize or adhere to them cannot endure.

The principles of moral and ethical conduct that have formed the basis for all civilizations come to us, in part, from the centuries-old Seven Noahide Laws. The Noahide Laws are actually seven commandments given to man by G-d, as recorded in the Old Testament. These commandments include prohibitions against murder, robbery, adultery, blasphemy, and greed, as well as the positive order to establish courts of justice.

Through the leadership of Rabbi Menachem Schneerson and the worldwide Lubavitch movement, the Noahide Law—and standards of conduct duly derived from them—have been promulgated around the globe....

In recognition of Rabbi Schneerson's vital efforts, and in celebration of his 87th birthday, the Congress, by House Joint Resolution 173, has designated April 16, 1989, and April 6, 1990, as "Education Day, U.S.A." and has authorized and requested the President to issue an appropriate proclamation in observance of these days.

NOW, THEREFORE, I, GEORGE BUSH, President of the United States of America, do hereby proclaim April 16, 1989, and April 6, 1990, as Education Day, U.S.A. I invite Governors from every State and Territory, community leaders, teachers, and all Americans to observe these days through appropriate events and activities.

Note the *detailed* listing of the Seven Noahide Laws.

Public Law 102-14

On March 20, 1991, both houses of Congress passed a joint resolution, known as Public Law 102-14,[3] declaring the Rebbe MHM's birthday as "Education Day, U.S.A," honoring his 90th year. It was

3. "Public Law 102-14, H.J. Res.104, 102nd Congress, 1st session"

signed by the Speaker of the House, the President of the Senate (Vice Pres. of the U.S.) and then by President George Bush. It reads as follows:

JOINT RESOLUTION

To designate March 26, 1991, as 'Education Day, U.S.A.'

***Whereas** Congress recognizes the historical tradition of ethical values and principles which are the basis of civilized society and upon which our great Nation was founded;*

***Whereas** these ethical values and principles have been the bedrock of society from the dawn of civilization, when they were known as the Seven Noahide Laws;*

***Whereas** without these ethical values and principles the edifice of civilization stands in serious peril of returning to chaos;*

***Whereas** society is profoundly concerned with the recent weakening of these principles that has resulted in crises that beleaguer and threaten the fabric of civilized society;*

***Whereas** the justified preoccupation with these crises must not let the citizens of this Nation lose sight of their responsibility to transmit these historical ethical values from our distinguished past to the generations of the future;*

***Whereas** the Lubavitch movement has fostered and promoted these ethical values and principles throughout the world;*

***Whereas** Rabbi Menachem Mendel Schneerson, leader of the Lubavitch movement, is universally respected and revered and his eighty-ninth birthday falls on March 26, 1991;*

***Whereas** in tribute to this great spiritual leader, 'the Rebbe', this, his ninetieth year will be seen as one of 'education and giving', the year in which we turn to education and charity to return the world to the moral and ethical values contained in the Seven Noahide Laws; and*

***Whereas** this will be reflected in an international scroll of honor* *signed by the President of the United States and other heads of state: Now, therefore, be it*

***Resolved** by the Senate and House of Representatives of the United States of America in Congress assembled, That March 26, 1991, the start of the ninetieth year of Rabbi Menachem Schneerson, leader of the worldwide Lubavitch movement, is designated as 'Education Day, U.S.A.' The President is requested to issue a proclamation calling upon the people of the United States to observe such day with appropriate ceremonies and activities.*

President Bill Clinton

President Clinton's proclamations designated the Rebbe MHM's birthday as "Education and Sharing Day, U.S.A." His 1998 proclamation reads in part:

A firm believer in nurturing both mind and heart, the Lubavitcher Rebbe, Rabbi Menachem M. Schneerson, devoted his life to helping young people realize their potential and become visionary leaders and thinkers, as well as concerned, caring, and productive citizens. He established more than 2,000 educational and social institutions in more than 40 States and nearly 60 countries. He was deeply committed to fostering civic pride and moral integrity along with professional success....

Now, Therefore, I, William J. Clinton, President of the United States of America, by virtue of the authority vested in me by the Constitution and laws of the United States, do hereby proclaim April 7, 1998, as Education and Sharing Day, U.S.A. I invite Government officials, educators, volunteers, and all of the people of the United States to observe this day with appropriate activities, programs, and ceremonies.

President George W. Bush

President Bush's proclamation of 2002 honored the Rebbe MHM's 100th birthday:

Rabbi Menachem Mendel Schneerson, the Lubavitcher Rebbe, demonstrated the power of teaching others to pursue both intellectual and moral excellence. For this purpose, he established education and outreach centers that offered social services and humanitarian aid. Today, more than 1,400 of these centers operate in 35 countries, meeting the spiritual and material needs of local communities. These beacons of hope communicate the Rebbe's ideal of living a meaningful life through personal responsibility and dedication to those in need. On the 100th anniversary of the Rebbe's birth, I encourage Americans to help teach our young people. In classrooms, homes, and places of worship, we can all make a lasting impact on individual children and thus invest in our country as a whole.

Now, Therefore, I, George W. Bush, President of the United States of America, by virtue of the authority vested in me by the Constitution and laws of the United States, do hereby proclaim March 24, 2002, as Education and Sharing Day, U.S.A., 2002. I call on all Americans to help our young people better understand the importance of character and develop a true love of learning. By doing so, we invest in thefuture of our country.

President George W. Bush, 2003

For the past 19 years, on the anniversary of his birth, we have recognized the contributions of the Rabbi Menachem Mendel Schneerson, the Lubavitcher Rebbe who was instrumental in establishing numerous educational, social, and rehabilitative institutions. The Rebbe believed that education is critical in cultivating the moral character of students. He lived what he said: "A single good deed on your part could transform the world."

President Barack Obama, 2009

Knowledge alone will not bring the future our children deserve. Our schools and community institutions must also help each child develop a moral compass....

Few have better understood or more successfully promoted these ideas than Rabbi Menachem Mendel Schneerson, the Lubavitcher Rebbe, who emphasized the importance of education and good character. Through the establishment of educational and social service institutions across the country and the world, Rabbi Schneerson sought to empow-

er young people and inspire individuals of all ages. On this day, we raise his call anew.

In Russia Too?

We have seen how much the concepts of the Seven Noahide Laws and the related concepts of fairness, goodness and kindness that the Rebbe MHM has promoted have permeated the U.S. government. But now they are also being spoken about openly in Russian political circles.

Our colleague Rabbi Berel Haskelevich has reported on this. He writes the following:[4]

Who has not heard of *Pravda*, the mighty Party newspaper, the official propaganda mouthpiece of the forever ruling Communist Party of the Soviet Union. The old saying was, 'there's not a word of pravda (truth) in Pravda', but the paper used to sell tens of millions of copies daily for many reasons:

(a) People needed paper to wrap the butter or cheese bought at the milk store, or meat at another store;

(b) There was nothing else to read: other papers copied the carefully tailored news from Pravda;

(c) Pravda was the official source of information about government decrees: all the new ordinances were carefully outlined in Pravda;

(d) those readers who were brilliant enough to read "between the lines" could, through careful analysis, get a glimpse of what goes on behind the scenes, what changes can be expected in the official thinking. A sentence in Pravda was like a briefing from the Politburo.

However, in the Era of Moshiach things change so quickly that we sometimes miss the quick flashes on the video screen. Pravda is no longer Communist, and no longer the official giant-paper. It is politically aligned with the rightist Russian nationalist forces, and makes a modest living through gossip, etc. Until recently, I was convinced: I will not find one word of truth in Pravda.

4. Printed in Beis Moshiach Magazine no. 574, 5 MarCheshvan, 5767, pp. 18-19. Available online at www.beismoshiach.org

That was until September 19. As if accidentally, I saw a link to Pravda.ru (on-line)[5] that read, 'Rebbe Schneerson against Belykh and Gozman.' Did they chose the title for sensational reasons? I did not know. I just clicked on it.

It was a political commentary written by one Andrey Scherbakov. He deplored the political platform of the leaders of the SPS (the Union of the Rightist Forces) Nikita Belykh and Leonid Gozman,…and Gozman's exposed atheism, which pushed away many religious people from the extreme rightist political movement.

So what does Pravda suggest instead?

> "We would add," writes A. Shcherbakov (commenting on a lead article by V. Tretyakov, Chief-Editor of the popular Moscow paper 'Moscow News') "that among the rightists of America today the ideas of reforming the political structures made by the…spiritual leader of Lubavitch Chasidim Rebbe Menachem Mendel Schneerson…whose birthday is proclaimed by a decision of Congress as Education Day USA, are gaining more and more popularity.
>
> "Although Rebbe Schneerson was an Orthodox Jew, his ideas are supported by many rightist Christian politicians. In the opinion of this prominent thinker, the future development of America is impossible without education and revival of morality, and fascism and democracy are extremes that must be avoided."

Shcherbakov then quotes [some of the Rebbe MHM's ideas] from Simon Jacobson's book (in Russian) and continues:

> "From such wells of wisdom…should the ideologues of SPS draw their ideas for their program documents. And Leonid Gozman, who so much loves to refer to the practice of American democracy, should rather know, as Rebbe Schneerson reminds us, that the sessions of the U.S. Congress begin with an appeal to G-d. Since there are pre-

5. (Pravda, Sept 18, 2006. http://www.pravda.ru/politics/parties/sps/197428-sps_programm-0).

> sented people of different confessions there, this is not a traditional prayer, but a specially prepared text."

As a result of the Rebbe MHM's campaign, the concept that the Seven Noahide Laws must be the foundation of a civilized society has been expressed by many other governments throughout the world including France, Belgium, England, Japan, Ghana, Poland, Austria, New-Zealand and Brazil as well as the Druze leaders in the Middle East.

The Mathematical Papers

The following mathematical paper, "The Location of Eigenvalues and Eigenvectors of Complex Matrices," was originally written in German by Rabbi Yisroel Aryeh Leib. In 1975 the Rebbe MHM gave the manuscript to Prof. Pesach (Paul) Rosenbloom to translate and prepare for publication. It was published in the Journal of Approximation Theory in 1978. (See Chapter II, section 2 for the story behind the publication.)

Rabbi Yisroel Aryeh Leib wrote the paper under the name "Mark Gurari," the name on his passport. (It was common in those days for people trying to leave Russia to use an alias.) Professor Rosenbloom told me that he assumed that it was written in the early 1930s when Rabbi Yisroel Aryeh Leib was with the Rebbe MHM in Berlin, since it was written in German. He also told me that he asked the Rebbe MHM if he should mention that the author was the brother of the Rebbe MHM. But the Rebbe MHM told him not to since the paper had to stand on its own merits

The mathematics in this paper is on a very high level and may not even be accessible to some mathematicians. Thus the paper itself is followed by a mathematical commentary on it written by Dr. Neil Miller. He wrote it in two parts-a short, general commentary and a long, detailed commentary. It was written specifically for this book and is published here for the first time.

JOURNAL OF APPROXIMATION THEORY **22**, 119–149 (1978)

The Location of Eigenvalues and Eigenvectors of Complex Matrices

M. GURARI*

Communicated by Oved Shisha

Received May 3, 1976

1. Introduction and Summary of Results

It is of considerable interest, both for theoretical purposes and for the applications, to obtain information about the location of the eigenvalues and eigenvectors of a matrix. There is an extensive literature on the location of the eigenvalues. (See, for example, the survey by Householder [11].) Somewhat less is known about the location of eigenvectors.

One important result is the Perron theorem which states that a positive matrix has a positive eigenvector belonging to a positive eigenvalue (Perron [17]; Frobenius [6, 7]). This has given rise to a considerable literature (see Brauer [4], Seneta [20]). Of course, many theorems on the effect of perturbations give information on the eigenvectors of a matrix close to a given matrix (Kato [14]).

Many results on the eigenvalues and eigenvectors of matrices can be extended to infinite-dimensional spaces. We may mention, for example, the Jentzsch theorem on integral operators with positive kernel [12], and its generalizations. (See Ostrowski [16], Krein and Rutman [15].)

In this paper we present several results on the location of the eigenvalues and eigenvectors of complex matrices, together with some extensions to infinite-dimensional sequence spaces. For example, we can obtain a result of the form (Theorem 12):

Let $C = (C_{jk})$, $0 \leqslant j, k \leqslant N$, be a matrix such that

$$C_{jk} = r_{jk} \exp(i\theta_{jk}), \qquad r_{jk} \geqslant 0, \qquad -\pi < \theta_{jk} \leqslant \pi,$$

* At the time of Mark Gurari's death on May 8, 1952, he was in the Department of Theoretical Physics at the University of Liverpool. A manuscript in German on the above subject was found among his papers. We have prepared this paper from his manuscript, extended and simplified some results, and put the material in relation to other published work. In the manuscript the author refers to a uniqueness theorem related to Theorem 12, and to analogs for integral equations. Unfortunately, these results are apparently lost. *Paul C. Rosenbloom, Department of Mathematics, Teachers College, Columbia University, New York, New York 10027.*

and set

$$a = \max_{j\geq 1, k\geq 0} (r_{jk}/r_{0k}), \qquad \theta = \max_{j,k\geq 0} | \theta_{jk} |,$$

$$b = \max_{j\geq 0} \left(\sum_{k=1}^{N} r_{jk}/r_{j0} \right).$$

If $0 < \theta \leqslant \pi/8$ and

$$a < F(b, \theta), \tag{1}$$

then C has an eigenvector z in the set

$$S(3\theta)\colon z_0 = 1, \qquad | \arg z_j | < 3\theta \qquad \text{for} \quad 0 < j \leqslant N.$$

Our proof gives $F(b, \theta) = \sin(4\theta) \cos(\theta/2)/b$. We have not tried to obtain a very sharp result, but have been concerned in getting an F which is explicit and easy to compute, and which lends itself to extension to the infinite-dimensional case.

If $a = 0$, then $C_{jk} = 0$ for $jk \neq 0$, so that we may call C a *border matrix*. Condition (1) says that C has a *dominant border*. While there is a considerable literature on matrices with a dominant *main diagonal*, little seems to have been done on matrices of the above type.

If C is a border matrix, then we may easily show that:

(a) C has the eigenvalue $\lambda = 0$ with multiplicity $N - 1$, and the corresponding $(N - 1)$-dimensional eigenspace defined by

$$z_0 = 0, \qquad \sum_{1}^{N} C_{0k} z_k = 0.$$

(b) The roots of the quadratic equation

$$\lambda^2 - C_{00}\lambda - d = 0, \qquad d = \sum_{k=1}^{N} C_{0k} C_{k0}\,,$$

are also eigenvalues, and have the eigenvectors

$$z_0 = \lambda, \qquad z_j = C_{j0} \qquad \text{for} \quad j > 0.$$

A border matrix is also a matrix of rank at most 2, but for our purposes the above representation in a particular coordinate system seems more convenient.

By applying known results in perturbation theory (see Kato [14], Rosenbloom[18]), we can also obtain sufficient conditions for the uniqueness of an eigenvalue in a specified region, and information on the location of the corresponding eigenvector. For instance, we obtain a result of the form: If

$$\left| \sum_{k=1}^{N} C_{jk} C_{k0} \right| \leqslant | C_{00} | \, | C_{j0} | \, \delta, \qquad \text{for } j \geqslant 0,$$

and

$$C_{0j} C_{j0} / C_{00}^2 \geqslant 0 \qquad \text{for all } j, \text{ and } d \neq 0,$$

and

$$\delta < B(d/C_{00}^2),$$

then C has a unique eigenvalue in the half-plane $R(\lambda/C_{00}) \geqslant \frac{1}{2}$ and a corresponding eigenvector x in the region

$$| \arg(x_j/C_{j0}) | \leqslant \alpha \qquad \text{for } j \geqslant 0,$$

where

$$\sin \alpha = B_1(d/C_{00}^2) \, \delta, \qquad 0 \leqslant \alpha < \pi/2.$$

Here B and B_1 are explicitly computable functions of d/C_{00}^2.

Of course, we can apply perturbation theory to obtain similar results for nearly positive matrices. Combining perturbation theory with the results of Ostrowski [16] (see also Birkhoff [2], Hopf [10]), we obtain results of the type:

If $C_{jk} = r_{jk} \exp(i\theta_{jk})$ and $r_{jk} > 0$, $| \theta_{jk} | \leqslant \theta < \pi/2$ for $j, k \geqslant 0$, and $0 < \gamma < \pi/2$, and

$$2 \sin(\theta/2) \leqslant B_2 \sin \gamma,$$

then C has an eigenvector in $S(\gamma)$.

Here B_2 is an easily computable function of the r_{jk}.

If $R = (r_{jk})$ is positive, and λ_R is the positive eigenvalue of R, then there is a certain constant $N_R < 1$ such that for $N_R \lambda_R < r < \lambda_R$ and θ sufficiently small, the matrix C has a unique eigenvalue in $| \lambda | \geqslant r$. There will be an eigenvector of C, belonging to this eigenvalue, in the region $S(\gamma)$. Also C has no other eigenvector in $S(\gamma)$. For N_R we can use Ostrowski's sharpening of Birkhoff's bound. Again all bounds are computable from the data R, r, and γ.

If R is nonnegative but some power R^m is positive, so that R belongs to the class of power-positive matrices studied by Brauer [3], then we can obtain similar results.

In many applications we are dealing with large matrices, or matrices depending on parameters, or families of matrices. It is then important to find comparatively simple functions of the elements, in terms of which we can obtain the desired information about the eigenvalues and eigenvectors. Thus the practical significance of our results is perhaps the identification of such computable functions of the data, and the orders of magnitude of the bounds we obtain.

We note, finally, that the approaches of Kantorovich *et al.* [13], and Krein and Rutman [15], may lend themselves to extensions to infinite-dimensional spaces.

2. The Perron Theorem

We begin by recalling the Perron theorem.[1] A matrix $C = (C_{jk})$, $1 \leqslant j, k \leqslant n$, is called *nonnegative*, $C \geqslant 0$, if $C_{jk} \geqslant 0$ for all j, k, and is called *positive*, $C > 0$, if $C_{jk} > 0$ for all j, k. We defined similarly the concepts of nonnegative and positive vectors.

Perron's Theorem [17]. *If $C > 0$, then C has a positive eigenvector belonging to a positive eigenvalue λ.*

(a) *If μ is any other eigenvalue, then $|\mu| < \lambda$.*

(b) *The eigenspace of λ is one-dimensional.*

(c) *There is no other eigenvalue which has a nonnegative eigenvector.*

We shall denote this eigenvalue by λ_C. It can also be characterized in terms of a variational problem. For any $x > 0$, let

$$\tau(x) = \min_j \left(\sum_k C_{jk} x_k / x_j \right). \tag{2}$$

Then the maximum of $\tau(x)$ for $x > 0$ and

$$\| x \| = \max_j | x_j | = 1$$

is attained, and this maximum is λ_C.

Frobenius [6, 7] extended Perron's theorem to certain classes of nonnegative matrices and characterized those nonnegative matrices which have more than one eigenvalue of maximum modulus. The extension is especially simple for the class of matrices introduced by Brauer [3]. He calls a matrix C *power-positive* if some power C^m is positive.

[1] *Editor's note*: In the original manuscript, the Perron theorem is rediscovered. The proof is similar to the one given in Bellman [1], and ascribed to unpublished work of Bohnenblust. We have presented Gurari's argument in a way which brings out some additional points of interest.

THEOREM 1. *If $C \geqslant 0$ and $C^m > 0$, then the maximum of $\tau(x)$ for $x > 0$, $\| x \| = 1$, is attained. The maximum is a positive eigenvalue λ_C, and is attained for a positive eigenvector ξ belonging to λ_C.*

Proof. We note first that $\tau(x)$ may be characterized as the maximum of the real numbers τ such that

$$Cx - \tau x \geqslant 0.$$

Since $Cx - \tau(x)x \geqslant 0$ and $C \geqslant 0$, we infer that

$$C(Cx - \tau(x)x) \geqslant 0,$$

that is,

$$C^2x - \tau(x)\, Cx \geqslant 0,$$

so that

$$\tau(Cx) \geqslant \tau(x).$$

It follows that

$$\tau(C^m x) \geqslant \tau(x).$$

For any positive matrix A, we define

$$\gamma(A) = \min_{j,k} A_{jk}. \tag{3}$$

It is then trivial that if $y \geqslant 0$ and $A > 0$, and $u = Ay$, then

$$\min u_j \geqslant \gamma(A) \| y \|.$$

We note also that for any matrix A, we have

$$\| A \| = \max_{\|x\| \leqslant 1} \| A \| = \max_j \sum | A_{jk} |.$$

Now let $y = C^m x$, $z = y / \| y \|$. Then we have

$$\min y_j \geqslant \gamma(C^m) \| x \|$$

and

$$\| y \| \leqslant \| C^m \| \| x \|,$$

so that

$$\min z_j \geqslant \gamma(C^m) / \| C^m \|$$

and

$$\tau(z) = \tau(y) \geqslant \tau(x).$$

Therefore the supremum of $\tau(x)$ on the set of $x \geqslant 0$, $\| x \| = 1$, is the same as its supremum on the subset where $\min_j x_j \geqslant \gamma(C^m)/\| C^m \|$. Since τ is continuous on this subset, it attains its maximum λ there at some vector ξ. If $\epsilon = \| C\xi - \lambda\xi \| > 0$ and $y = C^m(C\xi - \lambda\xi)$, then $\min y_j \geqslant \gamma(C^m)\epsilon > 0$.

But since $y = C(C^m\xi) - \lambda C^m\xi$, this implies that $\tau(C^m\xi) > \lambda$, which contradicts the definition of λ. Hence $\epsilon = 0$ and $C\xi = \lambda\xi$. Thus all $\xi > 0$ for which $\tau(\xi) = \lambda$ are eigenvectors of C.

The function

$$\mu(C) = \gamma(C)/\| C \|$$

seems to be a natural measure of the positivity of a matrix and arises frequently in the sequel. In the course of the argument, we proved

COROLLARY 1a. *If $C \geqslant 0$ and $C^m > 0$, and ξ is the positive unit vector which maximizes τ, then $\min_j \xi_j \geqslant \mu(C^m)$.*

For the sake of completeness we prove that properties (a)–(c) in Perron's theorem hold also for nonnegative power-positive matrices. Let C' be the transpose of C, and let η be a positive unit eigenvector belonging to λ. If μ is any eigenvalue of C other than λ and z is an eigenvector belonging to μ, let $| z |$ be the vector with the components $| z_j |$, $1 \leqslant j \leqslant n$. Let $A = C^m$. We have

$$A'\eta = \lambda^m\eta \qquad \text{and} \qquad Az = \mu^m z.$$

It follows that

$$| \mu |^m | z | \leqslant A | z |$$

and

$$A | z | - | \mu |^m | z | \neq 0$$

unless z is a scalar multiple of a nonnegative vector, and then μ must also be nonnegative. Consequently, except in this case, we have

$$\eta \cdot (A | z | - | \mu |^m | z |) > 0,$$

that is,

$$(\lambda^m - | \mu |^m)(\eta \cdot | z |) > 0,$$

so that

$$| \mu | < \lambda.$$

In the exceptional case we may assume $z \geqslant 0$, $\mu \geqslant 0$. Then we obtain

$$0 = \eta \cdot (Cz - \mu z) = (\lambda - \mu)(\eta \cdot z),$$

and therefore $\mu = \lambda$.

Finally if z is any eigenvector of C belonging to λ, let

$$u = z - \frac{(\eta \cdot z)}{\eta \cdot \xi}\,\xi,$$

so that

$$Cu = \lambda u, \qquad \eta \cdot u = 0.$$

If z is not a scalar multiple of ξ, then $u \neq 0$. If u were a scalar multiple of a nonnegative vector, then $|\eta \cdot u|$ would be positive. Hence we find that

$$A|u| - \lambda^m |u| \geqslant 0, \qquad A|u| - \lambda^m |u| \neq 0,$$

and therefore

$$\eta \cdot (A|u| - \lambda^m |u|) > 0.$$

But since $\eta \cdot A|u| = \lambda^m |u|$, we have arrived at a contradiction. Therefore $u = 0$, and z is a scalar multiple of ξ.

The same argument shows that λ can be characterized by another extremal problem.

THEOREM 2. *If $C \geqslant 0$ and $C^m > 0$ and for $x > 0$*

$$\sigma(x) = \max_j \left(\sum C_{jk} x_k / x_j\right), \tag{4}$$

then

$$\lambda = \min_{x>0} \sigma(x) = \sigma(\xi).$$

Proof. Let $\mathbf{1}$ be the vector with all components equal to 1. It is sufficient to look for the minimum of σ on the set S_1 of all $x > 0$ such that $\| x \| = 1$ and $\sigma(x) \leqslant \sigma(\mathbf{1})$. The argument of Theorem 1 shows that $\sigma(C^m x) \leqslant \sigma(Cx) \leqslant \sigma(x)$, and that if $\sigma(C^m x) \leqslant \sigma(\mathbf{1})$, $x > 0$, and $\| x \| = 1$, then

$$x_j \geqslant \gamma(C^m)/\sigma(\mathbf{1}) \qquad \text{for all } j. \tag{5}$$

Thus it suffices to look for the minimum of σ on the subset S_2 of those x in S_1 which satisfy (5). Again if the minimum is attained at $v \in S_2$ and $\sigma(v)v - Cv \neq 0$, then $\sigma(C^m v) < \sigma(v)$, which contradicts the minimum property of v.

We note

COROLLARY 2a. *If $C \geqslant 0$ and $C^m > 0$, then*

$$\lambda_C \geqslant \max_j C_{jj},$$

and

$$\min_j \sum_k C_{jk} \leqslant \lambda_C \leqslant \max_j \sum_k C_{jk} = \| C \|.$$

Proof. The first estimate follows from $\lambda_C \geqslant \tau(\delta^{(j)})$, where $\delta^{(j)}$ is the vector with components δ_{jk}. The second follows from $\tau(\mathbf{1}) \leqslant \lambda_C \leqslant \sigma(\mathbf{1})$.

If τ^* and σ^* are the functions corresponding to τ and σ for the transposed matrix C', then we have the inequalities

$$\tau(x) \leqslant \sigma^*(y) \qquad \text{and} \qquad \sigma(x) \geqslant \tau^*(y)$$

for any positive x and y.

We wish to sharpen property (c) of Perron's theorem.

THEOREM 3. *If $C \geqslant 0$ and $C^m > 0$, and*

$$Cz = \mu z, \quad z_1 = 1, \quad |\arg z_j| \leqslant \pi/2 \quad \text{for } j > 1,$$

then $\mu = \lambda$ and $z > 0$.

Proof. Let η be the vector such that

$$\eta > 0, \quad \|\eta\| = 1, \qquad \text{and} \qquad C'\eta = \lambda\eta.$$

By property (b), the conclusion follows if $\mu = \lambda$. If $\mu \neq \lambda$, then from $\eta \cdot Cz = \lambda(\eta \cdot z)$ we obtain $\eta \cdot z = 0$. But

$$R(\eta \cdot z) = \eta_1 z_1 + \sum_2^n \eta_j R(z_j) > 0,$$

which is a contradiction.

Following Ostrowski [16], if $y > 0$ and x is any real vector, we define $m(x; y)$ and $M(x; y)$ as the upper and lower bounds, respectively, of m and M such that

$$my \leqslant x \leqslant My.$$

Then we have

$$\tau(x) = m(Cx; x), \qquad \sigma(x) = M(Cx; x),$$
$$\min x_j = m(\lambda; \mathbf{1}), \qquad \max x_j = M(x; \mathbf{1}).$$

Birkhoff [2] introduced into the study of positive matrices the projective metric

$$\theta(x, y) = \log(M(x; y)/m(x; y)) \qquad (x, y > 0)$$

of Hilbert [8] (see also Busemann and Kelley [5]). If C is a positive matrix, then we have

$$\max_{x,y>0} \theta(Cx, Cy) = \log T_C = \Delta_C,$$

where

$$T_C = \max_{j,k,r,s} (C_{jk}C_{rs}/C_{rk}C_{js});$$

(Ostrowski [16, p. 87]), and

$$\max \theta(Cx, Cy)/\theta(x, y) = \frac{T_C^{1/2} - 1}{T_C^{1/2} + 1} = N_C$$

(Birkhoff [2, Lemma 1; p. 221]).

We can use these relations to sharpen the considerations of Theorem 1. If $C^m > 0$ and $x > 0$ and $T = T(C^m)$, then

$$M(C^{m+1}x - \tau(x)\, C^m x;\, C^m x) \leqslant Tm(C^{m+1}x - \tau(x)\, C^m x;\, C^m x).$$

If $M = M(C^{m+1}x - \tau(x)\, C^m x;\, C^m x)$ and η is the positive eigenvector of C' as in the proof of Theorem 3, then we have

$$C^{m+1}x - \tau(x)\, C^m x \leqslant MC^m x,$$

so that, with $\lambda = \lambda_C$,

$$\lambda^m(\lambda - \tau(x))(\eta \cdot x) \leqslant M\lambda^m(\eta \cdot x),$$

and

$$\lambda - \tau(x) \leqslant M \leqslant Tm(C^{m+1}x - \tau(x)\, C^m x;\, C^m x).$$

We infer that

$$C^{m+1}x - \tau(x)\, C^m x \geqslant T^{-1}(\lambda - \tau(x))\, C^m x,$$

which yields

$$\tau(C^m x) \leqslant \tau(x) + T^{-1}(\lambda - \tau(x)),$$

or

$$\lambda - \tau(C^m x) \leqslant (1 - T^{-1})(\lambda - \tau(x)).$$

If $v = mq + r$, $0 \leqslant r \leqslant m$, then it follows that

$$\begin{aligned} \lambda - \tau(C^v x) &\leqslant \lambda - \tau(C^{mq} x) \\ &\leqslant (1 - T^{-1})^q(\lambda - \tau(x)), \end{aligned} \tag{6}$$

so that

$$\lim_{v \to \infty} \tau(C^v x) = \lambda.$$

Similarly, we find that

$$\sigma(C^m x) - \lambda \leqslant (1 - T^{-1})(\sigma(x) - \lambda). \tag{7}$$

In the course of the proofs of Theorems 1 and 2, we proved $\tau(Cx) \geqslant \tau(x)$, $\sigma(Cx) \leqslant \sigma(x)$, which imply that

$$\begin{aligned} \theta(C^m x;\, x) &\leqslant \sum_{k=0}^{m-1} \theta(C^{k+1}x;\, C^k x) \\ &\leqslant m\theta(Cx;\, x). \end{aligned}$$

But if ξ is the positive eigenvector of C such that $\| \xi \| = 1$ and $N = N(C^m)$, then

$$\begin{aligned}\theta(x; \xi) &\leqslant \theta(x; C^m x) + \theta(C^m x, \xi)\\ &\leqslant \theta(x, C^m x) + N\theta(x, \xi)\end{aligned}$$

so that

$$\theta(x, \xi) \leqslant (1 - N)^{-1} \theta(x, C^m x).$$

If x is normalized so that $m(x; \xi)\, M(x; \xi) = 1$, that is

$$\log M(x; \xi) = \theta(x, \xi)/2 = \theta/2,$$

then

$$\exp(-\theta/2)\xi \leqslant x \leqslant \exp(\theta/2)\xi,$$

and

$$\begin{aligned}\| x - \xi \| &\leqslant (\exp(\theta/2) - 1) \| \xi \|\\ &\leqslant (\sigma(x)/\tau(x))^k - 1,\end{aligned} \tag{8}$$

where

$$k = m(1 - N)^{-1}/2.$$

Thus we can estimate the distance from x to ξ in terms of the ratio $\sigma(x)/\tau(x)$.

THEOREM 4. *If $C \geqslant 0$ and $C^m > 0$ and $x > 0$, then*

$$\lim_{v\to\infty} \sigma(C^v x) = \lim_{v\to\infty} \tau(C^v x) = \lambda_C\,,$$

and we have inequalities (6) *and* (7) *on the rates of convergence of $\sigma(C^v x)$ and $\tau(C^v x)$ to λ_C, and* (8) *on the distance from x to the positive eigenvector.*

In the following we shall continue to denote by ξ and η the positive eigenvectors of C and C', respectively. If we set

$$\xi \cdot \eta = 1,$$

then we may still replace ξ and η by $a\xi$ and $a^{-1}\eta$, respecitvely, where a is any positive number. It will be convenient to postpone further normalization of ξ and η until later.

The transformation C' may be considered as the adjoint of C, operating on the dual space with the norm

$$\| y \|_1 = \sum | y_j |.$$

For future use we give the following modification of (8):

LEMMA 1. *If $x > 0$ and $\theta(x, \xi) \leqslant \epsilon$, then*

$$\begin{aligned}\| x - (\eta \cdot x)\xi \| &\leqslant (e^\epsilon - 1) \mid \eta \cdot x \mid \| \xi \| \\ &\leqslant (e^\epsilon - 1) \| \eta \|_1 \| \xi \| \| x \|.\end{aligned}$$

Proof. Let $z = x - (\eta \cdot x)\xi$. We have

$$\begin{aligned}M(x; \xi) &= (\eta \cdot x) + M(z; \xi), \\ m(x; \xi) &= (\eta \cdot x) + m(z; \xi),\end{aligned}$$

and $m(z; \xi) \leqslant \eta \cdot z = 0 \leqslant M(z; \xi)$.

Furthermore, we have $M(x; \xi) \leqslant e^\epsilon m(x; \xi)$. It follows that

$$M(z; \xi) - m(z; \xi) \leqslant M(z; \xi) - e^\epsilon m(z, \xi) \leqslant (e^\epsilon - 1)(\eta \cdot x).$$

Since

$$\| z \| \leqslant \max(M(z; \xi), -m(z; \xi)) \| \xi \|,$$

we obtain

$$\| z \| \leqslant (e^\epsilon - 1)(\eta \cdot x) \| \xi \|.$$

The following estimates are also useful.

LEMMA 2.

$$\begin{aligned}\| \xi \| \| \eta \|_1 &\leqslant 1/\mu(C), \\ \| \xi \| \| \eta \|_1 &\leqslant \exp((1 - N_C)^{-1} \theta(C\mathbf{1}, \mathbf{1})), \\ m(\eta; \mathbf{1}) &\geqslant \mu(C) \| \eta \|_1 .\end{aligned}$$

Proof. Since

$$1 = \eta \cdot \xi \geqslant m(\xi; \mathbf{1}) \| \eta \|_1 ,$$

the first estimate follows from Corollary 1a. An alternative estimate of $\| \xi \|/m(\xi; \mathbf{1}) = \exp(\theta(\xi; \mathbf{1})$ follows from

$$\begin{aligned}\theta(\xi; \mathbf{1}) &\leqslant \theta(\xi, C\mathbf{1}) + \theta(C\mathbf{1}; \mathbf{1}) \\ &\leqslant N_C\theta(\xi, \mathbf{1}) + \theta(C\mathbf{1}, \mathbf{1}),\end{aligned}$$

and this yields the second inequality. The third follows from

$$\lambda_C\eta = C'\eta,$$

which implies

$$\begin{aligned}\gamma(C) \| \eta \|_1 &\leqslant \lambda_C m(\eta; \mathbf{1}) \\ &\leqslant \| C \| m(\eta; \mathbf{1}).\end{aligned}$$

We note that $\theta(C\mathbf{1}, \mathbf{1}) = \log(\sigma(\mathbf{1})/\tau(\mathbf{1}))$, and $\sigma(\mathbf{1})/\tau(\mathbf{1})$ is the ratio of the bounds for λ_C given in Corollary 2a.

THEOREM 5. *If $\eta \cdot z = 0$, and $C > 0$, then*

$$\| C^v z \| \leqslant (34/\mu(C))(\lambda_C N_C)^v \| z \|.$$

Remark. Ostrowski gives bounds which imply that $\| C^v z \| = O((\lambda_C N_C)^v)$, but do not specify the constant implicit in this result.

Proof. If we set $m = m(\xi; \mathbf{1})$, then from

$$-(\| z \|/m)\xi \leqslant z \leqslant (\| z \|/m)\xi$$

we obtain

$$M(z; \xi) \leqslant \| z \|/m \leqslant (\| z \|/(\mu(C) \| \xi \|)) = a,$$

and

$$-m(z; \xi) \leqslant a.$$

If $k > 1$, then we have

$$x = ka\xi + z > 0$$

and

$$M(x; \xi) \leqslant (k+1)a, \qquad m(x; \xi) \geqslant (k-1)a,$$

so that

$$\theta(x; \xi) \leqslant (k+1)/(k-1) = t.$$

Now we obtain

$$\theta(C^v x; \xi) \leqslant N_C{}^v \theta(x; \xi) \leqslant t N_C{}^v = \epsilon.$$

From

$$C^v x = ka\lambda_C{}^v \xi + C^v z$$

and

$$\eta \cdot C^v z = 0$$

we infer, by Lemma 1, that

$$\| C^v z \| \leqslant (e^\epsilon - 1)\, ka\lambda_C{}^v \| \xi \|.$$

But $\epsilon < t$, so that we have

$$e^\epsilon - 1 \leqslant e^\epsilon \epsilon \leqslant e^t \epsilon.$$

If we now choose $k = 5$, we obtain the theorem.

3. Perturbation of Simple Eigenvalues and Eigenvectors

In this section we give results which we need on the perturbation of simple eigenvalues and eigenvectors. While these results are implicit, in principle, in the available literature, it is hard to find there explicit quantitative results. Rosenbloom [18] and Kato [14] have obtained results of the kind we want by quite different methods. Since their estimates are expressed in terms of different data, it is difficult to compare them. We shall work out here various estimates using Kato's method, based on the analysis of the resolvent (see Hille and Phillips [9]). This approach has the advantage that it can easily be extended to eigenvalues of higher multiplicity.

Suppose that C is a linear transformation of a complex Banach space $\mathfrak{X}$ into itself, and let

$$R(\lambda; C) = (\lambda - C)^{-1}$$

be the resolvent of C. We say that an eigenvalue λ_0 of C is *simple* if it is also an eigenvalue of the conjugate transformation C^* on the conjugate space $\mathfrak{X}^*$, the null-spaces of $C - \lambda_0$ and $C^* - \lambda_0$ are one-dimensional, and λ_0 is an isolated point of the spectrum of C. It follows that $R(\lambda; C)$ has a pole of order 1 at λ_0 and that its residue there is a projection P_0 onto the null-space of $C - \lambda_0$, and

$$CP_0 = P_0C = \lambda_0 P_0 .$$

Let x_0 and $x_0{}^*$ be eigenvectors of C and C^*, respectively, such that $x_0{}^*(x_0) = 1$. Thus P_0 can be expressed in the form

$$P_0 = x_0 \otimes x_0{}^*,$$

that is,

$$P_0 x = x_0{}^*(x)\, x_0 \qquad \text{for all} \quad x \in \mathfrak{X}.$$

If Ω is a domain with rectifiable boundary containing no eigenvalues on its boundary, then

$$P = \frac{1}{2\pi i}\int_{\partial\Omega} R(\lambda; C)\, d\lambda$$

is a projection onto the union of the eigenspaces corresponding to the portion of the spectrum of C contained in Ω. In particular, if Ω contains λ_0 and no other point of the spectrum of C, then $P = P_0$.

Suppose Ω is such a domain, and let

$$M = \max_{\lambda \in \partial\Omega} \| R(\lambda; C)\|.$$

If U is a bounded linear transformation of $\mathfrak{X}$ into itself, then we have

$$R(\lambda; C + U) = R(\lambda; C)(1 - UR(\lambda; C))^{-1}$$

and

$$R(\lambda; C + U) - R(\lambda; C) = R(\lambda; C + U)\, UR(\lambda; C),$$

by Hille and Phillips [9, p. 196–197].

Hence if

$$\| U \| \leqslant \delta < 1/M,$$

then we obtain

$$\| R(\lambda; C + U)\| \leqslant M/(1 - \delta M)$$

and

$$\| R(\lambda; C + U) - R(\lambda; C)\| \leqslant M^2\delta/(1 - \delta M).$$

If

$$P_U = \frac{1}{2\pi i}\int_{\partial\Omega} R(\lambda; C + U)\, d\lambda,$$

then

$$\| P_U - P_0 \| \leqslant BM^2\delta/(1 - \delta M),$$

where $B =$ (length of $\partial\Omega)/2\pi$. Consequently, if

$$\delta < 1/(M + BM^2),$$

then we obtain

$$\| P_U - P_0 \| < 1.$$

By Kato [14, p. 33], this implies that the rank of P_U is one-dimensional, so that $C + U$ has a unique eigenvalue $\lambda(U)$ in Ω, and

$$P_U x = x_U{}^*(x)\, x_U\,, \qquad x_U{}^*(x_U) = 1,$$

for all $x \in \mathfrak{X}$, where x_U and $x_U{}^*$ are the eigenvectors of $C + U$ and $(C + U)^*$, respectively. Furthermore, we have

$$(C + U)\, P_U = P_U(C + U) = \lambda(U)\, P_U\,.$$

But the formula

$$(\lambda(U) - \lambda_0)\, P_U = \frac{1}{2\pi i}\int_{\partial\Omega} (\lambda - \lambda_0)\, R(\lambda; C + U)\, d\lambda$$

$$= \frac{1}{2\pi i}\int_{\partial\Omega} (\lambda - \lambda_0)(R(\lambda; C + U) - R(\lambda; C))\, d\lambda$$

implies that

$$| \lambda(U) - \lambda_0 |\, \| P(U)\| \leqslant dBM^2\delta/(1 - \delta M)$$
$$\leqslant dM(1 + BM)\delta,$$

where

$$d = \max_{\lambda \in \partial\Omega} | \lambda - \lambda_0 |$$

and

$$| \lambda(U) - \lambda_0 | \leqslant dM(1 + BM)\delta$$

since

$$\| P(U) \| \geqslant 1.$$

Furthermore, we have

$$\| P_U(x_0) - P_0(x_0) \| = \| x_U^*(x_0)\, x_U - x_0 \| \leqslant M(1 + BM)\delta \, \| x_0 \|.$$

Similarly, we can show that

$$\| x_0^*(x_U)\, x_U^* - x_0^* \| \leqslant M(1 + BM)\delta \, \| x_0^* \|.$$

We summarize these results in

THEOREM 6. *Suppose that C is a linear transformation of $\mathfrak{X}$ into itself, that λ_0 is a simple eigenvalue of C, that x_0 and x_0^* are eigenvectors of C and C^*, respectively, belonging to λ_0 such that $x_0^*(x_0) = 1$, that Ω is a domain with boundary $\partial\Omega$ of length $2\pi B$ containing λ_0 and no other points of the spectrum of C, and that*

$$M = \max_{\lambda \in \partial\Omega} \| R(\lambda; C) \|.$$

Then for

$$\| U \| \leqslant \delta < 1/(M + BM^2) = 1/K,$$

the transformation $C + U$ has a unique eigenvalue $\lambda(U)$ in Ω, which is simple. This eigenvalue satisfies

$$| \lambda(U) - \lambda_0 | \leqslant Kd\delta,$$

where

$$d = \max_{\lambda \in \partial\Omega} | \lambda - \lambda_0 |.$$

There are eigenvectors of $C + U$ and $(C + U)^*$, respectively, belonging to $\lambda(U)$ in the spheres

$$\| x - x_0 \| \leqslant K\delta \, \| x_0 \| \quad \textit{and} \quad \| x^* - x_0^* \| \leqslant K\delta \, \| x_0^* \|. \tag{9}$$

We remark that we can always normalize x_0 and x_0^* so that $\| x_0^* \| = \| x_0 \| = \| P_0 \|^{1/2}$.

By minor modifications of the above argument, we can obtain similar results for unbounded regions Ω. For example, we have

COROLLARY 6a. *Suppose that C is a bounded linear transformation on $\mathfrak{X}$ to itself, and λ_0, x_0, and x_0^* are as above. Suppose also that there are no other points of the spectrum of C in $|\lambda| \geqslant r$, where $r < |\lambda_0|$, and that*

$$M = \max_{|\lambda|=r} \| R(\lambda; C)\|.$$

Then for

$$\| U\| \leqslant \delta < 1/(M + rM^2) = 1/K$$

the transformation $C + U$ has a unique eigenvalue $\lambda(U)$ in $|\lambda| \geqslant r$, which is simple. It satisfies

$$|\lambda(U) - \lambda_0| \leqslant (1 + 2|\lambda_0|K)\delta.$$

There are eigenvectors of $C + U$ and $(C + U)^*$, respectively, belonging to $\lambda(U)$ and satisfying (9).

For the proof, we take Ω to be the annulus $r < |z| < R$, and let $R \to \infty$. We note that

$$R(\lambda; C) = \sum_0^\infty C^k/\lambda^{k+1}$$
$$= \lambda^{-1} + C\lambda^{-2} + O(\lambda^{-3})$$

for $|\lambda| > |\lambda_0|$. This implies that

$$P_U - P_0 = -\frac{1}{2\pi i}\int_\Gamma (R(\lambda; C + U) - R(\lambda, U))\, d\lambda$$

and

$$(\lambda(U) - \lambda_0)\, P_U = U - \frac{1}{2\pi i}\int_\Gamma (\lambda - \lambda_0)(R(\lambda; C + U) - R(\lambda, U))\, d\lambda,$$

where Γ is the circle $|\lambda| = r$. The rest of the reasoning is as before.
Another variant of the argument yields

COROLLARY 6b. *Let $a > 0$ and suppose that*

$$M_j = \sup_{R\lambda=a} |\lambda|^j \| R(\lambda; C)\|, \qquad j = 0, 1.$$

Suppose also that the bounded transformation C has the simple eigenvalue λ_0, $R\lambda_0 > a$, and that the half-plane $R\lambda \geqslant a$ contains no other points of the spectrum of C. Then for

$$\| U\| \leqslant \delta < \frac{2a}{2aM_0 + M_1^2} = \frac{1}{K}$$

the transformation $C + U$ *has a unique eigenvalue* $\lambda(U)$ *in the half-plane* $R\lambda \geqslant a$, *and it is simple. It satisfies*

$$|\lambda(U) - \lambda_0| \leqslant (2M_1^{-1} + 1) K \| C \| \delta \tag{10}$$

if $\delta \leqslant \| C \|$. *There are eigenvectors of* $C + U$ *and* $(C + U)^*$, *respectively, belonging to* $\lambda(U)$ *and satisfying* (9).

This time we take Ω to be the portion of the circle $|\lambda| < r$ in the half-plane $R\lambda > a$. Again the contribution of the circular part $|\lambda| = r$ to the integral

$$P_U - P_0 = \frac{1}{2\pi i} \int_{\partial\Omega} (R(\lambda; C + U) - R(\lambda; C))\, d\lambda$$

approaches zero, so that we obtain

$$P_U - P_0 = \frac{1}{2\pi i} \int_{a-i\infty}^{a+i\infty} (R(\lambda; C + U) - R(\lambda; C))\, d\lambda.$$

From the identity

$$R(\lambda; C + U) - R(\lambda; C) = R(\lambda; C)(1 - UR(\lambda; C))^{-1}\, UR(\lambda; C)$$

it follows that

$$\| R(\lambda; C + U) - R(\lambda; C)\| \leqslant \frac{M_1{}^2\delta}{|\lambda|^2 (1 - M_0\delta)},$$

which yields

$$\| P_U - P_0 \| \leqslant \frac{M_1{}^2\delta}{2a(1 - M_0\delta)} \leqslant K\delta.$$

We set

$$D(\lambda) = R(\lambda; C + U) - R(\lambda; C).$$

From the identities

$$\lambda R(\lambda; C) = 1 + R(\lambda; C)C$$

and

$$\lambda R(\lambda; C + U) = 1 + R(\lambda; C + U)(C + U),$$

we derive

$$\begin{aligned} D(\lambda) &= \lambda^{-1}(D(\lambda)C + R(\lambda; C + U)U) \\ &= \lambda^{-1}D(\lambda)C + \lambda^{-2}(U + R(\lambda; C + U)(C + U)U). \end{aligned}$$

Consequently, we infer

$$\begin{aligned} (\lambda(U) - \lambda_0)\, P_U &= \frac{1}{2\pi i} \int_{\partial\Omega} (\lambda - \lambda_0)\, D(\lambda)\, d\lambda \\ &= \frac{1}{2\pi i} \int_{\partial\Omega} \lambda D(\lambda)\, d\lambda - \lambda_0(P_U - P_0) \\ &= J(C + U)\, U + (P_U - P_0)(C - \lambda_0), \end{aligned}$$

where

$$J = \frac{1}{2\pi i}\int_{\partial\Omega} \frac{R(\lambda;\, C+U)}{\lambda}\, d\lambda.$$

Again the integrand is $O(\lambda^{-2})$, is $|\lambda| \to \infty$, and J is independent of r for r sufficiently large. Hence we find that

$$J = \frac{1}{2\pi i}\int_{a-i\infty}^{a+i\infty} \frac{R(\lambda;\, C+U)}{\lambda}\, d\lambda.$$

Now the estimate

$$\| R(\lambda;\, C+U)\| \leqslant \| R(\lambda;\, C)\|/(1 - M_0\delta)$$

yields

$$| J | \leqslant \frac{M_1}{2a(1 - M_0\delta)},$$

from which we conclude the inequality.

Remark. We always have the relation

$$M_1 \leqslant 1 + \| C \|\, M_0 ,$$

but sometimes we can obtain a much sharper bound for M_1.

4. Perturbation of Power-Positive Matrices

We wish now to apply the results of the previous section to power-positive matrices. For this purpose we need to estimate the resolvent of such a matrix. If C is power-positive, we shall denote by λ_C its largest eigenvalue and by ξ and η the positive eigenvectors of C and C^*, respectively, normalized by the conditions

$$\eta \cdot \xi = 1, \qquad \| \eta \| = \| \xi \| = \| P_0 \|^{1/2},$$

where $P_0 = \xi \otimes \eta$ is the projection defined by

$$P_0 x = (\eta \cdot x)\xi \qquad \text{for all } x.$$

THEOREM 7. *If $C > 0$ and*

$$M(r) = \max_{|\lambda|=r} \| R(\lambda;\, C)\|,$$

then for $\lambda_C N_C < r < \lambda_C$, we have

$$M(r) \leqslant \mu(C)^{-1}((\lambda_C - r)^{-1} + B(r - \lambda_C N_C)^{-1}),$$

where

$$B = B(C) = 34(1 + \mu(C)^{-1}).$$

Proof. Let y be a given vector,

$$b = \eta \cdot y, \qquad z = (1 - P_0)y = y - b\xi,$$

and

$$(\lambda - C)x = y.$$

Then

$$x = (\lambda - \lambda_C)^{-1}\, b\xi + (\lambda - C)^{-1}\, z.$$

The estimate in Theorem 5 yields

$$\| R(\lambda; C)z \| \leqslant 34\mu(C)^{-1}(|\lambda| - \lambda_C N_C)^{-1} \| z \|.$$

Since Lemma 2 implies that

$$\| z \| \leqslant \| y \| + \| \xi \| \| \eta \| \| y \| \leqslant (1 + \mu(C)^{-1}) \| y \|,$$

we obtain the estimate for $M(r)$ stated above.

The minimum of $(\lambda_C - r)^{-1} + B(r - \lambda_C N_C)^{-1}$ is attained for $r = \alpha\lambda_C$, where $\alpha = (N_C + B^{1/2})/(1 + B^{1/2})$, from which we obtain

COROLLARY 7a. *For $\alpha = (N_C + B^{1/2})/(1 + B^{1/2})$, we have*

$$M(\alpha\lambda_C) \leqslant (1 + B^{1/2})^2/((1 - N_C)\, \mu(C)\, \lambda_C).$$

To deal with power-positive matrices, we use the identity

$$R(\lambda, C) = \left(\sum_{k=0}^{m-1} \lambda^k C^{m-1-k} \right) R(\lambda^m; C^m).$$

COROLLARY 7b. *If $C \geqslant 0$ and $C^m > 0$, and $N = N(C^m)$, $B' = B(C^m)$, and $\lambda_C N^{1/m} < r < \lambda_C$, then*

$$M(r) \leqslant \frac{\| C \|^m - r^m}{\| C \| - r} \frac{1}{\mu(C^m)} \frac{1}{\lambda_C{}^m - r^m} + \frac{B'}{r^m - \lambda_C{}^m N}.$$

For example, if $m = 2$, and $\alpha = (N + (B')^{1/2})/(1 + (B')^{1/2})$, then an easy computation yields

$$M(\alpha^{1/2}\lambda_C) \leqslant \frac{200 \| C \|}{(1 - N)(\mu(C^2)\, \lambda_C)^2}.$$

Application of Corollary 6a and Lemma 2 leads to

COROLLARY 7c. *If $C > 0$ and α is as in Corollary 7a, and*

$$\| U \| \leqslant \delta < 1/K,$$

where

$$K = M(1 + \lambda_C M),$$

$$M = \frac{100}{(1 - N_C)\,\mu(C)^2\,\lambda_C},$$

then $C + U$ *has a unique eigenvalue* $\lambda(U)$ *in the set* $|\lambda| \geqslant \alpha\lambda_C$, *and* $\lambda(U)$ *satisfies*

$$|\lambda(U) - \lambda_C| \leqslant (1 + 2\lambda_C K)\delta,$$

and this eigenvalue is simple. There are eigenvectors x *and* y *of* $C + U$ *and* $(C + U)^*$, *respectively, in the spheres*

$$\| x - \xi \| \leqslant K\delta \| \xi \|, \qquad \| y - \eta \|_1 \leqslant K\delta \| \eta \|_1 .$$

If $\delta < \mu(C)/K$, *then* $|\arg x_j| \leqslant \theta$, *where*

$$\sin\theta = K\delta/\mu(C), \qquad 0 \leqslant \theta < \pi/2.$$

We can apply Corollary 7b in a similar way to obtain a corresponding result for power-positive matrices.

We now wish to prove a generalization of Theorem 3. For this purpose we first derive a lemma.

LEMMA 3. *If* $C > 0$, $z \neq 0$, $|\arg z_j| \leqslant \gamma < \pi/2$ *for all* j, *and*

$$\| Cz - \mu z \| \leqslant \epsilon,$$

then

$$|\lambda_C - \mu| \leqslant \epsilon/(\| z \|\,\mu(C)\cos\gamma) = K\epsilon/\| z \|, \tag{11}$$

and

$$\| z - (\eta \cdot z)\,\xi \| \leqslant \frac{34(1 + K)\,\epsilon}{\mu(C)\,\lambda_C(1 - N_C)} = K_1\epsilon. \tag{12}$$

Proof. Let $z_j = r_j \exp(i\theta_j)$, $|\theta_j| \leqslant \gamma < \pi/2$ for all j. Since

$$\eta \cdot (Cz - \mu z) = (\lambda_C - \mu)(\eta \cdot z)$$

and

$$R(\eta \cdot z) = \sum \eta_j r_j \cos\theta_j$$
$$\geqslant m(\eta; 1)\,\| z \| \cos\gamma,$$

we obtain

$$|\lambda_C - \mu|\, m(\eta; 1)\,\| z \| \cos\gamma \leqslant \| \eta \|_1\, \epsilon,$$

and now Lemma 2 implies (11).

Now let $u = P_0 z = z - (\eta \cdot z)\xi$, so that

$$(C - \lambda_C)u = (C - \lambda_C)z = v.$$

Hence we have

$$u = -\sum_{k=0}^{\infty} \lambda^{-k-1} C^k v$$

and, by Theorem 5, we find that

$$\| u \| \leqslant 34 \| v \| / (\mu(C) \lambda_C (1 - N_C)).$$

Since

$$v = (C - \mu)z + (\mu - \lambda_C)z,$$

so that

$$\| v \| \leqslant \epsilon + K\epsilon,$$

we obtain the estimate (12).

THEOREM 8. *If $C > 0$, $\| U \| \leqslant \epsilon$, and if z is an eigenvector of $C + U$ such that*

$$\| z \| = 1 \qquad \text{and} \qquad | \arg z_0 | \leqslant \gamma < \pi/2 \quad \text{for all } j, \tag{13}$$

belonging to the eigenvalue μ, then

$$| \lambda_C - \mu | \leqslant K\epsilon$$

and

$$\| z - (\eta \cdot z)\xi \| \leqslant K_1 \epsilon,$$

where K and K_1 are as in Lemma 3.

In Corollary 7a we obtain conditions on U that $C + U$ have a *unique* eigenvalue in $| \lambda | \geqslant \alpha\lambda_C$, and then find that its eigenvector is close to ξ. In Theorem 8 we find that if $C + U$ has an eigenvector satisfying (13) and U is small, then the corresponding eigenvalue is close to λ_C and the eigenvector is close to a scalar multiple of ξ. Here U is not necessarily so small that Corollary 7a applies, and so there may very well be other eigenvalues in $| \lambda | \geqslant \alpha\lambda_C$.

In the next theorem we give a sufficient condition that $C + U$ have at least one eigenvector satisfying (13) and with a positive component. Again the condition may not be strong enough to ensure uniqueness of the corresponding eigenvalue.

THEOREM 9. *If $C > 0$ and $0 < \gamma < \pi/2$ and*

$$\epsilon < \mu(C)^{1/2} \sin \gamma / (2K_1),$$

where K_1 is as in Lemma 3, then for $\| U \| \leqslant \epsilon$, the matrix $C + U$ has an eigenvector z in the set

$$S(\gamma)\colon z_1 = 1, \qquad | \arg z_j | \leqslant \gamma \quad (\text{all } j).$$

Proof. For $0 \leqslant t \leqslant 1$, let $C(t) = C + tU$, and let t_0 be the least upper bound of the t in [0, 1] such that $C(\tau)$ has an eigenvector in $S(\gamma)$. Then t_0 is positive by Corollary 7c. If $t_0 < 1$, then $C(t_0)$ has an eigenvector z on the boundary of $S(\gamma)$. We may assume that $| \arg z_j | = \gamma$. Let $a = \eta \cdot z$. Then by Lemma 3, we have

$$| 1 - a\xi_1 | \leqslant K_1 \epsilon \| \xi \|$$

and

$$| z_j - a\xi_j | \leqslant K_1 \epsilon \| \xi \|.$$

Consequently, we infer that

$$\begin{aligned} | \xi_1 z_j - \xi_j | &= | \xi_1(z_j - a\xi_j) + (a\xi_1 - 1)\, \xi_j | \\ &\leqslant 2K_1 \epsilon \| \xi \| \\ &\leqslant 2K_1 \epsilon\, \mu(C)^{-1/2}\, \xi_j , \end{aligned}$$

by Lemma 2. This is impossible if ϵ satisfies the above inequality.

COROLLARY 9a. *If $A = (C_{jk} \exp(i\theta_{jk}))$, where $C > 0$, and $| \theta_{jk} | \leqslant \theta < \pi/2$ for all j, k and if*

$$2 \sin(\theta/2) < \mu(C)^{1/2} \sin \gamma/(2K_1 \| C \|),$$

then A has an eigenvector in $S(\gamma)$.

Proof. We set $U = A - C$ in Theorem 9.

By means of these methods, we can obtain similar results for power-positive matrices.

LEMMA 4. *If $C \geqslant 0$, $C^m > 0$, $\| z \| = 1$, $| \arg z_j | \leqslant \gamma < \pi/2$, and $N = N(C^m)$, and*

$$\| Cz - \mu z \| \leqslant \epsilon,$$

then

$$| \lambda_C - \mu | \leqslant \epsilon/(\mu(C^m) \cos \gamma) = K(C^m)\epsilon,$$

and

$$\begin{aligned} \| z - (\eta \cdot z)\, \xi \| &\leqslant \left(\sum_0^{m-1} \frac{\| C \|^k}{\lambda_C{}^k} \right) \frac{34(1 + K(C^m))\, \epsilon}{\lambda_C \mu(C^m)(1 - N)} \\ &= K_1(C^m)\, \epsilon. \end{aligned}$$

THEOREM 10. *If* $C \geqslant 0$, $C^m > 0$, *and* $0 < \gamma < \pi/2$, *and*

$$\epsilon < \sin \gamma / (2\mu(C^m)^{1/2} K_1(C^m)),$$

then for $\| U \| \leqslant \epsilon$, *the matrix* $C + U$ *has an eigenvector in* $S(\gamma)$.

5. MATRICES WITH DOMINANT BORDER

In this section it will be convenient to have the indices in our vectors and matrices run from 0 to N. We shall begin by determining the eigenvalues and eigenvectors of a *border matrix* C, i.e., a matrix such that $C_{jk} = 0$ for $jk \neq 0$

If $Cz = \lambda z$, $z \neq 0$, and $\lambda \neq 0$, then for $j > 0$ we have

$$z_j = C_{j0} z_0 / \lambda$$

so that

$$\lambda^2 z_0 = \lambda \sum C_{0k} z_k = (C_{00}\lambda + d)\, z_0\,,$$

where

$$d = \sum_1^N C_{0j} C_{j0}\,.$$

Since $z \neq 0$, we must have $z_0 \neq 0$, and therefore λ is a root of the quadratic polynomial

$$Q(\lambda) = \lambda^2 - C_{00}\lambda - d,$$

and

$$\lambda = (C_{00} \pm (C_{00}^2 + 4d)^{1/2})/2.$$

Incidentally, it is easy to prove that

$$\det(\lambda - C) = \lambda^{N-1} Q(\lambda).$$

If $\lambda = 0$, then z is in the $(N - 1)$-dimensional subspace

$$z_0 = 0, \qquad \sum_1^N C_{0k} z_k = 0.$$

Similarly, we easily compute

$$R(\lambda; C)\, y = (\lambda Q(\lambda))^{-1}\, x,$$

where

$$x_0 = \lambda^2 y_0 + \lambda \sum_1^N C_{0k} y_k ,$$

$$x_j = \lambda C_{j0} y_0 + C_{j0} \sum_1^N C_{0k} y_k + y_j Q(\lambda)$$

$$= \lambda C_{j0} y_0 + C_{j0} \sum{}' C_{0k} y_k + (Q(\lambda) + C_{j0} C_{0j})\, y_j ,$$

for $j > 0$. Here $\sum'$ denotes the summation over all indices $k \neq 0, j$.

If $C_{j0} \neq 0$ for all j, then it is often convenient to normalize the matrix C. We transform by the diagonal matrix A defined by

$$A_{jj} = C_{j0} , \qquad A_{jk} = 0 \qquad \text{for} \quad j \neq k,$$

and set

$$A^{-1} C A = C_{00} \tilde{C}. \tag{14}$$

If x is an eigenvector of $\tilde{C}$ belonging to the eigenvalue μ, then Ax is an eigenvector of C belonging to the eigenvalue $C_{00}\mu$. The matrix $\tilde{C}$ is a border matrix with

$$\tilde{C}_{j0} = 1 \qquad \text{and} \qquad \tilde{C}_{0j} = C_{0j} C_{j0} / C_{00}^2 \qquad \text{for all } j.$$

Hence we first focus our attention on border matrices with $C_{j0} = 1$ and $C_{0j} \geqslant 0$ for all j. There is a unique positive eigenvalue λ_1 , a unique negative eigenvalue λ_2 , and an eigenvalue of multiplicity $N - 1$ at 0. Let us compute the other data needed in order to apply the results of Section 3.

The positive eigenvector ξ belonging to λ_1 has the components

$$\xi_j = \xi_0 / \lambda_1 \qquad \text{for} \quad j > 0,$$

and since $\lambda_1 > 1$, we have $\| \xi \| = \xi_0$. The positive eigenvector η of C^* has the coordinates

$$\eta_j = C_{0j} \eta_0 / \lambda_1 \qquad \text{for} \quad j > 0,$$

and

$$\| \eta \|_1 = \eta_0 (1 + d/\lambda_1) = \eta_0 \lambda_1 .$$

The normalization

$$\eta \cdot \xi = 1, \qquad \| \eta \|_1 = \| \xi \|,$$

leads to

$$\eta_0 = (1 + 4d)^{-1/4}, \qquad \xi_0 = (\eta_0 + \eta_0^{-1})/2. \tag{15}$$

Finally we have

$$| \lambda Q(\lambda) | \, \| R(\lambda; C) \| = \max(| \lambda |^2 + | \lambda | d, | \lambda | + (d - C_{0j}) + | Q(\lambda) + C_{0j} |).$$

On the line $R\lambda = \frac{1}{2}$, the perpendicular bisector of the segment $[\lambda_2, \lambda_1]$, we have $\lambda - \lambda^2 = |\lambda|^2 \geqslant |\lambda|/2$, so that

$$|Q(\lambda)| = |\lambda|^2 + d.$$

Therefore we have

$$\frac{|\lambda| + d}{|Q(\lambda)|} \leqslant \frac{2|\lambda|^2 + d}{|\lambda|^2 + d} \leqslant 2.$$

Furthermore, we see that

$$|Q(\lambda) + C_{0j}| = |\lambda|^2 + (d - C_{0j}) \leqslant |\lambda|^2 + d.$$

It follows that

$$\begin{aligned}|\lambda| + (d - C_{0j}) + |Q(\lambda) + C_{0j}| &\leqslant |\lambda| + d + |Q(\lambda)| \\ &\leqslant 3|Q(\lambda)| \leqslant 6|\lambda Q(\lambda)|,\end{aligned}$$

and conclude that

$$\| R(\lambda; C)\| \leqslant 6 \qquad \text{for} \quad R\lambda = \tfrac{1}{2}.$$

We note that for $t > 0$, $s - 0$,

$$\begin{aligned}s^2 + sd &\leqslant s^2 + (tds^2 + t^{-1}d)/2 \\ &= ((2 + td)\, s^2 + t^{-1}d)/2.\end{aligned}$$

If we choose t as the positive solution of

$$(2 + td)t = 1, \qquad \text{i.e.,} \quad t = (-1 + (1 + d)^{1/2})/d,$$

then we obtain

$$s^2 + sd \leqslant (s^2 + d)(2 + td)/2.$$

Hence for $R\lambda = \frac{1}{2}$, we have

$$|\lambda|\, \| R(\lambda; C) \leqslant \max((1 + (1 + d)^{1/2})/2, 3) = M_1, \tag{16}$$

which is sharper than the bound $1 + 6\|C\|$ obtained from the identity

$$\lambda R(\lambda; C) = 1 + R(\lambda; C)C.$$

We can now apply Corollary 6b. A little computation yields

THEOREM 11. *If C is a nonnegative border matrix with $C_{j0} = 1$ for all j, and if*

$$\| U\| \leqslant \delta < 1/(6 + M_1^2) = 1/K,$$

then $C + U$ *has a unique eigenvalue* $\lambda(U)$ *in the half plane* $R\lambda \geqslant \frac{1}{2}$. *This eigenvalue is simple, and satisfies*

$$| \lambda(U) - \lambda_1 | \leqslant \varphi(\| C \|)\delta,$$

where

$$\begin{aligned} \varphi(k) &= 1 + 35k & k \leqslant 25, \\ &= (k^2/2)(1 + 12k^{-1/2}) & k > 25. \end{aligned}$$

There is an eigenvector x *of* $C + U$, *belonging to* $\lambda(U)$, *such that*

$$\| x - \xi \| \leqslant K\delta \| \xi \|.$$

If $\lambda_1 K\delta < 1$, *then for all* j *we have*

$$| x_j | \geqslant \frac{(1 - \lambda_1 K\delta)}{\lambda_1(1 - K\delta)} \| x \|$$

and

$$| \arg x_j | \leqslant \alpha,$$

where

$$\sin \alpha = \lambda_1 K\delta, \qquad 0 \leqslant \alpha < \pi/2.$$

There is an eigenvector y *of* $(C + U)^*$, *belonging to* $\lambda(U)$, *such that*

$$\| y - \eta \|_1 \leqslant K\delta \| \eta \|_1 .$$

If we apply Theorem 11 to the matrix $\tilde{C}$ defined by (14), then we obtain

COROLLARY 11a. *Suppose that* C *is a border matrix and that* $C_{j0} \neq 0$ *and* $C_{j0}C_{0j}/C_{00}^2 \geqslant 0$ *for all* j. *If* V *is a matrix such that*

$$\sum_0^N | V_{jk}C_{k0} | \leqslant \delta | C_{00}C_{j0} | \qquad \text{for all } j,$$

where

$$\delta < 1/(6 + M_1^2) = 1/K,$$

and M_1 *is defined by* (16) *with*

$$d = \sum_1^N C_{0j}C_{j0}/C_{00}^2 ,$$

then $C + V$ *has a unique eigenvalue* $\lambda(V)$ *in the half-plane* $R(\lambda/C_{00}) \geqslant \frac{1}{2}$. *This eigenvalue is simple. If we set*

$$\lambda_1 = (1 + (1 + 4d)^{1/2})/2 \tag{17}$$

then there is an eigenvector x of $C + V$, *belonging to* $\lambda(V)$, *such that*

$$\left| \frac{x_0}{C_{00}} - 1 \right| \leqslant K\delta,$$

and

$$\left| \frac{x_j}{C_{j0}} - \frac{1}{\lambda_1} \right| \leqslant K\delta \qquad \text{for} \quad j > 0.$$

If $\lambda_1 K\delta < 1$, *and*

$$\sin \alpha = \lambda_1 K\delta, \qquad 0 \leqslant \alpha < \pi/2,$$

then

$$| \arg(x_j/C_{j0})| \leqslant \alpha \qquad \text{for all } j.$$

There is an eigenvector y of $(C + V)^*$, *belonging to* $\lambda(V)$, *such that*

$$| C_{00} y_0 - 1 | + \sum_1^N \left| C_{k0} y_k - \frac{C_{0k} C_{k0}}{\lambda_1 C_{00}^2} \right| \leqslant \lambda_1 K\delta.$$

Note that x and y satisfy

$$\max_j \left| \frac{x_j}{C_{j0}} \right| \leqslant \frac{(1 + K\delta)\, \lambda_1}{(1 - \lambda_1 K\delta)} \min_j \left| \frac{x_j}{C_{j0}} \right|,$$

and

$$\lambda_1(1 - K\delta) \leqslant \sum_0^N | C_{0k} y_k | \leqslant \lambda_1(1 + K\delta).$$

By an easy limiting process, or by imitating the above argument, we can obtain extensions of this result to certain infinite-dimensional spaces.

Let l_1 be the Banach space of absolutely convergent series y with the norm

$$\| y \|_1 = \sum_0^\infty | y_j |,$$

and let $l_\infty = l_1^*$ be its conjugate space, the set of all bounded sequences x with the norm

$$\| x \|_\infty = \sup | x_j |.$$

If $\{a_n\}$, $n \geqslant 0$, is a sequence of nonzero complex numbers, and A is the diagonal transformation defined by

$$(Ax)_j = a_j x_j \qquad \text{for all } j,$$

then we may denote by Al_1 and Al_∞, respectively, the transforms of l_1 and l_∞ by A.

COROLLARY 11b. *Suppose that C is a border matrix and that $C_{j0} \neq 0$ and $C_{j0}C_{0j}/C_{00}^2 \geqslant 0$ for all $j \geqslant 0$, and that*

$$0 < d = \sum_1^\infty C_{0j}C_{j0}/C_{00}^2 < +\infty.$$

Let A be defined by $(Ax)_j = C_{j0}x_j$ for $j \geqslant 0$.

Then if V is a matrix such that

$$\sum_0^\infty | V_{jk}C_{k0} | \leqslant \delta \, | C_{00}C_{j0} | \qquad (all\ j \leqslant 0),$$

where

$$\delta < 1/K, \qquad K = 6 + M_1^2,$$

then $C + V$, as a linear transformation on Al_∞, has a unique eigenvalue $\lambda(V)$ in the half plane $R(\lambda/C_{00}) \geqslant \frac{1}{2}$. This eigenvalue is simple, and is also a simple eigenvalue of $(C + V)^*$ on $A^{-1}l_1$. If $\lambda_1 K\delta < 1$, then there is an eigenvector x of $C + V$, belonging to $\lambda(V)$ such that

$$| \arg(x_j/C_{j0})| \leqslant \alpha \qquad for\ all\ j,$$

where

$$\sin \alpha = \lambda_1 K\delta, \qquad 0 \leqslant \alpha < \pi/2.$$

We can also obtain results like Theorem 9, which may not be strong enough to imply uniqueness. Suppose that

$$C_{jk} = r_{jk} \exp(i\theta_{jk}), \qquad r_{jk} \geqslant 0,$$

and

$$| \theta_{jk} | \leqslant \theta < \pi/2 \qquad \text{for all } j, k, \tag{18}$$

and that

$$\max_{j \geqslant 1, k \geqslant 0} (r_{jk}/r_{0k}) = a, \tag{19}$$

$$\max_{j \geqslant 0} \sum_1^N r_{jk}/r_{j0} = b. \tag{20}$$

Then a and b are measures of the dominance of the border of the matrix C.

THEOREM 12. *If C satisfies Eqs.* (18)–(20), *and*

$$\begin{aligned} \theta + \gamma &< \pi/2, \qquad 0 < \gamma, \\ \sin((\gamma/2) + 2\theta) &< \sin(3\gamma/2) - 2ab, \end{aligned} \tag{21}$$

then C has an eigenvector z in the convex set

$$S(\gamma)\colon z_0 = 1, \qquad |\arg z_j| < \gamma \quad (\textit{all } j).$$

Remarks. If $a = \theta = 0$, then C is a power-positive border matrix ($C^2 > 0$). If $s = \sin(\gamma/2)$, then

$$\begin{aligned}\varphi(\gamma) &= \sin(3\gamma/2) - \sin(\gamma/2)\\ &= 2s\cos\gamma = 2(s - 2s^3) > 0\end{aligned}$$

for $0 < \gamma < \pi/2$. Hence (16) is satisfied for sufficiently small θ and a. Thus (21) defines a class of matrices close to power-positive border matrices, which are sure to have an eigenvector in $S(\gamma)$. For γ close to 0, our result may be considered a perturbation of the Perron–Frobenius result, while if γ is close to $\pi/2$, the result is related to Theorem 3. Note also the maximum of φ is attained for

$$\sin(\gamma/2) = 1/6^{1/2}$$

and is $(8/27)^{1/2}$. Thus if $2ab < (8/27)^{1/2}$, then (21) is satisfied for some γ and for all sufficiently small θ.

Proof. For given γ, θ, and r_{jk} $(j, k \geqslant 0)$ satisfying (21), let θ_0 be the least upper bound of the numbers such that

$$0 \leqslant \theta_0 \leqslant \theta$$

and such that every matrix C with $|\theta_{jk}| \leqslant \theta_0$ (all j, k) has an eigenvector in $S(\gamma)$. Then $\theta_0 > 0$, and if $\theta_0 < \theta$, then there is a matrix C with $|\theta_{jk}| \leqslant \theta_0$ (all j, k) having an eigenvector z on the boundary of $S(\gamma)$.

Let

$$z_k = \rho_k \exp(i\alpha_k), \qquad \rho_k \geqslant 0,$$

for $k \geqslant 0$. We may assume that $\alpha_0 = 0$, and $\alpha_j = \gamma$, $l_j > 0$, and $|\alpha_k| \leqslant \gamma$ for $k > 0$. For any $m > 0$, we have

$$|\lambda|\,\rho_m \leqslant \sum_0^N r_{mk}\rho_k \leqslant a\sum_0^N r_{0k}\rho_k\,.$$

But

$$\begin{aligned}|\lambda| \geqslant R\lambda &= R\left(\sum_0^N C_{0k}z_k\right)\\ &\geqslant \sum r_{0k}\rho_k\cos(\theta + \gamma),\end{aligned}$$

and this implies that

$$\rho m \leqslant a/\cos(\theta + \gamma)$$

for $m > 0$.

We set

$$Z_m = \sum_1^N C_{mk} z_k ,$$

so that

$$\lambda = C_{00} + Z_0 ,$$
$$\lambda z_j = C_{j0} + Z_j .$$

The estimate

$$| Z_m | \leqslant \sum_1^N r_{mk} \rho_k \leqslant abr_{m0}/\cos(\theta + \gamma)$$

yields

$$| \arg(1 + z_m/C_{m0})| \leqslant \beta,$$

where

$$\sin \beta = ab/\cos(\theta + \gamma), \qquad 0 < \beta < \pi/2.$$

Consequently, we obtain

$$\gamma = \arg z_j = \arg((C_{j0} + Z_j)/(C_{00} + Z_0))$$
$$\leqslant \theta + \beta - (-\theta - \beta) = 2(\theta + \beta),$$

or

$$\sin((\gamma/2) - \theta) \leqslant \sin \beta,$$

that is

$$2 \sin((\gamma/2) - \theta) \cos(\theta + \gamma) < 2ab.$$

But the left-hand side is

$$\sin(3\gamma/2) - \sin(2\theta + (\gamma/2)),$$

so this inequality contradicts (21).

COROLLARY 12a. *If C satisfies the conditions*

$$C_{j0} \neq 0 \qquad (j \geqslant 0),$$
$$| \arg((C_{jk}C_{k0})/(C_{00}C_{j0}))| \leqslant \theta \qquad (j, k \geqslant 0),$$
$$| C_{00} | \, | C_{jk} | \leqslant a \, | C_{0k} | \, | C_{j0} | \qquad (j > 0, k \geqslant 0),$$

and

$$\sum_1^N | C_{jk}C_{k0} | \leqslant b \, | C_{00} | \, | C_{j0} | \qquad (j \geqslant 0),$$

and condition (20), *then C has an eigenvector z such that*

$$z_0 = C_{00} , \qquad | \arg(z_j/C_{j0})| < \gamma \qquad (j > 0).$$

Clearly Theorem 12 and its corollary can be extended in the obvious way to certain infinite-dimensional spaces.

References

1. R. Bellman, "Introduction to Matrix Analysis," McGraw–Hill, New York, 1960.
2. G. Birkhoff, Extensions of Jentzsch's theorem, *Trans. Amer. Math. Soc.* **85** (1957), 219–227.
3. A. Brauer, On the characteristic roots of power-positive matrices, *Duke Math. J.* **27** (1961), 439–446.
4. A. Brauer, On the characteristic roots of non-negative matrices, Ref. [19, pp. 3–38].
5. H. Busemann and P. J. Kelly, "Projective Geometries and Projective Metrics," Academic Press, New York, 1953.
6. G. Frobenius, Über Matrizen aus positiven Elementen, *Sitzungsber. Kgl. Preuss. Akad. Wiss.* (1908), 471–476; and (1909), 514–518.
7. G. Frobenius, Über Matrizen aus nicht-negativen Elementen, *Sitzungsber. Kgl. Preuss. Akad. Wiss.* (1912), 456–477.
8. D. Hilbert, Neue Begründung der Bolyai-Lobatschefskyschen Geometrie, *Math. Ann.* **57** (1903), 137–150.
9. E. Hille and R. S. Phillips, "Functional Analysis and Semi-Groups," Colloquium Publ., Vol. 31, Amer. Math. Soc., Providence, R.I., 1957.
10. E. Hopf, An inequality for positive linear operators, *J. Math. Mech.* **12** (1963), 683–692.
11. A. S. Householder, Localization of the characteristic roots of matrices, Ref. [19, pp. 39–60].
12. R. Jentzsch, Über Integralgleichungen mit positiven Kern, *J. Math.* **141** (1912), 235–244.
13. L. V. Kantorovich, B. Z. Vulich, and A. G. Pinsker, Functional analysis in semi-ordered spaces (Russian), *Gos. Izdat. Tekhniko-Teoret. Lit., Moscow,* 1950.
14. T. Kato, "Perturbation Theory for Linear Operators," Springer, Berlin, 1966.
15. M. G. Krein and M. A. Rutman, Linear operators leaving invariant a cone in a Banach space, *Uspehi Mat. Nauk* **23** (1948), 3–95. English translation: Amer. Math. Soc. Translation No. 26.
16. A. M. Ostrowski, Positive matrices and functional analysis, Ref. [19, pp. 81–101].
17. O. Perron, *Zm* Theorie der Matrices, *Math. Ann.* **64** (1907), 248–263.
18. P. C. Rosenbloom, Perturbation of linear operators in Banach spaces, *Arch. Math. (Basel)* **6** (1955), 89–101.
19. H. Schneider (Ed.), "Recent Advances in Matrix Theory," Univ. of Wisconsin Press, Madison, 1964.
20. E. Seneta, "Non-Negative Matrices, An Introduction to Theory and Applications," Wiley, New York, 1973.

A Commentary on Gurari's "The Location of Eigenvalues and Eigenvectors of Complex Matrices"

by Dr. Neil Miller

It is very important to find the eigenvalues and eigenvectors of a given matrix. Information about them is critical in the development of matrix theory and applications to many fields of science. Such applications play a central role in our daily lives, from airplanes to candle flames to animal populations.

The following example demonstrates how eigenvalues can make matrix multiplication easy. Let $A\mathbf{x} = \lambda\mathbf{x}$, so that A has an eigenvalue λ with eigenvector $\mathbf{x}$. Then

$$A^2\mathbf{x} = A(A\mathbf{x}) = A(\lambda\mathbf{x}) = \lambda(\lambda\mathbf{x}) = \lambda^2\mathbf{x}.$$

Moreover, for any positive integer n, $A^n\mathbf{x} = \lambda^n\mathbf{x}$, since each multiplication of A on the left-hand side of the above equation produces a factor of λ on the right-hand side.

The next result stems from the theory of nonnegative matrices. It is often applied to animal and plant populations, the subject of my Ph.D. thesis. It is **Perron's Theorem** (Gurari, p.122) which states that if the matrix $C > 0$ (that is, C is a **positive** matrix), then it has a positive eigenvalue λ with a positive eigenvector ξ with the properties (a), (b), and (c) as listed.

Gurari's main task is to generalize Perron's Theorem to larger classes of matrices. Unfortunately, an unconditional generalization, as well as a theorem that yields exact eigenvalues, is next to impossible. However, many of his results, especially those at the end of his article, confine the eigenvalues and eigenvectors to a small range of values, given certain conditions about the matrix. In particular, Theorem 1 (p.123) and the discussion on pp.124-5 prove Perron's Theorem for nonnegative power-positive matrices. A matrix C is **power-positive** if $C^m > 0$ for some $m > 0$. Clearly if C is positive, then C is power-positive for $m = 1$.

Some of the remaining theorems further generalize the first statement and Property (c) of Perron's Theorem. For example, Theorem 3 (p.126) adds that any other eigenvector, when multiplied by a scalar (**normalized**) so that its first component is 1, has at least one component with a negative real part. This result is recast in approximate form in Theorem 8 (p.139) for $E = C + U$ where C is positive and U is a small enough perturbation of C. In particular, if E is nonnegative and power-positive, then adding even an arbitrarily small amount to the zero entries of E perturbs it to a positive matrix $C > 0$. Therefore, any power-positive matrix E is a perturbation $(C + U)$ to which the perturbation theorems can be applied.

Theorem 9 (pp.139-40) assures that there exists an eigenvector with each component in a small region near the positive real axis. Corollary 9a(p.140) and Theorem 10(p.141) extend Theorem 9 to cases when C is not necessarily positive.

The above results are cleverly woven together to form Theorem 12 and Corollary 12a(p.148), the main result. They extend Theorem 9 for matrices that are close to a border matrix B, whose entries B_{ij} are all zero except its top row and leftmost column. This class includes ALL complex matrices whose top and left border elements are large enough in absolute value relative to the other elements. Moreover, both Theorem 12 and Corollary 12a can be applied to certain infinite-dimensional spaces (p.148).

Theorem 8 assures the uniqueness of the "positive" eigenvector whose existence is proven in Theorems 9-12.

Several of Gurari's intermediate results are listed below, for they contribute significantly to his theory:

- p.120: an exact formula for the eigenvalues and eigenvectors of a border matrix
- p.125: an exact formula for the largest eigenvalue of a power-positive matrix (Theorem 2), as well as a simple estimate of this eigenvalue (Corollary 2a)
- p.131: quantitative results on the perturbation of eigenvalues and eigenvectors, which can be extended to eigenvalues with multiplicities of more than 1; that is λ for which the rank of $(C-\lambda I)$ is less than $n-1$ where C is an $n\times n$ matrix. Such results were difficult to find elsewhere at the time this article was written.
- pp.133-5: Theorem 6 and its corollaries show that in certain cases, perturbing the matrix C does not greatly alter its eigenvalues and eigenvectors. Corollary 6b, which addresses the right half-plane, is especially useful in proving the later theorems.

Full Commentary on Gurari's "The Location of Eigenvalues and Eigenvectors of Complex Matrices"

by Dr. Neil Miller

This article is a running commentary of Gurari's work that expands on my previous paper. Its goals are to explain the concepts introduced in his article to the average mathematician, and to ease him through those proofs that are hard to understand. Lines quoted from Gurari's text, such as
"Let $C = (C_{jk})$, $0 \leq j, k \leq N$, be a matrix...":
are surrounded by quotation marks.

<u>Page 119</u>

The discussion in Section 1 is expanded in Section 5.

"Let $C = (C_{jk})$, $0 \leq j, k \leq N$, be a matrix...":

(C_{jk}) is the following matrix:

$$\begin{pmatrix} C_{00} & C_{01} & C_{02} & \cdots & C_{0,N-1} & C_{0N} \\ C_{10} & C_{11} & C_{12} & \cdots & C_{1,N-1} & C_{1N} \\ C_{20} & C_{21} & C_{22} & \cdots & C_{2,N-1} & C_{2N} \\ \vdots & \vdots & \vdots & & \vdots & \vdots \\ C_{N-1,0} & C_{N-1,1} & C_{N-1,2} & \cdots & C_{N-1,N-1} & C_{N-1,N} \\ C_{N0} & C_{N1} & C_{N2} & \cdots & C_{N,N-1} & C_{NN} \end{pmatrix}$$

The rows of C are numbered $0, 1, 2, ..., N$, and so are the columns of C. Each entry of C is called C_{jk}, where j and k are integers from 0 to N. Thus C_{jk} is in row j and column k of C. For example, if $j = 0$ and $k = N$, then $C_{jk} = C_{0N}$ is the entry in row 0 and column N in the upper-right corner of C above.

"$C_{jk} = r_{jk}exp(i\theta_{jk}), r_{jk} \geq 0, -\pi < \theta_{jk} \leq \pi$":

Each entry C_{jk} is a complex number, represented by a point on a two- dimensional graph (see below). Its radius r_{jk} is its distance from the origin (point O).

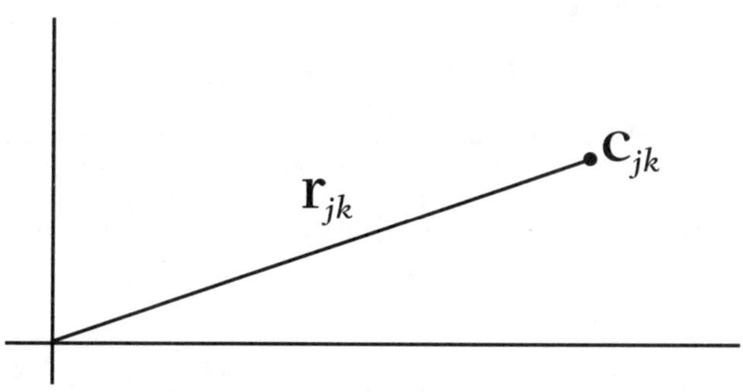

If the point is on Line A, directly to the right ("due east") of the origin, then its angle is zero. Otherwise, we define its angle θ_{jk} as the angle between it and Line A. The value of θ_{jk} is positive if the point lies in the upper half of the graph; otherwise it is considered negative. Since the value of a right angle is $\pi/2$, a point due north of the origin has angle $\pi/2$, while a point due south has angle $-\pi/2$. A point due west has angle π. Therefore, in the range $-\pi < \theta_{jk} \leq \pi$, θ starts due west at $-\pi$ and goes counterclockwise around the origin, until it returns to due west (π).

The variable i is the theoretical square root of -1. It is NOT a real number. The term $exp(i\theta_{jk})$ is defined as $e^{i\theta_{jk}} = cos(\theta_{jk}) + i * sin(\theta_{jk})$.

<ins>Page 120</ins>

"

$$a = \max_{j\geq 1, k\geq 0}(r_{jk}/r_{0k}),$$

$$\theta = \max_{j,k\geq 0}|\theta_{jk}|,$$

$$b = \max_{j\geq 0}(\sum_{k=1}^{N} r_{jk}/r_{j0})$$

":

The variable a is defined as the maximum value of r_{jk}/r_{0k} among all values of r_{jk}/r_{0k} for which $j \geq 1$ and $k \geq 0$.

The term $|\theta_{jk}|$ is θ_{jk} itself if θ_{jk} is nonnegative; otherwise the minus sign is removed. For instance, $|-\pi/4| = \pi/4$.

The expression

$$\sum_{k=1}^{N} r_{jk}/r_{j0}$$

is the sum $r_{j1}/r_{j0} + r_{j2}/r_{j0} + ... + r_{jN}/r_{j0}$. One of these exists for each $j \geq 0$, and b is the maximum of these sums.

"$F(b, \theta)$ is a function F of b and θ."

The eigenvector **z**: its components $z_0, z_1, ..., z_N$ are complex numbers, or points on the graph of the previous page. The **argument** of a complex number z_j, denoted $argz_j$, is defined as its angle θ_j on the graph. $S(3\theta)$ is the set of all **z** such that $z_0 = 1$ and for all j besides zero $|argz_j| < 3\theta$.

"If $a = 0$, then $C_{jk} = 0$ for $jk \neq 0$":

Proof: a was defined as

$$a = \max_{j\geq 1, k\geq 0}(r_{jk}/r_{0k}).$$

The quantities r_{jk} and r_{0k}, being radii or distances of points from the origin of the complex plane, are never negative. Thus $a = 0$ implies that for ALL j and k such that $j \neq 0$, $r_{jk} = 0$. Since the entries of C are $C_{jk} = r_{jk}exp(i\theta_{jk})$ (bottom of Page 119), it follows that $C_{jk} = 0$ unless $j = 0$, in which case $j * k = 0$. Therefore, $j * k \neq 0$

implies that $C_{jk} = 0$, proving the above statement. Please note that the jk in C_{jk} is really j, k whereas the jk in $jk \neq 0$ is the product $j * k$.

Condition (1) refers to the above equation $a < F(b, \theta)$ followed on the right by the equation marker "(1)".

"(a) C has the eigenvalue $\lambda = 0$ with multiplicity $N - 1$....":

This statement means that the determinant of the matrix $(\lambda I - C)$ is the polynomial $(\lambda - 0)^{N-1}(\lambda - x)(\lambda - y)$; that is, $\lambda^{N-1}(\lambda - x)(\lambda - y)$. If $y = x$ then C has the eigenvalue x with multiplicity two; otherwise x and y each have multiplicity one.

Proof: Since C is a border matrix, it has the form

$$\begin{pmatrix} C_{00} & C_{01} & C_{02} & \cdots & C_{0,N-1} & C_{0N} \\ C_{10} & 0 & 0 & \cdots & 0 & 0 \\ C_{20} & 0 & 0 & \cdots & 0 & 0 \\ \vdots & \vdots & \vdots & & \vdots & \vdots \\ C_{N-1,0} & 0 & 0 & \cdots & 0 & 0 \\ C_{N0} & 0 & 0 & \cdots & 0 & 0 \end{pmatrix}$$

Thus $(\lambda I - C) =$

$$\begin{pmatrix} \lambda - C_{00} & -C_{01} & -C_{02} & \cdots & -C_{0,N-1} & -C_{0N} \\ -C_{10} & \lambda & 0 & \cdots & 0 & 0 \\ -C_{20} & 0 & \lambda & \cdots & 0 & 0 \\ \vdots & \vdots & \vdots & & \vdots & \vdots \\ -C_{N-1,0} & 0 & 0 & \cdots & \lambda & 0 \\ -C_{N0} & 0 & 0 & \cdots & 0 & \lambda \end{pmatrix}$$

whose determinant is found, after an easy calculation, to be

$$\lambda^{N-1}(\lambda^2 - C_{00}\lambda - d)$$

where

$$d \equiv \sum_{k=1}^{N} C_{0k}C_{k0}.$$

Solving the quadratic polynomial in the above formula yields two solutions x and y (which might be complex numbers). Therefore C has the eigenvalue $\lambda = 0$ with multiplicity $N - 1$. QED

The $(N - 1)$-dimensional **eigenspace** is the set of vectors $\mathbf{z}$ such that $C\mathbf{z} = \mathbf{0}$. We prove below that this set is defined by

$$z_0 = 0, \sum_{1}^{N} C_{0k}z_k = 0 :$$

Proof: Expanding the equation $C\mathbf{z} = \mathbf{0}$, we obtain the system of equations:

$$C_{00}z_0 + C_{01}z_1 + C_{02}z_2 + ... + C_{0,N-1}z_{N-1} + C_{0N}z_N = 0$$

$$C_{10}z_0 = 0$$

$$C_{20}z_0 = 0$$

$$\vdots$$

$$C_{N-1,0}z_0 = 0$$

$$C_{N0}z_0 = 0$$

Assuming that all of C_{0k} and C_{j0} are nonzero, it is clear that the above system of equations is satisfied only for $z_0 = 0$ and

$$C_{01}z_1 + C_{02}z_2 + ... + C_{0,N-1}z_{N-1} + C_{0N}z_N = 0$$

by substituting $z_0 = 0$ in the top equation of the system. Thus the eigenspace is

$$z_0 = 0, \sum_1^N C_{0k}z_k = 0,$$

as required. QED

"(b) The roots of the quadratic equation.........
........have the eigenvectors

$z_0 = \lambda,$ $z_j = C_{j0}$ for $j > 0$"

Proof: Let the roots of the quadratic equation be λ_i, where $i = 1$ or 2. If $\mathbf{z} = [z_0, z_1, z_2, ..., z_{N-1}, z_N]^T = [\lambda_i, C_{10}, C_{20}, ..., C_{N-1,0}, C_{N0}]^T$, then $C\mathbf{z} =$

$$\begin{bmatrix} C_{00}\lambda_i + C_{01}C_{10} + C_{02}C_{20} + ... + C_{0N}C_{N0} \\ C_{10}\lambda_i \\ C_{20}\lambda_i \\ \vdots \\ C_{N-1,0}\lambda_i \\ C_{N0}\lambda_i \end{bmatrix}$$

$$= \begin{bmatrix} C_{00}\lambda_i + d \\ C_{10}\lambda_i \\ C_{20}\lambda_i \\ \vdots \\ C_{N-1,0}\lambda_i \\ C_{N0}\lambda_i \end{bmatrix}$$

$$= \begin{bmatrix} \lambda_i^2 \\ C_{10}\lambda_i \\ C_{20}\lambda_i \\ \vdots \\ C_{N-1,0}\lambda_i \\ C_{N0}\lambda_i \end{bmatrix}$$

since λ_i solves $\lambda_i^2 - C_{00}\lambda_i - d = 0$. However, this vector is simply λ_i times the original vector **z**. Therefore, the two vectors $z_0 = \lambda, z_j = C_{j0}$ for $j > 0$ are the eigenvectors that correspond to the quadratic roots λ_i. QED

"A border matrix is also a matrix of rank at most 2":

The **rank** of a matrix C is the number of rows of C that are linearly independent. In this case, assuming that all of C_{j0} and C_{0k} are nonzero, the top two rows are linearly independent, but all subsequent rows are multiples of the second row (indexed by $j = 1$). Thus the rank of C is 2. Obviously, if $C_{j0} = 0$ for all $j > 0$ or $C_{0k} = 0$ for all $k > 0$, then the rank of C is 1.

Page 121

The results on this page are stated here without proof. There is no need to derive them here.

$S(\gamma)$ is a set of vectors. This set consists of vectors, some of whose components are functions of γ.

Page 122

"terms of a variational problem...."

The word "variational" is not of major importance here.

From the equation

$$\text{“}\tau(x) = \min_j(\sum_k C_{jk}x_k/x_j)\text{''}$$

we can deduce that

$$\tau(x) = \min_j \frac{(Cx)_j}{x_j}$$

, because

$$(Cx)_j = \sum_k C_{jk}x_k.$$

"Then the maximum......"

This result will be proved as part of Theorem 1 on Page 123.

"more than one eigenvalue of maximum modulus...."

The **modulus** $|\lambda|$ of an eigenvalue λ is defined as follows: If λ is a real number then its modulus is its absolute value. Otherwise, its modulus is its distance from the origin of the complex plane pictured above. If $\lambda = re^{(i\theta)}$ then its modulus is r.

"power-positive if some power C^m is positive."

The matrix C^m is the matrix product $C * C * ... * C$ with m occurences of C. For example, $C^0 = I$, the identity matrix, and $C^1 = C$ itself.

Page 123

"Proof: We note...
... $Cx - \tau x \geq 0$."

We show, as follows, that $\tau(\mathbf{x})$, as defined above, is indeed the maximum τ such that $C\mathbf{x} - \tau\mathbf{x} \geq \mathbf{0}$ for the vector $\mathbf{x}$. From the definition of $\tau(\mathbf{x})$, $\tau(\mathbf{x}) \leq \frac{(Cx)_j}{x_j}$ for all j. Therefore, for each j, $(C\mathbf{x} - \tau(\mathbf{x}) * \mathbf{x})_j \geq (C\mathbf{x})_j - (C\mathbf{x})_j = 0$. Thus $\tau(\mathbf{x})$ satisfies the required inequality.

To show that $\tau(\mathbf{x})$ is the maximum number that does so, we let J be the number for which

$$\tau(\mathbf{x}) = \frac{(C\mathbf{x})_J}{\mathbf{x}_J};$$

that is, J is the j at which the expression

$$\frac{(Cx)_j}{x_j}$$

is minimal. If $\tau(\mathbf{x})$ is not the maximum, then $T \equiv \tau(\mathbf{x}) + E$ satisfies the inequality for some positive amount E. However, $(C\mathbf{x} - T(\mathbf{x}) * \mathbf{x})_J = (C\mathbf{x})_J - \tau(\mathbf{x}) * \mathbf{x}_J - E\mathbf{x}_J = (C\mathbf{x})_J - (C\mathbf{x})_J - E\mathbf{x}_J = -E\mathbf{x}_J < 0$ since $\mathbf{x}_J$ is not zero from the definition of $\tau(\mathbf{x})$ and the fact that

$$\tau(\mathbf{x}) = \frac{(C\mathbf{x})_J}{\mathbf{x}_J}.$$

Thus our assumption that $\tau(\mathbf{x})$ is not the maximum leads to the statement we have just disproved that T satisfies the inequality. Therefore, $\tau(\mathbf{x})$ IS the maximum. QED

"

$$C^2x - \tau(x)Cx \geq 0$$

"

Since $\tau(\mathbf{x})$ is a number, not a vector, $C\tau(\mathbf{x}) = \tau(\mathbf{x})C$. Thus the above equation follows from the previous equation $C(Cx - \tau(x)x) \geq 0$. QED

"$\tau(C\mathbf{x}) \geq \tau(\mathbf{x})$"

Proof: From the first equation on Page 123, we substitute $C\mathbf{x}$ for $\mathbf{x}$ and find that $\tau(C\mathbf{x})$ is the maximum number τ such that $C^2x - \tau Cx \geq 0$. But by the equation last proven, $\tau(\mathbf{x})$ is also a number τ that satisfies this inequality. Therefore, $\tau(\mathbf{x})$ can not exceed the maximum τ; that is, $\tau(C\mathbf{x})$. Thus $\tau(C\mathbf{x}) \geq \tau(\mathbf{x})$. QED

"$\tau(C^m\mathbf{x}) \geq \tau(\mathbf{x})$"

Proof: If we substitute $C\mathbf{x}$ for $\mathbf{x}$ in all equations from the top of Page 123 to this point, we arrive at the inequality $\tau(C^2\mathbf{x}) \geq \tau(C\mathbf{x})$. By substituting $C^2\mathbf{x}$ for

$\mathbf{x}$ we similarly obtain $\tau(C^3\mathbf{x}) \geq \tau(C^2\mathbf{x})$. We proceed in this manner until we get to $\tau(C^m\mathbf{x}) \geq \tau(C^{m-1}\mathbf{x})$. Now we combine these inequalities to form $\tau(C^m\mathbf{x}) \geq \tau(C^{m-1}\mathbf{x}) \geq ... \geq \tau(C^2\mathbf{x}) \geq \tau(C\mathbf{x}) \geq \tau(\mathbf{x})$, which yields the result $\tau(C^m\mathbf{x}) \geq \tau(\mathbf{x})$. QED

"$\min u_j \geq \gamma(A)||\mathbf{y}||$"

Proof: We know that $||\mathbf{y}||$ is defined as $\max y_l$. Since $\mathbf{u} = A\mathbf{y}$, it follows that for each j from 0 to N,

$$\begin{aligned} u_j = \sum_{k=0}^{N} A_{jk}y_k &\geq \sum_{k=0}^{N} \gamma(A)y_k \\ = \gamma(A)\sum_{k=0}^{N} y_k &\geq \gamma(A)max y_l \\ &\geq \gamma(A)||\mathbf{y}|| \end{aligned}$$

. The first "$\geq$" is true since by the definition of $\gamma(A), \gamma(A) \leq A_{jk}$ for each j and k. The next link of the statement is obtained by removing the constant $\gamma(A)$ from the sum. The next link, another "$\geq$", results from the fact that

$$\sum_{k=0}^{N} y_k \geq \max y_l$$

which is true since all entries y_l are nonnegative from the statement "It is trivial that if $\mathbf{y} \geq \mathbf{0}$...". Since $u_j \geq \gamma(A)||\mathbf{y}||$ for each j, it follows that $\min u_j \geq \gamma(A)||\mathbf{y}||$. QED

"

$$||A|| = \max_{||x|| \leq 1} ||A|| = \max_j \sum |A_{jk}|$$

"

The first equality in the above statement is a definition. However, I think the $||A||$ in the middle term is an innocent error; we replace it with $||A\mathbf{x}||$. Now we prove the second equality as follows:

For all $\mathbf{x}$ such that $|\mathbf{x}| \leq 1$:

$$\begin{aligned} ||A\mathbf{x}|| &= \max_j(\sum_k |A_{jk}x_k|) \\ &\leq \max_j(\sum_k |A_{jk}||x_k|) \\ &\leq \max_j(\sum_k |A_{jk}|) \end{aligned}$$

since each $|x_k|$ is at most 1. On the other hand, if we set $\mathbf{x} = \mathbf{1}$, the vector whose entries are all 1, then

$$\max_{||\mathbf{x}|| \leq 1} ||Ax|| \geq ||A\mathbf{1}|| = \max_j(\sum_k |A_{jk}|)$$

Thus we have proven that

$$\max_{||\mathbf{x}|| \leq 1} ||A\mathbf{x}|| \leq \max_j \sum |A_{jk}|$$

and also

$$\max_{||\mathbf{x}||\leq 1} ||A\mathbf{x}|| \geq \max_j \sum |A_{jk}|,$$

which together imply that

$$\max_{||\mathbf{x}||\leq 1} ||A\mathbf{x}|| = \max_j \sum |A_{jk}|.$$

Q.E.D.

"$\min y_j \geq \gamma(C^m)||x||$":

Proof:

$$\begin{aligned}\min_j y_j &= \min_j(\sum_k C^m_{jk} x_k) \\ &= \min_j(\sum_k |C^m_{jk}||x_k|) \\ &\geq \gamma(C^m)\sum_k |x_k| \geq \gamma(C^m)||\mathbf{x}||\end{aligned}$$

since C^m and $\mathbf{x}$ are nonnegative. Q.E.D.

"$||y|| \leq ||C^m||||x||$":

Proof: Since $\mathbf{y}$ is defined as $C^m\mathbf{x}$, we see that

$$\begin{aligned}||\mathbf{y}|| &= \max_j y_j = \max_j(\sum_k C^m_{jk} x_k) \\ &\leq \max_j(\sum_k C^m_{jk}) \max_j(|x_k|) \\ &\leq \max_j(\sum_k |C_{jk}|^m) \max_j(|x_k|) \\ &= ||C^m||||\mathbf{x}||\end{aligned}$$

The second line is true because $\max(AB) \leq \max(A)\max(B)$ for any quantities A and B. The third line follows from the second since $|C_{jk}|$ is the absolute value of C_{jk}. Q.E.D.

"$\min z_j \geq \gamma(C^m)/||C^m||$":

Proof:

$$\begin{aligned}\min_j z_j &= \min_j(y_j/||\mathbf{y}||) = (\min_j y_j)/||\mathbf{y}|| \\ &\geq \gamma(C^m)||\mathbf{x}||/||\mathbf{y}|| \\ &\geq \gamma(C^m)||\mathbf{x}||/||C^m||||\mathbf{x}|| \\ &= \gamma(C^m)/||C^m||\end{aligned}$$

In the first line, the term $||\mathbf{y}||$ can be brought outside the expression being minimized over j, since it is independent of j. The second line follows from the inequality $\min y_j \geq \gamma(C^m)||\mathbf{x}||$, and the third line from $||\mathbf{y}|| \leq ||C^m||||\mathbf{x}||$. Q.E.D.

"$\tau(z) = \tau(y) \geq \tau(x)$":

Proof: The equality on the left is true since

$$\begin{aligned}\tau(\mathbf{z}) &\equiv \min_j(\sum_k C_{jk}z_k/z_j) \\ &= \min_j \frac{\sum_k C_{jk}(y_k/||\mathbf{y}||)}{(y_j/||\mathbf{y}||)} \\ &= \min_j(\sum_k C_{jk}y_k/y_j).\end{aligned}$$

Since $\mathbf{y}$ was defined as $C^m\mathbf{x}$, the inequality on the right is merely the statement proven above that $\tau(C^m\mathbf{x}) \geq \tau(\mathbf{x})$. Q.E.D.

"Therefore the supremum...
...on the subset..."

Proof: Let us call X the set referred to in the above statement, Y its subset, and Z the set of all possible values of $\mathbf{z}$ as defined above. Since $C^m > 0$ and $\mathbf{x} \geq 0$, we have $\mathbf{y} = C^m\mathbf{x} \geq 0$. As $||\mathbf{y}|| > 0$, we see that for each $\mathbf{z}$, $\mathbf{z} \equiv \mathbf{y}/||\mathbf{y}|| \geq 0$. In addition,

$$\begin{aligned}||\mathbf{z}|| &= \max_j z_j = \max_j y_j/||\mathbf{y}|| \\ &= ||\mathbf{y}||/||\mathbf{y}|| = 1.\end{aligned}$$

Thus $\mathbf{z} \in X$. Also, from the statement $minz_j \geq \gamma(C^m)/||C^m||, \mathbf{z} \in Y$ too. Therefore $Z \subset Y \subset X$. That is, X contains all vectors in Y and maybe additional vectors not in Y. Thus, the supremum of τ over X is at least the supremum of the same function τ over Y; that is,

$$\sup_X \tau \geq \sup_Y \tau.$$

Similarly, since Y contains all vectors in Z and maybe more vectors,

$$\sup_Y \tau \geq \sup_Z \tau.$$

Therefore,

$$\sup_X \tau \geq \sup_Z \tau. \tag{1}$$

If we can prove that

$$\sup_Z \tau \geq \sup_X \tau, \tag{2}$$

then we can combine it with (1) to prove that

$$\sup_X \tau = \sup_Z \tau$$

and therefore use the two inequalities above (1) to show that

$$\sup_X \tau = \sup_Y \tau$$

as required. To prove Result (2), let us suppose that there is an $\mathbf{x}_*$ such that $\tau(\mathbf{x}_*)$ equals the supremum of τ over X. Then, by the statement "$\tau(z) = \tau(y) \geq \tau(x)$", $\tau(\mathbf{z}_*) \geq \tau(\mathbf{x}_*)$ where $\mathbf{z}_* \equiv \mathbf{y}_*/||\mathbf{y}_*||$ and $\mathbf{y}_* \equiv C^m\mathbf{x}_*$. Otherwise, there is a sequence $\mathbf{x_1}, \mathbf{x_2}, \mathbf{x_3}, ...$ such that the values $\tau(\mathbf{x_1}), \tau(\mathbf{x_2}), \tau(\mathbf{x_3}), ...$ converge to the supremum of τ over X. If we derive $\mathbf{z_n}$ from each of the above $\mathbf{x_n}$ for $n = 1, 2, 3, ...$ then for each n, $\tau(\mathbf{z_n}) \geq \tau(\mathbf{x_n})$. If the $\tau(\mathbf{z_n})$ converge to the supremum of τ over X, then we already have Result (2). Otherwise, there is an N such that

$$\tau(\mathbf{z_N}) \geq \sup_X \tau,$$

which implies (2). Q.E.D.

"Since τ is
continuous on this subset (Y), it attains its maximum λ there at some vector ξ."

Proof: Suppose that there is no vector $\xi \in Y$ such that

$$\tau\xi = \sup_Y \tau.$$

Then there is a sequence $\mathbf{x_1}, \mathbf{x_2}, \mathbf{x_3}, ...$ that converges to, say, ξ such that the values $\tau(\mathbf{x_1}), \tau(\mathbf{x_2}), \tau(\mathbf{x_3}), ...$ converge to the supremum of τ over Y. Now Y is the set of all vectors $\mathbf{x} \geq 0$, $||\mathbf{x}|| = 1$, and $\min x_j \geq \gamma(C^m)/||C^m||$. Thus Y is closed; that is,

$$\xi \in Y. \tag{3}$$

Moreover, because $\tau(\mathbf{x})$ is continuous, $\tau(\xi)$ is the supremum of τ over Y. From the above equation

$$\sup_Y \tau = \sup_X \tau,$$

we deduce that τ attains its maximum (over X) at ξ. Q.E.D.

"If ϵ...then $\min y_j \geq \gamma(C^m)\epsilon > 0$."

Proof: Since λ is the τ that corresponds to ξ, $C\xi - \lambda\xi \geq 0$ from the equation

$$Cx \quad - \quad \tau x \geq 0 \tag{4}$$

(5)

at the top of Page 123. Let K be the k for which $(C\xi - \lambda\xi)_k$ is greatest. Then $(C\xi - \lambda\xi)_K = \epsilon$. Thus for each j,

$$y_j = \sum_k (C^m)_{jk}(C\xi - \lambda\xi)_k.$$

Since $C^m \geq 0$ and one of the terms in the above summation involves K, we have $y_j \geq (C^m)_{jK}(C\xi - \lambda\xi)_k \geq \gamma(C^m)\epsilon$. Thus $\min y_j \geq \gamma(C^m)\epsilon$. Since $C^m > 0$, it follows that $\gamma(C^m) > 0$ as well as $\epsilon > 0$, completing the proof. Q.E.D.

Page 124

"But since $y = C(C^m\xi) - \lambda C^m\xi$, this implies that $\tau(C^m\xi) > \lambda$":

Proof: We just proved above that every entry y_j is > 0. From (3) above, with $\mathbf{x}$ replaced by $C^m\xi$, we obtain our result. Q.E.D.

"Thus all $\xi > 0$ for which $\tau(\xi) = \lambda$ are eigenvectors of C." That is to say: if $\xi > 0$ and $\tau(\xi) = \lambda$, then ξ is an eigenvector of C.

Proof: The above statement is clear from the fact just before it that $C\xi = \lambda\xi$.

"COROLLARY 1a"; i.e. $\min \eta_j \geq \gamma(C^m)/||C^m||$:

Proof: From (4) above, and the fact that Y is the subset of X for which $\min x_j \geq \gamma(C^m)/||C^m||$, we see immediately that $\min \eta_j \geq \gamma(C^m)/||C^m||$. Q.E.D.

"Let $|z|$ be the vector with the components $|z_j|, 1 \leq j \leq n$."

Note that $|\mathbf{z}|$ is different from $||\mathbf{z}||$, a *scalar* defined before as $max|z_j|$.

"We have $A'\eta = \lambda^m\eta$ and $Az = \mu^m z$."

Proof: η was defined as a positive eigenvector of C',, NOT of C. Since $A = C^m$, we have $A' = (C^m)' = (C * C * ... * C)'(m \text{ times}) = C' * C' * ... * C'(m \text{ times})$. Thus $A'\eta = C' * C' * ... * C'\eta = C' * ... * C'\lambda\eta = \lambda C' * ... * C'\eta = ... = \lambda^m\eta$. In the above equations, there are m terms C' in $C' * C' * ... * C'$ and $m-1$ terms C' in $C' * ... * C'$.λ, being a scalar, was brought outside the matrix-vector product $C' * ... * C'\eta$. In the same way, but without the argument that $A' = C' * C' * ... * C'$, we show that $A\mathbf{z} = \mu^m\mathbf{z}$. Q.E.D.

"It follows that $|\mu|^m|z| \leq A|z|$":

Proof: Since $A\mathbf{z} = \mu^m \mathbf{z}$, for each j, $A\mathbf{z}_j = \mu^m z_j$. Thus $\mu^m = (A\mathbf{z})_j / z_j$, so the jth component of the left-hand side of the above equation is

$$\begin{aligned} |\mu|^m|z_j| &= \frac{|(Az)_j|}{|z_j|}|z_j| = |(Az)_j| \\ &= |\sum_k A_{jk} z_k| \leq \sum_k |A_{jk}||z_k| \\ &= \sum_k A_{jk}|z_k| = A|z_j|, \end{aligned}$$

the jth component of the right-hand side, where we have used $|A_{jk}| = A_{jk}$ since $A = C^m$ is positive. Q.E.D.

"and $A|z| - |\mu|^m|z| \neq 0$ unless z is a scalar multiple of a nonnegative vector...":

Proof: Let $z_k \equiv r_k e^{i\theta_k}$. If $A|\mathbf{z}| - |\mu|^m|\mathbf{z}|$ is zero, then $A|\mathbf{z}| = |\mu|^m|\mathbf{z}|$. If we define $\mathbf{r}$ as the vector with components $r_k = |z|_k$, then for each k, $(A\mathbf{r})_k = |\mu|^m r_k$. However,

$$(A\mathbf{r})_k = \sum_j A_{kj} r_j,$$

which implies

$$\sum_j A_{kj} r_j = |\mu|^m r_k.$$

Since $A\mathbf{z} = \mu^m \mathbf{z}$, we take the absolute value of each component of both sides to obtain, for each k,

$$|\mu|^m|z_k| = |A\mathbf{z}|_k = |sum_j A_{kj} z_j|$$

$z_k = r_k e^{i\theta_k}$ yields

$$|\mu|^m r_k = |sum_j A_{kj} r_j e^{i\theta_j}|$$

which, from (?) above, becomes

$$\sum_j A_{kj} r_j = |sum_j A_{kj} r_j e^{i\theta_j}|.$$

Since $A = C^m$ is a positive matrix and each r_j is nonnegative, (?+1) can be true only under the following condition: all complex numbers $A_{kj} r_j e^{i\theta_j}$ for which $r_j > 0$ point in the same direction in the complex plane; that is, the angle θ_j is the same for all such j. In other words, $\mathbf{z}$ must be a scalar multiple of a nonnegative vector, where the scalar is the common angle θ_j. Q.E.D.

"unless z is a scalar multiple..., and then μ must also be nonnegative."

Proof: If $\mathbf{z}$ is a scalar multiple β of a nonnegative vector $\mathbf{y}$, then $C\mathbf{z} = \mu\mathbf{z}$ implies that, for each j such that $z_j \neq 0$ (and there exists such j unless $\mathbf{z} = 0$),

$$\begin{aligned}\mu &= \frac{(C\mathbf{z})_k}{z_k} = \frac{sum_j C_{kj} z_j}{z_k} \\ &= \frac{sum_j C_{kj} \beta y_j}{\beta y_k} \\ &= \frac{sum_j C_{kj} y_j}{y_k}\end{aligned}$$

in which each term in the numerator is nonnegative and the denominator y_k is positive. Thus $\mu \geq 0$.
In case $\mathbf{z} = 0$, we have $\mu = 0$ since $C\mathbf{z}$ and $\mu\mathbf{z}$ are both zero. Q.E.D.

"unless z is a scalar multiple..., and then μ must also be nonegative. Consequently, except in this case, we have

$$\eta * (A|z| - |\mu|^m|z|) > 0\text{":}$$

Proof: Since η is positive, $\eta_k > 0$ for each k. Moreover, let $\mathbf{q} \equiv (A|\mathbf{z}| - |\mu|^m|\mathbf{z}|)$. Then EXCEPT in the scalar-multiple case, there is some component k^* such that $q_k^* > 0$ while $q_k \geq 0$ for all other k. Then

$$\begin{aligned}\eta * (A|\mathbf{z}| - |\mu|^m|\mathbf{z}|) &= \eta * \mathbf{q} \\ &= \sum_k \eta_k q_k = \eta_k^* q_k^* + \sum_{k \neq k^*} \eta_k q_k\end{aligned}$$

with a positive first term and nonnegative sum. Thus $\eta * (A|z| - |\mu|^m|z|) > 0$. Q.E.D.

"that is, $$(\lambda^m - |\mu|^m)(\eta * |z|) > 0\text{":}$$

Proof: In order to prove the above statement, we show that $(\lambda^m - |\mu|^m)(\eta * |\mathbf{z}|)$ is the same as $\eta * (A|\mathbf{z}| - |\mu|^m|\mathbf{z}|)$ as follows:

$$\begin{aligned}&(\lambda^m - |\mu|^m)(\eta * |\mathbf{z}|) \\ &\quad= (\lambda^m\eta) * |\mathbf{z}| - \eta * (|\mu|^m|\mathbf{z}|) \\ &\quad= (A'\eta) * |\mathbf{z}| - \eta * (|\mu|^m|\mathbf{z}|) \\ &\quad= \eta * (A|\mathbf{z}|) - \eta * (|\mu|^m|\mathbf{z}|) \\ &\quad= \eta * (A|\mathbf{z}| - |\mu|^m|\mathbf{z}|)\end{aligned}$$

as required, where the fourth line of the above expression stems from the fact that $(A'\eta) * |\mathbf{z}|$ can be written as $< A'\eta, |\mathbf{z}| >=< \eta, A|\mathbf{z}| >= \eta * (A|\mathbf{z}|)$ since the matrix A' becomes $(A')' = A$ when brought to the other side of an inner product. Q.E.D.

"In the exceptional case we may assume $z \geq 0, \mu \geq 0$. Then we obtain

$$0 = \eta * (Cz - \mu z) = (\lambda - \mu)(\eta * z),$$

and therefore $\mu = \lambda$."

Proof: We know that $\mu \geq 0$ from the lines [Gurari] "unless z is a scalar multiple of a nonnegative vector, and then μ must also be nonegative...."
Now suppose $\mathbf{y} = 0$, where we have defined $\mathbf{z} = \beta\mathbf{y}$. Then $\mathbf{z} = 0$ and we have our result $\lambda = \mu = 0$. Thus we can assume that $\mathbf{y} \neq 0$. Since $C\mathbf{z} = \mu\mathbf{z}$, it is clear that $0 = \eta * (C\mathbf{z} - \mu\mathbf{z})$. Since we are treating the exceptional case, we write $\mathbf{z} \equiv \beta\mathbf{y}$ as before, where $\mathbf{y}$ is nonnegative. Next we show that the right equality in the above statement is true with z replaced by $\mathbf{y}$ as follows:

$$\begin{aligned}
&\eta * (C\mathbf{y} - \mu\mathbf{y}) \\
&\quad = \eta * C\mathbf{y} - \eta * \mu\mathbf{y} \\
&\quad = C'\eta * \mathbf{y} - \eta * \mu\mathbf{y} \\
&\quad = \lambda\eta * \mathbf{y} - \mu\eta * \mathbf{y} \\
&\quad = (\lambda - \mu)(\eta * \mathbf{y})
\end{aligned}$$

where the fourth line above is true since η was defined as "a positive unit eigenvector (of C') belonging to λ." Therefore, this last expression is also zero. Since $\mathbf{y} \neq 0$, there is a k^* such that $y_k^* > 0$. Since $\eta > 0$ and $\mathbf{y} \geq 0$, it follows that $(\eta * \mathbf{y}) > 0$; thus $\lambda = \mu$.
The same result holds for $\mathbf{z}$ instead of $\mathbf{y}$, since the nonzero scalar β is divided out of the above equality. Q.E.D.

The above statements state that either $|\mu| < \lambda$, or $\mu = \lambda$ in case $\mathbf{z}$ is a nonnegative eigenvector or a scalar multiple of one. This is merely properties (a) and (c) of Perron's theorem [Gurari, p.122] generalized to a power-positive matrix C.

The next step is to prove property (b). For the bottom of [Gurari, Page 124], please recall that $C\xi = \lambda\xi$.

Page 125

"If z is not a scalar multiple of η, then $u \neq 0$. If u were a scalar multiple of a nonnegative vector, then $|\eta * u|$ would be positive...":

Remarks: We now prove that $\mathbf{z}$ IS a multiple of η by arriving at a contradiction.

To prove that $\mathbf{u}$ being a multiple of a nonnegative vector would imply $|\mathbf{u} * \eta| > 0$, let $\mathbf{u} \equiv \beta\mathbf{y}$ where $\mathbf{y} \geq 0$. Since we assume that $\mathbf{z}$ is not a scalar multiple of η, we can say that $\mathbf{u} \neq 0$, so $\beta \neq 0$ and $\mathbf{y} \neq 0$. Thus for some $k^*, y_k^* > 0$. Since $\eta > 0$, it is clear that

$$\mathbf{y} * \eta = y_k^* \eta_k^* + \sum_{k \neq k^*} y_k \eta_k$$

in which the k^* term is positive and the sum is nonnegative. Therefore, we would have $|\mathbf{u} * \eta| = |\beta|\mathbf{u} * \eta > 0$.

Since $|\mathbf{u} * \eta| = 0$, we see that $\mathbf{u}$ is NOT a multiple of a nonnegative vector.

"$A|u| - \lambda^m|u| \geq 0, A|u| - \lambda^m|u| \neq 0$":

Proof:

$$\begin{aligned} A\mathbf{u} &= A\mathbf{z} - \frac{(\eta * \mathbf{z})}{(\eta * \xi)} A\xi \\ &= C^m \mathbf{z} - \frac{(\eta * \mathbf{z})}{(\eta * \xi)} C^m \xi \\ &= \lambda^m \mathbf{z} - \frac{(\eta * \mathbf{z})}{(\eta * \xi)} \lambda^m \xi \\ &= \lambda^m \mathbf{u}. \end{aligned}$$

Then for each component j,

$$\begin{aligned} (\lambda^m|\mathbf{u}|)_j &= (|\lambda^m \mathbf{u}|)_j = |\lambda^m u_j| \\ &= |(A\mathbf{u})_j| = |\sum_k A_{jk} u_k| \leq \sum_k |A_{jk}||u_k| \\ &= \sum_k A_{jk}|u_k| = (A|\mathbf{u}|)_j \end{aligned}$$

since all entries of A are positive. Thus we have proved the first half of our statement. Now suppose the second half is false; that is, for each j $(A|\mathbf{u}|)_j = \lambda^m|u_j|$. Letting $u_k \equiv r_k e^{i\theta_k}$ yields

$$\begin{aligned} |\sum_k A_{jk} r_k e^{i\theta_k}| &= |\sum_k A_{jk} u_k| \\ &= |(A\mathbf{u})_j| = |\lambda^m \mathbf{u}_j| = (|\lambda|^m |\mathbf{u}|)_j \\ &= (A|\mathbf{u}|)_j = \sum_k A_{jk}|u_k| = \sum_k A_{jk} r_k. \end{aligned}$$

Then θ_k is the same for each k, making $\mathbf{u}$ a scalar multiple of a nonnegative vector. This conclusion contradicts the above remark that $\mathbf{u}$ is not such a multiple. Thus the second half is true. Q.E.D.

"But...$\eta * A|u| = \lambda^m|u|$...":

I think the above statement may contain an innocent error; it probably means "$\eta * A|u| = \eta * \lambda^m|u|$...":

Proof:

$$\begin{aligned} \eta * A|\mathbf{u}| &= <\eta, A|\mathbf{u}|> \\ &= <A'\eta, |\mathbf{u}|> = <\lambda^m \eta, |\mathbf{u}|> \\ &= <\eta, \lambda^m|\mathbf{u}|> = \eta * \lambda^m|\mathbf{u}|. \end{aligned}$$

Q.E.D.

"But.....we have arrived at a contradiction. Therefore
$u = 0$, and z is a scalar multiple of ξ."

We have just shown that $\eta * (A|\mathbf{u}| - \lambda^m|\mathbf{u}|) = 0$, which contradicts the previous equation $\eta * (A|\mathbf{u}| - \lambda^m|\mathbf{u}|) > 0$. This contradiction uproots our previous statement that $\mathbf{u}$ is not a multiple of a nonnegative vector, which was based on our assumption that $\mathbf{u} \neq 0$ and $\mathbf{z}$ is not a multiple of ξ. Therefore $\mathbf{u} = 0$ and $\mathbf{z}$ IS a multiple of ξ.

From the statement [Gurari, p.124]
"Finally if z is any eigenvector of C belonging to λ"
we now see that any eigenvector of C and λ is a multiple of ξ itself. Thus the eigenspace (set of eigenvectors) of λ is one-dimensional, which is property (b) of Perron's theorem for power-positive C. Q.E.D.

"Proof....It is sufficient
to look for the minimum of σ on the set S_1 of all $x > 0$ such that $||x|| = 1$
and $\sigma(x) \leq \sigma(\mathbf{1})$."

Proof: We wish to show that the value $\mathbf{x}$ which minimizes $\sigma(\mathbf{x})$ has norm one and that $\sigma(\mathbf{x}) \leq \sigma(\mathbf{1})$. That $\sigma(\mathbf{x}) \leq \sigma(\mathbf{1})$ is clear; to show that $||\mathbf{x}|| = 1$, we let $\mathbf{x}^*$ be the minimum and $\mathbf{x} \equiv \mathbf{x}^*/||\mathbf{x}^*||$. Then $||\mathbf{x}|| = ||\mathbf{x}^*||/||\mathbf{x}^*|| = 1$ and

$$\begin{aligned}\sigma(\mathbf{x}) &= \max_j(\sum \frac{C_{jk}x_k}{x_j}) \\ &= \max_j(\sum \frac{C_{jk}x^*{}_k/||\mathbf{x}^*||}{x^*{}_j/||\mathbf{x}^*||}) \\ &= \max_j(\sum \frac{C_{jk}x^*{}_k}{x^*{}_j}) = \sigma(\mathbf{x}^*)\end{aligned}$$

which is the minimum. Thus we can find the minimum in S_1. Q.E.D.

"and...The argument of Theorem 1 shows that $\sigma(C^m x) \leq \sigma(Cx) \leq \sigma(x)$, and that if $\sigma(C^m x) \leq \sigma(\mathbf{1}), x > 0$, and $||x|| = 1$, then
$x_j \geq \gamma(C^m)/\sigma(\mathbf{1})$ for all j."

Proof: The proof of the first part of our statement is as follows:
We first show that $C\mathbf{x} - \sigma(\mathbf{x})\mathbf{x} \leq 0$. By definition,

$$\sigma(\mathbf{x}) \geq \frac{(Cx)_j}{x_j}$$

for all j. Therefore, for each j, $(C\mathbf{x} - \sigma(\mathbf{x}) * \mathbf{x})_j \leq (C\mathbf{x})_j - (C\mathbf{x})_j = 0$. Thus $\sigma(\mathbf{x})$ satisfies

$$(C\mathbf{x} - \sigma(\mathbf{x}) * \mathbf{x})_j \;\leq\; 0.$$

Now let J be the number for which

$$\sigma(\mathbf{x}) = \frac{(C\mathbf{x})_J}{\mathbf{x}_J};$$

that is, J is the j at which the expression

$$\frac{(Cx)_j}{x_j}$$

is maximal. If $\sigma(\mathbf{x})$ is not the minimum number that satisfies (?+2), then $S \equiv \sigma(\mathbf{x}) - E$ satisfies (?+2) for some positive amount E. However, $(C\mathbf{x} - S * \mathbf{x})_J = (C\mathbf{x})_J - \sigma(\mathbf{x}) * \mathbf{x}_J + E\mathbf{x}_J = (C\mathbf{x})_J - (C\mathbf{x})_J + E\mathbf{x}_J = E\mathbf{x}_J > 0$ since $\mathbf{x}_J$ is not zero from the definition of $\sigma(\mathbf{x})$ and the fact that

$$\sigma(\mathbf{x}) = \frac{(C\mathbf{x})_J}{\mathbf{x}_J}.$$

Thus our assumption that $\sigma(\mathbf{x})$ is not the minimum leads to the statement we have just disproved that S satisfies the inequality. Therefore, $\sigma(\mathbf{x})$ IS the minimum.

Next we prove that $C^2\mathbf{x} - \sigma(\mathbf{x})C\mathbf{x} \leq 0$. Since $\sigma(\mathbf{x})$ is a number, not a vector, we have $C\sigma(\mathbf{x}) = \sigma(\mathbf{x})C$. Thus the above equation follows from $C(C\mathbf{x} - \sigma(\mathbf{x})\mathbf{x}) \leq 0$.

Now we show that $\sigma(C\mathbf{x}) \leq \sigma(\mathbf{x})$. We substitute $C\mathbf{x}$ for $\mathbf{x}$ in Equation (?+2) and find that $\sigma(C\mathbf{x})$ is the minimum number σ such that $C^2\mathbf{x} - \sigma C\mathbf{x} \leq 0$. But by the equation last proven, $\sigma(\mathbf{x})$ is also a number σ that satisfies this inequality. Therefore, $\sigma(\mathbf{x})$ can not be smaller than the minimum σ above; that is, it can not be smaller than $\sigma(C\mathbf{x})$. Thus $\sigma(C\mathbf{x}) \leq \sigma(\mathbf{x})$ as required.

Finally we show that $\sigma(C^m\mathbf{x}) \leq \sigma(\mathbf{x})$. If we substitute $C\mathbf{x}$ for $\mathbf{x}$ in all equations in the last three paragraphs, we arrive at the inequality $\sigma(C^2\mathbf{x}) \leq \sigma(C\mathbf{x})$. By substituting $C^2\mathbf{x}$ for $\mathbf{x}$ we similarly obtain $\sigma(C^3\mathbf{x}) \leq \sigma(C^2\mathbf{x})$. We proceed in this manner until we get to $\sigma(C^m\mathbf{x}) \leq \sigma(C^{m-1}\mathbf{x})$. Now we combine these inequalities to form $\sigma(C^m\mathbf{x}) \leq \sigma(C^{m-1}\mathbf{x}) \leq ... \leq \sigma(C^2\mathbf{x}) \leq \sigma(C\mathbf{x}) \leq \sigma(\mathbf{x})$, which yields the result $\sigma(C^m\mathbf{x}) \leq \sigma(C\mathbf{x}) \leq \sigma(\mathbf{x})$.

STUCK FOR NOW To prove the second part of our statement, we assume that $\sigma(C^m\mathbf{x}) \leq \sigma(\mathbf{1}), \mathbf{x} > 0$, and $||\mathbf{x}|| = 1$. We show that $x_j \geq \gamma(C^m)/\sigma(\mathbf{1})$ for each j as follows: Since $\mathbf{x} > 0 and ||\mathbf{x}|| = 1$, there is a K such that $x_K = 1$. Then for any j, $\gamma(C^m) \leq C^m_{jK} = C^m_{jK}x_K$

$$\leq \sum_k C^m_{jk}x_k = \frac{\sum_k C^m_{jk}x_k}{x_j}x_j$$

$\leq \sigma(C^m\mathbf{x})x_j \leq \sigma(\mathbf{1})x_j$ from our above assumption. Dividing both sides by $\sigma(\mathbf{1})$ yields $\gamma(C^m)/\sigma(\mathbf{1}) \leq x_j$ for any j, and thus our result. Q.E.D.
ENDSTUCK

"Thus it suffices to look for the minimum of σ on the subset S_2 of those x in S_1 which satisfy (5)."

Proof: Let $\mathbf{x} \in S_1$ have the minimum $\sigma(\mathbf{x})$. Then $\sigma(C^m\mathbf{x}) \leq ... \leq \sigma(\mathbf{x}) \leq \sigma(\mathbf{1})$, so $\sigma(C^m\mathbf{x}) \leq \sigma(\mathbf{1})$. Thus $\mathbf{x} \in S_2$, so we need only search S_2 for the minimum. Q.E.D.

"which...if the minimum is attained at $v \in S_2$ and $\sigma(v)v - Cv \neq 0$, then $\sigma(C^m v) < \sigma(v)$, which contradicts the minimum property of v."

Proof: We start by proving that $\sigma(C^m\mathbf{v}) < \sigma(\mathbf{v})$. By the Equation (?+2) and our hypothesis that $\sigma(\mathbf{v})\mathbf{v}-C\mathbf{v} \neq 0$, there is a component k such that $\sigma(\mathbf{v})v_k-(C\mathbf{v})_k > 0$. Then by definition of σ,

$$(C\mathbf{v})_k < \sigma(\mathbf{v})v_k = \frac{(C\mathbf{v})_I}{v_I}v_k$$

where I is the component j which maximizes $(C\mathbf{v})_j/v_j$. We multiply each side by v_I to obtain $(C\mathbf{v})_k v_I < (C\mathbf{v})_I v_k$. By (?+2), for each component l, we have

$$(C\mathbf{v})_l \leq \sigma(\mathbf{v})v_l = \frac{(C\mathbf{v})_I}{v_I}v_l$$

so that $(C\mathbf{v})_l v_I \leq (C\mathbf{v})_I v_l$. Then

$$\begin{aligned} v_I(C^m * C\mathbf{v})_l &= v_I\sum_k C^m{}_{lk}(C\mathbf{v})_k \\ &= \sum_k C^m{}_{lk}(C\mathbf{v})_k v_I < \sum_k C^m{}_{lk}v_k(C\mathbf{v})_I \\ &= (C\mathbf{v})_I\sum_k C^m{}_{lk}v_k = (C\mathbf{v})_I(C^m\mathbf{v})_l \end{aligned}$$

where the inequality in the above set of equations results from $(C\mathbf{v})_k v_I < (C\mathbf{v})_I v_k$. In short, we just proved that $v_I(C^m * C\mathbf{v})_l < (C\mathbf{v})_I(C^m\mathbf{v})_l$; since $C^m * C = C^{m+1}$, we have $v_I(C^{m+1}\mathbf{v})_l < (C\mathbf{v})_I(C^m\mathbf{v})_l$. Dividing each side by $(C^m\mathbf{v})_l v_I$ yields

$$\frac{(C^{m+1}\mathbf{v})_l}{(C^m\mathbf{v})_l} < \frac{(C\mathbf{v})_I}{v_I}.$$

But since l is arbitrary and $C^{m+1} = C * C^m$, we have $\sigma(C^m\mathbf{v}) < \sigma(\mathbf{v})$, which contradicts the minimum property of $\mathbf{v}$ since $C^m\mathbf{v} > 0$, as $C^m > 0$ and $\mathbf{v} > 0$. Q.E.D.

To complete the proof of Theorem 2, we note that $C\xi = \lambda\xi$, or $C\xi_j = \lambda\xi_j$ for each j. Therefore, $\lambda = C\xi_j/\xi_j$ for each j. In other words, $\lambda = \sigma(\xi)$.

Also, since the hypothesis $\sigma(\mathbf{v})\mathbf{v} - C\mathbf{v} \neq 0$ resulted in a contradiction, Thus if $\mathbf{v}$ attains the minimum, then $C\mathbf{v} = \sigma(\mathbf{v})\mathbf{v}$, so $\sigma(\mathbf{v})$ is an eigenvalue with eigenvector $\mathbf{v}$. Since $\mathbf{v}$ is nonnegative, Property (c) of Perron's theorem [Gurari, p.122] tells us that $\sigma(\mathbf{v}) = \lambda$. By Property (b), $\mathbf{v}$ is (a scalar multiple of) ξ. Thus we conclude that

$$\lambda = \min_{\mathbf{x}>0}\sigma(\mathbf{x}) = \sigma(\xi)$$

which is Theorem 2. Q.E.D.

"Proof. The first estimate follows from $\lambda_C \geq \tau(\delta^{(j)})$, where $\delta^{(j)}$ is the vector with components δ_{jk} (= 1 for $k = j$ and 0 for all other k)."

Proof: By Theorem 1 [Gurari, p.123], for each component j $\lambda_C = \max(\tau\mathbf{x}) \geq \tau(\delta^{(j)})$. But the last term is just

$$\min_l \frac{\sum_k C_{lk}\delta^{(j)}{}_k}{\delta^{(j)}{}_l}$$

$$= \sum_k C_{jk}\delta^{(j)}{}_k = C_{jj}$$

Thus for each j, $\lambda_C \geq C_{jj}$ so that

$$\lambda_C \geq \max_j C_{jj}$$

. Q.E.D.

"vector...The second follows from $\tau(\mathbf{1}) \leq \lambda_C \leq \sigma(\mathbf{1})$."

Proof: By Theorem 1 [Gurari, p.123], $\lambda_C = \max(\tau\mathbf{x}) \geq \tau(\mathbf{1})$. On the other hand, Theorem 2 states that $\lambda_C = \min(\sigma\mathbf{x}) \leq \sigma(\mathbf{1})$. Thus $\tau(\mathbf{1}) \leq \lambda_C \leq \sigma(\mathbf{1})$. Therefore,

$$\begin{aligned} \lambda_C &\geq \tau(\mathbf{1}) = min_j(\sum_k C_{jk}\mathbf{1}_k/\mathbf{1}_j) \\ &= min_j(\sum_k C_{jk}). \end{aligned}$$

On the other hand,

$$\begin{aligned} \lambda_C &\leq \sigma(\mathbf{1}) = max_j(\sum_k C_{jk}\mathbf{1}_k/\mathbf{1}_j) \\ &= max_j(\sum_k C_{jk}), \end{aligned}$$

which completes the proof that

$$\min_j \sum_k C_{jk} \leq \lambda_C \leq \max_j \sum_k C_{jk}$$

and the last term is just the definition of ——C—— [Gurari, p.123]. Q.E.D.

<u>Page 126</u>

"If τ^*...
for any positive x and y."

Proof: First we show that λ_C, or λ for short, is an eigenvalue of C'; that is, $det(C' - \lambda I) = 0$. Indeed, since $\lambda > 0, \lambda$ is a real number, so $\lambda = \overline{\lambda}$, its complex conjugate. Therefore,

$$\begin{aligned} det(C' - \lambda I) &= det[(C - \overline{\lambda} I)'] = det(C - \overline{\lambda} I) \\ &= det(C - \lambda I) = 0 \end{aligned}$$

since λ is an eigenvalue of C. Thus every eigenvalue of C is an eigenvalue of C'. By substituting C' for C in the above proof and noting that $C'' = C$, we see that each eigenvalue of C' is also an eigenvalue of C. Therefore, λ_C is the maximal eigenvalue of both C and C', so by Theorems 1 and 2,

$$\begin{aligned} \tau(\mathbf{x}) &\leq \max(\tau) \leq \lambda_C \leq \min(\sigma^*) \\ &\leq \sigma^*(\mathbf{y}). \end{aligned}$$

Similarly,

$$\begin{aligned}\tau^*(\mathbf{y}) &\leq \max(\tau^*) \leq \lambda_C \leq \min(\sigma)\\ &\leq \sigma(\mathbf{x});\end{aligned}$$

that is, $\tau(\mathbf{x}) \leq \sigma^*(\mathbf{y})$ and $\tau^*(\mathbf{y}) \leq \sigma(\mathbf{x})$, as required. Q.E.D.

"Proof: Let η be the vector such that $\eta > 0, ||\eta|| = 1$, and $C'\eta = \lambda\eta$."

Remark: Since C is power-positive, so is C', so by Perron's theorem, the above conditions are satisfied for some vector η. By Property (b), η is the only vector that satisfies them.

"By property (b), the conclusion follows if $\mu = \lambda$."

Proof: We need only show that $\mathbf{z} > 0$. By the main part of Perron's theorem, λ has a positive eigenvector, its only eigenvector by property (b). But by the hypothesis of Theorem 3 and the fact that $\mu = \lambda, C\mathbf{z} = \mu\mathbf{z} = \lambda\mathbf{z}$. Thus $\mathbf{z}$ is the above eigenvector, so $\mathbf{z} > 0$. Q.E.D.

"By...If $\mu \neq \lambda$, then from $\eta * Cz = \lambda(\eta * z)$ we obtain $\eta * z = 0$."

Proof: Writing all dot products $x * y$ as inner products $< x, y >$, we have

$$\begin{aligned}\overline{\mu} < \eta, \mathbf{z} > &= < \eta, \mu\mathbf{z} >\\ &= < \eta, C\mathbf{z} >=< C'\eta, \mathbf{z} >\\ &= < \lambda\eta, \mathbf{z} >= \lambda < \eta, \mathbf{z} >\end{aligned}$$

so that $0 = (\overline{\mu} - \lambda) < \eta, \mathbf{z} >$. By taking the complex conjugate and noting that λ is real, we get

$$\begin{aligned}0 &= (\mu - \overline{\lambda})\overline{< \eta, \mathbf{z} >}\\ &= (\mu - \lambda)\overline{< \eta, \mathbf{z} >}\end{aligned}$$

which, together with $\mu \neq \lambda$, implies that $0 = \overline{< \eta, \mathbf{z} >}$ and thus $0 =< \eta, \mathbf{z} >$. Q.E.D.

$$\text{“}R(\eta * z) = \eta_1 z_1 + \sum_2^n \eta_j R(z_j) > 0\text{''}:$$

Note: $R(\eta * z)$ denotes the real part of the complex number $(\eta * z)$.

Proof: Since $\eta > 0$ and $z_1 = 1$, the left half (the equality) of the above statement is true. To prove the right half (the inequality), we notice that $\eta_1 z_1 > 0$. Also,

$|argz_j| \leq \pi/2$ for $j > 1$ means that each component z_j lies in the right half of the complex plane, so for each $j, R(z_j) \geq 0$ and thus $\eta_j R(z_j) \geq 0$. This proves the inequality. Q.E.D.

"which is a contradiction."

Proof: What we just proved, that $R(\eta * \mathbf{z}) > 0$, contradicts the earlier assumption that $\eta * \mathbf{z} = 0$, upon which was based the assumption that $\mu \neq \lambda$. Therefore, $\mu = \lambda$. The above statement "By property (b), the conclusion follows if $\mu = \lambda$" completes the proof of Theorem 3. Q.E.D.

"Then we have

$$\begin{aligned} \tau(x) &= m(Cx;x), \sigma(x) = M(Cx;x), \\ minx_j &= m(\lambda;\mathbf{1}), \max m_j = M(x;\mathbf{1})." \end{aligned}$$

Remark: Another innocent error. We replace λ with x.

Proof: $m(C\mathbf{x};\mathbf{x})$ is the upper bound of m's that satisfy $m\mathbf{x} \leq C\mathbf{x}$; that is, $C\mathbf{x} - m\mathbf{x} \geq 0$. But from the first equation of [Gurari, p.123], this upper bound is none other than $\tau(\mathbf{x})$. Therefore $\tau(\mathbf{x}) = m(C\mathbf{x};\mathbf{x})$.

The proof of the next statement is similar. $M(C\mathbf{x};\mathbf{x})$ is the lower bound of M's such that $C\mathbf{x} \leq M\mathbf{x}$; that is, $C\mathbf{x} - M\mathbf{x} \leq 0$. But from () this lower bound is exactly $\sigma(\mathbf{x})$. Therefore $\sigma(\mathbf{x}) = M(C\mathbf{x};\mathbf{x})$.

Now we show that $min\mathbf{x}_j = m(\mathbf{x};\mathbf{1})$. Since every component of $\mathbf{1}$ is 1, we have

$$\begin{aligned} m(\mathbf{x};\mathbf{1}) &= \max_m(m\mathbf{1} \leq \mathbf{x}) \\ &= \max_m((\forall j)m * 1_j \leq x_j) = \max_m((\forall j)m \leq x_j) \end{aligned}$$

which is obviously $minx_j$.

The proof of the last statement is like the above.

$$\begin{aligned} M(\mathbf{x};\mathbf{1}) &= \min_M(\mathbf{x} \leq M\mathbf{1}) \\ &= \min_M((\forall j)x_j \leq M * 1_j) = \min_M((\forall j)x_j \leq M) \end{aligned}$$

which is clearly $maxx_j$. Q.E.D.

$$\text{"}max_{x,y>0}\theta(Cx,Cy) = logT_C = \Delta''_C$$

Proof: We observe that

$$m(C\mathbf{x};C\mathbf{y}) = \max_m(mC\mathbf{y} \leq C\mathbf{x})$$

$$\begin{aligned} &= \max_m((\forall k) m * (C\mathbf{y})_k \leq (C\mathbf{x})_k) \\ &= \max_m((\forall k) m \leq \frac{(C\mathbf{x})_k)}{(C\mathbf{y})_k} \\ &= \min_k \frac{(C\mathbf{x})_k)}{(C\mathbf{y})_k} \end{aligned}$$

and

$$\begin{aligned} M(C\mathbf{x}; C\mathbf{y}) &= \min_M(C\mathbf{x} \leq MC\mathbf{y}) \\ &= \min_M((\forall j)(C\mathbf{x}_j \leq M * (C\mathbf{y})_j) \\ &= \min_M((\forall j)\frac{(C\mathbf{x})_j)}{(C\mathbf{y})_j} \leq M) \\ &= \max_j \frac{(C\mathbf{x})_j)}{(C\mathbf{y})_j}. \end{aligned}$$

Then

$$max_{x,y>0}\theta(Cx, Cy) \tag{6}$$

$$= \sup_{x,y>0} \log(\frac{M(Cx;Cy)}{m(Cx;Cy)} \tag{7}$$

$$= \log \sup_{x,y>0} \frac{M(Cx;Cy)}{m(Cx;Cy)} \tag{8}$$

$$= \log \sup_{x,y>0} \frac{\max_j (Cx)_j/(Cy)_j}{\min_r (Cx)_r/(Cy)_r} \tag{9}$$

$$= \log \sup_{x,y>0} \max_{j,r} \frac{(Cx)_j/(Cy)_j}{(Cx)_r/(Cy)_r} \tag{10}$$

$$= \log \sup_{x,y>0} \max_{j,r} \frac{(Cx)_j/(Cx)_r}{(Cy)_j/(Cy)_r} \tag{11}$$

$$= \log \max_{j,r} \sup_{x,y>0} \frac{(Cx)_j/(Cx)_r}{(Cy)_j/(Cy)_r} \tag{12}$$

$$= \log \max_{j,r} \max_{k,s} \frac{C_{jk}/C_{js}}{C_{rk}/C_{rs}} \tag{13}$$

$$= \log \max_{j,k,r,s} (C_{jk}C_{rs})/(C_{rk}C_{js}) = logT_C. \tag{14}$$

Equation (2nd) above follows from (1st) since log commutes with sup, as log is a monotone increasing function. Equation (3rd) is derived from the above statements about $m(Cx, Cy)$ and $M(Cx, Cy)$. To obtain (6th) from (5th) need only to show that max and sup commute: Let $f(j, r, x, y)$ be the function whose max and sup we evaluate. Given x and y, we define

$$N(x, y) \equiv \max_{j,r} f(j, r, x, y).$$

Similarly, given j and r, we let

$$s(j, r) \equiv \sup_{x,y>0} f(j, r, x, y)$$

and call

$$S \equiv \max_{j,r} s(j,r).$$

We now show that

$$\sup_{x,y>0} \max_{j,r} f(j,r,x,y) = \max_{j,r} \sup_{x,y>0} f(j,r,x,y);$$

that is,

$$\sup_{x,y>0} N(x,y) = S.$$

In other words, we prove that:
(1) $(\forall x, y > 0) N(x,y) \leq S$, and
(2) $(\forall \delta)(\exists x_\delta, y_\delta) s.t. N(x_\delta, y_\delta) + \delta > S$. The proof of (1) is easy: given $x, y > 0$, let j_0, r_0 be the j, r that maximize $f(j,r,x,y)$. Then by definition, $N(x,y) = f(j_0, r_0, x, y) \leq s(j_0, r_0) \leq S$. To prove (2), given δ, we let $\epsilon \equiv \delta$, $x_\delta \equiv x_\epsilon$, $y_\delta \equiv y_\epsilon$, and $f(j_1, r_1, x_\epsilon, y_\epsilon) + \epsilon > s(j_1, r_1) = S$ where $s(j_1, r_1)$ maximizes $s(j,r)$. Then

$$\begin{aligned} N(x_\delta, y_\delta) + \delta &= N(x_\epsilon, y_\epsilon) + \epsilon \\ &= \max_{j,r} f(j,r,x_\epsilon,y_\epsilon) + \epsilon \\ &\geq f(j_1, r_1, x_\epsilon, y_\epsilon) + \epsilon > s(j_1, r_1) = S, \end{aligned}$$

which proves (6th).

To arrive at (7th), given j and r we let L be the index m that maximizes C_{jm}/C_{rm}, and let l minimize C_{jm}/C_{rm}. That is, for all indices m, $C_{jl}/C_{rl} \leq C_{jm}/C_{rm} \leq C_{jL}/C_{rL}$. Then

$$(\forall m) \frac{C_{jl}}{C_{rl}} \leq \frac{C_{jm}}{C_{rm}}.$$

Multiplying both sides by C_{rl} yields

$$(\forall m) C_{jl} \leq \frac{C_{jm}}{C_{rm}} C_{rl}.$$

Now we prove that for all $x > 0$,

$$\frac{C_{jl}}{C_{rl}} \leq \frac{(Cx)_j}{(Cx)_r} \leq \frac{C_{jL}}{C_{rL}}$$

where $(Cx)_j = C_{j1}x_1 + C_{j2}x_2 + \ldots + C_{jn}x_n = \sum C_{jm}x_m$. After cross-multiplying, we rewrite the left half of (st) as

$$C_{jl} \sum_m C_{rm}x_m \leq C_{rl} \sum_m C_{jm}x_m.$$

But using the above expression of C_{jl} yields

$$\begin{aligned} C_{jl} \sum_m C_{rm}x_m &= \sum_m C_{jl}C_{rm}x_m \\ &\leq \sum_m \left(\frac{C_{jm}}{C_{rm}} C_{rl}\right) C_{rm}x_m \\ &= \sum_m C_{jm}C_{rl}x_m = C_{rl} \sum_m C_{jm}x_m \end{aligned}$$

which proves the left half. Similarly for L,

$$(\forall m)\frac{C_{jL}}{C_{rL}} \geq \frac{C_{jm}}{C_{rm}}.$$

Multiplying both sides by C_{rL} gives

$$(\forall m)C_{jL} \geq \frac{C_{jm}}{C_{rm}}C_{rL}.$$

After cross-multiplying, the right half of (st) becomes

$$C_{jL}\sum_m C_{rm}x_m \geq C_{rL}\sum_m C_{jm}x_m.$$

We use the expression of C_{jL} to obtain

$$\begin{aligned} C_{jL}\sum_m C_{rm}x_m &= \sum_m C_{jL}C_{rm}x_m \\ &\geq \sum_m (\frac{C_{jm}}{C_{rm}}C_{rL})C_{rm}x_m \\ &= \sum_m C_{jm}C_{rL}x_m = C_{rL}\sum_m C_{jm}x_m \end{aligned}$$

which completes the proof of the right inequality. Therefore, (st) is true, so

$$\sup_x((Cx)_j/(Cx)_r) = C_{jL}/C_{rL}$$

and

$$\begin{aligned} \sup_y \frac{1}{(Cy)_j/(Cy)_r} &= \frac{1}{\inf_y (Cy)_j/(Cy)_r} \\ &= C_{jl}/C_{rl}. \end{aligned}$$

Thus

$$\begin{aligned} \sup_{x,y>0} \frac{(Cx)_j/(Cx)_r}{(Cy)_j/(Cy)_r} & \\ &= \frac{C_{jL}/C_{rL}}{C_{jl}/C_{rl}} \\ &= \frac{\max_k C_{jk}/C_{rk}}{\min_s C_{js}/C_{rs}} \\ &= \max_{k,s} \frac{C_{jk}/C_{rk}}{C_{js}/C_{rs}} \\ &= \max_{k,s} \frac{C_{jk}/C_{js}}{C_{rk}/C_{rs}} \end{aligned}$$

which proves (7th). Q.E.D.

Page 127

$$\text{“}\max\theta(Cx,Cy)/\theta(x,y) = \frac{T_C^{1/2}-1}{T_C^{1/2}+1}\text{..”}$$

Proof: See “Extensions of Jentzsch's Theorem” by Garrett Birkhoff.

[Bir], p.219

For now we use the notation in [Bir].

“Since...$\theta(f,g)$ is invariant under
all projective transformations mapping the interval $0 < f_2/f_1 < +\infty$ onto itself.”

Proof: In other words, we must show that for any scalar t $\theta(tf,tg) = \theta(f,g)$. The proof is easy: let $f \equiv (f1, f2)$ represent the quotient $f2/f1$ and g represent $g2/g1$. Then

$$\begin{aligned}\theta(tf,tg) &= |Ln(\frac{tf2/f1}{tg2/g1})| = |Ln(\frac{f2/f1}{g2/g1})| \\ &= \theta(f,g).\end{aligned}$$

Q.E.D.

[Bir], p.220

“of... $y \to 1/y$ is an isometry for $\theta(y,y')$...”:

Proof: We must show that $y \to 1/y$ is an isometry for θ; that is, $\theta(1/y,1/y') = \theta(y,y')$, where this y' does NOT mean the derivative of y. But

$$\begin{aligned}\theta(1/y,1/y') &= |ln\frac{1/y'}{1/y}| = |ln(y/y')| = |-ln(y/y')| \\ &= |ln(y'/y)| = \theta(y,y')\end{aligned}$$

where we have used the fact that $|z| = |-z|$ for $z = ln(y/y')$. Q.E.D.

“order....we can assume $x > x'$ implies $xP \geq x'P$.”

Proof: This statement says that since $\theta(y',y) = \theta(1/y,1/y') = \theta(y,y')$, we can assume WITHOUT LOSS OF GENERALITY that P is such that $x > x'$ implies $xP \geq x'P$. To prove this contention, we observe that the only relevance of P in our final result is in the quotient $\theta(fP,gP)/\theta(f,g)$, where $f = f_2/f_1$ and $g = g_2/g_1$. By the above fact that $\theta(y',y) = \theta(y,y')$, we substitute $y' = fP$ and $y = gP$ to obtain

$$\frac{\theta(fP,gP)}{\theta(f,g)} = \frac{\theta(gP,fP)}{\theta(f,g)},$$

so the result is the same whether or not $f > g$ implies $fP \geq gP$. Q.E.D.

"$0P = b/d$, clearly b and d have the same sign...":

Proof: $0P = P(0) = (a * 0 + b)/(c * 0 + d) = b/d$. To show that b and d have the same sign, we use Birkhoff's previous statement that P "maps $0 < x < \infty$ onto a proper subinterval of $0 < y < \infty$" (top of p.220). Since P is a projective transformation, we may assume a, b, c, and d are all nonzero, so P is continuous at $x = 0$. Then P maps 0 to a point on the lower boundary of the above subinterval. This point, which is b/d as shown above, is therefore zero or greater. Since b and d are nonzero, $b/d > 0$, so that b and d have the same sign. Q.E.D.

"$0P = b/d$....since $xP = 0$ has no positive
solution, b and a have the same sign":

Proof: If there were a positive solution of $xP = 0$, or a value x such that $P(x) = 0$, then this x would contradict the fact that we have chosen P as a projective transformation "mapping the interval $0 < f_2/f_1 < \infty$ onto itself" (bottom of p.219).

Now suppose that b and a have different signs. Then $b/a < 0$, so that $-b/aispositiveandthusinthed$
b/a) = 0/(-cb/a + d) = 0, $whichcontradictsthefactthat$xP = $0hasnopositivesolution.Wehaveassumed$

$$= \frac{ax + b}{cx + d} = \frac{ax + b}{cx + cb/a} = a/c$$

for all x, and thus P would not be a projection. Q.E.D.

"solution...since $xP = \infty$ has no positive solution,
d and c have the same sign. Hence we can assume a, b, c, d *positive* in (3)."

Proof: Any positive value x such that $P(x) = \infty$ would, as above, contradict our choice of P as a projective transformation "mapping the interval $0 < f_2/f_1 < \infty$ onto itself" (bottom of p.219).

Now suppose that d and c have different signs. Then $d/c < 0$, so that $-d/c$ is positive and thus in the domain of P. But $P(-d/c) = (-ad/c + b)/0 = \infty$, which contradicts the fact that $xP = \infty$ has no positive solution. We have assumed that b is not ad/c, for we would otherwise have

$$P(x) = \frac{ax + b}{cx + d} = \frac{ax + ad/c}{cx + d} = a/c$$

for all x, so that P would not be a projection.

We have shown above that a, b, c, and d all have the same sign. If they are all negative, then let a', b', c', and d' be equal to $-a, -b, -c$, and $-d$, so that a', b', c', d' are all positive. Then

$$P(x) \quad = \quad \frac{ax + b}{cx + d} = \frac{-(ax + b)}{-(cx + d)}$$

$$= \frac{(-a)x + (-b)}{(-c)x + (-d)} = \frac{a'x + b'}{c'x + d'}.$$

Therefore, we can use the positive parameters a', b', c', d' instead of the negative ones a, b, c, d. By removing the $'$ mark from $a', b', c', d' > 0$ we see that we can assume $a, b, c, d > 0$. Q.E.D.

"(4) $\qquad y' = dy/dx = (ad - bc)/(cx + d)^2$":

Proof: Here y' stands for the derivative of y. In Equation (3) of [Bir], $y(x) = (ax + b)/(cx + d)$. Taking dy/dx yields the above result. Q.E.D.

"we can assume $ad > bc$."

Proof: Above we proved that $x > x'$ implies $P(x) \geq P(x')$; thus $P(x)$ (that is, $y(x)$) increases as x grows, so $dy/dx > 0$. From (4) of [Bir] and the fact that $c, d > 0$, it follows that $(ad - bc) > 0$; i.e. $ad > bc$. Q.E.D.

"We now consider...$d\theta(y)/d\theta(x)$
$= xdy/ydx$. By (3) and (4), this is $d\theta(y)/d\theta(x) = (ad - bc)x/(ax + b)(cx + d)$."

Proof: Let $\theta(y) \equiv \theta(y, z) = |lny - lnz|$ where z is a constant less than x and y, where x is as below. The derivative is $d\theta(y)/dy = 1/y$, so that $d\theta(y) = dy/y$. Similarly, $d\theta(x) = dx/x$, yielding $d\theta(y)/d\theta(x) = xdy/ydx$, the first equation. It readily leads to the final equation. Q.E.D.

"By the differential calculus, the *maximum* of this (i.e., the minimum of its reciprocal) occurs when $ac = bdx^{-2}$, or $x = (bd/ac)^{1/2}$."

Proof: We minimize the reciprocal which is $(ax+b)(cx+d)/(ad-bc)x$. Since we can assume $ad > bc$, $(ad-bc)$ is a positive constant. Then the problem reduces to minimizing $(ax+b)(cx+d)/x = acx + (bc+ad) + bd/x$. We set its derivative to zero; that is, $0 = ac - bdx^{-2}$ from which we get $x = (bd/ac)^{1/2}$ as a critical point. Since this x is positive, the second derivative $2bdx^{-3}$ is also positive. Therefore, x is the minimum of $(ax+b)(cx+d)/x$ and thus the maximum of our desired function $d\theta(y)/d\theta(x)$. Q.E.D.

"simplifying, we see that P *contracts* all hyperbolic distances, by a factor whose supremum is

$$(5) \qquad N(P) = \frac{\nu - 1}{\nu + 2(\nu)^{1/2} + 1},$$

where $\nu = (ad/bc) > 1$."

Proof: $N(P)$ is defined in (6) of [Bir] as $sup[\theta(fP, gP)/\theta(f, g)]$. We can replace fP, gP, f, g by y, y', x, x' respectively. Then $N(P) = sup[\theta(y, y')/\theta(x, x')] = sup[d\theta(y)/d\theta(x)]$ which is merely the maximum value of $d\theta(y)/d\theta(x) = (ad-bc)x/(ax+$

$b)(cx+d)$. Now we divide by $(ad-bc)$ to obtain

$$\frac{d\theta(y)/d\theta(x)}{ad-bc} = x/(ax+b)(cx+d)$$

and take the reciprocal to get

$$\begin{aligned}\frac{ad-bc}{d\theta(y)/d\theta(x)} &= (ax+b)(cx+d)/x \\ &= (a+\frac{b}{x})(cx+d).\end{aligned}$$

Now we substitute $x = (bd/ac)^{1/2}$, at which $d\theta(y)/d\theta(x)$ attains its maximum, into the above equation as follows:

$$\begin{aligned}\frac{ad-bc}{d\theta(y)/d\theta(x)} &= (a+\frac{b}{x})(cx+d) \\ &= (a+(\frac{acb}{d})^{1/2})((\frac{bdc}{a})^{1/2}+d) \\ &= (abcd)^{1/2}+bc+ad+(abcd)^{1/2} = ad+2(abcd)^{1/2}+bc \\ &= ((ad)^{1/2}+(bc)^{1/2})^2 = bc((\frac{ad}{bc})^{1/2}+1)^2 \\ &= bc(\nu^{1/2}+1)^2\end{aligned}$$

where $\nu \equiv (ad/bc)$ as in Equation (5) of [Bir]. Taking the reciprocal yields

$$\frac{d\theta(y)/d\theta(x)}{ad-bc}\frac{1}{bc(\nu^{1/2}+1)^2},$$

and multiplying by $(ad-bc)$ gives us

$$\begin{aligned}d\theta(y)/d\theta(x) &= \frac{ad-bc}{bc(\nu^{1/2}+1)^2} \\ &= \frac{ad/bc-1}{(\nu^{1/2}+1)^2} = \frac{\nu-1}{(\nu^{1/2}+1)^2} \\ &= \frac{\nu-1}{\nu+2(\nu)^{1/2}+1}\end{aligned}$$

as required. Since we have assumed that $ad > bc$, and that a, b, c, d are all positive, we get $\nu = (ad/bc) > 1$ which completes the proof of Equation (5) of [Bir]. Q.E.D.

"(6') $N(PP') \leq N(P)N(P')$."

Proof: P' is another projection. From (6) of [Bir],

$$\begin{aligned}N(PP') &= sup_{f,g}[\theta(fPP',gPP')/\theta(f,g)] \\ &= sup_{f,g}([\theta(fPP',gPP')/\theta(fP,gP)] * [\theta(fP,gP)/\theta(f,g)]) \\ &\leq sup_{f,g}[\theta(fPP',gPP')/\theta(fP,gP)] * \sup_{f,g}[\theta(fP,gP)/\theta(f,g)]\end{aligned}$$

$$
\begin{aligned}
&= sup_{q\equiv fP, r\equiv gP}[\theta(qP', rP')/\theta(q,r)] * \sup_{f,g}[\theta(fP, gP)/\theta(f,g)] \\
&\leq sup_{q,r}[\theta(qP', rP')/\theta(q,r)] * \sup_{f,g}[\theta(fP, gP)/\theta(f,g)] \\
&= N(P')N(P) = N(P)N(P').
\end{aligned}
$$

The reason for the last inequality is as follows: The supremum over $q \equiv fp, r \equiv gP$ only covers those quantities q and r that are the values of $P(f)$ and $P(g)$ for some f and g. However, the supremum over q, r covers *all* values q and r, and is thus equal to or greater than the other supremum. Q.E.D.

$$\text{“}(7) N(P) = \frac{\nu^{1/4} - \nu^{1/4}}{\nu^{1/4} + \nu^{1/4}} = \frac{2\sinh(\lambda/4)}{2\cosh(\lambda/4)} = \tanh\frac{\lambda}{4}.\text{”}$$

The first fraction in the above equation contains an innocent error; we replace it with

$$\frac{\nu^{1/4} - \nu^{-1/4}}{\nu^{1/4} + \nu^{-1/4}}.$$

Proof: From (5) of [Bir],

$$
\begin{aligned}
N(P) &= \frac{\nu - 1}{\nu + 2(\nu)^{1/2} + 1} \\
&= \frac{(\nu^{1/2} + 1)(\nu^{1/2} - 1)}{(\nu^{1/2} + 1)(\nu^{1/2} + 1)} \\
&= \frac{\nu^{1/2} - 1}{\nu^{1/2} + 1} \\
&= \frac{\nu^{1/4} - \nu^{-1/4}}{\nu^{1/4} + \nu^{-1/4}}
\end{aligned}
$$

as required, where we obtain the last equation from its preceding one by dividing the numerator and denominator by $\nu^{1/4}$. Now the definition $\lambda = \ln(ad/bc) = \ln(\nu)$ yields $\nu = e^{\lambda}$, so that $\nu^{1/4} = e^{\lambda/4}$ and thus

$$N(P) = \frac{2\sinh(\lambda/4)}{2\cosh(\lambda/4)} = \tanh\frac{\lambda}{4}$$

using the definitions of sinh, cosh, and tanh. Q.E.D.

“ **3. Convex cones.** Now let C be...”:

Now we prepare to prove the statement on the top of [Gur], Page 127. To avoid confusion, we replace C of [Bir] by R, L by R^n, and P by C. This equation uses only vectors $\mathbf{x}$ and $\mathbf{y}$ whose components satisfy $x_j \geq 0$ and $y_j \geq 0$ for all j. Therefore, we let R be the set of vectors $\mathbf{x}$ for which $0 \leq x_j \leq M$ for some large constant M. We let H be the hyperplane (analogy of a plane in $n-1$ dimensions) given in Example 1 of [Bir], p.221.

"Since R is a bounded closed convex set, *every line intersects R in a closed segment.*"

This statement is NOT true because of the definition of H. Rather, we apply the above statement as follows: The line $l(f,g)$ that connects points f and g, which lies in H, intersects R in a closed segment. This segment can be mapped onto $0 \le x \le \infty$; that is, to part or all of $0 \le x \le \infty$.

"mapped...so that $fA < gA$ by a *unique* affine transformation A."

Proof: Let a and b be the two places where $l(f,g)$ intersects the boundary of R. Let a be the one closer to f than to g, so the points appear in the order a, f, g, b or b, g, f, a. For any point u on $l(f,g)$, let $uA = A(u) \equiv d(u,a)/d(u,b)$ where $d(p1,p2)$ denotes the distance between the points $p1$ and $p2$. Then $0 = A(a) \le A(f) < A(g) \le A(b) = \infty$; that is, $0 = aA \le fA < gA \le bA = \infty$.

A is affine in the following sense: let $B(u-a) \equiv A(u)$. Then B maps from the line segment $[0, b-a]$ to $0 \le x \le \infty$, and is thus in a way a "linear" function. Then A is affine since its argument u is merely a translation by a of the argument $(u-a)$ of $B(u-a)$.

Now we prove that A is unique among affine functions such that $fA < gA$. Suppose A' is another affine transformation that maps $l(f,g)$ to $[0,\infty]$ and $fA' < gA'$. Since the points are ordered a,f,g,b or b,g,f,a, the requirement $fA' < gA'$ implies $A'(a) = 0, A'(b) = \infty$. If we define $B'(u-a) \equiv A'(u)$, then B' maps from $[0, b-a]$ to $0 \le x \le \infty$. Since only one "linear" function can do this, it follows that $B' = B$ and thus $A' = A$. Therefore A is unique. Q.E.D.

(CHECK WITH PROF. SIROVICH)

[Bir], p.221

We skip to the bottom of this page, and prove that $N(C;R)$ as defined in (9) is the same as the N_C on the top of [Gur], p.127 as follows:

Our main task is to prove that $\theta(f,g;R)$ (replacing C in [Bir] by R) is the same as $\theta(x,y)$ of [Gur]. Since $\theta(f,g)$ is an absolute value as defined in (2) of [Bir] we can, without loss of generality (WLOG), take the points in the order a,f,g,b, where $f \equiv (1-S)a + Sb$ and $g \equiv (1-T)a + Tb$, where $0 \le S < T \le 1$. Note that $S < T$ stems from the order a,f,g,b.

Now we show

$$\frac{T/s}{(1-T)/(1-S)} = \frac{\max_i g_i/f_i}{\min_i g_i/f_i}$$

so we can derive the fifth line of the equations below. If the line a,f,g,b lies along the boundary, then each point has a zero component, so that f and g are not positive. That is to say, in [Gur]'s notation, that x and y are not positive, which contradicts the assumption made in the definition of $\theta(x,y)$ ([Gur], bottom of p.126).

Thus we may assume that the line a,f,g,b does NOT lie along the boundary. Since a and b are boundary points of R, we may define $a \equiv (0, a_2, ..., a_n)$ and

$b \equiv (b_1, 0, b_3, ..., b_n)$. Here we have WLOG defined the first two coordinates to correspond to the zero components of a and b respectively.

The above definition of f and g yields $f_1 = Sb_1, f_2 = (1-S)a_2$, and $f_i = (1-S)a_i + Sb_i$ for all $i \geq 3$. g is the same, with the S's replaced by T's. Thus $g_1/f_1 = T/S, g_2/f_2 = (1-T)/(1-S)$, and

$$g_i/f_i = \frac{(1-T)a_i + Tb_i}{(1-S)a_i + Sb_i}$$

for $i \geq 3$, where we note that $a_i, b_i \geq 0$ for all i.

To derive the fifth line below, we only need to prove that $maxg_i/f_i = f_1/f_1 = T/S$ and $ming_i/f_i = f_2/f_2 = (1-T)/(1-S)$; that is, for all $i \geq 3$,

$$\frac{1-T}{1-S} \leq \frac{(1-T)a_i + Tb_i}{(1-S)a_i + Sb_i} \leq \frac{T}{S}.$$

To prove the first inequality above, we assume it is false and cross-multiply both sides to obtain $(1-T)[(1-S)a_i + Sb_i] > (1-S)[(1-T)a_i + Tb_i]$. Next we subtract $(1-T)(1-S)a_i$ from both sides to get $(1-T)Sb_i > (1-S)Tb_i$. We then add TSb_i to get $Sb_i > Tb_i$. If $b_i = 0$ then we obtain $0 > 0$, a contradiction. Otherwise we divide both sides by b_i to yield $S > T$, a contradiction of our above premise that $S < T$. The proof of the second inequality is similar. Assuming it false, we cross- multiply both sides to get $S[(1-T)a_i + Tb_i] > T[(1-S)a_i + Sb_i]$. Subtracting STb_i yields $S(1-T)a_i > T(1-S)a_i$; adding STa_i gives $Sa_i > Ta_i$. Now either $a_i = 0$ which yields the false statement $0 > 0$, or $a_i \neq 0$ in which case we divide by a_i and wind up with $S > T$, a contradiction.
Therefore, both inequalities hold and the fifth line follows from the fourth.

Then

$$\begin{aligned}
\theta(f,g;R) &= \theta(fA, gA) = \theta(A(f), A(g)) \\
&= \theta(\frac{d(f,a)}{d(f,b)}, \frac{d(g,a)}{d(g,b)}) \\
&= \theta(\frac{S}{1-S}, \frac{T}{1-T}) \\
&= \ln(\frac{T/s}{(1-T)/(1-S)} \\
&= \ln(\frac{\max_i g_i/f_i}{\min_i g_i/f_i} \\
&= \ln(\frac{\max_i y_i/x_i}{\min_i y_i/x_i} \\
&= \ln(\frac{(1/\min_i x_i/y_i)}{(1/\max_i x_i/y_i)} \\
&= \ln(\frac{\max_i x_i/y_i}{\min_i x_i/y_i} \\
&= \ln(M(x;y)/m(x;y)) = \theta(x,y).
\end{aligned}$$

In the sixth line of the above equations, we replace [Bir]'s notation f, g with [Gur]'s x, y. In the final line, we use the identities

$$M(x;y) = \max_i x_i/y_i, m(x;y) = \min_i x_i/y_i.$$

By substituting fC for f and gC for $g, \theta(fC, gC; R)$ of [Bir] is identical to $\theta(Cx, Cy)$ of [Gur]. Q.E.D.

" LEMMA 1. *If the transform CP of C under P has finite diameter* Δ *under* $\theta(f, g; C)$, *then*

(9a) $$N(P, C) = \tanh(\Delta/4) < 1."$$

This lemma is proved below. Once it is proven, [Gur]'s statement follows as shown after the proof of the lemma.

[Bir], p.222

Proof of LEMMA 1: We recall that f and g lie on the segment $s(a, b)$.

"image ... $s(aC, bC) \leq RC$: hence $\theta(aC, bC; R) \leq \Delta$.

Proof: If y is a point on the image segment $s(aC, bC)$, then $y = Cx$ for some x on the segment $s(a, b)$ between a and b. Since $a, b \in R$, the set of all nonnegative vectors, for each component $j, a_j \geq 0$ and $b_j \geq 0$. Also, for each j, either $a_j \geq x_j \geq b_j$ or $b_j \geq x_j \geq a_j$. Thus $x_j \geq 0$ for each j, so that $x \in R$. Thus $y \in RC$ which has been defined as the set $\{z : z = Cx \text{ for some } x \in R\}$. Therefore, $s(aC, bC) \subset RC$, which I think is the meaning of the statement $s(aC, bC) \leq RC$ quoted above.

It is clear that

$$\theta(aC, bC; R) \leq \sup_{y,z} \theta(yC, zC; R)$$

which is the definition of *Delta*. Q.E.D.

" $$\theta(fC, gC; R)/\theta(f, g; R) \leq \tanh(\Delta/4)."$$

Proof: Dropping the notation R, we obtain

$$\begin{aligned} \theta(fC; gC; R)/\theta(f, g; R) &\leq \sup[\theta(fC, gC), \theta(f, g)] \\ &= N(C) = \tanh\frac{\lambda}{4} = \tanh(\Delta/4). \end{aligned}$$

On the last line above, both equalities are derived from (6) and (7) of [Bir] respectively. The final equality results since both λ and Δ have the same definition. (see PROF. SIROVICH) Q.E.D.

Notes (to finish the proof of LEMMA 1): We show as follows that

$$\tanh(\Delta/4) \leq N(P, C) = \sup_{f,g} \theta(Cf, Cg)/\theta(f, g):$$

For any small quantity ϵ, we must prove that there is some f, g such that $\tanh(\Delta/4) < \theta(Cf, Cg)/\theta(f, g) + \epsilon$. For each $n \in 1, 2, ...$ we let $\epsilon \equiv 2^{-n}$ and let f_n, g_n be inverse images of x_n, y_n respectively on segment (c_n, d_n) of length a bit over $\Delta - 2^{-n}$. We prove below that f_n, g_n are the points f, g that satisfy the above inequality. LATER (see PROF. SIROVICH)

Q.E.D. for LEMMA 1.

Proof of [Gur]'s statement: To avoid confusion, we replace C of [Bir], (9) and (9a), by R and P by C, so that (9) now reads $N(C; R) = sup[\theta(fC, fC; R)/\theta(f, g; R)]$ and (9a) becomes $N(C, R) = \tanh(\Delta/4) < 1$. The Δ_C defined on [Gur], Page 126, as

$$max_{x,y>0}\theta(Cx, Cy) = logT_C = \Delta_C,$$

from which $T_C = e^{\Delta_C}$. Then Δ_C is in fact the diameter Δ of the transform RC. Thus, from Lemma 1,

$$\begin{aligned}
\max\theta(C\mathbf{x}, C\mathbf{y}) &= N(C, R) = \tanh(\Delta/4) \\
&= \tanh(\Delta_C/4) \\
&= \frac{e^{\Delta_C/4} - e^{-\Delta_C/4}}{e^{\Delta_C/4} + e^{-\Delta_C/4}} \\
&= \frac{e^{\Delta_C/2} - 1}{e^{\Delta_C/2} + 1} \\
&= \frac{T_C^{1/2} - 1}{T_C^{1/2} + 1}
\end{aligned}$$

as required. Q.E.D.

Back to [Gur]:

"$M(C^{m+1}x - \tau(x)C^m x; C^m x) \leq Tm(C^{m+1}x - \tau(x)C^m x; C^m x)$."

Proof: From the definitions on Page 126,

$$\begin{aligned}
\log T &= \log T(C^m) = \sup_{x,y>0} \theta(C^m x, C^m y) \\
&= \sup_{x,y>0} \log(M(C^m x; C^m y)/m(C^m x; C^m y)) \\
&= \log(\sup_{x,y>0}(M(C^m x; C^m y)/m(C^m x; C^m y)).
\end{aligned}$$

Taking the exponent of both sides, we see that

$$T = \sup_{x,y>0}(M(C^m x; C^m y)/m(C^m x; C^m y)).$$

In other words, for any $x, y > 0, T \geq M(C^m x; C^m y)/m(C^m x; C^m y)$. Now we change the notation in the above statement by replacing x with $Cx - \tau(x)x$ and y with x, where x is as on Page 127. We substitute the change in the above inequality to obtain

$$\begin{aligned} T &\geq \frac{M(C^m(Cx-\tau(x)x);C^mx)}{m(C^m(Cx-\tau(x)x);C^mx)} \\ &= \frac{M(C^{m+1}x-\tau(x)C^mx);C^mx)}{m(C^{m+1}x-\tau(x)C^mx);C^mx)}. \end{aligned}$$

Multiplying both sides by $m(C^{m+1}x-\tau(x)C^mx);C^mx)$ yields the required statement. Q.E.D.

"$C^{m+1}x-\tau(x)C^mx \leq MC^mx$":

Proof: From the definition of $M(C^{m+1}x-\tau(x)C^mx;C^mx)$ ([Gur], p.126), we have $C^{m+1}x-\tau(x)C^mx \leq MC^mx$. Q.E.D.

"so that, with $\lambda = \lambda_C$,

$$\lambda^m(\lambda-\tau(x))(\eta\cdot x) \leq M\lambda^m(\eta\cdot x)\text{"}:$$

Proof: η was defined in [Gur], Page 126 as the vector satisfying $\eta > 0, ||\eta|| = 1$, and $C'\eta = \lambda\eta$. We take the dot product (with notation (a,b) instead of $a\cdot b$) of both sides of the inequality $C^{m+1}x-\tau(x)C^mx \leq MC^mx$ with η to get $(\eta, C^{m+1}x - \tau(x)C^mx) \leq (\eta, MC^mx)$. The right-hand side is

$$\begin{aligned} (\eta, MC^mx) &= M(\eta, C^mx) = M((C^m)'\eta, x) = M((C')^m\eta, x) \\ &= M(\lambda^m\eta, x) = M\lambda^m(\eta, x) = M\lambda^m(\eta\cdot x), \end{aligned}$$

the right-hand side of the inequality we seek to prove. On the other hand, the left-hand side becomes

$$\begin{aligned} (\eta, C^{m+1}x-\tau(x)C^mx) &= (\eta, C^{m+1}x) - (\eta, \tau(x)C^mx) \\ &= (\eta, C^{m+1}x) - \tau(x)(\eta, C^mx) \\ &= ((C^{m+1})'\eta, x) - \tau(x)((C^m)'\eta, x) \\ &= ((C')^{m+1}\eta, x) - \tau(x)((C')^m\eta, x) \\ &= (\lambda^{m+1}\eta, x) - \tau(x)(\lambda^m\eta, x) \\ &= \lambda^{m+1}(\eta, x) - \lambda^m\tau(x)(\eta, x) \\ &= (\lambda^{m+1} - \lambda^m\tau(x))(\eta, x) \\ &= \lambda^m(\lambda-\tau(x))(\eta, x) \\ &= \lambda^m(\lambda-\tau(x))(\eta\cdot x) \end{aligned}$$

But this is the left-hand side of the statement to prove. Q.E.D.

"$\lambda^m(\lambda-\tau(x)) \leq M \leq Tm(C^{m+1}x-\tau(x)C^mx;C^mx)$."

Proof: To obtain the left inequality, we take the statement just proven and divide both sides by $\lambda^m(\eta\cdot x)$. The right inequality is merely $M \leq Tm$ as proven above.

$$\text{“}C^{m+1}x - \tau(x)C^m x \geq T^{-1}(\lambda - \tau(x))C^m x\text{”}:$$

Proof: By the definition of $m1 \equiv m(C^{m+1}x - \tau(x)C^m x; C^m x)$ we have $m1C^m x \leq C^{m+1}x - \tau(x)C^m x$. We divide the left and right sides of the inequality we just proved by T to get $T^{-1}(\lambda - \tau(x)) \leq m1$. Multiplying each side by $C^m x$ gives $T^{-1}(\lambda - \tau(x))C^m x \leq m1C^m x \leq C^{m+1}x - \tau(x)C^m x$, thereby proving our claim. Q.E.D.

This spot is in the middle of page 127 of the article by Gurari

Bibliography

[Bir]: Birkhoff, Garrett. Extensions of Jentzsch's Theorem. Transactions of the American Mathematical Society, Vol. 85, pp.219-227.

[Gur]: Gurari, M. The Location of Eigenvalues and Eigenvectors of Complex Matrices. Journal of Approximation Theory, Vol. 22, pp.119-149.

[Ost]: Ostrowski, A. M. Positive Matrices and Functional Analysis, pp.81-101 in Recent Advances in Matrix Theory. The University of Wisconsin Press, Madison, 1964.

Index

יחי אדוננו מורנו ורבינו מלך המשיח לעולם ועד

לזכרון
הוו"ה אי"א נו"מ ובעל מדות
הרב ישראל ארי' ליב ז"ל
בן כ"ק הרה"ג והרה"ח ומקובל רב פעלים לתורה ולמצות
ורבים השיב מעון רבי לוי יצחק נ"ע זי"ע שניאורסאהן

In memory of

Rabbi Yisroel Aryeh Leib Schneerson

The chosid who inspired
Moshiach and science research

And in memory of his brother
הוו"ה אי"א נו"מ ובעל מדות
הרב דובער הי"ד

Rabbi DovBer Schneerson

who was killed in the Holocaust
sanctifying Hashem's name

May Hashem avenge his blood

In memory of my father

הרב משה דובער ז״ל

בן הרב יעקב אליעזר ז״ל סילמאן

Rabbi Moshe DovBer Silman

Whose home was filled
with hundreds of Jewish books—בית מלא ספרים
and whose example inspired me to write my own book.

והקיצו ורננו שוכני עפר והוא בתוכם

May we be reunited with him
with תחיית המתים immediately

In honor of my mother

מרת רחל תי׳ בת שרה לאה ע״ה

Rebbitzen Rochel Silman

May she have long, healthy years לחיים נצחיים

and *nachas* from the whole family

May we all go together to greet
The Rebbe Melech HaMoshiach immediately!

שמעון אברהם בן רחל

הנאהבים והנעימים בחייהם...

לעילוי נשמת אבי מורי היקר
ר' יהודה ליב ע"ה בן ר' נתן נטע ע"ה
ולעילוי נשמת אמי מורתי היקרה
מרת אדל ע"ה בת ר' ישעי' ע"ה

נדבת בתם ומשפחתה

לעילוי נשמת מרת מעניא ע"ה בת ר' ישעי' אברהם ע"ה
ולעילוי נשמת אחותה שנהרגה בשואה על קידוש השם
מרת רבקה הי"ד בת ר' ישעי' אברהם ע"ה

...ובמותם לא נפרדו

והקיצו ורננו שוכני עפר והם בתוכם
בהתגלות הרבי מלך המשיח לעיני כל
תיכף ומי"ד ממ"ש ממ"ש!

In memory of

הרב משה דובער ז״ל
בן הרב יעקב אליעזר ז״ל סילמאן

Rabbi Moshe DovBer Silman

Who served the Jewish community of Minnesota
as Rabbi for twenty-five years
and is remembered with love almost forty years later

His home was always filled with *seforim*
and he was constantly learning from them

May his children, grandchildren and great grandchildren
always follow in his footsteps

And may we all go together
to greet Melech HaMoshiach *now*!

רחל בת שרה לאה
יעקב אליעזר בן רחל
דוד מאיר ירחמיאל בן סימה ומלכה נחמה בת רחל ומשפחתם
שמעון אברהם בן רחל ונעטל בת אדל ומשפחתם
יהודית ברכה בת רחל ומשפחתה
יעקב מענדל בן רייזל ונעמי אסתר בת רחל ומשפחתם

לזכות

שמעון אברהם בן רחל ונעטל בת אדל וביתם

חיים אלי-הו בן אסתר מרים מלכה ושרה נחמה בת נעטל וביתם

דניאל יונתן בן צבי׳ רות ושולמית בת נעטל וביתם

משה דובער בן נעטל וחנה בת טילא העניא וביתם

יהודה ליב בן נעטל

לוי יצחק בן נעטל ולאה בת שרה וביתם

מנחם מענדל בן נעטל

ישראל בן מרגולית וחנה בת נעטל וביתם

ישראל ארי׳ בן נעטל

חי׳ מושקא בת נעטל

In memory of our dear colleagues

Rabbi Dr. Naftali Berg ז"ל

and

Dr. Tsvi Yehuda Saks ז"ל

Their contributions to Moshiach & Science research
and their lectures at our conferences
will always be an inspiration to us.

והקיצו ורננו שוכני עפר והם בתוכם
ונלך ביחד לקבל פני הרבי מלך המשיח

The RYAL Institute

In honor of the 5th yartzeit of

ר׳ צבי יהודה ז״ל
בן יצחק אלחנן ע״ה סאקס

Dr. Tsvi Yehuda Saks
14 MarCheshvan, 5771

May his shining example continue to inspire
his loving family and friends
until the coming of Moshiach...

May it be immediately!

The Saks Family

Dedicated in loving memory of

הרב נפתלי צבי יהודה ז"ל
בן ראובן ע"ה בערג

Dr. Naftali Berg

A true chosid of the Rebbe
asking the Rebbe's advice on every step

One who was מהדר במצות and in particular
in his generosity with צדקה —
giving unstintingly to all who asked.

By living his life this way
he showed us the true דרך of עבודת השם

May we be זוכה to march with the רבי and our זיידע
immediately in our days.

אשתו לאה פייגא בת גיטל
שמעון ומרגולא רייקא רומאני ומשפחתם
אליעזר וצירל רחל טורין ומשפחתם
בן ציון ואיזה בעלה בערג ומשפחתם
משה ורבקה הינדא רייניץ ומשפחתם

לזכות ההתגלות המלאה של
הרבי מלך המשיח

For the full revelation of

The Rebbe Melech HaMoshiach

and for the **זכות** of our children and grandchildren.

May all of us together see the face of
The Rebbe Melech HaMoshiach *now*!

ועינינו תחזינה מלכותך

נפתלי הערץ בן רייזל ופייגא סאשא בת חנה רבקה עסטולין

In memory of my dear friend

ר׳ דוב בן בנימין ע״ה

Berel Kriegsman

Larry Lasher, Ph.D.

Berel Kriegsman (left) and Dr. Lasher (right) at the 2002 Moshiach & Science Conference

Dr. Larry Lasher is the Pioneer Mission's Science Chief at the NASA Ames Research center. Pioneer was earth's first space probe to the outer planets and became the first earth-made object to leave the solar system and go towards interstellar space. It has been commemorated by several U.S. postage stamps, and an image of Pioneer is also included in U.S. Passports.

Special Thanks to:

Mr. Joel Grae
Mr. Golan Ben-Ami
Rabbi Motti Anati
Mr. Ovadia Anatian
Mr. Danny Branover
Rabbi Mottel Deitsch
Rabbi Levi Garelik

לזכות
הרב שלום מרדכי הלוי בן רבקה ש׳ רובאשקין

Rabbi Shalom Mordechai Rubashkin

שיצא לשלום מה׳ שלום
תיכף ומי״ד ממ״ש

In memory of
my dear friend and *chavrusa*

ר׳ ראובן בן בנימין ע״ה

Reuven Amrami

A friend who lived by ״שלי שלך״

We spent many long nights together learning
from his vast library of *seforim*

Shimon Silman

לזכרון
הרב גבריאל נח הי״ד בן הרב נחמן שי׳
וריעתו מרת רבקה הי״ד בת הרב שמעון שי׳
הולצברג
שנהרגו על קידוש השם
והקיצו ורננו שוכני עפר והם בתוכם
ונלך ביחד לקבל פני הרבי מלך המשיח
תיכף ומי״ד ממ״ש!

In blessed memory of

ר׳ ליב אלי-הו בן יעקב דוד הכהן ע״ה

For his yartzeit, 3 Shevat

G-d willing, the publishing of this book should help illuminate
all remaining vestiges of darkness
and immediately reveal the true and complete Redemption
with the Rebbe Melech HaMoshiach.

Dedicated by his daughter Sarah